INTERNATIONAL SERIES OF MONOGRAPHS IN

PURE AND APPLIED MATHEMATICS

VOLUME 77

MATHEMATICAL ANALYSIS

G. YE. SHILOV

MATHEMATICAL ANALYSIS

A Special Course

TRANSLATED BY

J. D. DAVIS

ENGLISH TRANSLATION EDITED BY

D. A. R. WALLACE
Department of Mathematics
The University of Glasgow

PERGAMON PRESS
OXFORD · LONDON · EDINBURGH · NEW YORK
PARIS · FRANKFURT

Pergamon Press Ltd., Headington Hill Hall, Oxford
4 & 5 Fitzroy Square, London W.1

Pergamon Press (Scotland) Ltd., 2 & 3 Teviot Place, Edinburgh 1

Pergamon Press Inc., 122 East 55th St., New York 22, N.Y.

Pergamon Press GmbH., Kaiserstrasse 75, Frankfurt-am-Main

First edition 1965

Library of Congress Catalog Card No. 65-18619

A translation of the original volume
Математический анализ, специальный курс
(Matematicheskii Analiz, Spetsial'nyi Kurs)
Fizmatgiz, Moscow 1961

PREFATORY NOTE

The book was written as a textbook for University Faculties of Mathematics, in a special course of Mathematical analysis. Problems in the theory of functions of a real variable, the calculus of variations and integral equations are treated here from the single viewpoint of the theory of linear spaces. The reader requires a grasp of a general course of mathematical analysis as covered in a degree course.

CONTENTS

FOREWORD

THIS book has been written as a textbook for a special course of mathematical analysis (in brief, "Analysis III"). "Analysis III" is delivered as the third course of the mechanics–mathematics faculty of the Moscow State University, and dates from 1949; the introduction of such a course was initiated by academician A. N. Kolmogorov, who was the first lecturer. "Analysis III" is based on material from earlier separate courses on the theory of functions of a real variable, the calculus of variations and integral equations, and treats all this material from a single viewpoint which has its source in the theory of linear spaces.

The layout of the book is according to the following scheme. The first chapter gives a straightforward minimal account of set theory. The second chapter contains elements of the theory of metric and normed linear spaces. In the third chapter the calculus of variations is developed; it is presented here as a theory of differentiable functionals in normed linear spaces. The fourth chapter is devoted to the theory of the Lebesgue integral; the scheme given by F. Riesz is thought to be a better starting-point for the account than that of Lebesgue, inasmuch as it is more economical and leads more rapidly to the heart of the subject. The fifth chapter, "The Geometry of Hilbert Space" contains the theory of orthogonal resolutions and a geometrical treatment of integral equations. In the sixth chapter the relation between integration and differentiation is explained and the Stieltjes integral is constructed. In the seventh and last chapter an account is given of the theory of the Fourier Transform; we depart from tradition here in including some material, which, in view of its special importance in mathematical physics, should long ago have had a place in a course of mathematical analysis. To facilitate modifications in a course of lectures, the material of the last three chapters is widely diversified.

The logical dependence between chapters is given schematically as follows:

Sets

Metric spaces

Lebesgue integral | Calculus of variations

Hilbert space | Integration and differentiation

Fourier transform

It must be observed that the general viewpoint of functional analysis developed in this course does not constitute an end in itself, but only a means; the chief aim is an introduction to the field of classical mathematical analysis.

For the sake of brevity we limit ourselves within each theme to a discussion of only the most important questions, fully recognising that the reader may well be dissatisfied in some instances. The choice of material generally, and in particular for the last chapters, presented the author with great difficulties. Some interesting but somewhat digressional questions have been set as problems; they can be used as material for seminars.

The reader needs to be acquainted with a general course of mathematical analysis, such as "A short course" by A. Ya. Klinchin for example. The book can then be used for an independent study of the subject. Towards the end of the book the elementary properties of analytic functions are also assumed to be familiar.

The author is particularly grateful to M. G. Krain, O. A. Olyanik, and D. A. Rykov, who have read through the manuscript and with their criticisms have contributed greatly to its improvement. In the second edition the text has been revised, supplemented, and improved in places. Some new problems have also been added. The author is indebted to his many correspondents in Leningrad, Kazan, Baku, and other cities of the U.S.S.R. for their valuable criticisms of the first edition.

CHAPTER I

SETS

1. Sets, Subsets, Inclusions

When we consider a number of objects ("elements"), we use such terms as "totality", "aggregate", and "set". For example, one can speak of the set of students in an audience, the set of grains of sand on a beach, the set of vertices of a polygon, or the set of its sides. The examples specified have the property that in each of them the corresponding set is composed of a definite number of elements (which may, in some cases, be difficult to determine). We shall call such sets *finite*.

In mathematics it is often necessary to deal with sets which are not composed of a finite number of objects; the simplest examples are provided by the set of all natural numbers 1, 2, 3, ... and the set of all points of an interval. We shall call such sets *infinite*. We also add to the totality of sets the *empty set*—the set which contains no elements at all.

So, for example, as can be seen from Fig. 1, the set of real roots of the equation $\sin x/x = b$ is infinite when $b = 0$ (in this case it consists in all the values $x = \pm\pi, \pm 2\pi \ldots$), finite but non-empty

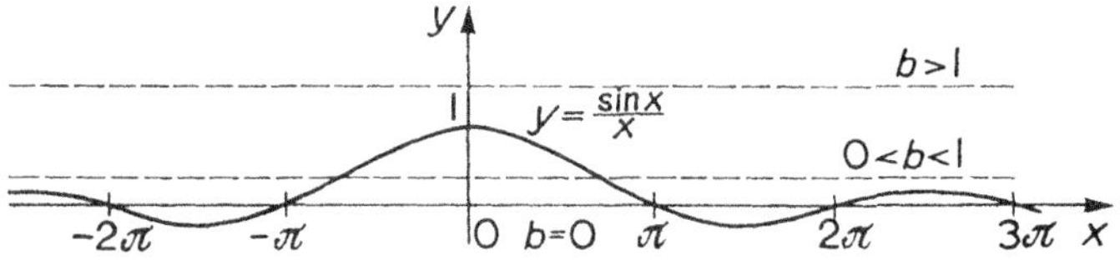

Fig. 1

when $0 < |b| \leq 1$ (the precise number of roots can be calculated for each b), and empty when $|b| > 1$ (no value of the function $\sin x/x$ exceeds 1 in absolute value, so the equation $\sin x/x = b$ has no roots at all when $|b| > 1$).

As a rule we shall denote sets by capital letters $A, B, C, \ldots$, and their elements by small ones. The notation $a \in A$ (or $A \ni a$) indicates that a is an element of the set A; $a \notin A$ (or $A \not\ni a$) means that a is not an element of the set A. The notation $A \subset B$ (or $B \supset A$) indicates that each element of the set A is an element of the set B; in this case the set A is called a *subset* of the set B. The largest subset of the set B is evidently the set B itself; the smallest is the empty set. Any other subset of B is known to contain some, but not all, of the elements of B. Each such subset is called a *proper* subset. The symbols $\in$, $\ni$, $\subset$, $\supset$ are called *inclusion* signs. If the inclusions $A \subset B$, $B \subset A$ both hold, then each element of the set A is an element of the set B, while conversely each element of B is an element of A; thus in this case A and B consist of the same elements, that is, they coincide with one another. This is denoted by the equation

$$A = B.$$

Sets can be formally denoted in the following distinct ways. The simplest consists in enumerating all the elements of a set, e.g. $A = \{1, 2, \ldots, n, \ldots\}$. An alternative form frequently used appeals to the properties of the elements of a set; thus, $A = \{x: x^2 - 1 < 0\}$ is the set of all x for which the inequality following the colon is satisfied.

2. Operations on Sets

We consider here three simple operations which can be carried out on sets: *union, intersection,* and *forming the complement.*

We describe first the operation of forming the union of sets. Let the sets $A, B, C, \ldots$ be given. We consider the totality of all elements, each of which belongs to *at least one* of $A, B, C, \ldots$ This aggregate is a new set, which is termed the *union* of the sets $A, B, C \ldots$

Thus the union of the set

$$A = \{6, 7, 8, \ldots\}$$

(all the natural numbers greater than 5) and the set

$$B = \{3, 6, 9, \ldots\}$$

(all the natural numbers that are multiples of 3) is the set

$$S = \{3, 6, 7, 8, 9, 10, \ldots\}$$

(all the natural numbers with the exception of 1, 2, 4, and 5).

We introduce now the operation of forming the intersection of sets. The *intersection* of the sets $A, B, C, \ldots$ is defined as the totality of elements belonging to *each* of the specified sets.

Thus, in the preceding example, the intersection of the sets

$$A = \{6, 7, 8, 9, 10, \ldots\}$$
$$B = \{3, 6, 9, 12, \ldots\}$$

is the set

$$D = \{6, 9, 12, \ldots\}.$$

It may happen that the sets $A, B, C, \ldots$ do not have a single element in common. In that case their intersection is the empty set, and the sets $A, B, C, \ldots$ are said to be non-intersecting or disjoint. For example, the sets of integers

$$A = \{1, 2\}, \quad B = \{2, 3\}, \quad C = \{1, 3\}$$

are disjoint (although when taken in pairs they have common elements).

Unions and intersections can be formed not only for finite collections of sets but also for infinite ones. For example, it is possible to form the union of the sets of points of all the straight lines in the plane that pass through a given point O. This union will obviously be the set of all points in the plane. The intersection of the specified sets will consist in the single point O.

The union S of sets $A, B, C, \ldots$ is sometimes called their *sum* and is written in the form $S = A + B + C + \ldots$; the intersection D is also termed the *product* and is denoted by $D = ABC \ldots$ There is some motivation for such "arithmetical" terminology. For instance, for any three sets A, B, C the equation

$$(A + B)\,C = AC + BC$$

holds. We shall give a proof of this equality as a simple but typical example of set-theoretic reasoning.

As we remarked, two sets are considered equal if each element of one of them is at the same time an element of the other. Thus we have to show that each element x of $(A + B) C$ (the left-hand side) is an element of $AC + BC$ (the right-hand side) and conversely that each element y of $AC + BC$ is contained in $(A + B)C$. First let x belong to $(A + B) C$. Being an element of the intersection of the sets $A + B$ and C, it must belong to each of them; thus we have

$$x \in A + B \quad \text{and} \quad x \in C.$$

Since x is contained in the union of A and B, it is certainly contained in one of them, say in A. But the inclusions $x \in A$, $x \in C$, imply $x \in A\,C$, whence $x \in AC + BC$. And if x is contained, not in A, but in B, then in the same way $x \in BC$, $x \in AC + BC$, as required. Conversely if y belongs to the sum $AC + BC$, then it belongs either to AC or to BC, $y \in BC$ say. But then $y \in B$ and $y \in C$; further it follows from $y \in B$ that $y \in A + B$ and finally that $y \in (A + B) C$. The case $y \in A\,C$ is treated analogously and the proof is complete.

It is to be observed, however, that by no means all arithmetical rules carry over to operations with sets. For example, we have for sets A, B, C, the formulae

$$A + A = A,$$
$$AA = A,$$
$$A + BC = (A + B)(A + C)$$

which are quite unlike the usual arithmetical equations. We suggest that the reader satisfy himself as to the accuracy of these formulae.

We shall indicate a few further symbols for sums and intersections of sets. For the union of sets we employ the symbols $\sum$ and $\cup$, so that, for example, the notation

$$S = \sum_{\nu=1}^{\infty} A_\nu \quad \text{or} \quad S = \bigcup_{\nu=1}^{\infty} A_\nu$$

denotes the union of the sets $A_1, A_2, \ldots, A_\nu, \ldots$

For the intersection of sets we employ the symbols $\prod$ and $\cap$, so that, for example

$$D = \prod_{\nu=1}^{\infty} A_\nu \quad \text{or} \quad D = \bigcap_{\nu=1}^{\infty} A_\nu$$

denotes the intersection of the sets $A_1, A_2, \ldots, A_\nu, \ldots$

We turn now to the operation of forming the complement.

If the set B is a subset of the set A, then the totality of those elements of A that do not belong to B is termed the *complement* of B with respect to A and is denoted by CB or $A - B$.

We note the obvious formula

$$(A - B) + B = A.$$

Note that for two arbitrary sets A and B, the result

$$(A + B) - B = A$$

is generally false; *it is true only when A and B have no common elements.*

Of more complicated results we remark the following, which will be of frequent occurrence:

$$C \sum_{\nu} B_{\nu} = \prod_{\nu} CB_{\nu}; \tag{1}$$

it can be read as follows: *the complement of a union of sets is the intersection of their complements.*

We shall give a proof of this result. Let $x \in C \sum_{\nu} B_{\nu}$; then $x \notin \sum_{\nu} B_{\nu}$; this implies that, for any ν, $x \notin B_{\nu}$, i.e. $x \in CB_{\nu}$; but then $x \in \prod CB_{\nu}$. Conversely, if $x \in \prod_{\nu} CB_{\nu}$, then, for any ν, $x \in CB_{\nu}$, i.e. $x \notin B_{\nu}$ for any ν; but then $x \notin \sum_{\nu} B_{\nu}$, i.e. $x \in C \sum_{\nu} B_{\nu}$, as required.

Operating again with C on both sides of equation (1) and letting $A_{\nu} = CB_{\nu}$, we obtain the result

$$\sum_{\nu} CA_{\nu} = C \prod_{\nu} A_{\nu}, \tag{2}$$

i.e. *the complement of an intersection of sets is the union of their complements.*

The results cited can be combined in the form of a general rule: the *complement symbol C* can be interchanged with either of the symbols $\sum$ and $\prod$ *provided that each of the latter is substituted for the other.*

3. Equivalence of Sets

We wish now to establish a rule by which it would be possible to compare different sets with respect to the number of elements in them.

For finite sets no problem arises here: by counting the elements of two finite sets A and B, we can immediately determine which is relatively the richer in elements. It is natural to call finite sets A and B equivalent if they have the same number of elements. However, this definition of equivalence does not carry over directly to the case of infinite sets. We shall now give it a form in which the extension to infinite sets will be immediate. To this end we observe that to establish the equivalence or non-equivalence of finite sets A and B there is no actual necessity to enumerate their respective elements. For example, if the set A is the audience in a hall and if B is the set of chairs in the same hall, then instead of counting the students and chairs separately, it is possible, by assigning to each student a free chair, to determine immediately and without any counting whether or not the specified sets are equivalent.

The procedure elaborated in the example given is, in abstract terms, the *establishing of a correspondence between the sets A and B.*

We introduce the following important definition. If each element of the set A somehow determines a unique element of the set B while in addition each element of B is determined by one and only one element of A, then there is said to be a *one–one correspondence* between the sets A and B. And in this case the sets A and B are called *equivalent*.

This new definition of equivalence applies to any sets, which need not be finite; thus, for example, the infinite set A of natural numbers 1, 2, ... is equivalent to the set B of negative integers $-1, -2, \ldots$, since a one–one correspondence between A and B is established by the rule: to each number $n \in A$ corresponds the number $-n \in B$.

In just the same way the set of natural numbers 1, 2, ... is equivalent to the set of all positive even integers 2, 4, ...; the equivalence between them is effected by the rule $n \to 2n$. In this example we see that a set can be equivalent to a proper subset of itself; it is obvious that such a situation can occur only with infinite sets.

The equivalence relation is denoted by the sign $\sim$. This relation is easily seen to be transitive: if $A \sim B$ and $B \sim C$, then $A \sim C$. If two sets are equivalent they are also said to be "equipotent", or to have the same "power".

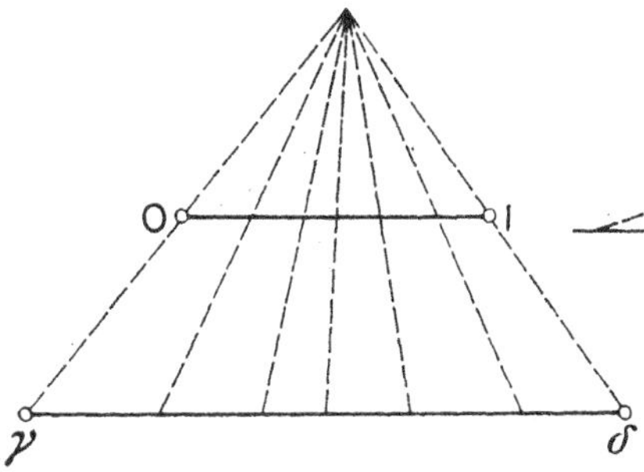

FIG. 2

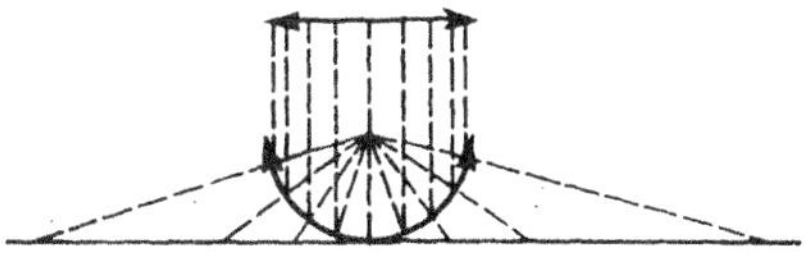

FIG. 3

The set of points of the closed interval [0, 1] is equivalent to the set of points of any other closed interval $[\nu, \delta]$; the correspondence can be established, for example, with the aid of a central projection as shown in Fig. 2. Similarly the sets of points of any two distinct closed intervals are equivalent.†

The set of points of an open interval is equivalent to a set of points on a straight line (Fig. 3).

It is not so easy to answer the question as to whether the set of points of a closed interval is equivalent to the set of points of an open interval.

There is the following general theorem, which contains as a particular application the answer to this question.

THEOREM (F. Bernstein, 1898). *If the set A is equivalent to a subset of the set B and the set B is equivalent to a subset of the set A, then the sets A and B are equivalent.*

Proof. Let us denote by B_1, a subset of the set B equivalent to the set A, and by A_1 a subset of the set A equivalent to B. In the one–one correspondence $B \sim A$, elements of B, correspond to certain elements in A, whose aggregate we shall denote by A_2 and so we obtain the inclusion chain,

$$A \supset A_1 \supset A_2$$

while $A_2 \sim A$ since $A_2 \sim B_1$, $B_1 \sim A$. If we show that $A \sim A_1$, the theorem will be proved (since $A_1 \sim B$). In the one–one map of A

† The closed interval $[\alpha, \beta]$ is determined by the inequality $\alpha \leqq x \leqq \beta$ (end points are included) and the open interval (x, β) by the inequality $\alpha < x < \beta$ (end points are not included).

onto A_2, the set $A_1 \subset A$ is mapped onto a certain set $A_3 \subset A_2$, the set $A_2 \subset A_1$ is mapped onto a set $A_4 \subset A_3$, the set $A_3 \subset A_2$ is mapped onto a set $A_5 \subset A_4$, and so on. Furthermore

the set $A - A_1$ maps onto $A_2 - A_3$
the set $A_1 - A_2$ maps onto $A_3 - A_4$
the set $A_2 - A_3$ maps onto $A_4 - A_5$

and so on.

From this it follows that the sets

$$A - A_1,\ A_2 - A_3,\ A_4 - A_5,\ A_6 - A_7, \text{ etc.}$$

are equivalent in pairs; the union of the sets, which are disjoint,

$$(A - A_1) + (A_2 - A_3) + (A_4 - A_5) + \cdots$$

is then equivalent to the union

$$(A_2 - A_3) + (A_4 - A_5) + (A_6 - A_7) + \cdots$$

We denote by D the intersection of the sets $A, A_1, A_2, \ldots$ The following equations then hold:

$$A = D + (A - A_1) + (A_1 - A_2) + (A_2 - A_3) + \cdots \qquad (1)$$
$$A_1 = D + (A_1 - A_2) + (A_2 - A_3) + (A_3 - A_4) + \cdots \qquad (2)$$

We shall prove the first equation. Let $a \in A$; we shall show that a is contained in the right-hand side of equation (1). If a is actually in each of the sets $A_1, A_2, \ldots$ then $a \in D$, and the assertion is proved. If, on the other hand, there are some A_n to which a does not belong, let A_k be the first such set, so that $a \in A_{k-1}$; but then $a \in A_{k-1} - A_k$ and is therefore contained in the right-hand side of (1). Conversely if a belongs to the right-hand side of (1), then evidently $a \in A$, since each term in the right-hand side is a subset of A.

Equation (2) is proved in exactly the same way.

Equations (1) and (2) can be written in the form

$$\begin{aligned} A = {} & [D + (A_1 - A_2) + (A_3 - A_4) + \cdots] \\ & + [(A - A_1) + (A_2 - A_3) + \cdots], \end{aligned} \qquad (3)$$

$$\begin{aligned} A_1 = {} & [D + (A_1 - A_2) + (A_3 - A_4) + \cdots] \\ & + [(A_2 - A_3) + (A_4 - A_5) + \cdots]. \end{aligned} \qquad (4)$$

The first square brackets on the right-hand sides of both equations contain the same set, while the second square brackets contain sets which were proved above to be equivalent. It is now easy to establish the equivalence of the sets A and A_1. To each point of the set $D + (A_1 - A_2) + (A_3 - A_4) + \cdots \subset A$ let correspond the same point in the set A_1; and to each point a of the set $(A - A_1) + (A_2 - A_3) + \cdots$ let correspond that point of the set $(A_2 - A_3) + \cdots$ which corresponds to a in virtue of the equivalence between these sets established above. Equations (3) and (4) show that this correspondence exhausts all the elements of the sets A and A_1. Thus the required one–one correspondence between A and A_1 is established.

Using the Bernstein theorem, it is easy to verify that the point-sets of open and closed intervals are equipotent. In fact a given closed interval $[\alpha, \beta]$ contains a set equivalent to the point-set of a given open interval (γ, δ) (any open interval) interior to $[\alpha, \beta]$ while the open interval (γ, δ) contains a set equivalent to the point set of the closed interval $[\alpha, \beta]$ (any interior closed interval). Applying the Bernstein theorem we have that $[\alpha, \beta] \sim (\gamma, \delta)$ as required.

4. Countable Sets

Definition. A set equivalent to the set of all natural numbers 1, 2, ..., is said to be a *countable set.*

Alternatively, a set is countable if all its elements can be numbered, thereby exhausting the natural numbers. We shall cite a few theorems concerning countable sets.

1. *Every infinite subset B of a countable set A is likewise countable.* In fact the elements of B can be renumbered according to their sequential order in A (in so doing, it will be necessary, since B is infinite, to use all the natural numbers).

2. *The union of a finite or countable number of countable sets is a countable set. Proof.* Let us consider first the case of two sets. Let $A = \{a_1, a_2, \ldots\}$ and $B = \{b_1, b_2, \ldots\}$ be countable sets. We write out all the elements of both sets in a single row as follows:

$$a_1, b_1, a_2, b_2, a_3, b_3.$$

Now all these elements can be renumbered according to their sequential order in the line. Of course an element occurring twice (i.e. one contained both in A and in B) acquires a number on its first occurrence and is omitted on its second. Consequently each element of the union of A and B is numbered as required.

The set of all integers $0, \pm 1, \pm 2, \ldots$ is countable since it is the union of two countable sets $1, 2, 3, \ldots$ and $0, -1, -2, \ldots$

The theorem is proved similarly for three, four, or in general, for any finite number of countable sets. For a countable number of countable sets, for instance for the aggregate of sets

$$\begin{aligned} A_1 &= \{a_{11}, a_{12}, \ldots, a_{1n}, \ldots\}, \\ A_2 &= \{a_{21}, a_{22}, \ldots, a_{2n}, \ldots\}, \\ &\ldots\ldots\ldots\ldots\ldots\ldots\ldots\ldots \\ A_k &= \{a_{k1}, a_{k2}, \ldots, a_{kn}, \ldots\}, \\ &\ldots\ldots\ldots\ldots\ldots\ldots\ldots\ldots \end{aligned}$$

the only difference will be that the rule for writing all the elements of the sets in a single line must be somewhat subtler, for example:

$$a_{11};\ a_{21}, a_{22}, a_{12};\ a_{31}, a_{32}, a_{33}, a_{23}, a_{13};\ a_{41}, a_{42}, a_{43}, a_{44}, a_{34}, a_{24}, a_{14};\ \ldots$$

the remainder of the proof is unchanged.

3. *The set of all rational numbers (i.e. numbers of the form p/q, where p and q are integers) is countable.*

In fact the set of all rational numbers is the union of the following countable sets:

(1) the set A_1 of all integers $n = 0, \pm 1, \pm 2, \ldots$;

(2) the set A_2 of all fractions of the form $n/2$, $n = 0, \pm 1, \pm 2, \ldots$;

(3) the set A_3 of all fractions of the form $n/3$, $n = 0, \pm 1, \pm 2, \ldots$;

. . . .

(k) the set A_k of all fractions of the form n/k, $n = 0, \pm 1, \pm 2, \ldots$

. . . .

The sets $A_1, A_2, \ldots A_k, \ldots$ constitute a countable set of sets; since each of them is countable, their union is also countable in virtue of theorem 2, as asserted.

4. If $A = \{a_1, \ldots a_k, \ldots\}$ and $B = \{b_1, \ldots b_n, \ldots\}$ *are countable sets, then the set of all pairs* (a_k, b_n) $(k, n = 1, 2, \ldots)$ *is also countable.*

In fact the set of all these pairs can be decomposed into a countable set of countable sets

$$\begin{aligned} A_1 &= \{(a_1, b_1), (a_1, b_2), \ldots, (a_1, b_n), \ldots\}, \\ A_2 &= \{(a_2, b_1), (a_2, b_2), \ldots, (a_2, b_n), \ldots\}, \\ &\ldots\ldots\ldots\ldots\ldots\ldots\ldots\ldots\ldots\ldots \\ A_k &= \{(a_k, b_1), (a_k, b_2), \ldots, (a_k, b_n), \ldots\} \\ &\ldots\ldots\ldots\ldots\ldots\ldots\ldots\ldots\ldots\ldots \end{aligned}$$

and by theorem 2 the union of these sets is a countable set.

This example is capable of a geometric interpretation: to the pair (a_k, b_n) corresponds the point in the plane with coordinates a_k, b_n, we see therefore that the set of all points in the plane both of whose coordinates are rational is countable.

5. *The set of all polynomials* $P(x) = a_0 + a_1 x + \cdots + a_n x^n$ *(of arbitrary degrees) with* ***rational*** *coefficients* $a_0, a_1, \ldots, a_n$ *is countable.* The set of all polynomials of the given form is the union of a countable set of sets A_n $(n = 0, 1, 2, \ldots)$ where A_n denotes the set of polynomials with degrees $\leq n$. Hence, in view of theorem 2, it is sufficient to show that each of the sets A_n is countable. For $n = 0$ it is a question of the countability of the rational numbers themselves, which was established in theorem 3. We shall now proceed by induction: we suppose that the set A_n has been proved countable, and go on to prove A_{n+1} countable.

Each element of the set A_{n+1} can be written in the form

$$Q(x) + a_{n+1} x^{n+1},$$

where $Q(x)$ is a polynomial of degree $\leq n$ with rational coefficients, i.e. an element of A_n, and a_{n+1} is a rational number.

The set of polynomials $Q(x)$ is by hypothesis countable, and so is the set of numbers a_{n+1}. Thus, corresponding to each element of the set A_{n+1}, it is possible to assign a pair $(Q(x), a_{n+1})$, in which each term runs through a countable set of values.

Note. It is clear that in the given instance, it is immaterial that we actually consider polynomials, i.e. linear combinations of powers of x. For example, linear combinations of trigonometric or other functions could equally well be considered. In general, instead of polynomials $a_0 + a_1 x + \cdots + a_n x^n$ with rational coefficients, we can consider any array $(a_0, a_1, \ldots a_n)$, each coordinate of which is an element of some countable set; the proof given above shows essentially that the set of all such arrays is also a countable set.

6. *The set of all algebraic numbers (i.e. zeros of polynomials with rational coefficients) is countable.*

By theorem 5 we can enumerate all polynomials with rational coefficients, and so they will form a sequence

$$P_1(x),\ P_2(x),\ \ldots,\ P_n(x),\ \ldots$$

But each of the specified polynomials possesses a certain finite number of zeros. Writing in a single line all the zeros of $P_1(x)$, then all the zeros of $P_2(x)$, etc., we are able to order the set of all algebraic numbers, as required.

Problems. 1. Show that the sets given below are countable:

(a) the set of all closed intervals $a \leqq x \leqq b$, where the end-points a and b are rational numbers.

(b) The set of all finite combinations of rational points in the plane.

2. Show that the following sets are either finite or countable:

(a) A set of disjoint intervals on a line.

(b) A set of figure eights in the plane, no two of which intersect.

(c) The set of points of discontinuity of a monotone function.

(d) A set M of real positive numbers, provided that all finite sums $\sum x_j, x_j \in M$, are bounded by a fixed number A.

Hints. (a) A rational point can be chosen in each interval. (b) A point with rational coordinates can be chosen in each half of a figure eight. (c) The discontinuities $[f(c-0), f(c+0)]$ of a monotone function $f(x)$ are disjoint. (d) Only finitely many points of M can lie outside any interval $[0, \epsilon]$.

Note. V. V. Grushin and V. P. Palamodov have established the same result for a set of non-intersecting figures in the plane which possess triple points (like the letter T), and also for a set of non-intersecting figures in space which contain saddle-points or incorporate Möbius strips as parts of themselves.

3. Resolve the set of natural numbers 1, 2, ... into a countable set of disjoint countable sets.

4. (Riddle). I. X, a mathematician, recently received a visit from his dear brothers N. In the entrance-hall they took off their hats and hung them on the stand. When they assembled to leave and began to put on their hats, it appeared, to their host's great confusion, that they were a hat short. Nobody had come into the entrance-hall during this time.

II. When the brothers N paid another visit to X (all with hats), they again hung their hats on the stand in the hall. When, on leaving, they started to put on their hats, it turned out that there was a hat too many. Both host and guests recollected definitely that until their arrival the hat stand had been quite empty.

III. On the next occasion, the guests put on their hats and left, and the host, having seen his guests to the street, returned to discover that all the hats were hanging on the stand.

IV. Finally, on the fourth occasion, the guests arrived hatless, and on their departure made use of the hats that remained from their last visit. When he had seen off his guests, the host again caught sight of all the hats on the hatstand,—the same number as were there before the guests' arrival.

What is the explanation of all these paradoxical events?

See the hint on p. 19.

5. Sets of the Power of the Continuum

It is found that there exist infinite sets the elements of which are incapable of being put into a sequence. Such sets are called uncountable. A typical example of an uncountable set is the *continuum*, the set of all points of an arbitrary closed interval.

Theorem 1 (G. Cantor, 1874). *The set of all points of the closed interval* $0 \leqq x \leqq 1$ *is uncountable.*

Proof. Suppose, on the contrary, that the set is countable and they can all be ordered in the sequence $x_1, x_2, x_n, \ldots$ Having got this sequence, we construct a sequence of closed sub-intervals as follows.

We partition the interval [0, 1] into three equal parts. Wherever the point x lies, it cannot belong simultaneously to all three of the intervals [0, 1/3] [1/3, 2/3] [2/3, 1] and so one of them can be selected which does not contain x_1 (neither as an interior point, nor as an end-point); we denote this interval by Δ_1. Further, we denote by Δ_2 one of the three equal parts of Δ_1 that does not contain x_2. When the intervals $\Delta_1 \supset \Delta_2 \supset \cdots \supset \Delta_n$ have been constructed in this way, we denote by Δ_{n+1} one of the subintervals formed by trisection of Δ_n that does not contain x_{n+1}, and so on. By a well-known theorem of analysis the infinite sequence of closed intervals $\Delta_1 \supset \Delta_2 \supset \ldots$ has a limit point ξ. This point ξ belongs to each of the Δ_n and so cannot coincide with any one of the points x_n. But this shows that the sequence $x_1, x_2, \ldots x_n, \ldots$ cannot exhaust the points of the interval [0, 1], contrary to the original hypothesis. The theorem is therefore proved.

We saw that all the rational numbers of the interval [0, 1] form a countable set. The remaining numbers of the interval are called irrational; for example, $\sqrt{1/2}$, $\pi/4$ etc. are irrational. We see now that the irrationals greatly outnumber the rationals; to be precise, the irrationals constitute a set known to be uncountable (otherwise, if the set of irrationals were countable, the set of all numbers in $0 \leqq x \leqq 1$, as the union of two countable sets, would also be countable). Furthermore, since the algebraic numbers (the zeros of polynomials with rational coefficients) also form a countable set (Section 4), the transcendentals, those numbers that are not zeros of polynomials with rational coefficients, constitute an uncountable set.

The argument put forward here proves, incidentally, the actual existence of transcendental numbers, which is by no means obvious *a priori*. *Every set equivalent to the set of all points of the closed interval* [0, 1] *is said to have the power of the continuum.*

We saw that the point-sets of an arbitrary closed interval $[a, b]$, an arbitrary open interval (α, β), and finally of the whole line $-\infty < x < \infty$, are equivalent to that of the closed interval [0, 1] and consequently have the power of the continuum. The following theorems facilitate the identification of new wide classes of sets with the power of the continuum.

The first of these theorems relates to an arbitrary infinite set:

THEOREM 2. *If a finite or countable set B is added to an infinite set A, there results a set equivalent to the original set A.*

For the proof, we shall extract at random from A a countable subset C, and let $D = A - C$. We have

$$A = D + C,$$
$$A + B = D + C + B.$$

Since the sets C and B are countable, their union $C + B$ is also a countable set; there exists therefore a one–one correspondence between C and $C + B$. Using this correspondence, and extending it by making the points of D self-corresponding points in it, we obtain the required one–one correspondence between the sets A and $A + B$†.

COROLLARY 1. *If a finite or countable set B is extracted from an infinite set Q, the remainder $A = Q - B$ is equivalent to Q, provided it again constitutes an infinite set.*

This follows immediately from the equation $Q = A + B$ on application of the theorem just proved to the set A.

COROLLARY 2. *The set of the irrationals has the power of the continuum; likewise the set of the transcendentals.*

Before turning to the theorems that follow, we shall consider the so-called dyadic representation of real numbers.

† This theorem, incidentally, implies as a corollary the equivalence of open and closed intervals, independently of Bernstein's theorem.

We restrict ourselves to the real numbers belonging to the closed interval [0, 1]. The point 1/2 divides this interval into two equal parts, which we denote by $\Delta_0 = [0, 1/2]$ and $\Delta_1 = [1/2, 1]$. The interval Δ_0 is divided into two equal parts by the point 1/4 we denote them by $\Delta_{00} = [0, 1/4]$ and $\Delta_{01} = [1/4, 1/2]$. Similarly the point 3/4 bisects Δ_1; we put $\Delta_{10} = [1/2, 3/4]$, $\Delta_{11} = [3/4, 1]$. Continuing the process of bisection we obtain eight intervals $\Delta_{000}, \Delta_{001}, \ldots, \Delta_{111}$ of length 1/8, sixteen intervals $\Delta_{0000}, \Delta_{0001}, \ldots$ Δ_{1111} of length 1/16, and so on. The end points of all these intervals are of the form $p/2^q$, where p and q are natural numbers; these points, which clearly form a countable set, are called dyadic rationals. The remaining points of the interval [0, 1] are called dyadic irrationals; the set which they form has the power of the continuum. We denote the aggregate of all the closed intervals of the system constructed by Δ.

For each point $\xi \in [0, 1]$, it is possible to find a sequence of closed intervals in Δ, successively embedded one in the other, with lengths respectively equal to $1/2$, $1/4$, ..., $1/2^n$ ..., and containing the point ξ. In fact ξ belongs to one of the intervals Δ_0, Δ_1; if it belongs to Δ_0, say, then it belongs to Δ_{00} or Δ_{01}, and so on. Thus for every ξ, we get:

$$\Delta_{\varepsilon_1} \supset \Delta_{\varepsilon_1 \varepsilon_2} \supset \Delta_{\varepsilon_1 \varepsilon_2 \varepsilon_3} \supset \cdots \supset \Delta_{\varepsilon_1 \varepsilon_2 \cdots \varepsilon_n} \supset \cdots \ni \xi \tag{1}$$

(the numbers ε_n are noughts or ones). Having the system of inclusions (1), we can identify with ξ the sequence of noughts and ones:

$$\varepsilon_1, \varepsilon_2, \varepsilon_3, \ldots, \varepsilon_n \ldots \tag{2}$$

The symbol (2) determines the *dyadic representation* of the real number ξ. Given the sequence (2), the number ξ itself is well-defined by the form:

$$\xi = \sum_{n=1}^{\infty} \frac{\varepsilon_n}{2^n}; \tag{3}$$

the partial sums of the series (3) are actually the left end-points of intervals appearing in the inclusions (1); it is clear that these left end-points form a monotone (non-decreasing) sequence tending to the value ξ.

It is indeed possible to commence with an arbitrary sequence (2) of zeros and ones and construct a number ξ from formula (3). It is easy to see that this number ξ will be the intersection of the closed intervals of the corresponding system (1). Thus all possible sequences of zeros and ones are involved in our construction.

If the point ξ is not a dyadic rational, it determines uniquely all the closed intervals occuring in (1): If, on the other hand, it is a dyadic rational, then it is a common end-point of two adjacent and equal closed intervals of the system Δ and at some stage of our process the choice between them will be arbitrary. If we take the one to the right, we shall have to take all the subsequent intervals to the left and all the subsequent numbers in (2) will be ones. While if we take the one to the left, we shall have to take all the subsequent intervals to the right and all the corresponding numbers in (2) will be zeros. It is easy to see that, conversely, if the terms in (2) from a certain point on are all zeros, or all ones, then the number ξ is a dyadic rational. This follows immediately from formula (3).

The set of dyadic irrationals is therefore in one–one correspondence with the set of all sequences of noughts and ones that contain an infinite number of each; the set of dyadic rationals is in one–one correspondence with the set of all sequences where the elements are all zeros from some point on, and also with the set where they are all ones from some point on. We can now turn to the next theorem.

THEOREM 3. *The set of all sequences consisting of zeros and ones has the power of the continuum.*

Proof. The set under consideration is the union of three sets: the set of sequences containing an infinite number both of ones and of zeros, the set of sequences containing only a finite number of zeros, and the set of sequences containing only a finite number of ones. As we have shown, the first of these is equivalent to the set of all dyadic irrationals and therefore has the same power as the continuum; the other two sets are countable, since they are equivalent to the set of dyadic rationals. In virtue of theorem 1, the set of all sequences of zeros and ones itself has the power of the continuum, as required.

THEOREM 4. *The set of all increasing sequences of natural numbers* (4) *has the power of the continuum.*

$$(0 <)\, k_1 < k_2 < \cdots < k_n < \cdots \tag{4}$$

Proof. Each sequence (4) determines a sequence of zeros and ones in which ones occupy the positions numbered $k_1, k_2, \ldots, k_n \ldots$ and zeros the remaining positions. This clearly constitutes a one–one correspondence between the set of all increasing sequences of natural numbers and the set of all sequences consisting in zeros and ones. We have shown the second set to have the power of the continuum; the theorem follows.

THEOREM 5. *The set of all sequences of natural numbers,*

$$m_1, m_2, \ldots, m_n, \ldots \tag{5}$$

(not necessarily increasing) *has the power of the continuum.*

Proof. Each sequence of natural numbers (5) determines an increasing sequence,

$$k_1 = m_1, \quad k_2 = m_1 + m_2, \ldots,$$
$$k_n = m_1 + m_2 + \cdots + m_n, \cdots$$

It is obvious that this constitutes a one–one correspondence between the set of all sequences of natural numbers and the set of increasing sequences. As we have shown, the second set has the power of the continuum; it follows that the first set has the same power.

THEOREM 6. *The set Z of all sequences of real numbers*

$$\xi = (\xi_1, \xi_2, \ldots, \xi_n, \ldots)$$

has the power of the continuum.

Proof. By theorem 5 each value ξ_n determines a sequence of natural numbers

$$\xi_n \to (p_{n1}, p_{n2}, \cdots, p_{nk}, \cdots)$$

and hence the symbol ξ determines an array

$$\begin{matrix} p_{11}\, p_{12} \cdots p_{1k} \cdots \\ p_{21}\, p_{22} \cdots p_{2k} \cdots \\ \cdots\cdots\cdots\cdots \\ p_{n1}\, p_{n2} \cdots p_{nk} \cdots \\ \cdots\cdots\cdots\cdots \end{matrix}$$

But all the elements of this array can be written in a simple sequence (cf. Section 4)

$$p_{11};\ p_{21}, p_{22}, p_{12};\ p_{31}, p_{32}, p_{33}, p_{23}, p_{13};$$
$$p_{41}, p_{42}, p_{43}, p_{44}, p_{34}, p_{24}, p_{14}, \cdots$$

Thus ξ determines a sequence of natural numbers. Conversely, it is clear that each sequence of natural numbers can be obtained in this way from some ξ. We see that the totality of symbols ξ is equivalent to the aggregate of all sequences of natural numbers, and so, by theorem 2, has the power of the continuum.

The proof of an analogous theorem goes through, with corresponding simplifications, in the case where ξ is defined by only a finite, and not a countable number of coordinates:

$$\xi = (\xi_1, \dots, \xi_n).$$

If we assume that each of the coordinates $\xi_1, \dots, \xi_n$ runs through the real line, we obtain the result that the *set of all points* in n-dimensional space *has the power of the continuum for any n.* In particular, the set of all complex numbers (or, what amounts to the same thing, the set of points in the plane) has the power of the continuum.

Note. There is no need whatever to regard each coordinate ξ_n as just a real number: the theorem holds if ξ_n runs through any set with the power of the continuum.

THEOREM 7. *The set $C(a, b)$ of all continuous functions $f(x)$ defined on the closed interval $[a, b]$ has the power of the continuum.*

Proof. Let $r_1, r_2, \dots, r_n, \dots$ be a sequence of all the rational points of the closed interval $[a, b]$. We identify with each continuous function $f(x)$ a sequence of real numbers—the values of the function $f(x)$ at the points $r_1, r_2, \dots, r_n, \dots$

$$f(x) \sim \{f(r_1), f(r_2), \dots, f(r_n), \dots\} = \{f(r_n)\}.$$

Under this correspondence two distinct functions $f(x)$ and $g(x)$ will determine distinct sequences $\{f(r_n)\}$ and $\{g(r_n)\}$, since if two continuous functions coincide at all rational points they coincide everywhere. Thus the set $C(a, b)$ of all continuous functions can be considered equivalent to some subset of the set of all numerical sequences. On the other hand, the set of all numerical sequences has by theorem 6 the power of the continuum, and is therefore equivalent to the subset of $C(a, b)$ comprising just constants. By Bernstein's theorem (Section 3), $C(a, b)$ is equivalent to the set of all numerical sequences, and consequently shares with it the power of the continuum.

Problems. 1. Show that the set of all continuous functions $f(x, y)$ defined on the square $|x| \leqq 1$, $|y| \leqq 1$, has the power of the continuum.

2. The function $f(x)$ $(a \leqq x \leqq b)$ is called a Baire function of the first class if it is the limit of a sequence of continuous functions:

$$f(x) = \lim_{n \to \infty} f_n(x), \qquad f_n(x) \in C(a, b).$$

Show that the power of the set of all Baire functions of the first class is equal to that of the continuum.

Hint to problem 4, Section 4. The set of brothers, N, is a countable set.

6. Sets of Higher Powers

If the sets A and B are not equivalent, but one of them, say A, is equivalent to some subset of B, the set B is said to have a higher power than A. Thus a countable set has a higher power than any finite set, and the continuum a higher power than a countable set.

There exist sets of higher power than the continuum. Furthermore, given a set of a certain power, it is always possible to construct a set of higher power using the following theorem:

Theorem (G. Cantor, 1878). *Let a set A be given and let B be the aggregate of all subsets of A. Then the set B is of higher power than the set A.*

Proof. We assume the existence of a one–one correspondence between the sets A and B, in other words, we assume that to each *element* $x \in A$ corresponds a uniquely determined *subset* A_x of A. The element x may or may not belong to the subset A_x; the first possibility is realised, for example, by A_x coinciding with the whole set A, and the second when A_x is the empty subset. We shall call elements of the first kind "good" and elements of the second kind "bad". We collect together all the bad elements x, i.e. those that do not belong to the corresponding subsets A_x, and let Z denote their aggregate. In virtue of the one–one correspondence between the subsets of the set A and its elements some element ζ must correspond to the subset Z. We examine two possibilities: ζ is either a good element or a bad one. If is good, it belongs to the corresponding subset, i.e. $\zeta \in Z$. But by construction Z consists only in bad elements, so the first possibility is excluded.

If ζ is a bad element, it does not belong to the corresponding subset: $\zeta \notin Z$. But by construction Z contains all the bad elements; hence the second possibility is excluded.

We have obtained a contradiction can be neither good, nor bad. Hence the original premise, the equivalence of the set A and the set B of all subsets of A, must be false; these sets cannot be equivalent. Since A itself is obviously equivalent to a subset of B (the one with singleton subsets of A as elements), we deduce that B has a higher power than A. The theorem is proved.

Examples. 1. It is easy to calculate that a finite set of n elements has just 2^n distinct subsets.

2. The set of all subsets of a countable set obviously coincides with the set of all sequences of distinct rational numbers, and therefore has the power of the continuum.

3. The set of all subsets of the continuum can be represented as a certain set of functions defined on the closed interval [0, 1]. Namely, to the subset A of this closed interval corresponds the function $f_A(x)$, equal to 1 when $x \in A$ and 0 when $x \notin A$ (the characteristic function of the set A). The set of all such functions therefore has a power greater than that of the continuum. And certainly the set of all functions on the closed interval $[a, b]$ has a power higher than that of the continuum. We recall that the set of continuous functions on a closed interval has the power of the continuum (Section 5, theorem 7).

Concluding remark

The fundamental ideas of set theory were first formulated at the end of the 19th century in the works of George Cantor (German mathematician 1845–1918), and since that time have pervaded various regions of mathematics, consummating its language quite remarkably. For a more detailed account we recommend *Mengenlehre* by F. Hausdorff (De Gruyter, Berlin, 1935) English translation *Set theory* (Chelsea, New York, 1957) and A. A. Fraenkel's *Foundation of set theory* (Amsterdam, 1958).

CHAPTER II

METRIC SPACES

1. Definition and Examples of Metric Spaces. Isometry

One of the most important concepts in mathematical analysis is that of the limiting process; it underlies such fundamental analytic operations as differentiation and integration.

A number x is said to be the limit of a sequence of real numbers x_n if the distance between x_n and x, i.e. the modulus of the difference $x - x_n$ tends to zero as $n \to \infty$. Thus the concept of the passage to the limit is based on the possibility of measuring the distance between points on the real line. Similarly the concept of the limiting process in the plane or in multidimensional space is based on the possibility of measuring the distance between points of the corresponding sets. We introduce the further concept of a metric space; this term will apply to an aggregate of objects to which are assigned mutual "distances" that fulfil certain intrinsic conditions. These "distances" facilitate the investigation of the properties of the limiting process "in its pure form", i.e. independently of the characteristics of the particular elements.

1. *Definition.* An arbitrary set M of elements ("points") $x, y, \ldots$ is termed a *metric space* if: (1) there is a rule that for any two points x, y determines a number $\varrho(x, y)$ ("the distance from x to y"), (2) this rule fulfils the following requirements (axioms):

(1) $\varrho(y, x) = \varrho(x, y)$ for any x and y (the symmetry of distance);

(2) $\varrho(x, y) > 0$ when $x \neq y$; $\varrho(x, x) = 0$ for any x;

(3) $\varrho(x, z) \leqq \varrho(x, y) + \varrho(y, z)$ for any x, y, z (the triangle inequality).

Examples. 1. Any set M on the real line R, constitutes a metric space with distance-function $\varrho(x, y) = |x - y|$. Similarly a set M in the plane R_2 or in three-dimensional space R_3 is a metric space if we consider as the distance between points the usual geometric distance; for points $x = (\xi_1, \xi_2, \xi_3)$, $y = (\eta_1, \eta_2, \eta_3)$ in R_3

$$\varrho(x, y) = \sqrt{(\xi_1 - \eta_1)^2 + (\xi_2 - \eta_2)^2 + (\xi_3 - \eta_3)^2}.$$

The triangle inequality (axiom 3) is in this instance the usual geometric inequality: the third side of a triangle does not exceed the sum of the other two sides.

Similarly, in n-dimensional space R_n, the distance between points $x = (\xi_1, \dots, \xi_n)$, $y = (\eta_1, \dots, \eta_n)$ can be defined by the formula

$$\varrho(x, y) = \sqrt{\sum_{j=1}^{n} (\xi_j - \eta_j)^2}, \tag{1}$$

so that any set M in n-dimensional space constitutes a metric space with distance-function (1).

Axioms 1 and 2 are evidently satisfied here. To verify that axiom 3 is satisfied we apply Cauchy's inequality†

$$\left|\sum_{j=1}^{n} a_j b_j\right| \leqq \sqrt{\sum_{j=1}^{n} a_j^2}\sqrt{\sum_{j=1}^{n} b_j^2},$$

which holds for any $a_1, \dots, a_n, b_1, \dots, b_n$. Substituting in this inequality $a_j = \xi_j - \eta_j$, $b_j = \eta_j - \zeta_j$, we find that

$$\begin{aligned}
\varrho^2(x, z) &= \sum_{j=1}^{n} (\xi_j - \zeta_j)^2 = \sum_{j=1}^{n} [(\xi_j - \eta_j) + (\eta_j - \zeta_j)]^2 \\
&= \sum_{j=1}^{n} (\xi_j - \eta_j)^2 + 2\sum_{j=1}^{n} (\xi_j - \eta_j)(\eta_j - \zeta_j) + \sum_{j=1}^{n} (\eta_j - \zeta_j)^2 \\
&\leqq \sum_{j=1}^{n} (\xi_j - \eta_j)^2 + 2\sqrt{\sum_{j=1}^{n} (\xi_j - \eta_j)^2}\sqrt{\sum_{j=1}^{n} (\eta_j - \zeta_j)^2} + \\
&\quad + \sum_{j=1}^{n} (\eta_j - \zeta_j)^2 = \left(\sqrt{\sum_{j=1}^{n} (\xi_j - \eta_j)^2} + \sqrt{\sum_{j=1}^{n} (\eta_j - \zeta_j)^2}\right)^2 \\
&= [\varrho(x, y) + \varrho(y, z)]^2, \text{ as required.}
\end{aligned}$$

† We give a proof of this inequality. We put $A = \sum a_j^2$, $B = \sum b_j^2$, $C = \sum a_j b_j$ we have to prove

$$C^2 \leqq AB \tag{*}$$

this inequality will be satisfied if the quadratic polynomial

$$P(\lambda) = A\lambda^2 + 2C\lambda + B$$

does not have distinct real zeros. But

$$P(\lambda) = \sum a_j^2 \lambda^2 + 2\sum a_j b_j \lambda + \sum b_j^2 = \sum (a_j \lambda + b_j)^2,$$

so that $P(\lambda)$ cannot have more than one real zero, namely $\lambda = -b_1/a_1 = \dots = -b_n/a_n$. Thus the inequality (*) holds.

2. In problems of analysis, spaces occur in which the elements are functions (functional spaces).

The choice of a particular metric in functional spaces depends on the requirements of the problem. When there is a geometric distance, it is clear that those elements that are close together must be considered close in terms of the metric. In analysis it is necessary for the most part to proceed the other way about: it is clear from the conditions of the problem what elements it is natural to regard as close together, and a definition of distance is introduced accordingly.

For instance, it is often natural to regard continuous functions $x(t)$ and $y(t)$ $(a \leqq t \leqq b)$ as close together if $\max\limits_{a \leqq t \leqq b} |x(t) - y(t)|$ is small. This quantity can be taken as a definition of the distance between $x(t)$ and $y(t)$; such a definition obviously satisfies axioms 1–3, and consequently any set M of functions, defined and continuous on the closed interval $[a, b]$ constitutes a metric space with metric

$$\varrho(x, y) = \max_{a \leqq t \leqq b} |x(t) - y(t)|. \tag{2}$$

3. In some cases (for example, in the calculus of variations), which involve functions that have derivatives up to the kth order, it is natural to consider close together those elements $x(t)$ and $y(t)$, for which the values, not only of the functions themselves, but also of their derivatives up to order k, are close together. The distance formula

$$\varrho(x, y) = \max_{a \leqq t \leqq b} \{|x(t) - y(t)|,\ |x'(t) - y'(t)|,\ \ldots,\ |x^{(k)}(t) - y^{(k)}(t)|\} \tag{3}$$

meets this requirement. A set of functions $x(t)$ with continuous derivatives up to order k evidently constitutes a metric space with the metric (3).

4. In other cases (for example, in the theory of integral equations), it is natural to regard functions $x(t)$ and $y(t)$ as close together if they are close in the integral sense, i.e. if

$$\int_a^b |x(t) - y(t)| \mathrm{d}t$$

is small. It is natural in this case to define distance by the formula

$$\varrho(x, y) = \int_a^b |x(t) - y(t)|\,\mathrm{d}t. \tag{4}$$

The axioms for a metric space are then clearly satisfied.

5. Sometimes it is necessary to define the proximity of functions, not linearly in terms of their difference, but in terms of some power of their difference; distance can then be given by the formula

$$\varrho^p(x, y) = \int_a^b |x(t) - y(t)|^p\,\mathrm{d}t. \tag{5}$$

For $p \geqq 1$, this definition satisfies the axioms for a metric space, though (with the exception of the simple cases $p = 1$ and $p = 2$) the verification of axiom 3 becomes rather involved; we shall not give it here.†

Thus the definition of a metric space is sufficiently elastic to meet the very diverse concrete requirements of mathematical analysis. In reviewing our whole course, we shall be convinced of the accuracy of this observation.

The metric space of *all* functions continuous on the closed interval $a \leqq t \leqq b$, with metric defined by (2), is denoted by $C(a, b)$. The metric space of *all* functions continuous on $[a, b]$, and having continuous derivatives up to order k, with metric defined by (2), is denoted by $D_k(a, b)$; we put $D_0(a, b) = C(a, b)$. The metric space of all functions continuous on $[a, b]$ together with the metric (5) is denoted by $C_p(a, b)$.

2. The inequality expressed in axiom 3 can be generalized for the case of any m elements $x_1, x_2, \dots, x_m$, in the form

$$\varrho(x_1, x_m) \leqq \varrho(x_1, x_2) + \varrho(x_2, x_3) + \cdots + \varrho(x_{m-1}, x_m).$$

This inequality follows from successive applications of axiom 3:

$$\varrho(x_1, x_m) \leqq \varrho(x_1, x_2) + \varrho(x_2, x_m) \leqq \varrho(x_1, x_2) + \varrho(x_2, x_3) + \\ + \varrho(x_3, x_m) \leqq \cdots$$

We observe the following simple property of metrics, which may be termed the "*quadrangle inequality*": for any four points x, y, z, u of a metric space, we have

$$|\varrho(x, z) - \varrho(y, u)| \leqq \varrho(x, y) + \varrho(z, u). \tag{1}$$

† Cf. Chapter IV, Section 5, art. 3.

Geometrically, this means that the difference between a pair of non-covertical sides of a quadrangle never exceeds the sum of such a pair. The proof follows from the inequalities

$$\varrho(x, z) \leqq \varrho(x, y) + \varrho(y, u) + \varrho(u, z),$$
$$\varrho(y, u) \leqq \varrho(y, x) + \varrho(x, z) + \varrho(z, u),$$

if from the first we subtract $\varrho(y, u)$ and from the second $\varrho(x, z)$. When $z = u$ the quadrangle inequality reduces to the *second triangle inequality*.

$$|\varrho(x, z) - \varrho(y, z)| \leqq \varrho(x, y), \tag{2}$$

which also has frequent application.

3. In the theory of sets the concept of *equivalence* played an essential part. Equivalent sets, i.e. sets in one–one correspondence with each other, were from the pure set-theoretic point of view absolutely identical, although they were composed of completely different elements. It was shown, for example, that the set of points of a closed interval and the set of functions defined and continuous on that interval have the same power, and so, in set-theoretic terms, there is no distinction whatever to be drawn between these sets.

But if we are considering two sets that constitute metric spaces (and interest us in this capacity), set-theoretic equivalence is not a sufficient criterion for us to consider them identical as spaces, since they may possess quite distinct metric relations. For instance, the metric spaces formed by the points of the closed interval $[a, b]$, and the functions continuous on this interval, though set-theoretically equivalent, are dissimilar in respect of their metrics, because in the first space distances between elements are bounded by the constant $b - a$, while in the second they are not bounded at all; many other differences could also be indicated. The following definition then suggests itself:

Two metric spaces are said to be *isometric* if it is possible to set up a one–one correspondence between their elements that preserves the distance between corresponding pairs of elements.

In other words, if M and M' are isometric spaces and to elements $x, y \in M$ correspond elements $x', y' \in M'$, then $\varrho(x, y) = \varrho(x', y')$.

For example, the spaces $C(0, 1)$ and $C(0, 2)$ of functions continuous on the closed intervals [0, 1] [0, 2], respectively are isometric. The correspondence between their elements can be expressed in the formula

$$C(0, 1) \ni x(t) \leftrightarrow y(t) = x\left(\frac{t}{2}\right) \in C(0, 2).$$

It is easily seen that this correspondence is one–one and distance-preserving as required.

2. Open Sets

1. The aggregate of all points x of a metric space M whose distance from a given point x_0 is less than a fixed magnitude $r > 0$, so that

$$\varrho(x, x_0) < r,$$

is termed a *sphere* (more precisely, an *open sphere*) of radius r; the point x_0 is said to be the *centre* of this sphere. The aggregate of all points x that satisfy the inequality

$$\varrho(x, x_0) \leqq r,$$

is termed a *closed sphere* of radius r. Finally, the set of all points situated precisely at a distance r from the point x_0, so that

$$\varrho(x, x_0) = r,$$

constitutes a *spherical surface* of radius r with centre x_0. We now give the following important definition.

A set U in a metric space M is said to be an *open set* or *region*, if each point $x_0 \in U$ is an *interior point* of U, i.e. if it is the centre of some open sphere contained in U (the radius of which will, in general, depend on x_0).

Thus the open sphere

$$U = \{x : \varrho(x, x_1) < r\}$$

with centre at some point x_1 is an open set. To see this, let $x_0 \in U$ so that $\varrho(x_0, x_1) = \theta < r$. We consider a sphere U_0 with centre x_0 and radius $r_0 < r - \theta$, and show that U_0 is contained in U. In fact, for any $x \in U_0$, we have by the triangle inequality

$$\varrho(x, x_1) \leqq \varrho(x, x_0) + \varrho(x_0, x_1) < r_0 + \theta < r - \theta + \theta = r,$$

as required.

Operations with open sets. The union of any number of open sets is evidently also an open set.

The intersection of any *finite number* of open sets is an open set. For let the point x_0 belong to the open sets $U_1, U_2, \ldots, U_m$, and let it be contained in the first of these together with a sphere of radius r_1 (with centre x_0), in the second together with a sphere of radius r_2, etc.; then the sphere with centre x_0; radius $\min(r_1, \ldots, r_m)$, is contained in each of the sets $U_1, \ldots, U_m$, and is therefore contained in their intersection as well.

For an infinite number of open sets, the foregoing reasoning breaks down, since the minimum (more precisely, the exact lower bound) of an infinite set of positive numbers may be zero. And in fact the intersection of the infinite aggregate of open sets

$$U_n = \left\{x : \varrho(x, x_0) < \frac{1}{n}\right\} \qquad (n = 1, 2, \ldots)$$

contains only those points x for which $\varrho(x, x_0) = 0$, i.e., by axiom 2, only the point x_0; this intersection is not, therefore, an open set.

2. On the real axis $-\infty < x < \infty$, every open interval (α, β) (bounded or unbounded) is evidently an open set. The union of a finite or countable number of open intervals (α_ν, β_ν) $(\nu = 1, 2, \ldots)$ without common points is also an open set. We shall show that *every open set U on the real axis is the union of a finite or countable number of disjoint open intervals.*

We consider an arbitrary point $x \in U$. By definition, U contains together with x some sphere, i.e. some open interval on the real axis that contains x. We now construct the largest open interval that contains x and is contained entirely in U.

We denote by S the set of points that lie to the right of x and do not belong to U. If S is empty the entire half-line (x, ∞) is contained in U. If S is non-empty it possesses an exact lower bound ξ. This point ξ cannot belong to U, since any point in U has a neighbourhood which lies entirely in U and cannot therefore contain a single point of S. In particular $\xi \neq x$. It is also clear that the whole open interval (x, ξ) is contained in U.

We proceed similarly with points lying to the left of x and obtain an open interval (η, x), contained in U, the left-hand end of which does not belong to U (the interval may be the whole of the half-line $(-\infty, x)$).

Thus, given the point $x \in U$, we have formed the interval (η, ξ) belonging to U and such that its end-points (of which either or both may lie at infinity) do not belong to U. Such an interval is said to be a *component open interval of the open set U.*

If two component intervals (η_1, ξ_1), (η_2, ξ_2) have a common point x_0, they coincide; for the inequality $\xi_1 < \xi_2$, say, is impossible, since ξ_1 must, on the one hand, as an interior point of interval (x_0, ξ_2), belong to U, and on the other, as an end-point of the interval (x_0, ξ_1), cannot belong to U.

The whole set U is therefore the union of disjoint component intervals. This union is at most countable, since in each of the component open intervals of U, a rational point can be chosen, and the rationals form a countable set. Our proof is therefore complete.

Problems. 1. If a set E on the line is covered by an arbitrary system of open intervals, an (at most countable) subsystem can be extracted which also covers E.

Hint. Of the open intervals that constitute the covering, select those with rational end-points, and discard one by one those intervals in the covering that contain a selected interval.

2. If a set E in the plane is covered by an arbitrary system of open discs, an (at most countable) subsystem can be extracted which also covers E.

Hint. Cf. problem 1.

3. A metric space M is said to possess a *countable base* if there exists a countable system of open sets $U_1, U_2, \ldots, U_n, \ldots$, such that for any point $x \in M$ and any region U that contains x, there exists k such that $x \in U_k \subset U$. Show that a theorem analogous to those of problems 1 and 2 holds in any metric space with a countable base.

4. Prove that the set E_0 of interior points of any set E is open (provided it is non-empty).

5. Prove that the aggregate of all open sets on the line has the power of the continuum.

Hint. Use theorem 6, Chapter I, Section 5.

3. Convergent Sequences and Closed Sets

1. We shall say that a sequence of points $x_1, x_2, \ldots, x_\nu, \ldots$ of a metric space M *converges to a point x* of that space if

$$\lim_{\nu \to \infty} \varrho(x, x_\nu) = 0,$$

or alternatively, that the sequence $x_1, x_2, \ldots$ converges to x if, given any sphere with centre x, it is the case that all points of the sequence from some point onwards lie inside it.

The point x is said to be the *limit* of the sequence $x_1, x_2, \ldots, x_\nu, \ldots$ and is denoted by $\lim x_\nu$.

It is not difficult to show that if the limit x exists, it is determined *uniquely*. In fact, if we had

$$\lim_{\nu\to\infty} \varrho(x, x_\nu) = 0, \quad \lim_{\nu\to\infty} \varrho(y, x_\nu) = 0,$$

then for any given $\varepsilon > 0$, we could find a number N such that the inequalities

$$\varrho(x, x_\nu) < \frac{\varepsilon}{2}, \quad \varrho(y, x_\nu) < \frac{\varepsilon}{2} \quad (\nu \geqq N)$$

would hold, and so, by the triangle inequality

$$\varrho(x, y) \leqq \varrho(x, x_\nu) + \varrho(y, x_\nu) \leqq \frac{\varepsilon}{2} + \frac{\varepsilon}{2} < \varepsilon.$$

Since in the inequality obtained ε is arbitrarily small,

$$\varrho(x, y) = 0$$

whence, in virtue of axiom 2,

$$x = y,$$

as required.

Examples. 1. Let a metric space M in n-dimensional space R_n (Section 1, example 1) be such that the distance between the points $x = (\xi_1, \ldots, \xi_n)$, $y = (\eta_1, \ldots, \eta_n)$ is given by the formula

$$\varrho(x, y) = \sqrt{\sum(\xi_j - \eta_j)^2}.$$

We shall explain what is meant by the convergence of the sequence

$$x_\nu = (\xi_1^{(\nu)}, \xi_2^{(\nu)}, \ldots, \xi_n^{(\nu)}) \quad (\nu = 1, 2, \ldots)$$

to the point

$$x = (\xi_1, \ldots, \xi_n).$$

Since

$$\varrho(x, x_\nu) = \sqrt{\sum(\xi_j - \xi_j^{(\nu)})^2},$$

$\varrho(x, x_\nu)$ tends to zero as $\nu \to \infty$ if and only if *all numerical sequences* $\xi_1^{(\nu)}, \xi_2^{(\nu)}, \ldots, \xi_n^{(\nu)}$ $(\nu = 1, 2, \ldots)$ tend to the limits $\xi_1, \xi_2, \ldots, \xi_n$ respectively as $\nu \to \infty$. Briefly, convergence in R_n is convergence with respect to all coordinates.

2. The convergence of a sequence of functions $x(t) \in C(a, b)$ to a function $x(t)$ means that, as $\nu \to \infty$,

$$\varrho(x, x_\nu) = \sup_{a \leq t \leq b} |x(t) - x_\nu(t)| \to 0.$$

In analysis such convergence is termed *uniform convergence.*

3. The convergence of a sequence of functions $x_\nu(t)$ in the space $C_p(a, b)$ to a function $x(t)$ means that, as $\nu \to \infty$,

$$\varrho^p(x, x_\nu) = \int_a^b |x(t) - x_\nu(t)|^p \, dt \to 0.$$

In analysis such convergence is termed *convergence in the mean of order p*, or if p is fixed, simply *convergence in the mean.*

Every uniformly convergent sequence is clearly also convergent in the mean for any p.

But it is easy to construct a sequence of functions, convergent in the mean for any p, but not uniformly convergent. For instance, let $x_\nu(t)$ be a function taking values between zero and unity, non-zero only in an open interval Δ_ν, of length less than $1/\nu$, and taking the value 1 at some point of this interval. Clearly

$$\int_a^b x_\nu^p(t) \, dt < \frac{1}{\nu},$$

so that the sequence is *convergent in the mean to zero.* But $\max x_\nu(t) = 1$ for any, ν, so the sequence does not *converge uniformly to zero.* It can be shown that this sequence does not converge uniformly to any function at all. Moreover the intervals Δ_ν can be so chosen that the sequence fails to converge for a single value of t.

LEMMA. *If $x_\nu \to x$, $y_\nu \to y$, then $\varrho(x_\nu, y_\nu) \to \varrho(x, y)$, (distance is a continuous function of its arguments).*

Proof. By the quadrangle inequality (Section 1, art. 2)

$$|\varrho(x, y) - \varrho(x_\nu, y_\nu)| \leq \varrho(x, x_\nu) + \varrho(y, y_\nu).$$

The right-hand side tends to zero as $\nu \to \infty$, as required.

2. A point x of a metric space M is said to be a *limit point* of a set $F \subset M$ if there exists a sequence $x_1, x_2, \ldots$ of (distinct) points of F convergent to the point x.

Another definition of a limit point, obviously equivalent to the above, runs as follows: a point x is a limit point of a set F if any sphere with centre x contains points of F (distinct from x).

The subset $F \subset M$ is said to be *closed* if it contains all its limit points.

Examples. 1. The interval $a \leqq x \leqq b$ on the real line is closed, but the interval $a \leqq x < b$ is not, since it does not contain its limit point b.

2. In any metric space, the sphere

$$U = \{x : \varrho(x, x_0) \leqq r\}$$

is a closed (and is therefore termed a closed set sphere).

For let us take any point x_1, not belonging to U, so that $\varrho(x_1, x_0) = r_1 > r$. We show that the sphere, centre x_1 radius $1/2\,(r_1 - r)$, contains no points of U; for if it contained such a point, then, denoting it by z, we should have

$$\varrho(x_0, x_1) \leqq \varrho(x_0, z) + \varrho(x_1, z) \leqq r + \frac{1}{2}(r_1 - r) < r_1$$

in contradiction to our construction. It follows that x_1 cannot be a limit point of the set U.

Closed sets in a metric space M bear a close relation to open sets of the space. We have the following theorem:

THEOREM. *A set U, complementary to a closed set F with respect to a metric space M, is always open. A set F, complementary to an open set U, is always closed.*

Proof. Let F be a closed set and U its complement; we shall show that U is open. We consider an arbitrary point $x_0 \in U$; we have to show that there exists a sphere defined by an inequality of the form

$$\varrho(x, x_0) < r,$$

and contained entirely in U.

If this is not the case, we must assume that any sphere with centre x_0 contains points of F. But then, according to the second definition of a limit point, x_0 is a limit point of F, and we must have $x_0 \in F$, which contradicts the hypothesis $x_0 \in U$. U is therefore open.

We turn now to the second part of the theorem. Let U be an open set and F its complement; we shall show that F is closed. Any point x_0 that belongs to U is the centre of some sphere contained in U and cannot, therefore, be a limit point of F. So the limit points of F can lie only in F itself, and consequently F is closed. The theorem is proved.

Recalling Section 2, art. 2, we obtain a general characterisation of all closed sets on the line $-\infty < x < \infty$: *every closed set on the line can be represented as the complement of a finite or countable aggregate of disjoint open intervals.* These intervals, which are the component open intervals of the complement of the closed set, are called the *contiguous open intervals* of the closed set.

Using the known properties of open sets in a metric space and the relation we have just found between open and closed sets, we can prove that the *union of a finite number of closed sets and the intersection of any number of closed sets are again closed sets.*

For let closed sets F_ν be given (ν running through some set of indices), and let $U_\nu = C F_\nu$ be the complementary open sets. By formula (1), Section 2, Chapter I, we have:

$$C \sum_\nu F_\nu = \prod_\nu C F_\nu = \prod U_\nu .$$

If ν runs through a finite set of indices, $\prod U_\nu$ is, in virtue of Section 2, an open set; hence the set $\sum F_\nu$, which is the complement of $\prod U_\nu$, is closed. Moreover, by (2), Section 2, Chapter I

$$C \prod_\nu F_\nu = \sum_\nu C F_\nu = \sum_\nu U_\nu . \tag{5}$$

The set $\sum U_\nu$ is always open; hence its complement $\prod F_\nu$ is always closed, as required.

3. A set A contained in a metric space M is said to be *everywhere dense in M* if each point $b \in M$ is the limit of a sequence of points $a_n \in A$ (not necessarily distinct). In other words, A is everywhere dense in M if any sphere whose centre is a point $b \in M$ contains a point $a \in A$.

Thus the set of rationals is everywhere dense on the line $-\infty < x < \infty$. By a well-known theorem of Weierstrass, every continuous function $f(x)$ on the closed interval $[a, b]$ can be expressed as the limit of a uniformly convergent sequence of polynomials; hence the set of all polynomials is everywhere dense in the space $C(a, b)$.

The property of being everywhere dense possesses a peculiar "transitivity": if a set A is everywhere dense in a space M and M in turn is everywhere dense in a larger space P, then A, considered as a subset of P, is everywhere dense in P. For since M is everywhere dense in P we can find for a given $x \in P$ and $\varepsilon > 0$

a point $b \in M$ such that $\varrho(b, x) < \varepsilon/2$, and since further, A is everywhere dense in M we can find a point $a \in A$ such that $\varrho(a, b) < \varepsilon/2$. By the triangle inequality $\varrho(a, x) \leqq \varrho(a, b) + \varrho(b, x) < \varepsilon$. Thus in any sphere of radius ε with centre at a point $x \in P$ there exists a point $a \in A$, as required.

Example. Let us show that the totality of trigonometric polynomials $T(x) = \sum_{k=0}^{n} (a_k \cos kx + b_k \sin kx)$ constitutes a set everywhere dense in the space $C_p(-\pi, \pi)$. It is well-known from analysis that every continuous function $f(x)$ on $C_p(-\pi, \pi)$ which possesses a piecewise-continuous derivative and satisfies the condition $f(-\pi) = f(\pi)$ has a Fourier series development

$$f(x) = \frac{a_0}{2} + \sum_{n=1}^{\infty} a_n \cos n x + b_n \sin n x,$$

which converges uniformly, i.e. with respect to the metrics of both the spaces $C(-\pi, \pi)$ and $C_p(-\pi, \pi)$.

Thus the trigonometric polynomials form a set A which is everywhere dense in the set M of continuous functions specified. Further, any continuous function $\varphi(x)$ on the closed interval $[-\pi, \pi]$ that satisfies the condition $\varphi(-\pi) = \varphi(\pi)$ is the limit of a uniformly convergent sequence of functions $f_n(x)$ in M, for instance, of the polygonal lines which coincide with $\varphi(x)$ at the points $k\pi/n$ $(k = 0, \pm 1, \ldots, \pm n)$. Finally any continuous function $g(x)$ is the limit of a sequence of continuous functions $\varphi_n(x)$ with $\varphi_n(-\pi) = \varphi_n(\pi)$ which converges in the metric of the space $C_p(-\pi, \pi)$; for example, we can put $\varphi_n(x) = g(x)$ for $|x| \leqq \pi - 1/n$, $\varphi_n(-\pi) = \varphi_n(\pi) = 0$, and define $\varphi_n(x)$ on the remaining intervals $(-\pi, -\pi + 1/n)$, $(\pi - 1/n, \pi)$ linearly. In virtue of the transitivity mentioned, we get that the set of trigonometric polynomials is everywhere dense in the space $C_p(-\pi, \pi)$, as required.

4. Let there be given an arbitrary subset A of a metric space M; we denote by $\bar{A}$ the union of the set A with all its limit points (not contained in it). If A is a closed set, then $\bar{A} = A$; and conversely, if $\bar{A}$ coincides with A, then A contains all its limit points and is therefore closed. In the general case the set $\bar{A}$ is said to be the *closure* of the set A. From the definition of closure it follows easily that a given point $b \in M$ belongs to the set $\bar{A}$ if and only if there exists in any sphere with centre b a point $a \in A$ (possibly coincident

with b). In particular, it is clear that A is everywhere dense in its own closure; and conversely if A is everywhere dense in some set Q, then $Q \subset \bar{A}$.

Utilising this last observation, we shall show that *the closure of an arbitrary set $A \subset M$ is always closed.* In other words, the closure $\bar{\bar{A}}$ of the set $\bar{A}$ coincides with $\bar{A}$ itself. We know by hypothesis that A is everywhere dense in $\bar{A}$ and that $\bar{A}$ is everywhere dense in $\bar{\bar{A}}$; hence in accordance with art. 3 A is everywhere dense in A and therefore $\bar{\bar{A}}$ is contained in the closure of A, i.e. $\bar{\bar{A}} \supset \bar{A}$; but this means that $\bar{A}$ is closed.

We observe further that every closed set F that contains A must, in addition, contain all the limit points of A, and hence the whole set $\bar{A}$. Since, as we have shown, $\bar{A}$ is closed, it can be characterised as *the smallest closed set that contains the set A.*

Examples. 1. The closure of the set A of all rational points on the real line is the aggregate of all the points (rational and irrational) on the line.

2. The closure in the space $C(a, b)$ of the set of all polynomials $P(x) = a_0 + a, x \ldots + a_n x^n$ is the whole space $C(a, b)$.

Problems. 1. Given any set A, we denote by A' the set of its limit points. Construct a set A on the line, such that $A'' = (A')'$ is non-empty but A''' is empty.

2. Prove that the set A' (cf. problem 1) is closed, whatever the set A.

3. Given that A' is countable, show that A is countable (assuming A to be a subset of the line).

4. Prove that the result of problem 3 holds in any metric space with a countable base (Section 2, problem 3).

5. A point x on the line is said to be an *accumulation point* of an uncountable set A if every neighbourhood of x contains an uncountable subset of A. Show that every uncountable set A has points of accumulation; moreover, almost all its points, except perhaps for a countable set, are accumulation points.

Hint. A point which is not an accumulation point of the set A can be covered by an interval with rational end-points and containing an at most countable number of points of the set.

6. Show that the result of problem 5 is valid in any metric space with a countable base.

7. Prove that the set M of points in the plane lying on the unit circle Γ with centre the origin and having polar angle coordinates $1, 2, \ldots, n, \ldots$ is everywhere dense on Γ.

Hint. If an arc $\Delta_0 \subset \Gamma$ does not contain any points of the set M, then neither do the arcs $\Delta_1 = \Delta_0 - 1$, $\Delta_2 = \Delta_0 - 2$, etc. The arcs $\Delta_0, \Delta_1, \Delta_2, \ldots,$

of equal length, and lying on Γ, cannot be mutually disjoint. If, say, the arcs Δ_k and Δ_{k+m} intersect, then the union of $\Delta_k, \Delta_{k+m}, \Delta_{k+2m}, \ldots$ covers the whole of Γ, which is impossible.

8. The quantity

$$\varrho(x, A) = \inf_{y \in A} \varrho(x, y)$$

is said to be the distance of the point x from the set A. Show that for a closed set A the relations

$$\varrho(x, A) = 0, \quad x \in A,$$

are equivalent; but if A is not closed, this is not the case.

9. Show that for any set A the aggregate of points x for which $\varrho(x, A) < \varepsilon$ is open, while the aggregate of points y for which $\varrho(y, A) \leqq \varepsilon$ is closed.

10. Show that every closed set F on the line is the intersection of a countable set of open sets. Similarly for a metric space with a countable base.

Hint. Put $U_n = \{x : \varrho(x, F) < 1/n\}$.

11. Given two disjoint closed sets F_1, F_2, construct two disjoint open sets U_1, U_2, such that $U_1 \supset F_1$, $U_2 \supset F_2$.

Hint. Put $U_1 = \bigcup_{x \in F_1} \{y : \varrho(x, y) < 1/2\varrho(x, F_2)\}$; similarly for U_2.

12. Show that every open set is the union of a countable set of closed sets (on the line or in a space with a countable base).

13. Show that the projection of a plane, closed, bounded set on the line is a closed set. Is it necessary to assume boundedness?

Hint. As an example, consider the projection of a rectangular hyperbola on an asymptote.

14. A metric space M is said to be *separable* if it contains a countable set A which is dense in it (so that the closure of A coincides with the whole of M). Show that the existence of a countable base is equivalent to separability.

15. Show that the spaces $C(a, b)$ and $D_m(a, b)$ are separable.

Hint. The set of polynomials with rational coefficients can be taken as a countable, everywhere dense set.

4. Complete Spaces

1. A sequence of points $x_1, x_2, \ldots, x_\nu, \ldots$ in a metric space M is said to be *fundamental* if, for any $\varepsilon > 0$, there exists a number N, such that for $\mu, \nu > N$,

$$\varrho(x_\nu, x_\mu) \leqq \varepsilon.$$

In such instances, we shall use the abbreviation

$$\lim_{\nu, \mu \to \infty} \varrho(x_\nu, x_\mu) = 0.$$

For example, any convergent sequence is fundamental. For by the triangle inequality,

$$\varrho(x_\nu, x_\mu) \leqq \varrho(x_\nu, x) + \varrho(x, x_\mu),$$

and if $x_\nu \to x$, then for sufficiently large μ, ν, the right-hand side becomes smaller than any preassigned ε.

If M is the real line with the usual metric, the concept of a fundamental sequence of points coincides with the classical concept of a fundamental numerical sequence. In the theory of real numbers there is a criterion due to Cauchy according to which every fundamental numerical sequence converges.

Cauchy's criterion is not valid in a general metric space.

We consider the open interval (0, 1); it represents a metric space with the usual metric for the real line. The sequence $1/2, 1/3, \ldots, 1/n, \ldots$ is clearly fundamental in this space, but does not converge in it. Cauchy's criterion is not, therefore, applicable in the metric space (0, 1).

2. Thus in a general metric space it is not possible to employ Cauchy's criterion. If the criterion turns out to be valid in certain particular metric spaces, this is due to the special properties of these spaces. We distinguish the class of such spaces by means of the following definition:

A metric space M is said to be *complete* if, in it, every fundamental sequence converges.

Examples. 1. We show that the n-dimensional space R_n with metric

$$\varrho(x, y) = \sqrt{\sum (\xi_j - \eta_j)^2}$$

is complete. Let $x_\nu = (\xi_1^{(\nu)}, \ldots, \xi_n^{(\nu)})$ be a fundamental sequence. Since

$$|\xi_j^{(\nu)} - \xi_j^{(\mu)}|^2 \leq \sum_j |\xi_j^{(\nu)} - \xi_j^{(\mu)}|^2 = \varrho^2(x_\nu, x_\mu),$$

the numerical sequence $\xi_j^{(\nu)}$ $(\nu = 1, 2, \ldots)$ is fundamental for every fixed $j = 1, 2, \ldots, n$, and as such has some limit ξ_i. The numbers $\xi_1, \xi_2, \ldots, \xi_n$ determine a vector $x \in R_n$. Since

$$|x - x_\mu|^2 = \sum_i [\xi_i - \xi_i^{(\mu)}]^2 \to 0 \quad \text{as} \quad \mu \to \infty,$$

the vector x is the limit of the given fundamental sequence. Hence every fundamental sequence in the space R_n converges in it.

2. The space $C(a, b)$ is complete. In fact, if the sequence of functions $y_n(x) \in C(a, b)$ is fundamental, then as $\mu, \nu \to \infty$,

$$\sup_{a \leq x \leq b} |y_\nu(x) - y_\mu(x)| \to 0. \tag{1}$$

We fix x; in virtue of (1), the numbers $y_n(x)$ form a fundamental numerical sequence, which by the classical Cauchy criterion must converge. Let $y(x)$ be the limit of $y_\nu(x)$ as $\nu \to \infty$. Letting $\nu \to \infty$ in the inequality

$$\sup_{a \leqq x \geqq b} |y_\nu(x) - y_\mu(x)| \leqq \varepsilon \quad (\nu, \mu > N = N(\varepsilon))$$

We obtain

$$\sup_{a \leqq x \leqq b} |y(x) - y_\mu(x)| \leqq \varepsilon \quad (\mu > N = N(\varepsilon)). \tag{2}$$

Hence $y(x)$ is the limit of a uniformly convergent sequence of continuous functions $y_n(x)$ and by a well-known theorem of analysis is therefore continuous. It follows further from (2) that $\varrho(y, y_\mu) \to 0$, so that in the space $C(a, b)$ every fundamental sequence is convergent; $C(a, b)$ is therefore a complete space.

3. A closed subset A of a complete metric space M, when considered as an independent metric space (with a metric borrowed from M), is itself complete. For every fundamental sequence $y_\nu \in A$ converges in M (since M is complete), and its limit belongs to A since A is closed. Conversely if a subset $A \subset M$ is known to be a complete metric space, then A is closed in M. For if A were not closed in M we could find a sequence $\{x_n\}$ in A that converged to some point $y \in M - A$. But the sequence $\{x_n\}$ is fundamental (since the metric in A is borrowed from M), and since A is complete, it must possess a limit in A. We should then have a sequence with two distinct limits, one in A, the other outside A, which is impossible.

4. The space $C_p(a, b)$ is not complete for any p. For the proof we consider a sequence of continuous functions $y_\nu(x)$, taking values between 0 and 1, and, as $\nu \to \infty$, tending uniformly to 0 on every open interval $(a, c - \varepsilon)$, and to 1 on every open interval $(c + \varepsilon, b)$ (where c is a fixed point between a and b). This is a Cauchy sequence, since

$$\int_a^b |y_\nu(x) - y_\mu(x)|^p \mathrm{d}x = \int_a^{c-\varepsilon} + \int_{c-\varepsilon}^{c+\varepsilon} + \int_{c+\varepsilon}^{b} \leqq \varepsilon + 2\varepsilon + \varepsilon = 4\varepsilon$$

for sufficiently large μ, ν. But *at the same time the sequence* $y_\nu(x)$ *is not convergent in the mean to any continuous function.*

To prove the last assertion we make the following observation. *If a sequence of functions* $f_\nu(x)$ $(\nu = 1, 2, \ldots)$ *is convergent in the mean over an interval* $\Delta = \{a \leqq x \leqq b\}$ *to a continuous function* $f(x)$, *and also converges uniformly to a function* $\varphi(x)$ *in some interval* $\delta = \{c \leqq x \leqq d\}$ *interior to* Δ, *then in the interval* δ, $\varphi(x) \equiv f(x)$ *identically*. For in the space $C_p(c, d)$ we have the relations

$$\varrho^p(f_\nu, f) = \int_c^d |f_\nu(x) - f(x)|^p \, dx \leqq \int_a^b |f_\nu(x) - f(x)|^p \, dx \to 0,$$

$$\varrho^p(f_\nu, \varphi) = \int_c^d |f_\nu(x) - \varphi(x)|^p \, dx \leqq \max_{x \in \delta} |f_\nu(x) - \varphi(x)|^p (d - c) \to 0.$$

And in virtue of the uniqueness of limits (Section 2), we have $f(x) \equiv \varphi(x)$.

If we suppose that the sequence $y_1(x), y_2(x), \ldots, y_\nu(x), \ldots$ constructed above is convergent in the mean to some continuous function $f(x)$, then by what we have proved we must have $f(x) = 0$ for $a \leqq x < c$, $f(x) = 1$ for $c < x \leqq b$. But there is obviously no value $f(c)$ for which the function $f(x)$ will be continuous over the closed interval $a \leqq x \leqq b$.

Problems. 1. Taking the metric

$$\varrho(x, y) = |\tan^{-1} x - \tan^{-1} y|$$

on the line $\{-\infty < x < \infty\}$, verify that all the axioms for a metric space are satisfied. Is this space complete?

Answer. The space is incomplete; the sequence $x_n = n$ $(n = 1, 2, \ldots)$ is fundamental but has no limit.

2. Show that the space $D_m(a, b)$ is complete for all m.

3. Is the space of all numerical sequences

$$x = (\xi_1, \xi_2, \ldots, \xi_n, \ldots),$$

where $\xi_n \to 0$ as $n \to \infty$, with metric given by

$$\varrho(x, y) = \max_n |\xi_n - \eta_n|$$

complete?

Answer. Yes.

4. Consider the following three spaces of functions on the real line

(a) all bounded continuous functions;

(b) all continuous functions for which $\lim_{|x| \to \infty} f(x) = 0$;

(c) all continuous functions, each identically zero outside some open interval.

Take as a metric in these spaces

$$\varrho(f, g) = \sup_x |f(x) - g(x)|.$$

Are the given spaces complete?

Answer. Complete in cases (a) and (b), incomplete in (c).

3. Lemma concerning closed spheres. The completeness of the real line is used in analysis to prove a well-known lemma about embedded closed intervals: a sequence of embedded closed intervals has a common point. An analogous result holds in a complete metric space with closed spheres substituted for closed intervals:

Lemma concerning closed spheres: *In a complete metric space, a sequence of embedded closed spheres,*

$$U_\nu = \{y : \varrho(y, y_\nu) \leqq r_\nu\}, \quad \nu = 1, 2, \ldots,$$

the radii r_ν of which tend to zero as $\nu \to \infty$, has a common point.

Proof. For all μ, the centre $y_{\nu+\mu}$ of the sphere $U_{\nu+\mu}$ lies in the sphere U_ν, so that

$$\varrho(y_\nu, y_{\nu+\mu}) \leqq r_\nu,$$

hence the sequence of centres $y_1, y_2, \ldots, y_\nu, \ldots$ is fundamental. Since the space, M say, is complete, this sequence has a limit y_0. Since the sphere U_ν is a closed set in M (Section 3, art. 2, example 2) and the points $y_\nu, y_{\nu+1}, \ldots$ belong to it, $y_0 = \lim_{\mu\to\infty} y_{\nu+\mu}$ also belongs to it; the number ν here is arbitrary, hence y_0 belongs to all the spheres U_ν, as asserted.

In Chapter I, using a theorem on embedded closed intervals, we proved that the point-set of the closed interval [0, 1] was uncountable. We can now reason analogously for a wide class of complete metric spaces.

First we give the following definition: a point x_0 of a metric space M is said to be *isolated* if there exists a sphere $\varrho(x, x_0) \leqq \delta$ containing no points of M with the exception of the point x_0 itself. Thus if M is some point-set on the line $-\infty < x < \infty$ with the usual metric, $x_0 \in M$ is an isolated point if there exists an open interval containing x_0, but not containing any other point of M.

THEOREM 1. *Every complete metric space M without isolated points is uncountable.*

Proof. Let us suppose, on the contrary, that the points of M form a countable set. Then they can all be arranged in a sequence $x_1, x_2, \ldots, x_n, \ldots$ Let y_1 be a point of M distinct from x_1; we denote by U_1 a sphere with centre y_1 not containing (either within, or on the boundary) the point x_1. Since y_1 is not isolated, this sphere contains other points of M. Let y_2 be an interior point of U_1, distinct from x_2; we denote by U_2 a sphere with centre y_2, contained entirely in U_1 and not containing the point x_2. Continuing this process, we obtain a sequence of spheres $U_1 \supset U_2 \supset \ldots$, the radii of which tend to zero, and in which the sphere U_n does not contain the point x_n. The common point ξ of all the U_n, which exists in virtue of the lemma just proved, cannot coincide with any of the points $x_1, x_2, \ldots, x_n \ldots$, in view of our construction. Hence the sequence $x_1, x_2, \ldots$ cannot exhaust the whole space M, contrary to our hypothesis; the theorem is proved.

We saw above (example 3) that a closed set in a complete metric space is itself a complete metric space. Hence in a complete metric space any closed set without isolated points is uncountable.

Note. If we reject the hypothesis that the space M contains no isolated points, the theorem ceases to be true. Any countable closed set (e.g. a convergent sequence) on the line provides a suitable counter-example when considered as an independent metric space.

We consider now a closed set F on the real axis. It is evident that every isolated point of F is a common end-point of two open intervals contiguous to F, or, what amounts to the same thing, is a common end-point of two component open intervals of the complementary open set.

A closed set without isolated points is said to be *perfect*. We saw (Section 3) that every closed set on the line can be obtained by extracting some set of disjoint open intervals; it is now apparent that a perfect set can be obtained by the extraction of some set of open intervals, not merely disjoint, but also without common end-points.

Any closed interval $[a, b]$ will serve as an example of a perfect set. But it is easy to construct a perfect set that does not contain a single closed interval as a subset. In particular, we consider the construction of the so-called *Cantor set.*

The Cantor set on the closed interval $[0, 1]$ is constructed as follows. We first extract the open interval $(1/3, 2/3)$ of length $1/3$, which forms the middle third of the whole closed interval. Then we proceed similarly with each of the two remaining closed intervals $[0, 1/3]$, $[2/3, 1]$, i.e. we extract the middle third from each, the open interval $(1/9, 2/9)$ from $[0, 1/3]$, and $(7/9, 8/9)$ from $[2/3, 1]$. A similar procedure obtains for each of the four remaining closed intervals $[0, 1/9]$, $[2/9, 1/3]$, $[2/3, 7/9]$, $[8/9, 1]$, and the process is continued indefinitely. As a result, we are left with a closed set called the Cantor set. Since the intervals extracted in the construction had no common end-points, the Cantor set is perfect. And since of every pair of adjacent open intervals with vertices at the points $p/3^q$, $(p + 1)/3^q$, $(p + 2)/3^q$, one at least was extracted, the Cantor set cannot contain a single closed interval. Nevertheless, in virtue of the theorem just proved, the Cantor set is uncountable.

More precisely, the power of the Cantor set is the power of the continuum. This is a consequence of the following theorem, which applies to any perfect set on a closed interval:

THEOREM 2 (G. Cantor). *Every perfect set on the closed interval* $[a, b]$ *has the power of the continuum.*

Proof. We shall assume that the points a, b are exact bounds of the set F; then, since F is closed, we have $a \in F$, $b \in F$.

If the perfect set F contains even one closed interval $[\alpha, \beta]$, the assertion of the theorem is obvious. We shall consider the case when F contains no closed intervals at all. In this case, between any two open intervals contiguous to F there is another contiguous open interval, and so the number of contiguous open intervals is infinite. Since F contains no isolated points, the intervals contiguous to F are not only disjoint, but also have no end-points in common.

The points of F can be divided into two classes, the first class containing points which are end-points of open intervals contiguous to F, termed *points of the first type*, and the second containing all the remaining points, termed *points of the second type*. Since the aggregate of open intervals contiguous to a perfect set is countable, the set of points of the first type is also countable. By theorem 1, the whole set F is uncountable, whence it follows that points of the second type always exist (which was previously not at all obvious) and form an uncountable set. We now use another construction, which is independent of theorem 1 and is based on a special property of the line; it will again be evident that points of the second type exist, and we shall see furthermore that they constitute a set with the power of the continuum.

We establish an order-preserving one–one correspondence between the set of open intervals contiguous to F and the set of dyadic rationals on the closed interval $[0, 1]$: if the contiguous open interval Δ' lies to the right of Δ'', the corresponding dyadic rationals are connected by the inequality $r'' < r'$. For example, such a map can be constructed as follows. We first map the contiguous open interval Δ_1 of greatest length (if there are several of these, we choose one at random) onto 1/2. Of all the contiguous open intervals lying to the left of Δ_1, we choose the greatest Δ_2 (with the same stipulation if there are several of these) and map it onto the point 1/4; similarly of those lying to the right of Δ_1, we choose the greatest Δ_3 and map it onto 3/4. On each of the four resulting closed intervals (to the left of Δ_2, between Δ_2 and Δ_1, between Δ_1 and Δ_3m, and to the right of Δ_3) we choose the greatest open interval and map the intervals chosen onto the numbers 1/8, 3/8, 5/8, 7/8, respectively. Order is clearly preserved under this mapping. Continuing this process indefinitely ,we ultimately arrive at the desired order-preserving map between the set of contiguous open intervals of F and the set of dyadic rationals [0, 1].

We now extend this map, on the one hand to the points of F, on the other to the dyadic irrationals of the closed interval [0, 1]. Let $\xi \in [0, 1]$ be a dyadic irrational. It divides the dyadic rationals in [0, 1] into two classes; a left K_l and a right K_r. The set of all contiguous open intervals of F is likewise divided under the correspondence into two classes, D_l and D_r. We denote by η_l the exact upper bound of the points contained in open intervals of the class D_l, and by η_r the exact lower bound of the points contained in intervals of D_r.

Between each rational point of [0, 1] and the point ξ, other rational points can be found; hence between each contiguous open interval of F that lies to the left of η_l and the point η_r itself can be found other contiguous open intervals; similarly, between each contiguous open interval of F that lies to the right of η_r and η_l itself can be found other contiguous open intervals. Hence the points η_l and η_r are neither interior nor end-points of contiguous open intervals; *they are therefore points of the second type.*

We can now state that $\eta_l = \eta_r$, for if this were not the case we could find a contiguous open interval separating η_l and η_r, but for this reason belonging neither to D_l nor to D_r, which is impossible. We denote the common value of η_l and η_r by η and map η onto the selected dyadic irrational ξ. The principle of the correspondence $\xi \to \eta$ is thus established. If $\xi' \neq \xi''$, a dyadic rational can be found between ξ' and ξ''; the corresponding contiguous open interval will separate the corresponding points η' and η'', which are therefore distinct. Hence the correspondence between the dyadic irrationals on [0, 1] and some subset of F is one–one, i.e. F contains a subset F^*, equivalent to this set of dyadic irrationals, and therefore has the power of the continuum, which completes the proof.

It is clear from the proof that the set F^* contains only points of the second type. We claim *that every point of the second type is the image of some dyadic irrational* ξ. For the point η determines a division of the aggregate of all open intervals contiguous to F into two classes; a left D_l (intervals to the left of η) and a right D_r (intervals to the right of η). At the same time the set of dyadic rationals is divided into classes K_l and K_r. The number $\xi = \sup K_l = \inf K_r$ is easily seen to be a dyadic irrational and to have as its image precisely the point η.

The transition from the closed interval [0, 1] to the set F can now be represented as a continuous deformation in which every dyadic rational point is extended to a whole open interval and every dyadic irrational point remains a point, the order of corresponding intervals and points being preserved.

Finally, a closed set in a metric space is said to be *nowhere dense* if it does not contain a single sphere. An arbitrary set A in a metric space is said to be *nowhere dense* if its closure A is nowhere dense.

Problems. 1. Show that every closed set (on the line or in a space with a countable base) is the sum of a perfect set and an at most countable set.

Hint. The perfect component is the set of accumulation points of the given closed set. Cf. problems 5, 6 of Section 3.

2. Prove that a complete metric space cannot be represented as the countable sum of its nowhere dense subsets.

Hint. Use the method of proof of theorem 1.

5. Theorem of the Fixed Point

1. Let us suppose that we have a function, defined on a metric space M, that maps each point y of the space onto a point $z = A(y)$ of the same space. A is then said to be *a reflection of the space* M *into itself*.

In analysis it is often necessary to deal with distinct reflections of functional spaces. For instance, if $f(x, y)$ is a given continuous function of its arguments in the region $a \leq x \leq b$, $-\infty < y < \infty$, it can be used to construct a reflection of the space $C(a, b)$ into itself by means of such formulae as

$$A[y(x)] = f(x, y(x)),$$

$$A[y(x)] = y_0 + \int_{x_0}^{x} f(\xi, y(\xi))\,d\xi \quad \text{(where } x_0, y_0 \text{ are fixed)}.$$

Every point y that is mapped onto itself by a reflection A (i.e. for which $A\,y = y$) is said to be a *fixed point* of the reflection A. Many problems in analysis concerning the existence of solutions to different equations reduce to the question as to whether certain reflections have fixed points. For example, a theorem on the existence of a solution to the differential equation

$$\frac{\mathrm{d}y}{\mathrm{d}x} = f(x, y) \tag{1}$$

with initial conditions $x = x_0$, $y = y_0$, essentially concerns the existence of a fixed point in the reflection

$$A\,y(x) \equiv y_0 + \int_{x_0}^{x} f(\xi, y(\xi))\,\mathrm{d}\xi,$$

since equation (1) with the given initial conditions is equivalent to the equation

$$y(x) = y_0 + \int_{x_0}^{x} f(\xi, y(\xi))\,\mathrm{d}\xi.$$

There are many general theorems that establish the existence of a fixed point in a reflection A on the strength of particular assumptions about A. We give here one of the simplest theorems of this kind, which gives not only the existence but also the uniqueness of a fixed point, admittedly with quite strong limitations imposed on the reflection under consideration.

Definition. A reflection A of a metric space M into itself is said to be *compressive* if for any two points $y, z \in M$, the inequality

$$\varrho(A\,y, A\,z) \leqq \theta\,\varrho(y, z),$$

holds, where θ is a fixed positive number smaller than 1.

THEOREM. *A compressive reflection A of a complete metric space M into itself has a fixed point, which is, moreover, unique.*

Proof. Proceeding from an arbitrary point $y_0 \in M$, we construct a sequence of points

$$y_1 = Ay_0,\ y_2 = Ay_1 = A^2y_0,\ \ldots,\ y_\nu = Ay_{\nu-1} = A^\nu y_0,\ \ldots$$

We contend that this sequence is fundamental in M, since for any ν

$$\varrho(y_\nu, y_{\nu+1}) = \varrho(A^\nu y_0, A^{\nu+1} y_0) \leqq \theta\,\varrho(A^{\nu-1} y_0, A^\nu y_0) \leqq \cdots$$
$$\leqq \theta^\nu \varrho(y_0, y_1)$$

and hence

$$\begin{aligned}\varrho(y_\nu, y_{\nu+\mu}) &\leqq \varrho(y_\nu, y_{\nu+1}) + \varrho(y_{\nu+1}, y_{\nu+2}) + \cdots + \varrho(y_{\nu+\mu-1}, y_{\nu+\mu}) \\ &\leqq \theta^\nu \varrho(y_0, y_1) + \theta^{\nu+1} \varrho(y_0, y_1) + \cdots + \theta^{\nu+\mu-1} \varrho(y_0, y_1) \\ &\leqq (\theta^\nu + \theta^{\nu+1} + \cdots + \theta^{\nu+\mu-1} + \cdots) \varrho(y_0, y_1) = \frac{\theta^\nu}{1-\theta} \varrho(y_0, y_1); \quad (2)\end{aligned}$$

by choosing ν sufficiently large, we can make this quantity as small as we please. Since M is complete, there exists a limit

$$y = \lim_{\nu \to \infty} y_\nu .$$

We show that y is a fixed point. We have

$$\varrho(A\, y, y_\nu) = \varrho(A\, y, A\, y_{\nu-1}) \leqq \theta\, \varrho(y, y_{\nu-1}) \to 0,$$

whence it follows that the sequence y_r converges to $A\, y$. Since the limit is unique, $A\, y = y$, as required. It remains to show that the fixed point obtained is the only fixed point of the transformation A. Let us assume that z is a second fixed point, so that in addition to the equation $A\, y = y$, we have $A\, z = z$. Then

$$\varrho(y, z) = \varrho(A\, y, A\, z) \leqq \theta\, \varrho(y, z).$$

If $\varrho(y, z) > 0$, we can cancel by $\varrho(y, z)$ and we obtain the contradiction: $1 \leqq \theta$. Hence we must have $\varrho(y, z) = 0$, $y = z$, i.e. a second fixed point, distinct from y, cannot exist. The theorem is proved.

Note. It is useful to estimate the distance from some point y_r to the fixed point y. To do this, we proceed to the limit as $\mu \to \infty$ in inequality (2):

$$\varrho(y_\nu, y_{\nu+\mu}) \leqq \frac{\theta^\nu}{1-\theta} \varrho(y_0, y_1) \quad (2d)$$

using the continuity of the distance-function (Section 3), we get:

$$\varrho(y_\nu, y) \leqq \frac{\theta^\nu}{1-\theta} \varrho(y_0, y_1). \quad (3)$$

This is the estimate that interests us. In actual problems it permits us to assess in advance the number of steps necessary to calculate y with a given accuracy.

Setting $\nu = 0$ in (3), we get

$$\varrho(y_0, y) \leqq \frac{1}{1-\theta} \varrho(y_0, y_1);$$

this inequality provides us with a bound on the distance between the original point y_0 and the fixed point.

2. We shall now demonstrate the application of this theorem to the problem of the existence and uniqueness of a solution to the differential equation

$$\frac{dy}{dx} = f(x, y) \tag{4}$$

with initial conditions $x = x_0$, $y = y_0$. As we remarked above, this problem concerns the existence and uniqueness of a fixed point in the reflection

$$A\,y(x) \equiv y_0 + \int_{x_0}^{x} f(\xi, y(\xi))\,d\xi . \tag{5}$$

The function $f(x, y)$ is assumed to be continuous in the region $a \leqq x \leqq b$, $-\infty < y < \infty$. The point x_0 is an interior point of the closed interval $[a, b]$.

The reflection (5) is defined in the metric space $C(a, b)$ of all continuous functions on the closed interval $[a, b]$. This space is, as we have seen, complete; we have only to ascertain under what conditions the operator A is compressive.

To this end, we estimate the distance in the space $C(a, b)$ between the elements that result from operating with A on the elements $y(x)$, $z(x)$:

$$\varrho(A\,y, A\,z) = \max_{a \leqq x \leqq b} |A\,y(x) - A\,z(x)|$$
$$= \max_{a \leqq x \leqq b} \left| \int_{x_0}^{x} [f(\xi, y(\xi)) - f(\xi, z(\xi))]\,d\xi \right| .$$

We suppose that the function $f(x, y)$ satisfies the inequality

$$|f(x, y_1) - f(x, y_2)| \leqq K |y_1 - y_2| \tag{6}$$

for any $x \in [a, b]$ and for any values of y_1, y_2, with K a fixed constant. The inequality (6) is termed the *Lipschitz condition*. If the Lipschitz condition is satisfied, we then have for any $\xi \in [a, b]$.

$$|f(\xi, y(\xi)) - f(\xi, z(\xi))| \leqq K |y(\xi) - z(\xi)| \leqq K\,\varrho(y, z)$$

and hence

$$\varrho(A\,y, A\,z) \leqq \max_{a \leqq x \leqq b} \int_{x_0}^{x} K\,\varrho(y, z)\,d\xi = K(b - a)\,\varrho(y, z).$$

We see that the operator A is compressive if the interval $[a, b]$ containing the point x_0 is sufficiently small, so that

$$K(b - a) = \theta < 1.$$

In this case we can apply the theorem of the fixed point, with the result: *if the function $f(x, y)$ satisfies the Lipschitz condition* (6), *then equation* (4) *with the initial conditions $y(x_0) = y_0$ has a solution $y = y(x)$ in some neighbourhood of the point x_0, which is moreover unique.* Thus we have proved a fundamental existence and uniqueness theorem for the solution of the differential equation (4).

By a similar procedure it is also possible to prove the continuous dependence of the solution on the initial conditions. We agree first to call the reflections A, B *ε-proximate* if, given $\varepsilon > 0$, the inequality $\varrho(A x, B x) < \varepsilon$ holds for any x in the space M.

LEMMA. *In a complete metric space M let two compressive reflections, A, B, be given, so that*

$$\varrho(A x, A y) \leqq \theta_A \varrho(x, y), \quad \varrho(B x, B y) \leqq \theta_B \varrho(x, y),$$

and let $\theta = \max(\theta_A, \theta_B) < 1$. Then if A and B are ε-approximate their fixed points will be situated at a distance not exceeding $\varepsilon/(1 - \theta)$ apart.

Proof. Let y_0 be the fixed point of the reflection A. In accordance with the general theory, the fixed point of B can be obtained as the limit y_ω of the sequence $y_0, y_1 = B y_0, y_2 = B y_1, \ldots$, where, by what we have proved

$$\varrho(y_0, y_\omega) \leqq \frac{\varrho(y_0, y_1)}{1 - \theta}.$$

But since A and B are ε-approximate, $\varrho(y_0, y_1) = \varrho(A y_0, B y_0) < \varepsilon$, whence $\varrho(y_0, y_\omega) < \varepsilon/(1 - \theta)$ as required.

The theorem on the continuous dependence of the solution of a differential equation on its initial conditions can be formulated as follows.

THEOREM. *If equation* (1) *is considered on the closed interval $[a, b]$ of length less than $1/K$, where K is the constant of the Lipschitz condition for $f(x, y)$ then for any $\varepsilon > 0$, the inequality $|y_0 - y_1| < \varepsilon$ implies the inequality*

$$\max_{a \leqq x \leqq b} |y_0(x) - y_1(x)| \leqq \frac{\varepsilon}{1 - K(b - a)};$$

here $y_j(x)$ *denotes the solution that satisfies the initial condition* $y_j(x_0) = y_j$ $(j = 0, 1)$.

Proof. We consider the reflections

$$A\, y(x) = y_0 + \int_{x_0}^{x} f(\xi, y(\xi))\, d\xi, \quad B\, y(x) = y_1 + \int_{x_0}^{x} f(\xi, y(\xi))\, d\xi.$$

They are both compressive, with the same parameter $\theta = K(b - a)$. If $|y_0 - y_1| < \varepsilon$ the reflections A, B are obviously ε-proximate. By the lemma, the distance between their fixed points does not exceed $\varepsilon/(1 - \theta) = \varepsilon/[1 - K(b - a)]$, as required.

Problems. 1. Formulate and prove by the fixed point method an existence and uniqueness theorem for the solution of the system of differential equations

$$\frac{d y_j}{d x} = f_j(x, y_1, \dots, y_n) \quad j = 1, 2, \dots, n.$$

Hint. The metric space M is composed of "vector functions" $y = (y_1(x), \dots, y_n(x))$ with the metric

$$\varrho(y, z) = \max_{a \leqq x \leqq b} \{|y_1(x) - z_1(x)|, \dots, |y_n(x) - z_n(x)|\}.$$

2. The reflection A on the half-line $1 \leqq x \leqq \infty$ maps each point x onto $x + 1/x$. Is the reflection compressive? Does it have a fixed point?

Answer. Although the reflection A diminishes distance, so that $\varrho(A\,x, A\,y) < \varrho(x\,y)$, there is no value $\theta < 1$ that satisfies the inequality $\varrho(A\,x, A\,y) \leqq \theta\,\varrho(x, y)$ (for all x, y). The reflection is not compressive, and it has no fixed point.

6. Completion of a Metric Space

In Cantor's theory of real numbers, irrationals are defined with the help of fundamental sequences†. With the usual metric the rationals form an *incomplete* metric space, and according to Cantor, the process of constructing the real numbers can be regarded as the process of constructing a complete metric space $\overline{M}$ that includes the space M. Cantor's method, properly generalized, allows the inclusion of *any* incomplete metric space M in some complete metric space $\overline{M}$.

Theorem. (F. Hausdorff, 1914). *Let* M *be a metric space (in general incomplete). There exists a complete metric space* $\overline{M}$, *called*

† Cf. for example, V. Nyemitski, M. Sludskaya, A. Cherkassov, *Course of Mathematical Analysis*, vol. 1, M–L., 1957.

the completion of the space M, which possesses the following properties:

(a) *M is isometric with some subspace $M_1 \subset \overline{M}$,*

(b) *M_1 is dense in M.*

Any pair of spaces $\overline{M}_1$, $\overline{M}_2$ that satisfy conditions (a) *and* (b) *are isometric with each other.*

Proof. We shall call two fundamental sequences $\{y_\nu\}$, $\{z_\nu\}$, in the space M, *equivalent* if $\lim_{\nu\to\infty} \varrho(y_\nu, z_\nu) = 0$. For example, every pair of sequences in M that converge to the same limit are equivalent, but no pair that converge to different limits are equivalent. Two fundamental sequences that are equivalent with a third are also equivalent with each other. Hence all fundamental sequences that can be constructed in M can be divided into classes, such that all the sequences belonging to a given class are equivalent with one another, and no sequence that is not a number of that class is equivalent with any sequence belonging to it. From such classes, which we shall denote by $Y, Z, \ldots$, we can construct the new space $\overline{M}$. All that requires definition is the distance between classes Y, Z. We define it by the formula

$$\varrho(Y, Z) = \lim_{\nu\to\infty} \varrho(y_\nu, z_\nu), \tag{1}$$

where $\{y_\nu\}$ is any fundamental sequence in the class Y, and $\{z_\nu\}$ any one in the class Z. Of course it is necessary in the first place to make sure that the given limit exists and is independent of the choice of the sequences $\{y_\nu\}$, $\{z_\nu\}$ in the classes Y, Z. By the quadrangle inequality (Section 1, art. 2)

$$|\varrho(y_\nu, z_\nu) - \varrho(y_{\nu+\mu}, z_{\nu+\mu})| \leqq \varrho(y_\nu, y_{\nu+\mu}) + \varrho(z_\nu, z_{\nu+\mu}),$$

hence the numbers $\varrho(y_\nu, z_\nu)$ form a Cauchy sequence. The limit $\lim_{\nu\to\infty} \varrho(y_\nu, z_\nu)$, therefore exists. If $\{y'_\nu\}$, $\{z'_\nu\}$ are other fundamental sequences in the classes Y, Z, we find, on applying the quadrangle inequality again, that

$$|\varrho(y_\nu, z_\nu) - \varrho(y'_\nu, z'_\nu)| \leqq \varrho(y_\nu, y'_\nu) + \varrho(z_\nu, z'_\nu) \to 0,$$

so that the sequence $\varrho(y'_\nu, z'_\nu)$ has the same limit as the sequence $\varrho(y_\nu, z_\nu)$. Thus the definition of distance between classes does not depend on the choice of fundamental sequences in these classes.

We must now verify that the quantity

$$\varrho(Y, Z) = \lim_{\nu\to\infty} \varrho(y_\nu, z_\nu)$$

satisfies axioms 1–3, Section 1.

Axiom 1: $\varrho(Y, Z) = \varrho(Z, Y)$ holds by construction

Axiom 2: $\varrho(Y, Z) > 0$ for $Y \neq Z$, $\varrho(Y, Y) = 0$.

First of all we have $\varrho(Y, Y) = 0$ by our construction of the function ϱ, since we can put $y_\nu = z_\nu$ in formula (1).

Suppose now that $\varrho(Y, Z) = 0$. This means that for arbitrary fundamental sequences $\{y_\nu\}$, $\{z_\nu\}$, in the classes Y, Z respectively, $\lim\,(y_\nu, z_\nu) = 0$. But then the sequences $\{y_\nu\}$, $\{z_\nu\}$ are equivalent and the classes Y, Z must coincide. Hence, if $\varrho(Y, Z) = 0$, $Y = Z$; it follows that for $Y \neq Z$, we must have $\varrho(Y, Z) > 0$, as required.

Axiom 3: $\varrho(Y, U) \leqq \varrho(Y, Z) + \varrho(Z, U)$. Let $\{y_\nu\}$, $\{z_\nu\}$, $\{u_\nu\}$, be fixed fundamental sequences in the classes Y, Z, U, respectively. The required inequality is obtained by passing to the limit in

$$\varrho(y_\nu, u_\nu) \leqq \varrho(y_\nu, z_\nu) + \varrho(z_\nu, u_\nu).$$

We now show that all the assertions formulated above in the theorem on completion hold for the space $\overline{M}$.

(1) *$\overline{M}$ contains a subset M_1 isometric with the space M.* We map each element $y \in M$ onto the corresponding class $Y \subset \overline{M}$, that contains the sequence $y, y, y, \ldots$ (i.e. the class of all sequences convergent to y). If under this mapping the point y corresponds to the class Y and the point z to the class Z, then

$$\varrho(Y, Z) = \lim \varrho(y, z) = \varrho(y, z).$$

And it follows that the aggregate of the classes Y is a subset of the space $\overline{M}$ isometric with M.

(2) *M_1 is dense in $\overline{M}$.* Let Y be an arbitrary class in $\overline{M}$ and let $\{y_\nu\}$ be a fundamental sequence in Y. We consider the sequence of classes $Y_1, Y_2, \ldots, Y_\mu, \ldots$, where Y_μ is determined by the sequence $(y_\mu, y_\mu, y_\mu, \ldots)$, i.e. it is the image of the element y_μ in the map $M \to M_1$.

For a given $\varepsilon > 0$, we can find a number μ_0 such that, for $\mu > \mu_0$, we have $\varrho(y_\mu, y_{\mu+p}) \leqq \varepsilon$. We shall then have

$$\varrho(Y, Y_\mu) = \lim_{\nu\to\infty} (y_\nu, y_\mu) \leqq \varepsilon.$$

But this signifies that the class Y is the limit of the classes Y_μ.

Since Y_μ belongs by construction to the set M_1, it follows that M_1 is dense in $\overline{M}$.

(3) *$\overline{M}$ is a complete space.* Let $Y_1, Y_2, \ldots$ be a fundamental sequence of elements of $\overline{M}$. For each class Y_ν we find a class $Z_\nu \subset M_1$ such that $\varrho(Y_\nu, Z_\nu) < 1/\nu$, and let $z_\nu \in M$ be the element corresponding to Z_ν. We claim that the sequence $\{z_\nu\}$ is fundamental in the space M. For

$$\begin{aligned}\varrho(z_\nu, z_\mu) = \varrho(Z_\nu, Z_\mu) &\leqq \varrho(Z_\nu, Y_\nu) + \varrho(Y_\nu, Y_\mu) + \varrho(Y_\mu, Z_\mu)\\ &\leqq \varrho(Y_\mu, Y_\mu) + \frac{1}{\nu} + \frac{1}{\mu} \to 0 \quad \nu \to \infty, \quad \mu \to \infty.\end{aligned}$$

The fundamental sequence $\{z_\nu\}$ determines a certain class $Z \subset \overline{M}$; we shall show that the class Z is the limit in $\overline{M}$ of the sequence Y_ν. For a given $\varepsilon > 0$, we have for sufficiently large $\nu \geqq \nu_0$:

$$\varrho(Z, Y_\nu) \leqq \varrho(Z, Z_\nu) + \varrho(Z_\nu, Y_\nu) \leqq \lim_{\mu\to\infty} \varrho(z_\mu, z_\nu) + \frac{1}{\nu} < \varepsilon.$$

Hence every fundamental sequence $Y_\nu \subset \overline{M}$ has a limit in $\overline{M}$, as required.

(4) *Any metric space $\overline{\overline{M}}$ that has the properties* (1)–(3) *is isometric with the space $\overline{M}$.*

For let M_1, M_2, be subsets of the spaces $\overline{M}, \overline{\overline{M}}$, respectively, that are isometric with the space M and consequently isometric with each other. We must extend this isometry from M_1 and M_2 to the spaces $\overline{M}, \overline{\overline{M}}$. Let us take any element $Y \subset \overline{M}$ and consider a sequence of elements $Y_\nu \subset M_1$ that converges to Y. The corresponding sequence $Z_\nu \subset M_2$ is always fundamental, since in virtue of the isometry between M_1 and M_2 the distances between elements of the sequence Z_ν are the same as those between the corresponding elements of the sequence Y_ν. Since $\overline{\overline{M}}$ is complete, it contains an element $z = \lim_{\nu\to\infty} z_\nu$. We map this element onto the chosen element $\overline{Y} \subset \overline{M}$. It is uniquely determined since equivalent sequences in $\overline{M}_1$ correspond to equivalent sequences in M_2 and the replacement of a sequence Y_ν by an equivalent one produces, under the mapping, the replacement of the sequence Z_ν by an equivalent one. The correspondence indicated is one–one and exhausts the elements of $\overline{M}$ and $\overline{\overline{M}}$. It remains for us to show that it is isometric. Let the elements $Y, Y' \in \overline{M}$ correspond to elements $Z, Z' \in \overline{\overline{M}}$, with

$$Y = \lim Y_\nu, \quad Y' = \lim Y'_\nu \quad (Y_\nu, Y'_\nu \in M_1).$$

If $Z_\nu, Z'_\nu \in M_2$ are the images of Y_ν, Y'_ν, then $\varrho(Z_\nu, Z'_\nu) = \varrho(Y_\nu, Y'_\nu)$ and in virtue of the lemma on the continuity of the distance-function (Section 3, art. 1)

$$\varrho(Z, Z') = \lim_{\nu\to\infty} \varrho(Z_\nu, Z'_\nu) = \lim_{\nu\to\infty} \varrho(Y_\nu, Y'_\nu) = \varrho(Y, Y'),$$

as required. Our proof is now complete.

Note 1. Let us suppose that a given metric space M is a subset of another complete metric space M^*. Then we can take as the completion of M the closure $\overline{M}$ of the set M in the space M^*. For $\overline{M}$, being a closed subset of the complete space M^*, is itself a complete space; hence it contains M as a dense subset. It thus fulfils the conditions of the theorem just proved, and can therefore serve as the completion of the space M.

Note 2. The space $C_p(a, b)$ of continuous functions on the closed interval $[a, b]$ with distance given by

$$\varrho^p(y, z) = \int_a^b |y(x) - z(x)|^p dx$$

is incomplete (we saw this in Section 4, art. 1). By the theorem just proved it has a completion $\overline{C}_p(a, b)$. The question naturally arises: is it possible to attach some concrete significance to the elements of the space $\overline{C}_p(a, b)$, which are abstractly determined by theorem 1; can they be interpreted in the form of some functions or other? It turns out that this can be done, though not very easily; we shall postpone the consideration of this question until Chapter IV, when we shall have the necessary tools to resolve it.

7. Continuous Functions and Compact Spaces

1. *Definitions and Elementary Properties*

A function $f(x)$, defined on a metric space M and taking numerical values, is said to be *continuous at the point* x_0 if, for any $\varepsilon > 0$, there exists $\delta > 0$ such that the condition

$$\varrho(x, x_0) < \delta$$

implies

$$|f(x) - f(x_0)| < \varepsilon .$$

As in classical analysis, a second definition equivalent to the first is possible: the function $f(x)$ is continuous at the point x_0 if for any sequence $x_n \to x_0$, we have $f(x_n) \to f(x_0)$. A function which is continuous at each point of the set M is said to be *continuous* on M.

One of the simplest examples of a continuous function in a metric space is the distance from a point x to a fixed point x_0. The continuity of this function follows from the second triangle inequality (Section 1, art. 2):

$$|\varrho(x', x_0) - \varrho(x'', x_0)| \leqq \varrho(x', x'') .$$

The usual properties of functions continuous on the line, which are familiar in analysis, are easily carried over to the case of functions continuous on metric spaces. Thus the sum, difference, and product of two continuous functions are also continuous functions. The quotient of two continuous functions is also continuous at all points where the denominator is non-zero.

We shall establish the following important properties of continuous functions on a metric space:

LEMMA. *If $f(x)$ is a continuous function, then the sets*

$$F_1 = \{x\colon f(x) \leqq A\},$$
$$F_2 = \{x\colon f(x) \geqq A\}$$

are closed for any A, and the sets

$$U_1 = \{x\colon f(x) < A\},$$
$$U_2 = \{x\colon f(x) > A\}$$

are open for any A.

Proof. We shall show that the set U_1 is open. Let $x_0 \in U_1$ so that $f(x_0) < A$, and let us put $\varepsilon = A - f(x_0)$. Since the function $f(x)$ is continuous at the point x_0, there exists a sphere $\varrho(x_0, x) < \delta$ in which the inequality $|f(x) - f(x_0)| < \varepsilon$ holds. Within this sphere

$$f(x) < f(x_0) + \varepsilon = A ,$$

and hence the whole sphere belongs to the set U_1. Since x_0 is an arbitrary point of U_1, it follows that U_1 is an open set. Its complement is the set F_2, which is therefore closed. The proof proceeds similarly for the sets U_2 and F_1; or with the substitution of $2A - f(x)$ for $f(x)$, the problem reduces to the foregoing one.

Conversely, *if it is known that for some function $f(x)$ defined on a metric space M, each of the sets*

$$U_1 = \{x\colon f(x) < A\}, \quad U_2 = \{x\colon f(x) > A\}$$

is open for any A, then the function $f(x)$ is continuous.

For in this case, for any point $x_0 \in M$ and any $\varepsilon > 0$, we can form the sets

$$U_1 = \{x\colon f(x) < f(x_0) + \varepsilon\}, \quad U_2 = \{x\colon f(x) > f(x_0) - \varepsilon\},$$

which are stipulated to be open. The intersection of these sets

$$U = \{x\colon f(x_0) - \varepsilon < f(x) < f(x_0) + \varepsilon\} = \{x\colon -\varepsilon < f(x) - f(x_0) < \varepsilon\}$$

is also open. It obviously contains the point x_0 and with it some sphere $U_\delta(x_0) = \{x : \varrho(x, x_0) < \delta\}$. Within this sphere the inequality

$$|f(x) - f(x_0)| < \varepsilon,$$

holds, but this also denotes the continuity of the function $f(x)$ at the point x_0.

A corresponding proposition can be formulated for the sets $F_1 = \{x\colon f(x) \leqq A\}$, $F_2 = \{x\colon f(x) \geqq A\}$; *if, for a function $f(x)$, each of these sets is closed for any A, then $f(x)$ is continuous.* The proof can be obtained by taking complements in the above.

Functions that are defined on a metric space which is itself composed of functions, such as the spaces $C(a, b)$, $D_n(a, b)$, etc. are generally termed *functionals*.

Problems. 1. Are the following functionals on the space $C(a, b)$ continuous?

(a) $F(y) = y(a)$;

(b) $F(y) = \max |y(x)|$;

(c) $F(y) = \max y(x)$;

(d) $F(y) = \int_a^b y(x)\,dx$;

(e) $F(y) = \begin{cases} 0, \text{ if } y(x) \text{ assumes a single negative value,} \\ 1/2, \text{ if } y(x) \equiv 0, \\ 1, \text{ if } y(x) \geqq 0 \text{ and } y(x) \not\equiv 0. \end{cases}$

Answer. Continuous in cases (a)–(d), but not in (e).

2. Are the following functionals on the space $D_1(a, b)$ continuous?

(a) $F(y) = y(a)$;

(b) $F(y) = \int_a^b \sqrt{1 + y'^2}\, dx$.

Answer. Yes.

3. Is the function $\varrho(x, B) = \inf_{y \in B} \varrho(x, y)$ (Section 3, problem 8) continuous?

Answer. Yes.

4. Is the function $d(x, B) = \sup_{y \in B} \varrho(x, y)$ continuous?

Answer. Yes.

2. *Compact Spaces*

The analogy with the properties of continuous functions on a closed interval is not always maintained under the transition to a general metric space. For example, a continuous function on the closed interval $[a, b]$ is always bounded on this interval and attains its upper and lower bounds. A continuous function on a sphere of radius r in a metric space M may be unbounded or may not attain its bounds (an example is given in one of the problems of the present paragraph). For specified properties of continuous functions to hold in a given metric space M, the space must suffer the imposition of further restrictions.

A metric space M is said to be *locally compact* if every infinite subset $A \subset M$ contains a fundamental sequence.

Thus every infinite subset A of the interval $a < x < b$ has by the well-known Bolzano–Weierstrass theorem, a limit point in $[a, b]$ and therefore contains a fundamental sequence; we see that the interval $M = (a, b)$ is a locally compact metric space.

A locally compact metric space need not be complete, as we saw with the open interval (a, b) in the preceding example. A metric space that is both locally compact and complete is said to be *compact*.

An independent definition of compactness can be given: a metric space M is compact if every infinite subset contained in it contains a *convergent* sequence.

A typical example of a compact space is the closed interval $a \leqq x \leqq b$ on the real line with the usual metric.

The properties of continuous functions in being bounded and in attaining their bounds are directly connected with the compactness of the sets on which they are defined.

THEOREM 1. *Every continuous function $f(x)$, defined on a compact space M, is bounded.*

Proof. Let us suppose that $f(x)$ is unbounded. Then for any integer n we can choose a point $x_n \in M$ for which $|f(x_n)| > n$.

By hypothesis the sequence of points $x_1, x_2, \ldots, x_n, \ldots$ must contain a convergent subsequence: rejecting a subset of these points if necessary, we can assume that the sequence itself converges to some point $x_0 \in M$. By the continuity of $f(x)$, there exists a neighbourhood of x_0, defined say, by the inequality $\varrho(x, x_0) < \delta$, in which $|f(x_0) - f(x)| < 1$ or $|f(x)| < |f(x_0)| + 1$. On the other hand this neighbourhood contains points of the sequence $x_1, x_2, \ldots$ with subscripts as great as we please; at these points $f(x)$ assumes values as large as we please. The contradiction obtained shows that $f(x)$ cannot be unbounded, but must in fact be bounded, as required.

THEOREM 2. *Every continuous function $f(x)$ defined on a compact space M, attains on M its exact upper (and lower) bound.*

Proof. Let b be the exact upper bound of $f(x)$. For any integer n we can find a point x_n satisfying the inequality

$$0 \leqq b - f(x_n) < \frac{1}{n}.$$

Let us suppose that the function $f(x)$ fails to attain the value b anywhere in M. Then the function

$$\varphi(x) = \frac{1}{b - f(x)}$$

is continuous on M and is bounded (by theorem 1). But as x runs through the sequence of points x_n, the denominator tends to zero. Hence the function $\varphi(x)$ cannot be bounded, and our assumption must have been untenable. It follows that $f(x)$ attains the value b at some point M.

The converse results are also true: if a metric space M is not compact, there exist continuous functions defined on M and either unbounded or, though bounded, not attaining their bounds. One of the problems below is devoted to this question. Thus the condition "M is compact" is both necessary and sufficient for the validity of theorems 1 and 2.

THEOREM 3. *Every continuous function $f(x)$, defined on a compact space M, is uniformly continuous on it, in other words, for any $\varepsilon > 0$, we can find $\delta > 0$ such that $\varrho(x, y) < \delta$ entails $|f(x) - f(y)| < \varepsilon$.*

Proof. Supposing the contrary to be true, we should be able for some $\varepsilon = \varepsilon_0$ to find sequences x_n, y_n, such that

$$\varrho(x_n, y_n) < \frac{1}{n}, \quad |f(x_n) - f(y_n)| \geqq \varepsilon_0.$$

By hypothesis the sequence x_n contains a subsequence convergent to the point x_0; discarding some subset of the points if necessary, we can suppose that the sequence x_n itself converges to x_0. Then the sequence y_n also converges to x_0. From some point on, x_n and y_n will lie in a neighbourhood of x_0 which is such that $|f(x) - f(x_0)| < \varepsilon_0/2$. But then

$$|f(x_n) - f(y_n)| \leqq |f(x_n) - f(x_0)| + |f(x_0) - f(y_n)| < \frac{\varepsilon_0}{2} + \frac{\varepsilon_0}{2} = \varepsilon_0$$

contrary to our construction. The theorem is therefore proved.

3. *Conditions for Compactness*

We shall now obtain some conditions that can be conveniently applied to establish the compactness of particular metric spaces.

We can suppose, without loss of generality, that a given metric space M is contained in some complete space P, since we can always take the completion of M as the space P.

A set $B \subset P$ is said to be an *ε-net for the set $M \subset P$* if each point x of the set M is within a distance ε of some point $y \in B$.

THEOREM 1 (F. Hausdorff, 1914). *A set M contained in a metric space P is locally compact (in terms of the metric of P) if and only if, for any $\varepsilon > 0$, P contains a finite ε-net for M.*

Proof. Let M be locally compact and let $\varepsilon > 0$ be given; we shall show that a finite ε-net for M exists. We take an arbitrary point $x_1 \in M$. If all the remaining points of M lie at a distance $\leqq \varepsilon$ from x_1, the point x_1 itself constitutes an ε-net for M and the construction is completed. If there are points of M situated at a distance $> \varepsilon$ from x_1, we choose an arbitrary point x_2 from among them. If now each point of M is within a distance ε of either x_1 or x_2, the points x_1, x_2 form a finite ε-net for M and the construction is complete; otherwise we continue the construction. Each new point x_n that is selected in the process is at a distance exceeding ε from each of its predecessors $x_1, x_2, \ldots, x_{n-1}$. Hence if the process were to continue indefinitely, we should have an infinite subset

$x_1, x_2, \ldots, x_n, \ldots$ of M that did not contain a single fundamental sequence, in contradiction to the local compactness of M. Since M is locally compact, the process must terminate after a finite number of steps; but the ending of the process signifies the construction of a finite ε-net for M.

Conversely let a finite ε-net for M exist in the space P for every $\varepsilon > 0$; we shall show that M is locally compact. We consider an arbitrary infinite subset $A \subset M$, and we have to find a fundamental sequence in A. As the first point of this sequence, we take any point $x_0 \in A$. Using the condition of the theorem with $\varepsilon = 1$ we can cover A with a finite number of spheres of radius 1; amongst these there must be one, which we denote by U_1, that contains an infinite subset $A_1 \subset A$. From A_1 we select any point $x_1 \neq x_0$. Using the condition of the theorem with $\varepsilon = 1/2$ we can cover A_1 with a finite number of spheres of radius $1/2$; amongst these there is a sphere U_2 that contains an infinite subset $A_2 \subset A_1$. We choose any point $x_2 \in A_2$, distinct from x_0 and x_1. Continuing in a similar fashion, we get a chain of infinite subsets $A \supset A_1 \supset \cdots \supset A_\nu \supset \cdots$ (where each set A_1 is contained in a sphere U_ν of radius $1/\nu$) and a sequence of distinct points $x_0, x_1, x_2, \ldots, x_\nu, \ldots$, where $x_\nu \in A_\nu$. We claim that the sequence $x_0, x_1, x_2, \ldots$ is fundamental. For $\mu < \nu$ we have $U_\mu \supset A_\mu \supset A_\nu$, so that $\varrho(x_\mu, x_\nu) < 2/\mu$. This quantity tends to zero as $\mu \to \infty$, and the sequence $x_0, x_1, \ldots$ is consequently fundamental, as asserted.

As an application of this criterion we show that *any bounded infinite set in n-dimensional Euclidean space $P = E_n$ is locally compact*. For a sphere in the space P that contains the bounded set M contains, for any m, only a finite number of points all the coordinates of which have the form $k/2^m$, where k is an integer, and the set of these points evidently constitutes, for sufficiently large m, an ε-net for M.

We now remark another simple criterion for local compactness: *a set M in a metric space P is locally compact if, for any $\varepsilon > 0$, a locally compact set B_ε (possibly infinite) can be found that is an ε-net for M.*

The proof of this criterion is very simple. We assert that for a given ε the finite $\varepsilon/2$-net Z for the set $B_{\varepsilon/2}$, which exists in virtue of the local compactness of $B_{\varepsilon/2}$, is an ε-net for the set M. For according to the stipulation, given an arbitrary point $x \in M$, there exists a point $y \in B_{\varepsilon/2}$ such that $\varrho(x, y) \leqq \varepsilon/2$, and a point $z \in Z$

such that $\varrho(y, z) \leqq \varepsilon/2$, but then $\varrho(x, z) \leqq \varrho(x, y) + \varrho(y, z) \leqq \varepsilon$, as asserted. Thus, for any $\varepsilon > 0$, M possesses a finite ε-net, and is therefore locally compact.

As an application of this criterion, we show *that the completion $\overline{M}$ of any locally compact set M is compact.* In fact, since the set M is dense in $\overline{M}$, it constitutes an ε-net for $\overline{M}$ for any ε. But M is stipulated to be locally compact, and so $\overline{M}$ is locally compact; since $\overline{M}$ is also complete, it is compact, as required.

THEOREM 2. *A locally compact subset M of a complete metric space P is compact if and only if it is closed in P.*

Proof. If the subset M is closed in the complete metric space P, it is itself a complete metric space (Section 4, art. 2, example 3). If it is in addition, locally compact, it is by definition compact. Conversely let $M \subset P$ be compact; then M is a locally compact set and we have only to show that it is closed. This follows from the same result of Section 4 (art. 2, example 3) in conjunction with the result that, being compact, M is a complete space.

Combining theorems 1 and 2, we get:

THEOREM 3. *A set M in a complete metric space P is compact if and only if it is closed in P and for every $\varepsilon > 0$, P possesses a finite ε-net for M.*

In particular any closed sphere in n-dimensional Euclidean space is compact. Every continuous function, defined on such a sphere, is bounded and attains its exact upper and lower bounds.

Problems. 1. Given a compact space Q, find a countable, everywhere dense set of points.

(*Hint.* Consider the union of all finite $1/m$-nets for Q, $m = 1, 2, \ldots$)

2. Show that from any system of open sets $\{G\}$ that together cover a compact space Q it is possible to select a finite subsystem $G_1, \ldots, G_m$, that also covers Q.

Hint. If from a given covering of the compact space Q it is impossible to extract a finite covering, then it is also impossible to extract a finite covering of some sphere Q_n, one of finitely many spheres of radius $1/2^n$ that together cover Q. Consider the limit point of the set of centres of the spheres Q_n.

3. Show that the descending chain $F_1 \supset F_2 \supset \ldots$ of non-empty closed subsets of a compact space has a non-empty intersection.

Hint. Pass over to complements and use problem 2.

4. If a sequence of continuous functions $f_1(x) \leqq f_2(x) \leqq \cdots$ converges on a compact space Q to a continuous function $f(x)$ it converges uniformly on that space (Dini's theorem).

Hint. For a given $\varepsilon > 0$ and a fixed point x_0 find a number n_0 such that $0 \leqq f(x_0) - f_{n_0}(x_0) \leqq \varepsilon$. There exists a neighbourhood of x_0 in which $0 \leqq f(x) - f_{n_0}(x) \leqq 3\varepsilon$ and hence also $0 \leqq f(x) - f_n(x) \leqq 3\varepsilon$ for all $n \geqq n_0$. Then use the result of problem 2.

5. An aggregate of continuous functions $\{f(x)\} = A$ on a compact space Q is said to be *uniformly equi-continuous* if for any $\varepsilon > 0$ there exists $\delta > 0$ such that the inequality $\varrho(x', x'') < \delta$ implies $|f(x') - f(x'')| < \varepsilon$ for any $f \in A$; it is said to be *uniformly bounded* if there exists a constant C such that $|f(x)| < C$ for any $f \in A$. Show that the aggregate A is a compact space with respect to the metric $\varrho(f, g) = \max\limits_{x \in Q} |f(x) - g(x)|$ if and only if it is uniformly equi-continuous and *uniformly bounded* (Arzela's theorem).

Hint. Include A in the space of all bounded (even discontinuous) functions on Q with the metric $\varrho(f\, g) = \sup |f(x) - g(x)|$. For a given δ-net on Q the set of piecewise-constant functions that assume constant values, equal to integral multiples of ε not exceeding C in absolute value, on spheres of radius δ with centres at points of the net form a finite 3ε-net for A. (To eliminate many-valuedness at points common to several spheres, choose one possible value at random.)

6. Show that the functional

$$F(y) = \int_0^{1/2} y(x)\, \mathrm{d}x - \int_{1/2}^{1} y(x)\, \mathrm{d}x$$

is continuous on the space C (0, 1); show that the exact upper bound of its values on a closed unit sphere in the space is equal to 1, but that this bound is not attained at any point of the sphere.

7. Given a non-compact metric space M construct on it a continuous unbounded function.

Hint. Each point x_k of a sequence $x_1, x_2, \ldots$ from which a convergent subsequence cannot be extracted is situated at a positive distance r_k from the set of all remaining points of the sequence. Every function $f(x)$, continuous on each of the spheres $\{\varrho(x, x_k) \leqq 1/2\, r_k\}$, is equal to zero on their boundaries, and outside the spheres is continuous on M.

8. Given a reflection of a compact space into itself which satisfies the condition $\varrho(A\,x, A\,y) < \varrho(x, y)$ for $x \neq y$, show that it possesses a unique fixed point.

Hint. The minimum of the continuous function $\varrho(A\,x, x)$ cannot be positive.

9. Show that a compact space cannot be reflected isometrically onto a part of itself (V. A. Rokhlin).

Hint. Supposing the contrary to be true, it is possible to find a point at a positive distance r_0 from the image of the space. Reiterating the reflection, construct a sequence in which the mutual distances between points are never less than r_0.

10. Construct a locally compact set on the plane which is isometric with a part of itself.

Hint. Consider the set of points with polar coordinates $\varrho = 1, \varphi = 0, 1, 2, \ldots$

11. Let A, B be isometric reflections of the compact space Q into itself. We define the distance between A and B by the formula

$$\varrho(A, B) = \max_{x \in Q} \varrho(A x, B x). \tag{1}$$

Show that the set of all isometric reflections of Q into itself, metricised by (1), is compact (B. L. Van der Waerden).

Hint. For a given finite ε-net on Q, observing that there exist only finitely many reflections of a finite set onto itself, construct a finite 2ε-net in the space of isometric reflections.

4. *Functions of Several Variables*

We shall sometimes come across continuous functions with several arguments that vary over a metric space M. For the sake of definiteness let us consider a function of a pair of points, y, z. A real-valued function $f(y, z)$ the arguments of which belong to a metric space M, is said to be continuous at $y = y_0$, $z = z_0$, if for any $\varepsilon > 0$ there exists $\delta > 0$ such that the inequalities

$$\varrho(y, y_0) < \delta, \quad \varrho(z, z_0) < \delta \tag{2}$$

imply

$$|f(y, z) - f(y_0, z_0)| < \varepsilon. \tag{3}$$

A function continuous over all pairs y_0, z_0, is said to be continuous everywhere.

An example of a continuous function of a pair of points y, z is the distance-function $\varrho(y, z)$. For by the quadrangle inequality (Section 1, art. 2).

$$|\varrho(y, z) - \varrho(y_0, z_0)| \leqq \varrho(y, y_0) + \varrho(z, z_0), \tag{4}$$

which can be made less than a given $\varepsilon > 0$ by taking $\varrho(y, y_0) < \varepsilon/2$, $\varrho(z, z_0) < \varepsilon/2$.

There is no need to reiterate the whole theory, expounded in paragraphs 1–2, for functions of several variables. In fact every function of several variables which vary over a metric space M can be represented as a function of a single variable which varies over some new metric space M'. For the sake of simplicity we shall consider the case of two variables, but we first define the *product of metric spaces*.

Let metric spaces M, N be given. We consider the set of all possible formal pairs $\{x, y\}$, where $x \in M$, $y \in N$, and define the distance between such pairs by the formula

$$\varrho(\{x_1, y_1\}, \{x_2, y_2\}) = \varrho(x_1, x_2) + \varrho(y_1, y_2), \tag{5}$$

where $\varrho(x_1, x_2)$, $\varrho(y_1, y_2)$ denote distances in the spaces M, N, respectively.

It is easily verified that the distance defined by this rule satisfies the axioms of Section 1. The set of all pairs $\{x, y\}$ together with this metric is said to be *the product of the metric spaces M, N*; it is denoted by $M \times N$.

Let us consider a function $f(x, y)$, the first argument of which runs through the space M, the second through the space N. It can evidently be regarded as a function of a single argument which runs through the space $M \times N$. In particular, a function $f(x, y)$, the arguments of which run through one and the same space M, can be regarded as a function of a single argument which runs through the space $M \times M$. If $f(x, y)$ is a continuous function of the arguments x, y in the sense indicated above, the corresponding function on the space $M \times M$ will evidently be continuous in the usual sense (art. 1). Thus the theory of continuous functions of two variables reduces to the theory of continuous functions of a single variable.

Problems. 1. Let the spaces M, N, be locally compact; show that their product $M \times N$ is also locally compact.

2. Let the spaces M, N be complete; show that their product $M \times N$ is also complete.

3. Give an example where a space M is locally compact and a space N is complete, but the product $M \times N$ is neither locally compact nor complete.

4. If metric spaces M, N are infinite, there exists a function $f(x, y)$, continuous in each of its arguments separately (with the other argument fixed), but not continuous on the product space $M \times N$.

8. Normed Linear Spaces

1. *Linear Spaces*

In analysis the operation of passing to the limit is most frequently encountered in combination with other operations, among which linear operations—the addition of elements and their multiplication by numbers—are greatly in evidence. These operations themselves are studied in linear algebra. We recall a fundamental definition connected with them—the definition of a linear space.†

A set E of elements $x, y, \ldots$ is said to be a *linear space* if there are defined on it operations of addition and multiplication by (real or complex) numbers, that satisfy the eight axioms to be listed below:

The first group of axioms (1–4) describes the properties of addition:

1. $x + y = y + x$ (*commutativity of addition*)
2. $(x + y) + z = x + (y + z)$ (*associativity of addition*)
3. *There exists an element* 0 *such that* $x + 0 = x$ *for any* $x \in E$.
4. *For each* $x \in E$ *the equation* $x + y = 0$ *has a solution.* The element y is said to be the *inverse* of the element x.

It is easily verified that the elements 0 and y, the existences of which are required by axioms 3 and 4, are determined uniquely.

The next group of axioms (5–8) provides a link between the operations of addition and multiplication by numbers (scalars). We denote by $\mathscr{E}$ the set of (real or complex) numbers by which multiplication is admissible.

5. $\lambda(\mu x) = (\lambda \mu) x \quad (\lambda \in \mathscr{E}, \mu \in \mathscr{E}, x \in E)$
6. $1x = x$
7. $\lambda(x + y) = \lambda x + \lambda y$
8. $(\lambda + \mu) x = \lambda x + \mu x$

It can be shown that $0 \cdot x = 0$ and that the inverse of a given x is the element y obtained by multiplying y by -1.

A set L of elements of a linear space E is said to be a *subspace* of E if the operations on elements of L of addition and multiplication by numbers always produce elements of L. The smallest subspace consists of the single element 0, the largest is the whole space E.

An example of a linear space is the aggregate R_n of arrays of n numbers

$$x = (\xi_1, \ldots, \xi_n) \tag{1}$$

† A detailed account can be found, for example, in: G. Ye. Shilov, Introduction to the theory of linear spaces, State Technical Publishing House, 1956 (2 issues), Chapter II on. English translation *An Introduction to the Theory of Linear Spaces* (Prentice-Hall, London, 1961).

with operations effected "coordinate by coordinate": if $x = (\xi_1, \ldots, \xi_n)$, $y = (\eta_1, \ldots, \eta_n)$, we define

$$\begin{aligned} x + y &= (\xi_1 + \eta_1, \ldots, \xi_n + \eta_n), \\ \alpha x &= (\alpha \xi_1, \ldots, \alpha \xi_n). \end{aligned} \tag{2}$$

This space is said to be *n-dimensional*, i.e. between any $n + 1$ elements $x^{(1)}, \ldots, x^{(n+1)}$, there exists a dependence relation

$$C_1 x^{(1)} + \cdots + C_{n+1} x^{(n+1)} = 0, \ \sum_{j=1}^{n+1} C_j^2 > 0 \tag{3}$$

and there exists also a set of n elements $e^{(1)}, \ldots, e^{(n)}$ for which no such dependence relation holds.

The functional spaces $C(a, b)$, $D_n(a, b)$, $C_p(a, b)$ (Section 1) are also linear spaces (with the natural operations). In contrast to R_n these spaces are infinite-dimensional.

We recall further the important concept of an isomorphism between linear spaces. Two linear spaces, E' and E'', are said to be *isomorphic* if there exists between them a one–one correspondence that preserves the operations of addition and scalar multiplication; this means that if a vector $x' \in E'$ corresponds to a vector $x'' \in E''$, and $y' \in E'$ corresponds to $y'' \in E''$, then the vector $x' + y' \in E'$ corresponds to the vector $x'' + y'' \in E''$, and for any $\lambda \in \mathscr{E}$ the vector $\lambda x' \in E'$ corresponds to the vector $\lambda x'' \in E''$.

Thus any two n-dimensional spaces are isomorphic; each of them is isomorphic with the n-dimensional coordinate space described above.

2. *Normed Linear Spaces*

The properties of a linear space are combined with the metric properties, to which we have so far devoted this chapter, in the definition of a normed linear space (a complete normed linear space is called a Banach space):

A set E of elements $x, y, \ldots$ is said to be a *normed linear space* if:

(1) it is a linear space;

(2) it is a metric space, and one in which

(3) the distance-function $\varrho(x, y)$ and the linear operations are connected by the following conditions:

(a) distance is invariant under displacement, i.e.

$$\varrho(x + z, y + z) = \varrho(x, y) \quad \text{for any} \quad x, y, z;$$

(b) distance is subject to the "homogeneity condition":

$$\varrho(\lambda x, 0) = |\lambda| \varrho(x, 0)$$

for any number λ and any element x.

From axiom (a) it follows that the distance between two points is equal to the distance of their difference from zero:

$$\varrho(x, y) = \varrho(x - y, y - y) = \varrho(x - y, 0).$$

It is therefore sufficient to know the distance of any element from zero. The distance $\varrho(x, 0)$ is said to be the *norm* (or length) of the element x and is denoted by $\|x\|$ or $|x|$. From the metric axioms and the properties (3) it is easy to deduce that the norm of any element satisfies the conditions:

(α) $\|x\| > 0$ for $x \neq 0$, $\|0\| = 0$;

(β) $\|\lambda x\| = |\lambda| \cdot \|x\|$;

(γ) $\|x + y\| \leqq \|x\| + \|y\|$.

Conversely, if a norm is defined on some linear space E, i.e. each element x determines a number $\|x\|$ such that conditions (α) – (γ) are satisfied, a metric on E can be introduced by the formula

$$\varrho(x, y) = \|x - y\|$$

and E becomes a normed linear space. In general a linear space with a norm that satisfies conditions (α) – (γ) *is also said to be a normed linear space.*

The metric spaces $C(a, b)$, D (a, b), $C(a, b)$ $(p = 1, 2)$ considered in Section 1 of this chapter, are normed linear spaces.

For in the space $C(a, b)$ of continuous functions on the closed interval $a \leqq x \leqq b$, distance was given by the formula

$$\varrho(y, z) = \max_{a \leqq x \leqq b} |y(x) - z(x)|.$$

Axiom (3) is evidently satisfied here. The norm of an element y is defined by the formula

$$\|y\| = \varrho(y, 0) = \max_{a \leqq x \leqq b} |y(x)|.$$

In the space D (a, b) of functions with continuous derivatives up to order m on the closed interval $[a, b]$, distance was given by the formula

$$\varrho(y, z) = \max_{a \leqq x \leqq b} \{|y(x) - z(x)|, \ldots, |y^{(m)}(x) - z^{(m)}(x)|\}.$$

It is easily seen that both the properties required in axiom (3) are satisfied here. The norm of an element y is defined by

$$\|y\| = \max_{a \leqq x \leqq b} \{|y(x)|, |y'(x)|, \ldots, |y^{(m)}(x)|\}.$$

In the space $C_p(a, b)$ $(p = 1, 2)$ of continuous functions on $[a, b]$ with distance given by

$$\varrho^p(y, z) = \int_a^b |y(x) - z(x)|^p \, dx$$

the two requirements of axiom (3) are again satisfied (the second because we have ϱ^p and not ϱ on the left). The norm of an element y is defined by

$$\|y\|^p = \int_a^b |y(x)|^p \, dx.$$

The set determined by the inequality

$$\|x\| \leqq 1,$$

is a sphere of radius 1 with centre at the zero of the space E in question. It is said to be the *unit sphere* of the space E.

The unit sphere (as also any sphere in a normed linear space) *is a convex set.* In general a set M in a linear space E is said to be convex if it contains, together with any two points x, y, all the points

$$z = \alpha x + \beta y, \quad \alpha + \beta = 1, \quad \alpha \geqq 0, \quad \beta \geqq 0,$$

or, in geometrical terms, contains the closed interval with end-points x, y. The convexity of the unit sphere follows immediately from the triangle inequality: if $\|x\| \leqq 1$, $\|y\| \leqq 1$, then

$$\|z\| = \|\alpha x + \beta y\| \leqq \alpha\|x\| + \beta\|y\| \leqq \alpha + \beta = 1.$$

We remark some general properties of normed linear spaces connected with the concept of convergence. Since the distance between the points x, y is defined as $\|x - y\|$, the convergence of a sequence x_n to the element x is incorporated in the relation

$$\lim_{n \to \infty} \|x - x_n\| = 0.$$

If $x_n \to x$, $y_n \to y$, *then* $x_n + y_n \to x + y$; for

$$\|x + y - (x_n + y_n)\| = \|(x - x_n) + (y - y_n)\|$$
$$\leqq \|x - x_n\| + \|y - y_n\| \to 0.$$

If $x_n \to x$, $\lambda_n \to \lambda$, *then* $\lambda_n x_n \to \lambda x$; for

$$\|\lambda_n x_n - \lambda x\| = \|\lambda_n (x_n - x) + (\lambda_n - \lambda) x\| \\ \leqq |\lambda_n| \|x_n - x\| + |\lambda_n - \lambda| \|x\| \to 0.$$

In conclusion we shall say a few words about linear isometries between normed linear spaces. Two isomorphic linear spaces can have distinct norms, as, for example, $C(a, b)$ and $C_p(a, b)$. If normed linear spaces are isomorphic as linear spaces and isometric as metric spaces, they are said to be *linearly isometric*. For example, the spaces $C(0, 1)$, $C(0, 2)$ and $C_p(0, 1)$, $C_p(0, 2)$ are linearly isometric. The required one–one correspondences can be given by the formulae

$$C(0, 1) \ni \varphi(x) \longleftrightarrow \varphi(2x) \in C(0, 2);$$

$$C_p(0, 1) \ni \varphi(x) \longleftrightarrow 2^{1/p} \varphi(2x) \in C_p(0, 2).$$

Instead of the precise term "linear isometry", the looser but brief "isomorphism" is often employed.

3. *Completion of a Normed Linear Space*

Like any other metric space, a normed linear space can be either complete or incomplete. In the latter case the space E can be completed by including it in a larger complete metric space $\overline{E}$, as in Section 7. It is to be observed that the completion of a normed linear space is not only a metric space but also a normed linear space. To establish this we must introduce linear operations into the completion and verify that axioms (1) and (3) are satisfied.

Each element X of the completion of a metric space E was defined by us as a symbol corresponding to a class of equivalent fundamental sequences of E.

We suppose now that E is a normed linear space. Then if we add two fundamental sequences $x_1, x_2, \ldots, x_n, \ldots$ and $y_1, y_2, \ldots, y_n, \ldots$ term by term, we get the sequence

$$x_1 + y_1, x_2 + y_2, \ldots, x_n + y_n, \ldots,$$

which is also fundamental, since

$$\|(x_n + y_n) - (x_m + y_m)\| \leqq \|x_n - x_m\| + \|y_n - y_m\|.$$

Substituting the equivalent sequences $\{x'_n\}$, $\{y'_n\}$ for the sequences $\{x_n\}$, $\{y_n\}$ we obtain the sequence of sums $\{x'_n + y'_n\}$, which is equivalent with the constructed sequence $\{x_n + y_n\}$, since

$$\|(x'_n + y'_n) - (x_n + y_n)\| \leqq \|x'_n - x_n\| + \|y'_n - y_n\|.$$

This fact permits the introduction of linear operations over the elements of the space $\overline{E}$ as follows.

We choose fundamental sequences $\{x_n\{$, $\{y_n\}$ from the classes X, Y respectively and consider as the sum of X and Y that class which contains the fundamental sequence $\{x_n + y_n\}$.

The preceding discussion affirms the correctness of this identification, and, in particular, the non-dependence of the result on the choice of the sequences $\{x_n\}$, $\{y_n\}$ in X, Y.

The product of the class X with a number λ is similarly defined by choosing a fundamental sequence $\{x_n\}$ in X and understanding as the class λX that class which contains the fundamental sequence $\{\lambda x_n\}$. We leave it to the reader to substantiate the correctness of this identification.

It is easily verified that the axioms 1–8 for a linear space are satisfied here; by their very definition the linear operations on classes reduce to the corresponding operations on elements of the original space. In particular, the class 0 consists of all sequences of the space E that converge to zero.

It only remains for us to verify that axiom 3 for a normed linear space is satisfied. The distance between classes X, Y is defined in Section 7 by

$$\varrho(X, Y) = \lim_{n\to\infty} \varrho(x_n, y_n),$$

where $\{x_n\}$, $\{y_n\}$ are any fundamental sequences in the classes X, Y, respectively. In particular, fixing the fundamental sequences $\{x_n\}$, $\{y_n\}$, $\{z_n\}$, in the classes X, Y, Z, we have

$$\begin{aligned}\varrho(X + Z, Y + Z) &= \lim_{n\to\infty} \varrho(x_n + z_n, y_n + z_n) \\ &= \lim_{n\to\infty} \varrho(x_n, y_n) = \varrho(X, Y),\end{aligned}$$

so that axiom (3a) is satisfied. Similarly

$$\varrho(\lambda X, 0) = \lim_{n\to\infty} \varrho(\lambda x_n, 0) = |\lambda| \lim_{n\to\infty} \varrho(x_n, 0) = |\lambda|\, \varrho(X, 0),$$

and so axiom (3b) is satisfied. Our assertion is therefore fully affirmed.

4. *Factor-Space*

Let L be a subspace of the linear space R (with no metric as yet). Two elements x, y are said to be *equivalent relative to* L if their difference $x - y$ belongs to L. If x, y are each equivalent individually to a third element z, they are equivalent to each other; for

$$x - y = (x - z) - (y - z) \in L.$$

Thus the whole space R can be partitioned into classes of mutually equivalent elements, i.e. x, y fall into the same class if and only if they are equivalent. The subspace L itself forms one of the classes; the class containing the element x_0 is the totality of sums $x_0 + l$, where l runs over all L. We shall denote classes of equivalent elements by X, Y, ...

We show that linear operations can be introduced into the set of classes X, Y, ... To define the sum of two classes X, Y, we take arbitrary elements x, y in these classes; the sum $z = x + y$ belongs to some class Z, which we shall regard as the sum, by definition, of classes X, Y. Given X, Y, this definition is unique: if x is replaced by an equivalent element $x = x + l$, $l \in L$, and y by an equivalent element $y = y + l'$, $l' \in L$, the sum $x + y$ is replaced by $x' + y' = (x + y) + (l + l')$, equivalent to $x + y$. The product of a class X by a number α is similarly defined: the class αX consists of all elements equivalent to the element αx, where x is any fixed element of the class X. All the axioms 1–8 of a linear space (art. 1) are satisfied here automatically, since they reduce to the corresponding axioms for the elements of R. In particular, the class L is the zero of the space of classes.

The linear space of classes constructed thus is said to be the *factor-space* of space R with respect to the subspace L and is denoted by R/L.

Now let R be a normed linear space and L a closed subspace of R. We can now define a norm on the factor-space R/L by putting

$$\|X\| = \inf_{x \in X} \|x\|.$$

Let us verify that the norm axioms are satisfied.

(a) Obviously $\|L\| = 0$ since $0 \in L$. We show that $\|X\| > 0$ if $X \neq L$. If $\|X\| = 0$, the class X contains a subsequence x_n for which $\|x_n\| \to 0$ as $n \to \infty$. If x is any element of the class X,

$x - x_n = l_n \in L$. But $x - x_n \to x$, and L is closed; hence $x \in L$, $X = L$, which contradicts the hypothesis.

(b) $\|\lambda X\| = \inf_{z \in \lambda X} \|z\| = \inf_{x \in X} \|\lambda x\| = |\lambda| \inf_{x \in X} \|x\| = |\lambda| \, \|X\|$.

(c) $\|X + Y\| = \inf_{z \in X+Y} \|z\| \leqq \inf_{x \in X,\, y \in Y} \|x + y\| \leqq \inf_{x \in X,\, y \in Y} \{\|x\| +$

$$+ \|y\|\} = \inf_{x \in X} \|x\| + \inf_{y \in Y} \|y\| = \|X\| + \|Y\|.$$

Thus R/L is here a normed linear space.

Now let R be also complete; then *R/L is also a complete space.* We prove this by considering a fundamental sequence of classes $X_1, X_2, \ldots, X_n, \ldots$ We find for a given k a number n_k such that $\|X_{n_k+m} - X_{n_k}\| < 1/2^k$ for any $m = 1, 2, \ldots$; in particular, $\|X_{n_k+1} - X_{n_k}\| < 1/2^k$. We choose an arbitrary element x_1 of the class X_{n_1}; since the class $X_{n_2} - X_{n_1}$ consists of all the differences $x - x_1$, where x runs over the class X_{n_2}, we can find an element $x_2 \in X_{n_2}$ such that $\|x_2 - x_1\| < 1$; and in the same way, we find an element $x_3 \in X_{n_3}$ such that $\|x_3 - x_2\| < 1/2$, and so on; the element x_{k+1} belongs to the class $X_{n_{k+1}}$ and $\|x_{k+1} - x_k\| < 1/2^{k-1}$. The sequence x_k is evidently fundamental and, since R is complete, converges to some element x. Let X be the class containing x; we have

$$\|X - X_{n_k}\| \leqq \|x - x_k\| \to 0 \quad \text{as} \quad k \to \infty.$$

The sequence of classes X_{n_k} is therefore convergent to the class X. But X_n is a fundamental sequence, so that it converges as a whole to the class X, as required.

Examples. 1. Let E be a linear space on which a seminorm is defined. This means that every element $x \in E$ corresponds to a number $\|x\|$ such that the norm axioms (β), (γ) are satisfied, (i.e. $\|\lambda x\| = |\lambda| \, \|x\|$ and $\|x + y\| \leqq \|x\| + \|y\|$), but (α) is not satisfied, i.e. there are elements $x \neq 0$ for which $\|x\| = 0$. Let us show how the space E can be "improved", i.e. transformed to a new space in which all three axioms (α), (β), (γ) are satisfied. By virtue of (β), (γ), the totality of elements x with $\|x\| = 0$ forms a subspace $Z \in$ E. We form the factor-space R/Z; as we know, its elements are the classes of elements equivalent relative to Z. All the elements of a class X have the same seminorm, since

$$\|x + l\| \leqq \|x\| + \|l\| = \|x\|,$$

$$\|x\| \leqq \|x + l - l\| \leqq \|x + l\| + \|l\| = \|x + l\|.$$

We define the norm of X as the common value of the seminorms of all its elements. Axioms (β), (γ) are satisfied by construction, since they are satisfied for the elements x. Axiom (α) is also satisfied; the norm of class L is zero, and if $\|X\| = 0$, all the elements $x \in X$ have zero norm, so that X coincides with Z. Thus a "genuine" normed space is obtained by forming the factor-space E/Z of E with respect to the subspace Z of elements with zero seminorm.

2. The completion of a normed linear space E (art. 3) can be interpreted as the formation of the factor-space of the space R of all fundamental sequences of elements of E with respect to the subspace Z of sequences convergent to zero.

The seminorm of space R is in fact

$$\|\{x_n\}\| = \lim \|x_n\|.$$

Elements with zero seminorm correspond to sequences convergent to zero. The classes of equivalent elements are the classes of coterminous fundamental sequences, and R/Z is the totality of these classes, normed in accordance with the rule indicated in example 1.

Problems. 1. Show that the triangle axiom in the definition of a normed linear space can be replaced by the convexity condition for the unit sphere.

2. An arbitrary, centrally symmetric, closed, convex set Q is taken on the plane with the origin of coordinates as an interior point. Show that there exists a norm for which Q is a unit sphere.

3. Let R be an n-dimensional normed space. Prove that the sequence $x_m = (\xi_1^{(m)}, \ldots, \xi_n^{(m)})$ $(m = 1, 2, \ldots)$ converges to zero if and only if each of the coordinates $\xi_j^{(m)}$ tends to zero as $m \to \infty$.

Hint. Since $\|x\| \leq \sum_{j=1}^{n} |\xi_j| \, \|e_j\|$ where $e_j = (0, 0, \ldots, 1, \ldots, 0)$ the 1 being in the jth position, convergence by coordinates implies convergence in the norm. To prove the converse it is enough to consider the sphere $\|x\| \leq 1$ and show that the coordinates of all its points are bounded by a fixed constant. But if for some sequence x_m, $\|x_m\| \leq 1$, we have $\max_j |\xi_j^{(m)}| = c_m \to \infty$, then $y_m = \dfrac{x_m}{c_m} \to 0$. At the same time, among the coordinates of each of the vectors y_m, there is one at least equal in absolute value to 1. It is possible to extract from the sequence y_m a subsequence that converges in all its coordinates, and hence also in the norm, to some vector $y_0 \neq 0$, which contradicts the given relation $y_m \to 0$.

4. Prove that a finite-dimensional subspace of a normed space R is always closed in R.

Hint. Use problem 3.

5. Let L be a closed subspace of a normed space R, not the whole of R. Show that the unit sphere of R contains a vector y the distance of which from all vectors of L exceeds 1/2.

Hint. Let $y_0 \in R - L$ and $d = \inf_{x \in L} |y_0 - x|$; further let $y_1 \in L$ be found such that $|y_0 - y_1| < 2d$. Then the vector $y = \frac{y_0 - y_1}{|y_0 - y_1|}$ satisfies the condition.

6. The constant 1/2 in problem 5 can be replaced by any constant smaller than 1, but cannot, in general, be replaced by 1. Consider as an example the subspace L of the space $C(-1, 1)$ composed of the functions $\varphi(x)$ for which

$$\int_{-1}^{0} \varphi(x)\, dx = \int_{0}^{1} \varphi(x)\, dx.$$

7. An infinite-dimensional normed linear space always contains a bounded, infinite set the elements of which are at mutual distances > 1/2 (F. Riesz).

Hint. Use problem 5.

Note. The result of this problem in conjunction with the results of Section 7 shows that the compactness of its bounded sets is a necessary and sufficient condition for the finite-dimensionality of a normed space.

8. F is a closed set in a complete linear normed space E and contains a closed interval $0 \leq t \leq \varepsilon$, $\varepsilon = \varepsilon(x_0)$, on each ray $t x_0$, $0 \leq t < \infty$. Show that F contains a sphere (I. M. Gelfand).

Hint. $E = \bigcap_{n=1}^{\infty} nF$; use problem 2 of Section 4, art. 5.

9. (continued). If the closed space F of problem 8 is centrally symmetric and convex, it contains a sphere centre O.

Note. Without the convexity condition the theorem is false even on a plane.

9. Linear and Quadratic Functions on a Linear Space

The simplest functions defined on normed linear spaces are linear functions.

A linear function can be defined on an unmetricised linear space as follows: a function $f(x)$, defined on a linear space E, is said to be *linear* (*a linear functional*) if it satisfies the conditions:

1° $f(x + y) = f(x) + f(y)$ for any $x, y \in E$;

2° $f(\lambda x) = \lambda f(x)$ for any $x \in E$ and any number λ.

By way of example we shall obtain the general form of a linear functional on an n-dimensional space E. Let $e_1, e_2, \ldots, e_n$ be a basis for the space E; for a given linear functional $f(x)$, we put

$$f(e_1) = \alpha_1, \ldots, f(e_n) = \alpha_n.$$

Then for any vector $x = \sum_{j=1}^{n} \xi_j e_j \in E$ we shall have

$$f(x) = f\left(\sum_{j=1}^{n} \xi_j e_j\right) = \sum_{j=1}^{n} \xi_j f(e_j) = \sum_{j=1}^{n} \alpha_j \xi_j .$$

Thus any linear functional on an n-dimensional space can be expressed relative to any basis as *a linear function of the coordinates of the vector* x.

Problems. 1. If $f(x)$ is a linear functional on a linear space E, the equation $f(x) = 0$ determines a subspace E_1 of E. Show that the factor space E/E_1 (Section 8, art. 4) is one-dimensional.

Hint. The class X of elements equivalent with respect to E_1 is a set on which $f(x)$ maintains a constant value. The map $x \to f(x)$ determines an isomorphism between the space E/E_1 and a one-dimensional space.

2. Linear functionals $f_1(x), \ldots, f_n(x)$ are said to be linearly independent if from $C_1 f_1(x) + \cdots + C_n f_n(x) \equiv 0$ it follows that $C_1 = \cdots = C_n = 0$. Let E_n be the subspace of a space E determined by the equations $f_1(x) = 0, \cdots, f_n(x) = 0$. Show that the factor space E/E_n is n-dimensional if $f_1, \ldots, f_n$ are linearly independent.

Hint. On each class X of elements equivalent with respect to E_n the functionals $f_1(x), \ldots, f_n(x)$ are constant. The map $x \to \{f_1(x), \ldots, f_n(x)\}$ of the space E into the n-dimensional space R_n determines an isomorphism between the factor space E/E_n and some subspace $R' \subset R_n$, which has, say, k-dimensions. If we had $k < n$, there would exist a linear dependence relation $C_1 f_1(x) + \cdots + C_n f_n(x) \equiv 0$; hence $k = n$, $R' = R_n$.

3. Let $f_1, \ldots f_n$, be linearly independent functionals on a linear space E and let E_n be the subspace described in problem 2. Show that any linear functional g that reduces to zero on E_n is a linear combination of the functionals $f_1, \ldots, f_n$ so that

$$g = \lambda_1 f_1 + \cdots + \lambda_n f_n .$$

Hint. g can be defined on the classes of elements equivalent with respect to E_n and therefore on the space E/E_n. Then use the general form of a linear functional on a finite-dimensional space.

4. Given any n linearly independent functionals $f_1, \ldots, f_n$, find n elements $x_1, \ldots, x_n$ such that $\det \| f_j(x_k) \| \neq 0$.

Hint. Choose them one by one at random from the n linearly independent classes of the n-dimensional space E/E_n (problem 2).

We observe that a linear function of the coordinates of a vector x in an n-dimensional space E is clearly a continuous function of x in the sense of Section 7. Hence a linear functional on a finite-dimensional space is always a continuous function.

On infinite-dimensional spaces there exist both continuous linear functionals, and those which are not continuous. Our discussion will be confined to the class of continuous linear functionals.

We give two important examples of continuous linear functionals on the space $C(a, b)$ of all continuous functions on the closed interval $[a, b]$:

(a) $F(y) = y(x_0)$, i.e. the functional F maps each point of the space $C(a, b)$ i.e. each continuous function $y(x)$, onto its value at a fixed point x_0 of the closed interval $[a, b]$.

(b) $F(y) = \int_a^b f(\xi)\, y(\xi)\, d\xi$, where $f(\xi)$ is a fixed continuous function of ξ.

We leave it to the reader to verify that these functionals have the required properties of continuity and linearity.

Notice that the linear functional $F(y) = y(x_0)$ is no longer continuous on space $C_p(a, b)$†.

We now cite some general properties of continuous linear functionals on a normed space.

Lemma 1. *A linear functional $f(x)$, continuous at the point $x = 0$, is bounded (in absolute value) on any sphere $\|x\| \leqq r$.*

For it follows from the continuity of the functional $f(x)$ at the point $x = 0$ that there exists a sphere $U = \{x: \|x\| \leqq \delta\}$ on which the values of the functional $f(x)$ are bounded by a given number ε. If now $\|x\| \leqq r$, then $(\delta/r)\, x \in U$, and hence

$$\left| f\left(\frac{\delta}{r} x\right) \right| \leqq \varepsilon .$$

But since the functional $f(x)$ is a linear, $f[(\delta/r)\, x] = \delta/r\, f(x)$, so that for $\|x\| \leqq r$

$$|f(x)| \leqq \frac{\varepsilon\, r}{\delta},$$

as required.

Lemma 2. *If a certain linear functional $f(x)$ is bounded on the unit sphere $\|x\| \leqq 1$ so that, say, $|f(x)| \leqq K$ on this sphere, then for any x the inequality*

$$|f(x)| \leqq K \|x\|$$

is satisfied, and the functional $f(x)$ is continuous on the whole space E.

† The functional $F(y)$ is defined on an incomplete space in this example. For an example with a complete space, see the Supplement at the end of this book.

For given any x, the ratio $x/\|x\|$ lies in the unit sphere; hence, by the boundedness condition,

$$f\left(\frac{x}{\|x\|}\right) = \frac{1}{\|x\|}\,|f(x)| \leqq K, \quad |f(x)| \leqq K\|x\|,$$

proving the first assertion of the lemma. The second assertion follows from the inequality

$$|f(x) - f(x_0)| = |f(x - x_0)| \leqq K|x - x_0|.$$

As a consequence of lemmas 1 and 2 we obtain the result that *the continuity of a linear functional at the single point $x = 0$ implies its continuity everywhere.* The exact upper bound of $|f(x)|$ on the unit sphere is called the *norm of the functional* f and is written as $\|f\|$. By lemma 2,

$$|f(x)| \leqq \|f\| \cdot \|x\|$$

for any $x \in E$.

Problem. A Linear functional $f(x)$ is bounded on the sphere $\|x - x_0\| \leqq r$. Show that it is continuous on the whole space E.

We often have to deal with linear functions of several arguments: bilinear, trilinear, etc. We now define a bilinear functional. A function $f(y, z)$ of a pair of arguments y, z, which vary over a linear space is said to be a *bilinear functional* if it is a linear functional of z for each fixed y and a linear functional of y for each fixed z.

We can easily obtain the general form of a bilinear functional on an n-dimensional space. Let $e_1, e_2, \ldots, e_n$ be a basis of the space and let $f(y, z)$ be a given bilinear functional. We determine n^2 numbers β_{jk} $(j, k = 1, 2, \ldots, n)$ by the equations

$$\beta_{jk} = f(e_j, e_k).$$

Now let $y = \sum_{j=1}^{n} \eta_j e_j$ and $z = \sum_{k=1}^{n} \zeta_k e_k$ be two arbitrary vectors. We evaluate $f(y, z)$

$$f(y, z) = f\left(\sum_{j=1}^{n} \eta_j e_j, \sum_{k=1}^{n} \zeta_k e_k\right)$$

$$= \sum_{j=1}^{n} \sum_{k=1}^{n} \eta_j \zeta_k f(e_j, e_k) = \sum_{j=1}^{n} \sum_{k=1}^{n} \beta_{jk} \eta_j \zeta_k, \qquad (1)$$

i.e. for the basis $\{e_j\}$ the functional $f(y, z)$ is expressed as a bilinear form in the coordinates of the vectors y, z.

In the infinite-dimensional case we shall consider bilinear functionals under additional assumptions of continuity. We recall that a function $f(y, z)$ is said to be continuous, say for $y = 0$, $z = 0$, if for any $\varepsilon > 0$ there exists $\delta > 0$ such that for $\|z\| < \delta$, $\|y\| \leqq \delta$, we have

$$|f(y, z)| < \varepsilon .$$

From this condition for a continuous bilinear functional we can deduce the following important estimate: *for any* y, z

$$|f(y, z)| \leqq C\|y\| \cdot \|z\| \tag{2}$$

where C *is a fixed constant.* For in virtue of the continuity of $f(y, z)$ at $y = 0$, $z = 0$, there exists a sphere, of radius δ, say, on which $|f(y, z)|$ does not exceed a given quantity, say 1. But then for any y, z

$$\left| f\left(\delta \frac{y}{\|y\|}, \delta \frac{z}{\|z\|}\right) \right| \leqq 1,$$

and consequently

$$|f(y, z)| \leqq \frac{1}{\delta^2} \|y\| \cdot \|z\| . \tag{3}$$

A function of y which is obtained by substituting y for the argument z in a bilinear functional $f(y, z)$, i.e. a function $f(y, y)$, is said to be a *quadratic functional.* It follows from (1) that a quadratic functional on an n-dimensional space can always be expressed in the form

$$f(y, y) = \sum_{j=1}^{n} \sum_{k=1}^{n} \beta_{jk} \eta_j \eta_k , \tag{4}$$

i.e. as a quadratic form in the coordinates of the vector y. In the general case a quadratic functional satisfies the inequality

$$|f(y, y)| \leqq C\|y\|^2, \tag{5}$$

which is obtained by substituting y for z in (2).

Trilinear functionals, cubic functionals, etc. can be constructed similarly.

Concluding remark

The emergence at the beginning of the twentieth century of the theory of metric spaces and its profound suffusion of mathematical analysis were anticipated by the whole preceding development of analysis. The fundamental concepts of the theory, including completeness, compactness, and separability were formulated in 1906 by M. Fréchet (French mathematician, b. 1878). The general definition of a normed linear space and of linear functionals on it was introduced in 1922 by Stephan Banach (Polish mathematician, 1892–1945), and Norbert Wiener (American Mathematician, 1894–1964). For further study we can recommend Hausdorff's *Theory of Sets* and Banach's *Course of Functional Analysis* (Kiev, 1948).

CHAPTER III

THE CALCULUS OF VARIATIONS

ONE of the central problems in the classical analysis of functions of one or several variables was that of discovering the extrema of differentiable functions. Extremum problems also play an important part in functional spaces, which we began to study in Chapter II. For example, the problem of determining the minimal surface of revolution bounded by two circles with a common axis (Fig. 4) can be interpreted as the problem of finding the extremum of the function

$$F(y) = 2\pi \int_a^b y \sqrt{1 + y'^2}\, \mathrm{d}x,$$

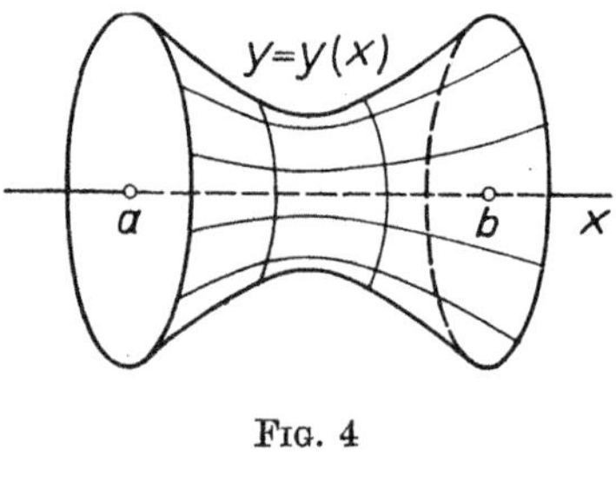

FIG. 4

the argument $y = y(x)$ of which is itself an element of a normed linear space (for example, the space $D_1(a, b)$ in the case considered).

The *calculus of variations* has as its aim a generalisation of the structures of classical analysis which will make possible the solution of similar extremum problems. More widely conceived, the calculus of variations is the analysis of infinitesimals (differential calculus) in infinite-dimensional spaces.

1. Differentiable Functionals

1. We first recall the definition of a differentiable function in classical analysis.

If a function $F(x) = F(x_1, x_2, \ldots, x_n)$ has continuous derivatives (in the usual sense) with respect to each of the arguments $x_1, \ldots, x_n$, then its increment from the point $x = (x_1, \ldots, x_n)$ to the point $x + h = (x_1 + h_1, \ldots, x_n + h_n)$ can be expressed in the form

$$\begin{aligned}\Delta F(x) = F(x+h) - F(x) &= \sum_{j=1}^{n} \frac{\partial F(x_j + \theta h_j)}{\partial x_j} h_j \\ &= \sum_{j=1}^{n} \frac{\partial F(x_j)}{\partial x_j} h_j + \sum_{j=1}^{n} \left[\frac{\partial F(x_j + \theta h_j)}{\partial x_j} - \frac{\partial F(x_j)}{\partial x_j} \right] h_j \\ &= \sum_{j=1}^{n} \frac{\partial F(x_j)}{\partial x_j} h_j + r(x, h) \quad (0 < \theta < 1). \end{aligned} \tag{1}$$

The first term on the right-hand side is the total differential of the function F; it represents an expression linearly dependent on the components of the displacement vector h. In virtue of the continuity of the partial derivatives $(\partial F)/(\partial x_j)$ the second term is *an infinitesimal of higher order than* $|h|$; this denotes that for any $\varepsilon > 0$ there can be found $\delta > 0$ such that, for $|h| < \delta$, the inequality

$$|r(x, h)| \leqq \varepsilon |h|$$

holds. If $F(x)$ possesses second-order derivatives $\partial^2 F/\partial x_i \, \partial x_k$ $(i, k = 1, 2, \ldots, n)$ in a neighbourhood of the point x, the quantity $r(x, h)$ is expressible in the form

$$r(x, h) = \sum \frac{\partial^2 F(x_1 + \theta h_1, x_2 + \theta h_2, \ldots, x_n + \theta h_n)}{\partial x_i \, \partial x_k} h_i h_k \qquad (0 < \theta < 1),$$

and if these second-order derivatives are bounded in the neighbourhood by a number N, a bound for $r(x, h)$ is given by

$$|r(x, h)| \leqq N n^2 |h|^2, \tag{1'}$$

which shows that $r(x, h)$ is then an infinitesimal of order not less than $|h|^2$.

Thus in the general case the first component in the sum (1) is predominant for sufficiently small displacements h; it is therefore said to be *the principal linear part of the increment of the function* $F(x)$. The general definition of a differentiable function on the n-dimensional space R_n is based on these properties. To wit, a function $F(x)$, defined on a set S in the space R_n, is said to be *differentiable at the point* $x_0 \in S$ if its increment between x_0 and the point $x_0 + h \in S$ can be expressed in the form

$$\Delta F(x) = F(x_0 + h) - F(x_0) = L(x_0, h) + r(x_0, h),$$

where $L(x_0, h)$ is a linear function of the displacement h, and $r(x_0, h)$ is an infinitesimal of higher order than h in the sense explained above.

2. This definition carries over to the case of a function defined on a normed linear space: a functional $F(y)$, defined on a normed linear space E, is said to be *differentiable at the point* $y_0 \in E$ if its increment between y_0 and the point $y_0 + h$ can be expressed in the form

$$\Delta F(y) = F(y_0 + h) - F(y_0) = L(y_0, h) + r(y_0, h),$$

where $L(y_0, h)$ is a continuous linear functional of the displacement h, and $r(y_0, h)$ is an infinitesimal of higher order than h; this denotes that for any $\varepsilon > 0$, $\delta > 0$ can be found, such that for $\varrho(0, h) < \delta$ we have $|r(y_0, h)| < \varepsilon \|h\|$.

We observe that there can only be one continuous linear functional $L(y_0, h)$ satisfying the conditions put forward. For if we had

$$\Delta F(y) = L_1(y_0, h) + r_1(y_0, h) = L_2(y_0, h) + r_2(y_0, h),$$

then by subtraction we should get

$$L_1(y_0, h) - L_2(y_0, h) = r_2(y_0, h) - r_1(y_0, h) = r(y_0, h),$$

where $r(y_0, h)$ is again an infinitesimal of higher order than h. The difference $L_1(y_0, h) - L_2(y_0, h) = L(y_0, h)$ represents a new continuous linear functional of h. For a given $\varepsilon > 0$, we can find $\delta > 0$, such that for $\|h\| < \delta$, the inequality

$$|r(y_0, h)| = |L(y_0, h)| < \varepsilon \|h\|$$

holds. Dividing through this last inequality by $\|h\|$, we find that on the unit sphere $\|h\| \leqq 1$ the values of the linear functional

$L(y_0, h)$ do not exceed ε, i.e. in absolute value. But since ε can be chosen arbitrarily small, $L(y_0, h) \equiv 0$, and $L_1(y_0, h) \equiv L_2(y_0, h)$, as required.

The linear functional $L(y_0, h)$, which we have just shown to be uniquely determined, is said to be *the differential* or, more commonly, *the variation of the functional F at the point* y_0 and is denoted by $\delta F(y_0, h)$.

It is well-known that a differentiable function $F(x_1, \ldots, x_n)$ of n variables has a derivative in every direction. This property carries over to differentiable functionals on any normed linear space:

LEMMA. *If a functional $F(y)$ is differentiable at $y = y_0$, then for any h the function $F(y_0 + t\,h)$, regarded as a function of t, is differentiable with respect to t in the usual sense at $t = 0$, and its derivative is equal to $L(y_0, h) = \delta F(y_0, h)$.*

Proof. The required derivative is the limit of the expressions

$$\frac{F(y_0 + th) - F(y_0)}{t} = \frac{\delta F(y_0, th) + r(y_0, th)}{t}$$

$$= \delta F(y_0, h) + \frac{r(y_0, th)}{t}. \tag{2}$$

By the condition of the lemma, for any $\varepsilon > 0$, there exists $\delta > 0$ such that, for $\|t\,h\| = |t|\,\|h\| < \delta$, we have the inequality

$$|r(y_0, th)| < \varepsilon\,\|th\| = \varepsilon\,|t|\,\|h\|.$$

It follows that the quotient

$$\frac{|r(y_0, th)|}{|t|} < \varepsilon\,\|h\|$$

and as $t \to 0$ can be made as small as we please. This shows that the expression on the left-hand side of (2) has the limit $\delta F(y_0, h)$ as $t \to 0$, as required.

Note. If the functional $F(y)$ is differentiable everywhere, then for fixed y, h, the function $F(y + t\,h)$ is differentiable at every t. For, putting $t = t_0 + \tau$, we have

$$F(y + th) = F(y + t_0 h + \tau h) = F(y_0 + \tau h) \quad (y_0 = y + t_0 h);$$

by what we have proved, this function of τ is differentiable with respect to τ at $\tau = 0$; but then the function $F(y + t\,h)$ is differentiable with respect to t at $t = t_0$.

3. *Examples.* 1. A continuous linear functional $F(y)$ is evidently always a differentiable function, since

$$F(y+h) - F(y) = F(h)$$

and the total increment of the functional reduces to a linear expression in h.

2. Let us consider the functional

$$F(y) = \int_a^b f(x, y(x))\, dx$$

on the space $C(a, b)$ of continuous functions on the closed interval $[a, b]$. The kernel $f(x, y)$ is assumed to be continuous and to have continuous first and second partial derivatives (in the usual sense) as a function of its arguments in the region $a \leqq x \leqq b$, $-\infty < y < \infty$. We give the functional $F(y)$ an increment in which the functional argument $y(x)$ receives a displacement $h(x)$:

$$\begin{aligned} \Delta F(y) &= F(y+h) - F(y) \\ &= \int_a^b \{f[x, y(x) + h(x)] - f[x, y(x)]\}\, dx. \end{aligned} \tag{3}$$

In accordance with the definition of $f(x, y)$ we have

$$f(x, y+h) - f(x, y) = \frac{\partial f}{\partial y} h + r(x, y, h),$$

where, for fixed x, y, $r(x, y, h)$ is an infinitesimal of order higher than h. In each bounded (in respect of y) region, the second derivatives of the function $f(x, y)$ are bounded in absolute value by a number, N say, and by what has gone before, $r(x, y, h)$ is bounded according to the inequality

$$|r(x, y, h)| \leqq N|h|^2.$$

Hence the integral (3) assumes the form

$$\Delta F(y) = \int_a^b \frac{\partial f(x, y)}{\partial y} h(x)\, dx + \varepsilon(y),$$

where the function $\varepsilon(y)$ is bounded in absolute value by the number

$$\int_a^b |r(x, y, h(x))|\,dx \leqq N\int_a^b |h^2(x)|\,dx \leqq N(b-a)\max|h^2(x)|$$
$$\leqq N(b-a)\,\|h\|^2.\ \dagger$$

We see that the increment of the functional $F(y)$ is split into a principal linear part and an infinitesimal of higher order. Thus it is differentiable on the space $C(a, b)$ and its variation has the form

$$\delta F(y, h) = \int_a^b \frac{\partial f(x, y)}{\partial y}\, h(x)\,dx.$$

3. We consider the functional

$$F(y) = \int_a^b f(x, y(x), y'(x))\,dx$$

on the space $D_1(a, b)$ of continuous functions on the closed interval $[a, b]$ with continuous derivatives of the first order. The kernel $f(x, y, y')$ is assumed to be a continuous function, defined on the region $a \leqq x \leqq b$, $-\infty < y < \infty$, $-\infty < y' < \infty$, and having continuous derivatives up to the second order. We give the functional $F(y)$ an increment in which the argument y goes from $y = y(x)$ to $y = y(x) + h(x)$, with $L \in D_1(a, b)$:

$$\begin{aligned}\Delta F(y) &= F(y+h) - F(y)\\ &= \int_a^b \{f[x, y(x) + h(x), y'(x) + h'(x)] - f(x, y, y')\}\,dx\\ &= \int_a^b \left[\frac{\partial f}{\partial y} h(x) + \frac{\partial f}{\partial y} h'(x)\right] dx + \int_a^b r(x, y, y', h, h')\,dx.\end{aligned}$$

The first component is evidently linear in the displacement $h(x)$. If N denotes the upper bound of the second derivatives of the function f with respect to the arguments y, y' in some region bounded in respect of y, y', and further, since we know that

$$\|h\| = \max\{|h(x)|, |h'(x)|\},$$

† Here and elsewhere $\{f(x)\}^2$ is written as $f^2(x)$.

in space $D_1(a, b)$, on writing $\mu = \|h\|$, the second component will be governed by the inequality (cf. unequality (1')):

$$\int_a^b |r(x, y, h, y', h')|\,dx \leq 4N \int_a^b \mu^2\,dx = 4N\mu^2(b-a).$$

The expression obtained is of the second order of smallness with respect to $\|h\|$. We see that the functional $F(y)$ is differentiable on the space $D_1(a, b)$ and its variation has the form

$$\delta F(y, h) = \int_a^b \left[\frac{\partial f}{\partial y} h(x) + \frac{\partial f}{\partial y'} h'(x)\right] dx.$$

4. Similarly the functional

$$F(y) = \int_a^b f[x, y(x), \ldots, y^{(m)}(x)]\,dx$$

is differentiable on the space $D_m(a, b)$; its variation has the form

$$\delta F = \int_a^b \left[\frac{\partial f}{\partial y} h(x) + \cdots + \frac{\partial f}{\partial y^{(m)}} h^{(m)}(x)\right] dx.$$

Here again we require the existence and continuity of the nth derivatives (with respect to all arguments) of the function f.

5. We can also consider functionals which have as argument a function of several variables. For definiteness, consider the functional

$$F(z) = \iint_G f(x, y, z(x, y), z_x(x, y), z_y(x, y))\,d\,x\,dy, \qquad (4)$$

where we have written for brevity

$$z_x = \frac{\partial z}{\partial x}, \quad z_y = \frac{\partial z}{\partial y}.$$

It is natural to consider it on the space $D_1(G)$ of functions $z(x, y)$, continuous together with their first partial derivatives on a region G. The norm in this space is given by the formula

$$\|z\| = \max_{(x, y) \in G} \left\{|z(x, y)|, \left|\frac{\partial z(x, y)}{\partial x}\right|, \left|\frac{\partial z(x, y)}{\partial y}\right|\right\}.$$

Assuming that the function f has continuous second derivatives with respect to all its arguments, we form an expression for the increment in the functional F corresponding to an increment $h(x, y)$ in the functional argument $z(x, y)$:

$$\Delta F = \iint\limits_G [f(x, y, z + h, z_x + h_x, z_y + h_y) - f(x, y, z, z_x, z_y)]\, dx\, dy$$

$$= \iint\limits_G \left[\frac{\partial f}{\partial z} h + \frac{\partial f}{\partial z_x} h_x + \frac{\partial f}{\partial z_y} h_y\right] dx\, dy$$

$$+ \iint\limits_G r(x, y, z, z_x, z_y, h, h_x, h_y)\, dx\, dy.$$

The first component is linear in h; a bound for the second is given, for $\|h\| = \mu$, by the inequality

$$\iint\limits_G |r(x, y, z, z_x, z_y, h, h_x, h_y)|\, dx\, dy \leqq 9N \iint\limits_G \mu^2 dx dy = 9N\mu^2 |G|,$$

where N denotes the upper bound of the absolute values of the second derivatives of f, and $|G|$ is the area of the region G. The functional (4) is therefore differentiable on the space $D_1(G)$, and its variation has the form

$$\delta F(z, h) = \iint\limits_G \left[\frac{\partial f}{\partial z} h + \frac{\partial f}{\partial z_x} h_x + \frac{\partial f}{\partial z_y} h_y\right] dx\, dy.$$

6. We now consider a functional dependent on several functional arguments, say

$$F(y_1, \ldots, y_n) = \int_a^b f(x, y_1(x), \ldots, y_n(x), y_1'(x), \ldots, y_n'(x))\, dx.$$

It is natural to consider it on the space $D_1^{(n)}$ which has as elements the vector functions $y = (y_1(x), \ldots y_n(x))$ $(a \leqq x \leqq b)$. We define the norm of a vector function $(y = y_1(x), \ldots, y_n(x))$ by the formula

$$\|y\| = \max_{a \leqq x \leqq b} \{|y_1(x)|, \ldots, |y_n(x)|, |y_1'(x)|, \ldots, |y_n'(x)|\}.$$

It is easily verified that all the axioms for a normed linear space are satisfied.

Assuming that f has bounded (by some number N) second derivatives with respect to $y_1, \ldots, y_n$, we evaluate the increment of the functional F corresponding to an increment $h = (h_1, \ldots, h_n)$

in the vector function $y = (y_1, \ldots, y_n)$:

$$\Delta F(y_1, \ldots, y_n) = \int_a^b f(x, y_1 + h_1, \ldots, y_n' + h_n')\,dx - \int_a^b f(x, y_1, \ldots, y_n')\,dx$$
$$= \int_a^b \left(\frac{\partial f}{\partial y_1} h_1 + \cdots + \frac{\partial f}{\partial y_n'} h_n'\right) dx + \int_a^b r(x, y_1, \ldots, y_n', h_1, \ldots, h_n')\,dx.$$

The first term in the result obtained is linear in $h = (h_1, \ldots, h_n)$. A bound for the second term is given by the inequality

$$\int_a^b |r(x, x_1, \ldots, y_n', h_1, \ldots, h_n')|\,dx \leqq n^2 N \int_a^b \mu^2\,dx = n^2 N (b - a)\,\mu^2$$

for $\|h\| < \mu$. This quantity is of the second order in μ; hence the functional F is differentiable and its variation has the form

$$\delta F(y, h) = \int_a^b \left[\frac{\partial f}{\partial y_1} h_1 + \cdots + \frac{\partial f}{\partial y_n'} h_n'\right] dx.$$

Problems. 1. Determine whether the following functionals are differentiable:

(a) $F(y) = y(a)$ on the space $C(a, b)$;
(b) $F(y) = y(a)$ on the space $D_1(a, b)$;
(c) $F(y) = y^2(a)$ on the space $C(a, b)$;
(d) $F(y) = \sqrt{1 + y'^2(a)}$ on the space $D_1(a, b)$;
(e) $F(y) = |y(a)|$ on the space $C(a, b)$.

Answer. (a)–(d) are differentiable, (e) not.

2. Show that if the functional $F(y)$ is differentiable, so is the functional: $F^2(y)$. Give the variation of $F^2(y)$.

Answer. $\delta F^2(y, h) = 2F(y)\,\delta F(y, h)$.

4. Just as in classical analysis, the remainder $r(y, h)$ in some cases admits of a further interpretation. We suppose that the remainder $r(y, h)$ of an increment of a differentiable functional $F(y)$, obtained by extracting the principal linear part, can be resolved into a quadratic functional and a new remainder with

higher than second-order smallness

$$r(y, h) = \frac{1}{2} Q(y, h, h) + r_2(y, h),$$

so that, for any $\delta > 0$, we can find $\varepsilon > 0$ such that, for $\|h\| < \delta$,

$$|r_2(y, h)| < \varepsilon\, |h|^2.$$

In this case the quadratic functional $Q(y, h, h)$ is said to be the second differential or *second variation of the functional* and the functional $F(y)$ is said to be *twice differentiable*. The second differential, like the first, is uniquely determined.

Functionals $F(y)$ of integral type, which we considered above, e.g.

$$F(y) = \int_a^b f(x, y, y')\, \mathrm{d}x, \tag{1}$$

on the space $D_1(a, b)$, are twice differentiable if the integrand f has continuous derivatives up to the third order. Expressions for the second variation of such functionals are easily obtained by expanding f in its Taylor expansion for the argument $y + h$ as far as the third-order terms. Thus, for the functional

$$F(y) = \int_a^b f(x, y)\, \mathrm{d}x$$

the second variation takes the form

$$\delta^2 F(y, h) = \int_a^b f_{yy}(x, y)\, h^2\, \mathrm{d}x; \tag{2}$$

for the functional (1),

$$\delta^2 F(y, h) = \int_a^b [f_{yy}\, h^2 + 2 f_{yy'}\, h\, h' + f_{y'y'}\, h'^2]\, \mathrm{d}x; \tag{3}$$

for the functional

$$F(y) = \int_a^b f(x, y, y', \ldots, y^{(m)})\, \mathrm{d}x$$

we have

$$\delta^2 F = \int_a^b [f_{yy}\, h^2 + \cdots + f_{y^{(k)} y^{(l)}}\, h^{(k)}\, h^{(l)} + \cdots + f_{y^{(m)} y^{(m)}}\, (h^{(m)})^2]\, \mathrm{d}x; \tag{4}$$

for the functional

$$F(z) = \iint\limits_G f(x, y, z, z_x, z_y)\,dx\,dy$$

we have similarly

$$\delta^2 F = \iint\limits_G [f_{zz} h^2 + f_{zz_x} h h_x + \cdots + f_{z_y z_y} h_y^2]\,dx\,dy; \qquad (5)$$

finally for the functional

$$F(y_1, \ldots, y_n) = \int\limits_a^b f(x, y_1, \ldots, y_n, y_1', \ldots, y_n')\,dx$$

we get

$$\delta^2 F = \int\limits_a^b [\sum f_{y_i y_k} h_i h_k + \sum f_{y_i y_k'} h_i h_k' + \sum f_{y_k' y_i'} h_k' h_i']\,dx. \quad (6)$$

Problems. 1. Establish the uniqueness of the second variation.

2. Show that a quadratic functional is differentiable and find its first and second variations.

Answer. $\delta A(y, y; h) = A(y, h) + A(h, y)$, $\delta^2 A(y, y; h, h) = 2A(h, h)$.

3. Find the second variation of the functional $e^{F(y)}$, where $F(y)$ is a twice differentiable functional.

Answer. $\delta^2 e^{F(y)} = \{[\delta F(y, h)]^2 \delta^2 F(y, h, h)\} e^{F(y)}$.

2. Extrema of Differentiable Functionals

1. Let us consider some differentiable functional F on a normed linear space E. We set ourselves the task of finding the points y at which the functional F attains extremal values—maxima or minima.

By definition, F has a relative minimum at a point y_0 if there exists a neighbourhood of y_0 (a sphere with centre y_0) within which the inequality

$$F(y) \geqq F(y_0)$$

is satisfied. If y_0 has a neighbourhood within which the converse inequality

$$F(y) \leqq F(y_0)$$

is satisfied, F is said to have a relative maximum at y_0. In analysis the relative extrema of differentiable functionals are determined by equating their differentials with zero. We shall show that a similar principle applies to differentiable functionals on normed linear spaces.

LEMMA. *At any extremum y_0 of a differentiable functional $F(y)$, the first variation $\delta F(y_0, h)$ of F is equal to zero for any displacement h.*

Proof. For an arbitrarily assigned h we consider the function $F(y_0 + t h)$ of the variable t. This function is differentiable with respect to t and takes on an extremal value at $t = 0$. Hence its derivative becomes zero at $t = 0$. But by the lemma of Section 1 this derivative is equal to $\delta F(y_0, h)$. Thus for any h the expression $\delta F(y_0, h)$ is equal to zero, as required.

Any point y_0 at which the first variation $\delta F(y_0, h)$ of a functional $F(y)$ vanishes for any h is said to be a *stationary point* of the functional.

We have then to determine at which points y_0 the variation $\delta F(y_0, h)$ vanishes (identically in h), and we shall then have the stationary points of the functional. But we have further to identify which of all the stationary points are maxima and minima, in which we are interested.

2. If the functional $F(y)$ is twice differentiable, we can turn to its second variation in investigating this last question. Since the first variation of the functional vanishes at a stationary point y_0, the increment of the functional under a displacement h from y_0 is expressible in the form

$$\Delta F = \frac{1}{2} \delta^2 F(y_0, h) + r_1(y_0, h),$$

where the quantity $r_1(y_0, h)$ is, for any $\varepsilon > 0$, bounded on the sphere $\|h\| < \delta(\varepsilon)$ by the inequality

$$|r_1(y_0, h)| \leqq \varepsilon \|h\|^2.$$

This bound can be expressed in the form of an equation

$$r_1(y_0, h) = \theta\, \varepsilon \|h\|^2, \quad \text{where} \quad -1 \leqq \theta \leqq 1.$$

Hence for $\|h\| < \delta(\varepsilon)$ the expression for the increment of the functional reduces to the form

$$\Delta F(y_0, h) = \frac{1}{2} \delta^2 F(y_0, h) + \theta\, \varepsilon \|h\|^2. \tag{1}$$

We can now formulate a necessary and sufficient condition for a minimum:

(a) *If y_0 is a point at which a functional $F(y)$ is a minimum, then $\delta^2 F(y_0, h_0) \geqq 0$ for any h_0*

(b) *If at a stationary point y_0 the inequality*

$$\delta^2 F(y_0, h) \geqq C\|h\|^2 \quad (C > 0 \text{ fixed}), \tag{1'}$$

is satisfied, y_0 determines a minimum of the functional $F(y)$.

For the proof of assertion (a) we suppose the contrary to be true, i.e. we suppose that for some h_0, $\delta^2 F(y_0, h_0) < 0$. We choose $\varepsilon < \frac{|\delta^2 F(y_0, h_0)|}{2\|h_0\|^2}$ and put $h = t h_0$, where t is so small that $\|h\| = |t| \, \|h_0\| < \delta(t)$. Then

$$\begin{aligned} \Delta F(y_0, h) &= \frac{1}{2} \delta^2 F(y_0, th_0) + \theta \, \varepsilon \|th_0\|^2 \\ &= t^2 \left[\frac{\delta^2 F(y_0, h_0)}{2} + \theta \, \varepsilon \|h_0\|^2\right] < 0, \end{aligned}$$

so that the functional $F(y)$ does not have a minimum at the point y_0.

To prove assertion (b) we choose $\varepsilon < C/2$ and find a lower bound for the increment of the functional $F(y)$ on a sphere of radius $\delta(\varepsilon)$, centre y_0. Using (1') in (1), we get

$$\begin{aligned} \Delta F(y_0, h) &= \frac{1}{2} \delta^2 F(y_0, h) + \theta \, \varepsilon \|h\|^2 \\ &\geqq \|h\|^2 \left(\frac{C}{2} + \theta \, \varepsilon\right) > 0 \quad \text{for} \quad h \neq 0, \end{aligned}$$

and hence the functional $F(y)$ has a minimum at the point y_0.

In general condition (b) cannot be weakened by substituting what would seem the natural condition $\delta^2 F(y_0, h) > 0$ for all h. A counter example to this condition is the functional

$$F(y) = \int_0^1 x \, y^2(x) \, dx - \int_0^1 y^3(x) \, dx = \int_0^1 y^2(x) \, [x - y(x)] dx$$

on the space $C(0, 1)$. The point $y(x) \equiv 0$ is a stationary point for this functional, and the second variation

$$\delta^2 F(0, h) = \int_0^1 x \, h^2(x) \, dx$$

is positive for each function $h(x) \neq 0$. But in any neighbourhood of zero the functional also assumes negative values; for a given $\varepsilon > 0$, it is enough to take as $y(x)$ any non-negative function that is positive at $x = 0$, does not exceed $\varepsilon - x$ for $x < \varepsilon$, and vanishes for $x \geqq \varepsilon$.

Problems. 1. Show that a linear functional, not identically zero, has no extrema.

2. Show that the extremum theory of the functional

$$F(y) = f(y(x))|_{x=a}$$

on the space $C(a, b)$ coincides with the usual extremum theory for a function $f(\xi)$.

3. Show that the extremum theory of the functional

$$F(y) = f(y(a), \quad y(b))$$

on the space $C(a, b)$ coincides with the usual extremum theory for a function of two variables $f(\xi, \eta)$.

3. By way of illustration we shall analyse the extremum problem for a functional of type

$$F(y) = \int_a^b f(x, y(x))\, dx.$$

This functional has a first variation (p. 83)

$$\delta F(y, h) = \int_a^b \frac{\partial f(x, y)}{\partial y} h(x)\, dx.$$

Let $y_0 = y_0(x)$ be an extremal point. The expression $L\, F(y_0, h)$ must vanish at this point for any function $L(x)$. We have the equation

$$\int_a^b \frac{\partial f(x, y_0)}{\partial y} h(x)\, dx = 0. \tag{2}$$

It will be shown below that such an equation can only hold provided that

$$\frac{\partial f(x, y_0)}{\partial y} \equiv 0 \quad \text{identically in } x. \tag{3}$$

If this equation is solved for y_0, it will in general yield one or several functions of x, and these are the only elements of the space $C(a, b)$ that give extrema of the functional under consideration.

The second variation of the functional $F(y)$ has the form (p. 87)

$$\delta^2 F(y_0, h) = \int_a^b \frac{\partial^2 f(x, y_0)}{\partial y^2} h^2(x)\,dx.$$

If it is non-negative for all $h(x)$, then clearly $f_{yy} \geqq 0$. The inequality $f_{yy} \geqq 0$ is therefore a necessary condition for a minimum of the functional F. On the other hand if we have $f_{yy}(x, y_0(x)) > 0$ for all x, then a stationary point is a minimum of the functional, since

$$\Delta F(y_0, h) = \int_a^b \Big[\frac{1}{2} f_{yy}(x, y_0(x))\, h^2(x) + \frac{1}{6} f_{yyy}(x, y_0(x) + \theta(x)\, h(x))\, h^3(x)\Big]\,dx$$

$$= \int_a^b h^2(x) \Big[\frac{1}{2} f_{yy} + \frac{1}{6} f_{yyy}\, h\Big]\,dx > 0$$

for sufficiently small h.

4. We turn now to equation (2). We have to show that if it holds for all continuous $h(x)$, then equation (3) holds. To do this we employ a lemma which will be of repeated use to us:

LEMMA. *If, for a continuous function $A(x)$, the equation*

$$\int_a^b A(x)\, h(x)\,dx = 0, \tag{4}$$

holds for any continuous function $h(x)$, then $A(x) \equiv 0$.

Proof. If $A(x) \not\equiv 0$, there exists an interior point x_0 of the closed interval $[a, b]$ at which $A(x_0) \neq 0$. For definiteness let $A(x_0) = c > 0$ and let $U(x_0) = \{x: |x - x_0| < \delta\}$ be a neighbourhood of x_0 in which the inequality $A(x) > c/2$ is satisfied. We consider any non-negative function $h(x)$ that vanishes outside the neighbourhood $U(x_0)$ and is positive at $x = x_0$. Then it is evident that

$$\int_a^b A(x)\, h(x)\,dx = \int_{|x - x_0| < \delta} A(x)\, h(x)\,dx > \frac{c}{2} \int_{|x - x_0| < \delta} h(x)\,dx > 0,$$

which contradicts the original hypothesis. The lemma is proved.

Note 1. The class of functions $h(x)$ for which we must demand that equation (4) hold in order to validate the result can be considerably reduced; as is evident from the construction, we can impose on $h(x)$ any smoothness restrictions we like (right up to infinite differentiability). We can also assume $h(x)$ to vanish in the vicinity of the end-points of the interval. We observe moreover that the lemma remains valid in the analogous formulation when a bounded region G is substituted for the closed interval $[a, b]$ and in the case of several independent variables.

Note 2. We can easily convince ourselves by reviewing the formulations above that they hold in somewhat more general situations. The functional $F(y)$ can be defined, not on a whole space E, but on some subset $E' \subset E$ with the property that, together with any two points y, $y + h$, it contains all the points of the form $y + t h$, $-\infty < t < \infty$, in other words, it contains the whole line determined by the points y, $y + h$. A subset $E' \subset E$ which possesses this property is said to be a *linear manifold* in E. In problems to be considered presently, a functional, defined and differentiable on a whole space E, will be considered just on some linear manifold $E' \subset E$, and we shall seek those points $y_0 \in E'$ at which the functional assumes a value that is extremal relative to displacements in the linear manifold. To solve such a problem, we have to consider the variation $\delta F(y, h)$ only for $y \in E'$, and look for the points y_0 for which $\delta F(y_0, h)$ vanishes for any displacement h that does not take us outside the manifold E'. Similarly for the second variation $\delta^2 F(y_0, h)$.

3. Functionals of the Type $\int_a^b f(x, y, y')\,dx$

We discuss these functionals, which are frequently encountered in problems of mathematics and mechanics, in some detail.

1. As we saw in Section 1 (p. 84), the functional

$$F(y) = \int_a^b f(x, y, y')\,dx,$$

on the space $D_1(a, b)$ has the variation

$$\delta F(y, h) = \int_a^b \left[\frac{\partial f}{\partial y} h(x) + \frac{\partial f}{\partial y'} h'(x)\right] dx. \quad (1)$$

To find the extremal points of the functional F, we must equate its variation $\delta F(y, h)$ with zero. For the required functions $y = y(x)$, we can obtain several variants on the conditions, depending on the manifolds on which the functional is defined.

We consider first the case in which the functional F is defined on the set of functions $y(x)$ which take on fixed values $y(a)$, $y(b)$ at the points a, b. The function $h(x)$ must then vanish at the end-points of the interval $[a, b]$. This set is evidently a linear manifold in $D_1(a, b)$ and we can treat it in accordance with note 2 of Section 2.

We suppose further that the solution $y = y(x)$ sought possesses a continuous second derivative. (This restriction will subsequently be withdrawn.) Then the coefficients of both $h(x)$ and $h'(x)$ in the integral (1), where the required solution $y(x)$ has been substituted for y, will be differentiable functions of x. Integrating the second term by parts, we get:

$$\int_a^b \frac{\partial f}{\partial y'} h'(x)\,\mathrm{d}x = \frac{\partial f}{\partial y'} h(x)\Big|_a^b - \int_a^b \left[\frac{\mathrm{d}}{\mathrm{d}x} f_{y'}\right] h(x)\,\mathrm{d}x.$$

The first term on the right vanishes since $h(a) = h(b) = 0$. Hence the expression for the variation is modified to the form

$$\delta F(y, h) = \int_a^b \left[\frac{\partial f}{\partial y} - \frac{\mathrm{d}}{\mathrm{d}x}\frac{\partial f}{\partial y'}\right] h(x)\,\mathrm{d}x.$$

At the required extremal point the variation $\delta F(y, h)$ must vanish for any $h(x)$. It follows (as in Section 2 above) that the coefficient of $h(x)$ vanishes identically. And so for the unknown function $y = y(x)$ we have the equation

$$\frac{\partial f}{\partial y} - \frac{\mathrm{d}}{\mathrm{d}x}\frac{\partial f}{\partial y'} = 0 \tag{2}$$

(the Euler equation). Expanding the total derivative with respect to x, we can write this equation in the form

$$f_y - f_{xy'} - f_{yy'}\,y' - f_{y'y'}\,y'' = 0.$$

This is an ordinary differential equation of the second order, linear in the major derivative. It follows that *if an extremum of the functional F exists and is attained at a function $y = y(x)$ that* *possesses a derivative of the second order, then the function* $y = y(x)$ *satisfies Euler's equation.*

The general solution of Euler's equation, as of any second-order differential equation, contains two parameters C_1, C_2. Each independent solution that obtains for fixed C_1, C_2, is said to be an *extremal* of the functional F. By a suitable choice of the constants C_1, C_2 we can generally find an extremal that satisfies prescribed conditions $y(a) = y_a$, $y(b) = y_b$ (there may be several such extremals). If no solution of Euler's equation satisfies these conditions, this signifies that our extremum problem has no solution in the class of twice differentiable functions.

2. Consideration of the second variation of the functional F enables us to obtain further necessary conditions for an extremum which are also sufficient.

As we found in Section 1 (p. 87) the second variation has the form

$$\delta^2 F = \tfrac{1}{2} \int_a^b [f_{yy} h^2 + 2 f_{yy'} h h' + f_{y'y'} h'^2] \, dx.$$

The middle term can be transformed as follows:

$$\int_a^b f_{yy'} h h' \, dx = \int_a^b \frac{1}{2} f_{yy'} dh^2(x) = -\frac{1}{2} \int_a^b h^2 \frac{d}{dx} f_{yy'} \, dx,$$

and hence

$$\delta^2 F = \int_a^b [P(x) h^2 + \tfrac{1}{2} f_{y'y'} h'^2] \, dx, \qquad (3)$$

where $$P(x) = \frac{1}{2}\left(f_{yy} - \frac{d}{dx} f_{y,y'}\right).$$

We claim that for minimal points $y = y_0(x)$ the inequality

$$f_{y'y'}(x, y_0(x), y_0'(x)) \geqq 0$$

holds for any x in the interval $[a, b]$.

For suppose that at some point x_0, $a \leqq x_0 \leqq b$, the expression $f_{y'y'}(x, y_0(x), y_0'(y))$ is negative. Then it is negative in some neighbourhood U of x_0. Let $L_0(x - x_0) \in D_1(a, b)$ be a function that takes on values between 0 and 1 on the neighbourhood U, equals 1 at the point x_0, and vanishes outside U. We can always find an open interval of length, say δ, on which $h_0'(x - x_0) \geqq c > 0$. We consider the expression (3) for the second variation under a displacement $h_m(x) = h_0[m(x - x_0)]$, when $m \to \infty$. The first term is bounded in absolute value by the quantity

$$\int_a^b |P(x)| \, dx,$$

the second evidently tends to $-\infty$, since by hypothesis $f_{y'y'} < 0$ on the neighbourhood U and the derivative of the function $h_m(x)$ is known to exceed $m^2 c^2$ on an open interval of length δ/m in this neighbourhood. Hence for sufficiently large m, $\delta^2 F$ takes on negative values under the displacement $h_m(x)$, but then the functional F cannot have a minimum at the point y_0.

The inequality $f_{y'y'}(x, y_0(x), y_0'(x)) \geqq 0$ is therefore a necessary condition for the functional F to have a minimum at the stationary point $y_0(x)$. This condition is known as *Legendre's condition.*

The determination of convenient *sufficiency* conditions for a minimum is considerably more complex.

We cite here proof *Weierstrass' sufficiency condition.*

Suppose that an extremal $y = y(x)$ can be included in a "field of extremals", i.e. in a single-parameter family of extremals $y = y(x, \alpha)$, where $-\in < \alpha < \in$ and $y(x, 0) = y(x)$; the function $y(x, \alpha)$ is in addition differentiable with respect to α, $\partial y/\partial\alpha > 0$, and for distinct of α the curves $y = y(x, \alpha)$ on the interval $a \leqq x \leqq b$ do not intersect. Then, *if for all x, y in the region covered by the extremals $y = y(x, \alpha)$ the inequality*

$$f_{y'y'}(x, y, \tau) > 0,$$

is satisfied for all, the extremal $y = y(x)$ determines a relative minimum of the functional F on the space D_1; moreover, of all the curves $y = \varphi(x) \in D_1$ that for sufficiently small β satisfy the inequality

$$|y(x) - \varphi(x)| < \beta,$$

the functional F takes a minimum value on the curve $y = y(x)$ whatever the derivatives $\varphi'(x)$ may be. On the other hand, if, with the given conditions, the inequality

$$f_{y'y'}(x, y, \tau) < 0,$$

is satisfied, then the extremal $y = y(x)$ determines a relative maximum with the same properties.†

We give some concrete problems of geometry and mechanics that reduce to problems of determining the extrema of functionals of the given form.

(a) The functional

$$F(y) = \int_a^b \sqrt{1 + y'^2}\,dx$$

expresses the arc-length of the curve $y = y(x)$ for $a \leqq x \leqq b$. The extremum problem for this function can be formulated as follows: of all the curves $y = y(x)$ that join prescribed points $(a, y(a))$, $(b, y(b))$ a curve of least length can be found. In the given instance the function $f(x, y, y') = \sqrt{1 + y'^2}$ is independent of y, and so Euler's equation (2) assumes the form

$$\frac{d}{dx} f_{y'} = 0,$$

from which it follows that

$$f_{y'} = \frac{y'}{\sqrt{1 + y'^2}} = \text{const};$$

but then $y' = \text{const.}$ also, and hence the solution is given by a linear function $y = C_x + C_1$; the required curve is the straight line joining the given points.

It is clear that we are dealing here with a minimum; we now consider the form assumed by the Weiserstrass condition. The extremal $y = C\,x + C_1$ can evidently be included in the field

$$y = C\,x + C_1 + \alpha, \quad -\varepsilon < \alpha < \varepsilon.$$

Further,

$$f_{y'y'} = \frac{1}{(1 + y'^2)^{3/2}},$$

and hence

$$f_{y'y'}(x, y, \tau) = \frac{1}{(1 + \tau^2)^{3/2}} > 0,$$

so that the Weierstrass condition is satisfied.

† The proof can be found in M. A. Lavrent'ev and L. A. Lyusternik, *A course of variational calculus*, for example M.–L., Chapter 8, (1950).

(b) We pose the analogous for a surface given by the parametric equations

$$x = x(u, v); \quad y = y(u, v); \quad z = z(u, v). \tag{4}$$

It is well known that the arc-length of a curve $v = v(u)$ joining points A, B on the surface (4) can be expressed by the integral

$$F(v) = \int_a^b \sqrt{E(u, v) + 2F(u, v)\frac{\mathrm{d}v}{\mathrm{d}u} + G(u, v)\left(\frac{\mathrm{d}v}{\mathrm{d}u}\right)^2}\,\mathrm{d}u, \tag{5}$$

where E, F, G are the Gaussian coefficients for an element of arc. Putting the general case aside for the moment, we consider the case of a sphere given by the equation

$$x = \cos\varphi\cos\psi, \; y = \sin\varphi\cos\psi, \; z = \sin\psi \tag{6}$$

in spherical coordinates ($u = \varphi$, $v = \psi$). The Gaussian parameters take the form

$$\begin{aligned} G &= x_\varphi x_\varphi + y_\varphi y_\varphi + z_\varphi z_\varphi = \cos^2\psi, \\ F &= x_\varphi x_\psi + y_\varphi y_\psi + z_\varphi z_\psi = 0, \\ E &= x_\psi x_\psi + y_\psi y_\psi + z_\psi z_\psi = 1, \end{aligned}$$

and the functional (5) the form

$$S = \int_a^b \sqrt{1 + \cos^2\psi\left(\frac{\mathrm{d}\varphi}{\mathrm{d}\psi}\right)^2}\,\mathrm{d}\psi.$$

The Euler equation

$$F_\varphi - \frac{\mathrm{d}}{\mathrm{d}\psi}F_{\varphi'} = 0,$$

just as above, admits a first integral

$$F_{\varphi'} = \frac{\cos^2\psi\cdot\varphi'}{\sqrt{1 + \cos^2\psi\cdot\varphi'^2}} = C \leq 1$$

which can also be written in the form

$$\frac{\mathrm{d}\varphi}{\mathrm{d}\psi} = \frac{C}{\cos\psi\sqrt{\cos^2\psi - C^2}}.$$

The general integral of this equation is obtained by making the substitution $\tan\psi = t$:

$$\sin(\varphi + C_2)\, C_1 \tan\psi \quad \left(C_1 = \frac{C}{\sqrt{1 - C^2}}\right)$$

or

$$\sin\psi = \alpha \sin\varphi \cos\psi + \beta \cos\varphi \cos\psi,$$

where α, β are new constants. Reverting to rectangular coordinates in accordance with (6), we get

$$z = \alpha x + \beta y.$$

We have obtained the equation of a plane passing through the origin of coordinates; the curve in which it intersects the sphere is an arc of a great circle. Thus the lines of least length on the sphere; if they exist, are arcs of great circles.

We verify that the Weierstrass condition is satisfied. Provided that the two points selected are not diametrically opposite (so that $\cos\psi = 0$), the arc of the great circle joining them can evidently be included in a field of extremals. Further,

$$F_{\varphi'\varphi'} = \frac{\cos^2\psi}{[1 + \varphi'^2 \cos^2\psi]^{3/2}},$$

so that

$$F_{\varphi'\varphi'}(\psi, \varphi, \tau) = \frac{\cos^2\psi}{[1 + \tau^2 \cos^2\psi]^{3/2}} > 0,$$

and the Weierstrass condition is satisfied; thus the arc Γ of the great circle that joins the two given points really does determine a minimum length in respect of all the curves that join these points and pass sufficiently close to the curve Γ.

Before proceeding to the two remaining examples, we shall make a practical observation concerning the integration of Euler's equation when the function $f = f(x, y, y')$ is independent of the argument x, so that

$$f = f(y, y').$$

We shall regard y here as the independent variable, and x as a function of y subject to definition. Then the functional F reduces to the form

$$F = \int_{y_1}^{y_2} f\left(y, \frac{1}{x'}\right) x' \, \mathrm{d}y.$$

Now the Euler equation

$$g_x - \frac{\mathrm{d}}{\mathrm{d}y} g_{x'} = 0, \quad \text{where} \quad g = f\left(y, \frac{1}{x'}\right) x',$$

will have a first integral

$$g_{x'} = \text{const},$$

just as above, or what is the same thing,

$$x' f_{y'}(y, y')\left(-\frac{1}{x'^2}\right) + f(y, y') = f - y' f_{y'}(y, y') = C. \tag{7}$$

It remains to integrate the first-order equation obtained, and this can be done by quadratures since x is absent.

We now consider the following examples:

(c) An extremum of the functional

$$F(y) = 2\pi \int_a^b y\sqrt{1 + y'^2}\,dx$$

provides the solution to the following problem: to find the curve $y = y(x)$ joining prescribed points $(a, y(a))$, $(b, y(b))$ for which the area of the corresponding surface of revolution about the x-axis is least.

To solve Euler's equation, we apply the foregoing procedure. In this case the first integral (7) takes the form

$$y\sqrt{1 + y'^2} - y'\frac{yy'}{\sqrt{1 + y'^2}} = C,$$

or, what is the same thing,

$$y' = \sqrt{\left(\frac{y}{C}\right)^2 - 1}.$$

With the substitution $y = C \cosh t$, the equation is easily integrated to

$$y = C \cosh\left(\frac{x}{C} + C_1\right).$$

The curve $y = y(x)$ will be the one required if it belongs to this two-parameter family and passes through the given points $(a, y(a))$, $(b, y(b))$. For the sake of simplicity we shall suppose that $a = -b$, $y(a) = y(-a)$. Then $C_1 = 0$ and the whole family of extremals reduces to the single-parameter family

$$y = C \cosh\frac{x}{C}, \tag{8}$$

which is obtained from the catenary $y = \cosh x$ as the family of all possible projections with centre at the origin of coordinates. Depending on the position of the points $(a, y(a))$ and $(-a, y(-a))$ in the plane, three distinct possibilities can arise: the number of catenaries in the family (8) that pass through the points may be two, one, or none at all (Fig. 5, the pairs of points A_1, B_1; A_2, B_2; A_3, B_3 respectively). We shall consider the situation in respect of Weierstrass' sufficiency condition. We have

$$f_{y'y'} = 2\pi \frac{y}{(1 + y'^2)^{3/2}};$$

this quantity is positive for $y > 0$ and any $y' = \tau$. The upper of the two possible extremals that join the points A_1 and B_1 can always be included in a field by employing extremals of the family (8). The Weierstrass condition is thus satisfied, and the upper extremal therefore determines a relative minimum of the functional F. The lower extremal cannot be so included, and the Weierstrass condition is not satisfied. We must therefore leave open the question as to the nature of the extremum determined by it. A more precise investigation will show that the lower extremal yields neither a maximum not minimum.

In the case of a single extremal joining the points A_2, B_2, it can also be included in a field and so determines a relative minimum.

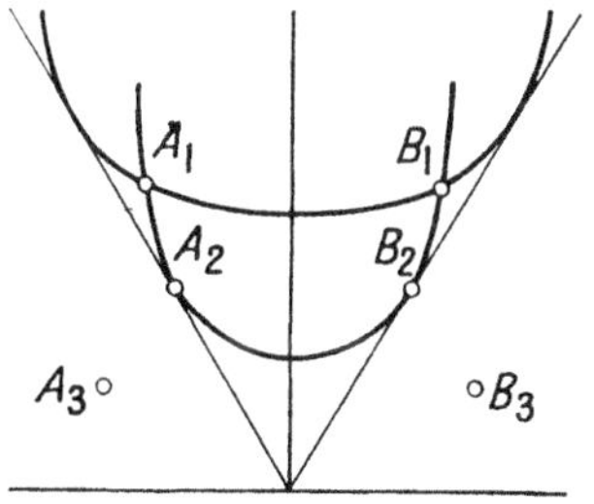

FIG. 5

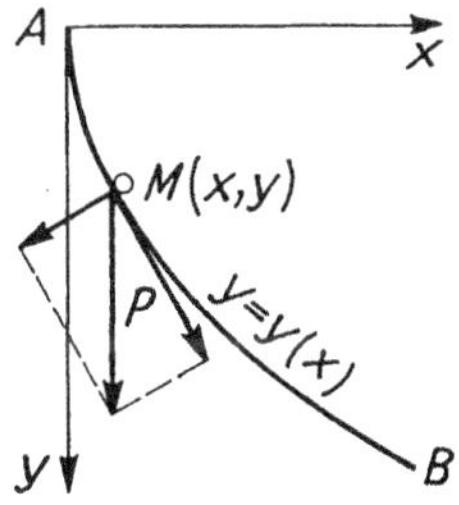

FIG. 6

In the third case there is no member of the class of twice differentiable curves joining the points A_3, B_3, at which the functional F attains a relative minimum.

(d) In 1696 J. Bernoulli proposed and solved the following problem, which became an important landmark in the development of the calculus of variations: *What is the curve that must join two points A, B in a vertical plane in order that a point mass constrained to slide along it under the force of gravity should pass from A to B in the least time?*

The required curve was termed by him the brachistochrone.

To commence the solution of the problem, we first find the time taken for the point M of mass m to slide along the prescribed curve from the first point to the second under the action of gravity and starting from rest. We choose the origin of coordinates at the first of the prescribed points and fix axes as shown in Fig. 6.

We show first of all that at a point with coordinates x, y the velocity of M is $v = \sqrt{2g\,y}$. To do this we resolve the gravitational force $P = mg$ into its normal and tangential components; the former plays no part in the motion, but the tangential component produces a tangential acceleration equal to $g\,\mathrm{d}y/\mathrm{d}s$. We have

$$\frac{\mathrm{d}v}{\mathrm{d}t} = g\frac{\mathrm{d}y}{\mathrm{d}s}, \quad \frac{\mathrm{d}s}{\mathrm{d}t} = v;$$

dividing these equations one by the other, we eliminate $\mathrm{d}s$ and $\mathrm{d}t$ and arrive at the equation

$$v\,\mathrm{d}v = g\,\mathrm{d}y.$$

Integrating this equation and taking into account the initial condition $y = 0$, $v = 0$, we get

$$v = \sqrt{2gy},$$

as required. We have further

$$\mathrm{d}t = \frac{\mathrm{d}s}{v} = \frac{\sqrt{1+y'^2}}{\sqrt{2gy}}\mathrm{d}x$$

and for the required duration, we obtain the expression

$$F(y) = \int_0^b \frac{\sqrt{1+y'^2}}{\sqrt{2gy}}\mathrm{d}x.$$

It is evident that $F(y) = F\big(y(x)\big)$ is a functional depending on the choice of the function $y(x)$. In this instance the function f is no longer twice differentiable with respect to y; nevertheless, a more precise investigation will disclose that here also F is differentiable, and its variation can be calculated in accordance with formula (1). Just as in the preceding example Euler's equation has a first integral

$$f - y' f_{y'} \equiv \frac{\sqrt{1 + y'^2}}{\sqrt{2gy}} - \frac{y'^2}{\sqrt{2gy}\sqrt{1 + y'^2}} = C,$$

or, what is the same thing,

$$y = \frac{C_1}{1 + y'^2} \quad \left(C_1 = \frac{1}{2gC^2}\right).$$

It is convenient here to go over to the parametric representation

$$y' = \tan\varphi .$$

As a result, we get

$$y = C_1 \cos^2\varphi, \quad y' = -C_1 \sin 2\varphi \cdot \varphi' = \tan\varphi,$$

so that

$$\varphi' = \frac{d\varphi}{dx} = -\frac{1}{C_1 \cos^2\varphi}$$

and hence

$$x = -\frac{C_1}{2}(2\varphi + \sin 2\varphi) + C_2 .$$

Replacing 2φ by $\pi - \theta$, we get a simpler parametric form of the solution:

$$x = a(\theta - \sin\theta) + b, \quad y = a(1 - \cos\theta),$$

where a, b are new constants.

Thus the extremals form a family of cycloids with cusps on the x-axis. A unique curve of the family is determined by the conditions of the problem $y(0) = 0$, $y(b) = C$, and this is the curve required.

Note 1. In the problems considered we have found curves $y = y_0(x)$ which determine a relative minimum of a functional $F(y)$. This is not quite an answer to the problem proposed: to find the curve $g = y_0(x)$ that determines the absolute minimum of the functional $F(y)$ (i.e. a minimum in respect not only of sufficiently close curves, but of all functions $y = y(x)$ for which $F(y)$ is defined). Of course the required minimum will also be a relative minimum if it is attained for some curve $y = y_0(x)$, and will then be yielded by our methods. But it may be that no smooth curve determines the absolute minimum. This is known to be the case, for example, in the problem of a minimal surface of revolution, if the prescribed points through which the generating curve must pass are sufficiently far apart. Moreover our results by no means exclude the possibility that, even in the "good" case when the prescribed points are close together, the extremal $y = y_0(x)$ joining them will determine a relative and not an absolute minimum (i.e. although the inequality $F(y) \geqq F(y_0)$ holds for curves close to $y = y_0(x)$, it may be that there are other curves for which $F(y) < F(y_0)$ and that there is no smooth curve for which $F(y)$ attains an absolute minimum). In fact this is not the case. There is a general theorem, called the Hilbert–Tonelli theorem, that guarantees the existence (in the class of rectifiable curves) of a solution to the extremal problem. However we cannot dwell here on the proof†.

Note 2. In the course of the discussion we *assumed* that the required solution $y = y(x)$ has a continuous second derivative. We can avoid making this assumption if we adopt a slightly different approach, which we now describe.

We recall that the variation of the functional

$$F(y) = \int_a^b f(x, y, y')\,\mathrm{d}x,$$

has the form

$$\delta F(y, h) = \int_a^b [f_y\, h(x) + f_{y'}\, h'(x)]\,\mathrm{d}x. \tag{9}$$

We transformed this expression, integrating the second term by parts in order to eliminate $h'(x)$ and deal only with $h(x)$. But we can adopt a different procedure integrating the first term by parts to eliminate $h(x)$ and leaving only $h'(x)$ to deal with. It turns out that with this method, not only is it unnecessary to assume the existence of y'', but we can even prove its existence. Thus, integrating the first term in (9) by parts, we find

$$\int_a^b \frac{\partial f(x, y, y')}{\partial y} h(x)\,\mathrm{d}x = g(x)\,h(x)\Big|_a^b - \int_a^b g(x)\,h'(x)\,\mathrm{d}x,$$

where $g(x)$ denotes the primitive of the function $\partial f(x, y, y')/\partial y$ (we remember that y is a function of x and hence so also is

† Cf. N. I. Akhiezer, *Lectures in the Calculus of Variations*, State Technical Publishing House, 1955, Chapter IV, Sections 33–36. English translation *The Calculus of Variations* (Blaisdell, New York, 1962).

$\partial f(x, y, y')/\partial y$. Since $h(a) = h(b) = 0$ the term without an integral sign vanishes, and we get

$$\delta F(y, h) = \int_a^b \left[- g(x) + \frac{\partial f}{\partial y'} \right] h'(x)\, \mathrm{d}x = 0.$$

It will be shown below that this equation holds for all permissible $h(x)$ only if

$$-g(x) + \frac{\partial f}{\partial y'} = \text{const}. \tag{10}$$

The function $\partial f(x, y, y')/\partial y'$ as a function of the variable x, is not in general differentiable. But in the given instance, equation (10) shows that it is differentiable together with $g(x)$.

Differentiating the left-hand side with respect to x, we arrive at *Euler's equation*:

$$-\frac{\partial f}{\partial y} + \frac{\mathrm{d}}{\mathrm{d}x}\frac{\partial f}{\partial y'} = 0.$$

A total derivative for the second term cannot be exhibited until the existence of y'' is proved. We show that *y'' exists wherever $f_{y'y'}(x, y, y')$ is non-zero.*

The derivative $\mathrm{d}/\mathrm{d}x\,(\partial f/\partial y')$ is the limit as $\Delta x \to 0$ of the expression

$$\frac{f_{y'}(x + \Delta x, y(x + \Delta x), y'(x + \Delta x)) - f_{y'}(x, y, y')}{\Delta x}$$

$$= \frac{\partial \overline{f_{y'}}}{\partial x} + \frac{\partial \overline{f_{y'}}}{\partial y}\frac{\Delta y}{\Delta x} + \frac{\partial \overline{f_{y'}}}{\partial y'}\frac{\Delta y'}{\Delta x},$$

where a bar denotes that the corresponding expression is evaluated for some intermediate values of its arguments. The bar is dispensed with as $\Delta x \to 0$ and the corresponding expression is considered at the original values of the arguments. In addition, $\Delta y/\Delta x$ tends to $y'(x)$. Since $\partial f'_y/\partial y'$ is stipulated to be non-zero, the expression $\Delta y'/\Delta x$ has a limit, and this signifies the existence of the second derivative.

It remains for us to prove the following lemma of Du Bois-Reymond:

If for some continuous function $A(x)$

$$\int_a^b A(x)\, h'(x)\, \mathrm{d}x = 0, \tag{11}$$

for any function $h(x) \in D_1(a, b)$ that vanishes at $x = a$ and $x = b$, then the function $A(x)$ is constant.

For the proof, we suppose that the function $A(x)$ is not constant and that there exist points x_1, x_2, say, for which $A(x_1) < A(x_2)$. We show that there exists a function $h(x) \in D_1(a, b)$, $h(a) = h(b) = 0$, for which equation (11) fails to hold. We take an arbitrary number C with value between $A(x_1)$ and $A(x_2)$. Since the function $A(x)$ is continuous, we can find disjoint open intervals $\Delta_1 \ni x_1$, $\Delta_2 \ni x_2$ such that for any $x' \in \Delta_1$, $x'' \in \Delta_2$

$$A(x') < C < A(x'').$$

As $h'(x)$ we take any continuous function that is positive on Δ_1, negative on Δ_2, and zero outside $\Delta_1 + \Delta_2$, and is such that

$$\int_a^b h'(x)\, \mathrm{d}x = \int_{\Delta_1} h'(x)\, \mathrm{d}x + \int_{\Delta_2} h'(x)\, \mathrm{d}x = 0.$$

the function $h(x)$ is then defined naturally by the formula

$$h(x) = \int_a^x h'(\xi)\, \mathrm{d}\xi;$$

evidently $h(x) \in D_1(a, b)$ and $h(a) = h(b) = 0$

We have further

$$\int_a^l [A(x) - C]\, h'(x)\, \mathrm{d}x = \int_{\Delta_1} [A(x) - C]\, h'(x)\, \mathrm{d}x + $$
$$+ \int_{\Delta_0} [A(x) - C]\, h'(x)\, \mathrm{d}x < 0,$$

since both terms are negative. But then

$$\int_a^b A(x)\, h'(x)\, \mathrm{d}x = \int_a^b [A(x) - C]\, h'(x)\, \mathrm{d}x + C \int_a^b h'(x)\, \mathrm{d}x$$
$$= \int_a^b [A(x) - C]\, h'(x)\, \mathrm{d}x < 0;$$

we see that for the given $h(x)$ equation (11) does not hold, as required.

Problems. 1. Find the extremals and investigate the conditions for solubility of the extremal problem for the following functionals:

(a) $\int_{-1}^{+1} \sqrt{y(1+y'^2)}\, dx, \quad y(-1) = y(1) = b > 0;$

(b) $\int_a^b \frac{1+y^2}{y'^2}\, dx, \quad y(a) = A, \quad y(b) = B.$

Answer. (a) One solution for $b = 1$, two for $b > 1$ (parabolas), and none for $b < 1$.
(b) Always one solution of the form $y = \sinh(C_1 x + C_2)$.

2. Analyse the extremal problem for the functionals:

(a) $\int_0^1 y'\, dx, \quad y(0) = 0, \quad y(1) = 1.$

(b) $\int_0^1 yy'\, dx, \quad y(0) = 0, \quad y(1) = 1.$

(c) $\int_0^1 xyy'\, dx, \quad y(0) = 0, \quad y(1) = 1.$

Answer. In cases (a) and (b) the value of the functional is independent of the choice of the function $y(x)$. In case (c) the variation of the functional does not vanish for any curve joining the given points; there is no extremum.

3. According to Fermat's principle, light travels in such a way that it traverses the distance between points A, B, in the least possible time. Assuming that the velocity of light in the earth's atmosphere varies linearly with altitude, find the form of a light ray. The curvature of the earth's surface may be ignored.

Answer. A circular arc.

4. Find an extremal of the functional

$$F(y) = \int_0^1 e^{y'} \tan y'\, dx, \quad y(0) = 0, \quad y(1) = 1.$$

Answer. $y = x$.

5. Prove the following generalisation of Du Bois-Reymond's lemma: If $A(x)$ is a continuous function and

$$\int_a^b A(y)\, h^{(n)}(x)\, dx = 0$$

for any function $h(x) \in D_n(a, b)$ that vanishes at $x = a$ and at $x = b$ together with its derivatives up to order $n - 1$, then $A(x)$ is a polynomial of degree $< n$.

Hint. Put

$$h^{(n)} x = p(x), \quad h(x) = \int_a^x \cdots \int_a^x p(x)\,\mathrm{d}x^n = \frac{1}{n!} \int_a^x p(\xi)\,(x - \xi)^{n-1}\,\mathrm{d}\xi$$

(Dirichlet's formula). The hypothesis of the theorem can now be expressed as follows: the integral

$$\int_a^b A(x)\,p(x)\,\mathrm{d}x$$

vanishes for any function $p(x)$ for which

$$\int_a^b x^k p(x)\,\mathrm{d}x = 0$$

$k = 0, 1, \ldots, n - 1$. Then apply the result of problem 3, Section 9, Chapter II.

4. Functionals of the Type $\int_a^b f(x, y, y')\,\mathrm{d}x$ (continued)

1. *Conditional extrema*

Apart from conditions of the form $y(a) = y(b)$, problems in the calculus of variations sometimes involve additional conditions of the form

$$G(y) = \int_a^b g(x, y, y')\,\mathrm{d}x = C \quad \text{(where } C \text{ is a given constant).}$$

Such a problem is that of Dido, which seeks the curve $y = y(x)$, $y(a) = 0$, $y(b) = 0$, that for a prescribed length $L > b - a$ bounds the greatest area. Here the subject of investigation is the extremum of the functional

$$F(y) = \int_a^b y\,\mathrm{d}x$$

under the restrictions $y(a) = y(b) = 0$ and the supplementary condition

$$G(y) = \int_a^b \sqrt{1 + y'^2}\,\mathrm{d}x = L.$$

The problem can be formulated abstractly as follows: *find an extremum of the differentiable functional $F(y)$ at a curve of the manifold determined by the equation*

$$G(y) = C,$$

where $G(y)$ is some other differentiable functional.

For the solution of this problem we make the additional assumption that the required point is not a stationary point of the functional G. In general, therefore, a separate study must be made of the stationary points of G. Fundamental to the solution of the problem is the fact that, while the variation of the functional F must vanish as before at the required extremal point, it no longer has to do so for all possible displacements h, but only for those h that yield the invariant value of the functional G. More precisely, we make the following assertion: *at the required extremal point, any vector h that satisfies the equation*

$$\delta G(y, h) = 0,$$

must also satisfy the equation

$$\delta F(y, h) = 0.$$

Let us suppose, on the contrary, that for some $h = h_0$ we have

$$\delta G(y, h_0) = 0, \quad \delta F(y, h_0) = A \neq 0.$$

Then for any t, $|t| \leqq 1$, it will be the case that

$$\delta G(y, t\,h_0) = 0, \quad \delta F(y, t\,h_0) = tA \neq 0 \quad \text{for} \quad t \neq 0.$$

In general the displacement $t h_0$ will lead us off the surface $G(y) = C$, and we shall have $G(y + t\,h_0) \neq C$ and with the aid of a suitable number s we modify the displacement $t_0 h$ so that the new displacement $t\,h_0 + s\,h_1$ satisfies the equation

$$G(y + t\,h_0 + s\,h_1) = C.$$

It can be shown that for all sufficiently small t the number s exists and is an infinitesimal of higher order than t, so that $s/t \to 0$ as $t \to 0$.

For let us write

$$G(y + t\,h_0 + s\,h_1) = G(y).$$

Since G is a differentiable functional, we have

$$G(y + t\,h_0 + s\,h_1) = G(y) + \delta G(t\,h_0 + s\,h_1) + r(t\,h_0 + s\,h_1),$$

where $r(h)$ is an infinitesimal of higher order than $|h|$. And since $\delta G(h_0) = 0$, $\delta G(h_1) = b \neq 0$, we get

$$K(s, t) \equiv s\,b + r(t h_0 + s h_1) = 0. \tag{1}$$

The function $r(t\,h_0 + s\,h_1)$ is differentiable with respect to t and s together with $G(y + t\,h_0 + s\,h_1)$, and so, in accordance with the hypotheses,

$$\left.\frac{\partial r(t\,h_0 + s\,h_1)}{\partial t}\right|_{\substack{s=0\\t=0}} = \left.\frac{\partial r(t h_0 + s\,h_1)}{\partial s}\right|_{\substack{s=0\\t=0}} = 0.$$

We see that the function $K(s, t)$ satisfies the conditions of the implicit function theorem; from equation (1), s can be expressed as a well-defined function of t, vanishing at $t = 0$. This function of t is differentiable,
and so,

$$\left.\frac{\mathrm{d}s}{\mathrm{d}t}\right|_{t=0} = -\left.\frac{\partial K}{\partial t}\right|_{\substack{s=0\\t=0}} \div \left.\frac{\partial K}{\partial s}\right|_{\substack{s=0\\t=0}} = 0,$$

showing that s is an infinitesimal of higher order than t.

We have further that for the functional F

$$\begin{aligned} F(y + t\,h_0 + s\,h_1) - F(y) &= \delta F(y, t\,h_0 + s\,h_1) + \cdots \\ &= t\,\delta F(y, h_0) + s\,\delta F(y, h_1) + \cdots \\ &= t\,a + \cdots \end{aligned}$$

where dots stand for infinitesimals of higher order than t. This expression obviously has differing signs for positive and negative t (sufficiently small in absolute value); hence when G maintains the constant value G, the functional F cannot have an extremum at the point y.

Our assertion is therefore proved. We now deduce from it a rule for determining the required extremal point.

For this we employ a simple lemma from the general theory of linear functionals.

LEMMA. *If a linear functional $F(h)$ vanishes for any vector h_0 for which another linear functional $G(h)$ vanishes, then $F(h)$ is proportional to $G(h)$:*

$$F(h) = \lambda G(h) \quad (\lambda \text{ fixed}) \tag{2}$$

Proof. If $G(h) \equiv 0$, then $F(h) \equiv 0$ and equation (2) holds for any λ. Let $G(h) \not\equiv 0$ and let h_0 be a vector for which $G(h_0) = b \neq 0$.

Then for any h, a number t can be found such that

$$G(h - t\,h_0) = G(h) - t\,G(h_0) = 0.$$

This equation is obviously satisfied by $t = G(h)/G(h_0)$. By hypothesis

$$F(h - t\,h_0) = 0,$$

so that

$$F(h) = t\,F(h_0) = G(h)\,\frac{F(h_0)}{G(h_0)} = \lambda\,G(h),$$

where

$$\lambda = \frac{F(h_0)}{G(h_0)},$$

and the lemma is proved.

We have seen that the linear functional $\delta F(y, h)$ vanishes for any vector h for which the functional $\delta G(y, h)$ vanishes. By the lemma just proved, we have

$$\delta F = \lambda\,\delta G,$$

or

$$\delta(F - \lambda G) = 0.$$

Hence the required extremal point y is determined as one for which the functional

$$H = F - \lambda G$$

for some (unknown) λ has a stationary value (on the whole space). If we have a method for finding y from this condition, we shall get a stationary point $y(\lambda)$ for each λ; but the only relevant λ are those for which the corresponding point $y(\lambda)$ satisfies the equation

$$G(y(\lambda)) = C.$$

The rule obtained is analogous to the well-known principle of Lagrange multipliers in the theory of conditional extrema of functions of several variables.

Example. Find an extremum of the functional

$$F(y) = \int_a^b y\sqrt{1 + y'^2}\,\mathrm{d}x$$

subject to the conditions $y(a) = y_a$, $y(b) = y_b$, and

$$G(y) = \int_a^b \sqrt{1 + y'^2}\,\mathrm{d}x = C. \tag{3}$$

Here

$$H = F - \lambda G = \int_a^b (y - \lambda)\sqrt{1 + y'^2}\,\mathrm{d}x.$$

Putting $y - \lambda = z$, we get the extremum problem for the functional

$$H(z) = \int_a^b z\sqrt{1 + z'^2}\,\mathrm{d}x$$

subject to the conditions $z(a) = y_a - \lambda$, $z(b) = y_b - \lambda$. The solution, as we know, is an arc of a catenary; the number λ can be determined from condition (3), which fixes the length of this arc. The functional $G(y)$ has only one stationary point; it is the closed interval formed by the straight line joining the points (a, y_a), (b, y_b), and its length l is the corresponding value of $G(y)$. The problem evidently becomes meaningless with the condition $G(y) = l$.

Problems. **1.** Solve Dido's problem of finding the curve $y = y(x)$, $y(b) = y(a) = 0$, of prescribed length $L < b - a$, that together with the closed interval $a \leqq x \leqq b$ bounds the greatest area.

Answer. The arc of a circle.

2. Find the closed curve of prescribed length L that encloses the greatest area.

Hint. Use polar coordinates. Show that Euler's equation for the functional $F - \lambda G$ implies that the required curve has constant curvature.

Answer. A circle.

3. Find the solid of revolution of prescribed axial cross-sectional area that contains the least volume.

Answer. A cylinder.

4. Find the solid of revolution of prescribed lateral surface area that contains the greatest volume.

Answer. The solid of revolution of a circular segment about a chord.

5. Differentiable functionals $G_1(y), \ldots, G_n(y)$ are said to be independent at a point y_0 if their variations $\delta G_1(y_0, h), \ldots, \delta G_n(y_0, h)$ are linearly independent. Show that $y \to \{G_1(y), \ldots, G_n(y)\}$ is a mapping of a neighbourhood of the point $y_0 \in E$ on to a neighbourhood of the point $\{G_1(y_0), \ldots, G_n(y_0)\}$ in n-dimensional space if functionals $G_1, \ldots, G_n$ are independent at y_0.

Hint. Find n elements $h_1, \ldots, h_n$ such that $\det \|G_j(h_k)\| \neq 0$ (problem 4 of Section 9, Chapter II) and apply the implicit function theorem to the system of equations (ξ_j given, t_j unknown)

$$G_j(y_0 + t_1 h_1 + \cdots + t_n h_n) = G_j(y_0) + \xi_j \quad (j = 1, 2, \ldots, n).$$

6. Show that the extremum problem of the functional

$$F(y) = \int_a^b f(x, y, y')\,dx$$

with n supplementary conditions

$$G_1(y) = \int_a^b g_1(x, y, y') = C_1, \dots, \qquad G_n(y) = \int_a^b g_n(x, y, y')\,dx = C_n$$

reduces, under the hypothesis that the $G_j(y)$ are linearly independent, to the extremum problem of the functional

$$F - \lambda_1 G_1 - \dots - \lambda_n G_n = \int_a^b [f(x, y, y') - \lambda_1 g_1(x, y, y') - \dots - \lambda_n g_n(x, y, y')]\,dx.$$

Hint. Show by using problem 5 that at the required extremal point, the equation $\delta F(y, h) = 0$ must hold for any displacement h that satisfies the conditions $\delta G_1(y, h) = \dots = \delta G_n(y, h) = 0$. Then use the result of problem 3, Section 9, Chapter II.

2. *Problems with Free End-points*

We now consider the case where the required curve $y = y(x)$ is subject to boundary conditions of another kind, and instead of its end-points being fixed at points (a, y_a), (b, y_b), they can vary along a given curve. Such problems are of frequent occurrence in geometry and mechanics.

We first consider the case where the left end-point of the required curve is fixed as before, but only the abscissa b of the right is fixed.

As before the variation of the functional $F(y)$ has the form

$$\delta F(y, h) = \int_a^b [f_y h + f_{y'} h']\,dx,$$

but the function $h(x)$ is no longer obliged to vanish at $x = b$. Integrating the second term by parts, we get

$$\delta F(y, h) = f_{y'}\Big|_{x=b} h(b) + \int_a^b \left[f_y - \frac{\mathrm{d}}{\mathrm{d}x} f_{y'}\right] h\,\mathrm{d}x.$$

At the required extremal point the variation $\delta F(y, h)$ must vanish whatever the displacement function $h(x)$. If we consider first only those displacement functions for which $h(b) = 0$, we find as before that the required curve satisfies Euler's equation

$$f_y - \frac{\mathrm{d}}{\mathrm{d}x} f_{y'} = 0, \tag{1}$$

i.e. it is an extremal of the functional F. At the same time the variation of the functional reduces at the extremal point to the form

$$\delta F(y, h) = f_{y'}|_{x=b}\, h(b). \tag{2}$$

Since $h(b)$ is arbitrary, the extremum condition $\delta F = 0$ reduces to the equation

$$f_{y'}(b, y(b), y'(b)) = 0, \tag{3}$$

which must be satisfied by the required curve.

Example. A variant on the brachistochrone problem (p. 102) consists in determining the curve $y = y(x)$ along which a point mass, starting from rest at the origin of coordinates, must slide in order to reach the straight line $x = b$ in the least possible time. We recall that the functional $F(y)$ has the form

$$F(y) = \int_0^b \frac{\sqrt{1 + y'^2}}{\sqrt{2gy}}\, \mathrm{d}x.$$

The extremals of $F(y)$ that pass through the origin of coordinates are the cycloids

$$x = C(\theta - \sin\theta), \qquad y = C(1 - \cos\theta). \tag{4}$$

It is easily seen that condition (3) reduces here to the form

$$y'(b) = 0.$$

We must therefore choose the cycloid (4) that is orthogonal to the straight line $x = b$, i.e. the one on which y is a maximum at $x = b$. The y-coordinate attains its greatest value when $\theta = \pi$, and so we get for C

$$b = C\pi.$$

Thus the required curve is

$$x = \frac{b}{\pi}(\theta - \sin\theta), \quad y = \frac{b}{\pi}(1 - \cos\theta).$$

We turn now to the case in which the right end-point of the required curve is restricted to lie on a given curve $y = b(x)$. In this case the functional $F(y)$ has the form

$$F(y) = \int_a^\xi F(x, y, y')\,\mathrm{d}x,$$

where not only the function $y = y(x)$ but also the right end-point ξ of the interval of integration has to be determined. As the normed linear space on which the functional $F(y)$ is defined, we naturally take the space $D_1(a, b)$ of all functions with continuous derivatives on the closed interval $[a, b]$, which comprises all possible positions of the point ξ. The increment in $F(y)$, when the functional argument $y(x)$ is replaced by $y(x) + h(x)$ can be written in the form

$$\Delta F(y, h) = \int_a^\xi [f(x, y + h, y' + h') - f(x, y, y')\,\mathrm{d}x + \\ + \int_\xi^{\xi+\Delta\xi} f(x, y + h, y' + h')\,\mathrm{d}x. \tag{5}$$

The principal linear part of the first term is evaluated just as in the preceding case:

$$\delta_1 F(y, h) = \int_a^\xi \left(f_y - \frac{\mathrm{d}}{\mathrm{d}x} f_{y'}\right) h\,\mathrm{d}x + f_{y'}\Big|_{x=\xi} h(\xi). \tag{6}$$

The increment $\Delta\xi$ of the abscissa ξ, the quantity $h(\xi)$ and the gradient of the curve $y = b(x)$, along which the right end-point of the required curve is constrained to move, are connected by the relation (cf. Fig. 7)

$$[b'(\xi) - y'(\xi)]\,\Delta\xi = h(\xi),$$

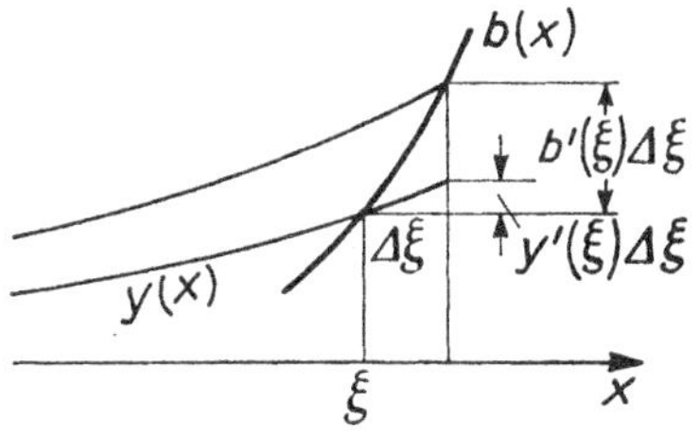

Fig. 7

and hence $\Delta\xi$ can be expressed linearly in terms of $h(\xi)$:

$$\Delta\xi = \frac{h(\xi)}{b'(\xi) - y'(\xi)}.$$

We shall assume that the fixed curve is not an extremal, so that $b'(\xi) \neq y'(\xi)$. We can now find the principal linear part of the second term in equation (5):

$$\delta_2 F(y, h) = \Delta\xi\, f(\xi, y(\xi), y'(\xi)) = \frac{h(\xi)}{b'(\xi) - y'(\xi)} f(\xi, y, y'). \tag{7}$$

Adding (6) and (7), we get

$$\delta F(y, h) = \delta_1 F(y, h) + \delta_2 F(y, h)$$

$$= \int_a^\xi \left(f_y - \frac{\mathrm{d}}{\mathrm{d}x} f_{y'}\right) h \,\mathrm{d}x + \left[f_{y'} + \frac{f}{b'(x) - y'(x)}\right]\Bigg|_{x=\xi} h(\xi).$$

At the required extremal point the variation $\delta F(y, h)$ must vanish for any displacement $h(x)$. If we consider first only those displacement functions for which $h(\xi) = 0$, the second member vanishes as before and the required curve must satisfy Euler's equation

$$f_y - \frac{\mathrm{d}}{\mathrm{d}x} f_{y'} = 0.$$

As before, this curve is one of the extremals of the functional F. For a general displacement function at the extremal point, we get the equation

$$\left[f_{y'} + \frac{f}{b'(x) - y'(x)}\right]\Bigg|_{x=\xi} h(\xi) = 0,$$

which reduces, since $h(\xi)$ is arbitrary, to the condition

$$f + [b'(x) - y'(x)]\, f_{y'}\,|_{x=\xi} = 0. \tag{8}$$

This relation imposes an additional restriction on the elements $y(x)$, $y'(x)$ of the required curve at a point of the curve $y = b(x)$, and it can now be fully determined.

Example. What curve determines the minimum distance between points A and B, where $A = A\,(0, 0)$ is fixed, and B can move along a prescribed curve $y = b(x)$?

Our functional here is

$$F(y) = \int_0^\xi \sqrt{1 + y'^2}\,\mathrm{d}x.$$

The extremals of this functional are straight lines: the first stipulation of the problem is met only by those lines that pass through the origin

$$y = k\,x.$$

Condition (8) acquires the form

$$\sqrt{1+k^2} + [b'(x) - k]\frac{k}{\sqrt{1+k^2}} = 0,$$

or, what is the same thing,

$$k\,b'(x) = -1.$$

This means that the required straight line $y = k\,x$ must intersect the curve $y = b(x)$ orthogonally.

Condition (8) is sometimes referred to as the condition of *transversality*; the required extremal must intersect the given curve *transversally*.

Note. We have considered only those cases where the right end-point of the required curve moves along a prescribed curve. If the same is true of the left end-point, we can show by the same procedure as above that the required curve is an extremal of the functional F, and that the condition of transversality must be satisfied by both the left and right end-points of this extremal.

Problems. **1.** If $f(x, y, y') = A(x, y)\sqrt{1+y'^2}$, the condition of transversality reduces to the condition for orthogonality.

2. Find the variation of the functional

$$F(y) = \int_0^1 y^3\,y'^2\,dx$$

under the single condition $y(0) = 1$.

Answer.

$$\delta F(y, h) = \int_0^1 [3\,y^2\,y'^2\,h + 2\,y^3\,y'\,h']\,dx$$

$$= \int_0^1 [3\,y^2\,y'^2 - 2\,(y^3\,y')']\,h\,dx + 2\,y^3(1)\,y'(1)\,h(1).$$

3. Find the variation of the functional

$$F(y) = \int_0^{x_0} [y^2 + y'^2]\,dx$$

under the conditions $y(0) = 0$, $y(x_0) = e^{2x_0}$.

Answer.

$$\delta F(y, h) = \int_0^{x_0} [2y - 2y''] h(x)\,dx + \frac{y^2(x_0) + y'^2(x_0)}{2e^{2x_0} - y'(x_0)} h(x_0).$$

4. Find the variation of the functional

$$F(y) = \int_{x_0}^{x_1} [y^2 + y'^2] \, ex$$

under the conditions $y(x_0) = \varphi(x_1)$, $y(x) = \psi(x_1)$.

Answer.

$$\delta F(y, h) = 2\int_{x_0}^{x_1} (y - y'') h\,dx + \frac{y^2(x_0) + y'^2(x_0)}{\varphi'(x_0) - y'(x_0)} h(x_0) +$$

$$+ \frac{y^2(x_1) + y'^2(x_1)}{\psi'(x_1) - y'(x_1)} h(x_1).$$

5. Piecewise-smooth extremals. Suppose that on the class of all piecewise-smooth curves $y = y(x)$ with fixed values $y(a)$, $y(b)$, and an angular point (i.e. a point of discontinuity of the derivative) at some $x = \xi$, an extremum of the functional $F(y) = \int_a^b f(x, y, y')\,dx$ is realised at $y_0(x)$. Prove that $y_0(x)$ is a solution of Euler's equation for $x < \xi$ and $x > \xi$, and that the expressions $f_{y'}(x, y_0(x), y_0(x))$ and $f(x, y_0, y_0') - y_0' f_{y'}(x, y_0', y_0)$ are continuous at $x = \xi$, (the Weierstrass–Erdmann principle).

Hint. If the angular point $(\xi, y(\xi))$ moves along a curve $\beta = \beta(\xi)$, the variations of the components $\int_a^\xi$ and $\int_\xi^b$ must be compensatory. Use the relation

$$\frac{h(\xi - 0)}{h(\xi + 0)} = \frac{\beta'(\xi) - y'(\xi - 0)}{\beta'(\xi) - y'(\xi + 0)},$$

which is evident from geometrical considerations.

6. Principle of reflection of extremals. With the hypotheses of the preceding problem, find a necessary condition for the irregularity of a piecewise-smooth extremal to lie on a prescribed curve $y = \beta(x)$.

Answer. The continuity of the expression $f_y (\beta' - y') + f$.

7. Principle of refraction of extremals. The curve $y = \beta(x)$ divides the plane into two parts A, B, one containing the point $(a, y(a))$, the other the point $(b, y(b))$. In the class of piecewise-smooth curves $y = y(x)$ with a single discontinuity on the curve $y = \beta(x)$, find the one for which the functional

$$\int_a^b f(x, y, y')\,dx, \quad f(x, y, y') = \begin{cases} g(x, y, y') & \text{for} \quad (x, y) \in A, \\ h(x, y, y') & \text{for} \quad (x, y) \in B. \end{cases}$$

attains an extremum.

Answer. The required curve in each of the regions A, B is the solution of the corresponding Euler equation. The condition

$$g_{y'}(\beta' - y'_A) + g = h_{y'}(\beta' - y'_B) + h$$

is satisfied on the boundary line.

5. Functionals with Several Unknown Functions

We consider a functional of the form

$$F(y) = \int_a^b f(x, y_1, \ldots, y_n, y'_1, \ldots, y'_n)\, dx \tag{1}$$

on the linear space $D_1^{(n)}(a, b)$ of vector functions $y = [y_1(x), \ldots, y_n(x)]$, defined on the closed interval $[a, b]$, and having continuous first-order derivatives; the norm in this space is given by the formula

$$\|y\| = \max_{a \leqq x \leqq b} \{|y_1(x)|, \ldots, |y_n(x)|, |y'_1(x)|, \ldots, |y_n(x)|\}.$$

If the function f has derivatives up to the second order with respect to all its arguments, then, as we saw in Section 2 (p. 86), the functional (1) is differentiable on the space $D_1^{(n)}$, and its variation has the form

$$\delta F(y, h) = \int_a^b \left[\frac{\partial f}{\partial y_1} h_1 + \ldots + \frac{\partial f}{\partial y_n} h_n + \frac{\partial f}{\partial y'_1} h'_1 + \cdots + \frac{\partial f}{\partial y'_n} h'_n\right] dx.$$

The displacement vector h here is the vector function $[h_1(x), \ldots, h_n(x)]$ in the same space $D_1^{(n)}$.

At the extremal point sought, the variation of the functional F vanishes for all h. In particular, if we set all the components of the displacement vector with the exception of one, $h_j(x)$, equal to zero, we obtain the equation:

$$\int_a^b \left[\frac{\partial f}{\partial y_j} h_j(x) + \frac{\partial f}{\partial y'_j} h'_j(x)\right] dx = 0. \tag{2}$$

We shall solve the extremal problem on the linear manifold of vector functions $y = [y_1(x), \ldots, y_n(x)]$ with prescribed boundary values

$$y(a) = [y_1(a), \ldots, y_n(a)], \quad y(b) = [y_1(b), \ldots, y_n(b)].$$

Then assuming that the required functions $y_1(x), \ldots, y_n(x)$ are twice differentiable with respect to x, and applying the procedure of Section 3, we derive from (2) the Euler equation

$$\frac{\partial f}{\partial y_j} - \frac{\mathrm{d}}{\mathrm{d}x} \frac{\partial f}{\partial y_j'} = 0. \tag{3}$$

The system of Euler equations (3) with $j = 1, 2, \ldots, n$ is a system of n second-order equations in n unknown functions. The general solution of such a system contains $2n$ arbitrary constants $C_1, \ldots, C_{2n}$; by choosing them appropriately, we can determine the solution that satisfies the boundary conditions.

We can, however, dispense with the existence hypothesis for the second derivative by adopting the same procedure as in Section 3 with the DuBois–Reymond lemma. Only here, in place of the restriction $\dfrac{\partial^2 f}{\partial y' \partial y'} \neq 0$ we shall have $\det \dfrac{\partial^2 f}{\partial y_j' \partial y_k'} \neq 0$.

Example. Equations of geodesics. Let us suppose that the square of the differential of arc on an n-dimensional surface L is given by the formula so that the arc-length of a curve joining points A, B

$$\mathrm{d}s^2 = \sum_{j,k=1}^{n} a_{jk}(u)\,\mathrm{d}u_j\,\mathrm{d}u_k,$$

is expressible in the form

$$S = \int_A^B \sqrt{\sum a_{jk}(u)\,\mathrm{d}u_j\,\mathrm{d}u_k}.$$

The coefficients $a_{jk}(u)$ are assumed differentiable with respect to each of the arguments $u_1, \ldots, u_n$, and the quadratic form $\sum a_{jk}(u)\,\mathrm{d}u_j\,\mathrm{d}u_k$ is positive definite.

We shall find the curves on which this functional has an extremal value. Taking the u_j to be expressible as functions of a parameter t, we get the system of Euler equations

$$\frac{1}{2\sqrt{g}} \sum \frac{\partial a_{jk}}{\partial u_l} u_j' u_k' - \frac{\mathrm{d}}{\mathrm{d}t} \frac{1}{\sqrt{g}} \sum a_{jl} u_j' = 0 \quad (l = 1, 2, \ldots, n), \tag{4}$$

where $g(u) = \sum_{j,k} a_{jk} u_j' u_k'$. This quantity becomes equal to x if t denotes arc-length, as we now suppose. Then equation (4) becomes

$$\frac{1}{2} \sum \frac{\partial a_{jk}}{\partial u_l} u_j' u_k' - \frac{\mathrm{d}}{\mathrm{d}t} \sum a_{jl} u_j' = 0 \quad (l = 1, 2, \ldots, n).$$

But on the other hand

$$\frac{\mathrm{d}}{\mathrm{d}t} \sum_j a_{jl} u_j' = \sum_{j,k} \frac{\partial a_{jl}}{\partial u_k} u_k' u_j' + \sum_j a_{jl} u_j''$$

$$= \frac{1}{2} \left[\sum_{j,k} \left(\frac{\partial a_{jl}}{\partial u_k} u_k' u_j' + \frac{\partial a_{kl}}{\partial u_j} u_j' u_k' \right) \right] + \sum_j a_{jl} u_j'',$$

and the equations can therefore be written in the form

$$\sum_j a_{jl} u_j'' = \frac{1}{2} \sum_j \sum_k \left(\frac{\partial a_{jk}}{\partial u_l} - \frac{\partial a_{jl}}{\partial u_k} - \frac{\partial a_{kl}}{\partial u_j} \right) u_j' u_k'. \tag{5}$$

Since the form $g = \sum a_{jk} u_j' u_k'$ is non-degenerate, $\det \| a_{jk} \| \neq 0$, and equation (5) can be solved for the u_j''. Denoting

$$\Gamma_{jk}^l = \frac{1}{2} \left[\frac{\partial a_{jk}}{\partial u_l} - \frac{\partial a_{jl}}{\partial u_k} - \frac{\partial a_{kl}}{\partial u_j} \right],$$

we get a system of equations of the form

$$u_m'' = \sum A_{lm} \Gamma_{jk}^l u_j' u_k'.$$

Applying general theorems on the existence and uniqueness of the solution to a second order system, we conclude that *through each point of the surface L* (more precisely, through each non-singular point, i.e. one at which the form g is non-degenerate) *and in each direction there passes a unique geodesic.*

Equations of motion of a system of point-masses. Let a system of n point-masses $m_1, \ldots, m_n$ be given. We denote the coordinates of the jth point by x_j, y_j, z_j. It is well known that the motion of the system can be expressed by Newton's system of equations

$$m_j \ddot{x}_j = F_{jx}, \quad m_j \ddot{y}_j = F_{jy}, \quad m_j \ddot{z}_j = F_{jz} \quad (j = 1, 2, \ldots, n), \tag{6}$$

where dots indicate differentiation with respect to time, and F_{jx}, F_{jy}, F_{jz} are the components of the force F_j acting on the jth point. We assume that the forces F_j possess a potential function $U = U(x_1, y_1, z_1, \ldots, x_n, y_n, z_n)$; this means that the following set of equations holds:

$$F_{jx} = -\frac{\partial U}{\partial x_j}, \quad F_{jy} = -\frac{\partial U}{\partial y_j}, \quad F_{jz} = -\frac{\partial U}{\partial z_j} \quad (j = 1, 2, \ldots, n).$$

The existence of a potential function U permits the evaluation of the work done by the forces acting on the system in displacements $\mathrm{d}x_1, \mathrm{d}y_1, \ldots, \mathrm{d}z_n$, as a "potential difference":

$$\sum F_{jx}\,\mathrm{d}x_j + F_{jy}\,\mathrm{d}y_j + F_{jz}\,\mathrm{d}z_j$$
$$= -\sum\left(\frac{\partial U}{\partial x_1}\mathrm{d}x_1 + \cdots + \frac{\partial U}{\partial z_n}\mathrm{d}z_n\right) = -\mathrm{d}U.$$

As we know, the function

$$T = \sum \frac{m_j}{2}(\dot{x}_j^2 + \dot{y}_j^2 + \dot{z}_j^2) = T(\dot{x}_1, \dot{y}_1, \dot{z}_1, \ldots, \dot{x}_n, \dot{y}_n, \dot{z}_n),$$

is termed the kinetic energy of the system. We introduce two important functionals,

$$J_1 = \int_a^b T(\dot{x}_1, \ldots, \dot{z}_n)\,\mathrm{d}t, \quad J_2 = \int_a^b U(t, x_1, \ldots, z_n)\,\mathrm{d}t.$$

Taking the initial position of the system $x_1(a), \ldots, z_n(a)$ and the final position $x_1(b), \ldots, z_n(b)$, to be fixed, we find the variations of both functionals J_1, J_2. Denoting the respective components of the displacement vector by $\delta x_1, \ldots, \delta z_n$, we have

$$\delta J_1 = \int_a^b \left[\frac{\partial T}{\partial \dot{x}_1}\,\delta\dot{x}_1 + \cdots + \frac{\partial T}{\partial \dot{z}_n}\delta\dot{z}_n\right]\mathrm{d}t$$

or, integrating each term by parts, and remembering that

$$\delta x_1(a) = \delta x_1(b) = \cdots = \delta z_n(a) = \delta z_n(b) = 0,$$

we get

$$\delta J_1 = -\int_a^b \left[\frac{\mathrm{d}}{\mathrm{d}t}\,\frac{\partial T}{\partial \dot{x}_1}\delta x_1 + \cdots + \frac{\mathrm{d}}{\mathrm{d}t}\,\frac{\partial T}{\partial \dot{z}_n}\delta z_n\right]\mathrm{d}t$$
$$= -\int_a^b [m_1\ddot{x}_1\,\delta x_1 + \cdots + m_n\ddot{z}_n\,\delta z_n]\,\mathrm{d}t.$$

Further

$$\delta J_2 = \int_a^b \left[\frac{\partial U}{\partial x_1}\,\delta x_1 + \cdots + \frac{\partial U}{\partial z_n}\,\delta z_n\right] \mathrm{d}t = -\int_a^b [F_{1x}\,\delta x_1 + $$
$$+ \cdots + F_{nz}\,\delta z_n]\,\mathrm{d}t.$$

In virtue of Newton's equations (6), we have

$$\delta J_1 = \delta J_2,$$

and therefore

$$\delta(J_1 - J_2) = 0.$$

We see that at the functions $x_1(t)$, ..., $z_n(t)$, *which describe the actual position of the system during the interval* $a \leqq t \leqq b$, *the functional*

$$J_1 - J_2 = \int_a^b (T - U)\,\mathrm{d}t$$

has a stationary value.

The fundamental problem in mechanics of a system of point masses thus turns out to be a problem in the variational calculus.

This fact was first discovered by W. Hamilton (in 1835), and the result given therefore bears the title of Hamilton's variational principle.

The function $L = T - U = L(x_1, \ldots, z_n, x_1, \ldots, z_1)$ is termed *Lagrange's function* for the system under consideration.

The motion of the system can often be expressed in terms of a smaller number of variables than the $3n$ functions $x_1, \ldots, z_n$, according to the number of degrees of freedom (i.e. $3n$ less the number of independent constraints). If r is the number of degrees of freedom, the position of the system is determined by r parameters—"generalised coordinates" $q_1, q_2, \ldots, q_r$. In particular, all the rectangular coordinates of points of the system can be expressed in terms of the parameters $q_1, q_2, \ldots, q_r$:

$$\left.\begin{aligned} x_j &= x_j(q_1, \ldots, q\ ),\\ y_j &= y_j(q_1, \ldots, q\ ),\\ z_j &= z_j(q_1, \ldots, q\ ), \end{aligned}\right\} \qquad j = 1, 2, \ldots, n.$$

It follows that

$$\dot{x}_j = \sum_{k=1}^{r} \frac{\partial x_j}{\partial q_k}\dot{q}_k, \quad \dot{y}_j = \sum_{k=1}^{r} \frac{\partial y_j}{\partial q_k}\dot{q}_k, \quad z_j = \sum_{k=1}^{r} \frac{\partial z_j}{\partial q_k}\dot{q}_k,$$

and hence the kinetic energy

$$T = \sum \frac{m_j}{2} (\dot{x}_j^2 + \dot{y}_j^2 + \dot{z}_j^2)$$

is some quadratic form in the "generalised velocities" q_j:

$$T = \sum_{j,k} a_{jk}\, \dot{q}_j\, \dot{q}_k,$$

the coefficients of which are functions of the generalised coordinates. Similarly the potential function $U(x_1, \ldots, z_n)$ is a function of the generalised coordinates

$$U = U(q_1, \ldots, q_n).$$

Lagrange's function $L = T - U$ now emerges as a function of $q_1, \ldots, q_r, \dot{q}_1, \ldots, \dot{q}_r$.

The conditions for stationary values of the functional

$$\int_a^b L\,\mathrm{d}t,$$

can, as always, be expressed as Euler equations, and now assume the form

$$\frac{\partial L}{\partial q_j} - \frac{\mathrm{d}}{\mathrm{d}t}\frac{\partial L}{\partial \dot{q}_j} = 0 \quad (j = 1, 2, \ldots, n); \tag{7}$$

this is the so-called *Lagrange system of equations of the second order.*

Since the kinetic energy of a system vanishes in a position of equilibrium, the conditions for equilibrium can easily be obtained from the equations of motion. Now the potential function is independent of the generalised velocities, and so we get as the conditions of equilibrium

$$\frac{\partial U}{\partial q_j} = 0,$$

i.e. an equilibrium position corresponds to a stationary value of the potential energy.

The equations (7) admit a first integral, termed the *energy integral*. To obtain it, we multiply each equation by $\mathrm{d}q_j = \dot{q}_j\,\mathrm{d}t$, and add the resulting equations, giving

$$\Sigma \frac{\partial T}{\partial q_j}\mathrm{d}q_j - \Sigma \frac{\partial U}{\partial q_j}\mathrm{d}q_j - \Sigma\, \dot{q}_j\, \mathrm{d}\left[\frac{\partial T}{\partial \dot{q}_j}\right] = 0.$$

Since

$$\dot{q}_j\,\mathrm{d}\left(\frac{\partial T}{\partial \dot{q}_j}\right) = \mathrm{d}\left(\dot{q}_j \frac{\partial T}{\partial \dot{q}_j}\right) - \frac{\partial T}{\partial \dot{q}_j}\mathrm{d}\dot{q}_j$$

and in virtue of the homogeneity of the form T in the q_j, which gives

$$\Sigma\, \dot{q}_j \frac{\partial T}{\partial \dot{q}_j} = 2T,$$

we get

$$\Sigma \frac{\partial T}{\partial q_j}\mathrm{d}q_j - \Sigma \frac{\partial U}{\partial q_j}\mathrm{d}q_j - 2\,\mathrm{d}T + \Sigma \frac{\partial T}{\partial \dot{q}_j}\mathrm{d}\dot{q}_j$$
$$= \mathrm{d}T - \mathrm{d}U - 2\,\mathrm{d}T = -(\mathrm{d}U + \mathrm{d}T) = 0,$$

and consequently

$$U + T = \text{const.}$$

Thus *the total energy of the system* (the sum of the potential and kinetic energies) *remains invariant during the motion.*

This last fact permits an easy proof of the following theorem on the stable equilibrium of a system:

THEOREM (Liouville). *If the potential function has a strict minimum at the point* $q^0 = (q_1^0, \ldots, q_r^0)$ *then for any* (sufficiently small) $\varepsilon > 0$, *there will exist* $\delta > 0$ *such that, if a kinetic energy of magnitude less than* δ *is imparted to the system when situated at rest at* $q_1^0, \ldots, q_r^0$, *then during the whole subsequent motion of the system the point* $q = (q_1, \ldots, q_r)$ *will remain within the neighbourhood* $|q - q^0| < \varepsilon$.

Proof. Since by hypothesis the function $U(q_1, \ldots, q_r)$ has a strict minimum at the point q^0, there will exist a sphere $|q - q^0| < \varepsilon$ at the boundary of which the inequality

$$U(q_1, \ldots, q_1) > U(q_1^0, \ldots, q_r^0) + \delta,$$

is everywhere satisfied, δ being a fixed positive number. If a kinetic energy $T \leqq \delta$ is imparted to the system when situated at rest at the point q^0, then in the subsequent motion the total energy $T + U$ of the system will remain constant and will not exceed

$U(q_1^0, \ldots, q_r^0) + \delta$. But since for $|q - q^0| = \varepsilon$, $U(q_1, \ldots, q_r)$ already exceeds the given magnitude, the total energy cannot fail to exceed $U(q_1^0, \ldots, q_r^0) + \delta$, whatever the value of T $(\geqq 0)$. Hence the point $(q_1, \ldots, q_r)$ must lie within the prescribed neighbourhood.

A more searching analysis, which we shall not give here† will show that in a sufficiently small neighbourhood $|q - q^0| \leqq \varepsilon$ it is possible to go over (by means of a linear transformation) from the coordinates $(q_1, \ldots, q_r)$ to new coordinates $(\tau_1, \ldots, \tau_r)$ so that, in the new coordinates, the equations of motion (with an accuracy up to infinitesimals of a higher order) have the form

$$\tau_k = \varepsilon_k \cos(\omega_k t + \alpha_k),$$

where $\varepsilon_k, \omega_k, \alpha_k$ are fixed numbers $(k = 1, 2, \ldots, n)$.

Note 1. It is of interest to observe that the problem of the motion of a mechanical system with n degrees of freedom can be treated in terms of the motion of a point along a geodesic on an n-dimensional surface $E =$ const., taken with a specially chosen metric.

For the functional corresponding to the kinetic energy can be expressed in the form

$$\int T\,\mathrm{d}t = \int \sqrt{E - U}\,\sqrt{T}\,\mathrm{d}t = \int \sqrt{(E - U) \sum a_{jk}(q)\,\dot{q}_j\,\dot{q}_k}\,\mathrm{d}t,$$

and under the condition $E =$ const. the extrema of this functional and Hamilton's functional are attained on one and the same curve.

Note 2. Hamilton's principle possesses a characteristic feature in that its formulation omits any assumption of finiteness in respect of the number of degrees of freedom. It can therefore be applied even to mechanical systems with infinitely many degrees of freedom, and in particular, to problems with a continuous mass distribution, provided the potential and kinetic energies can be calculated for these systems. (The applicability of Hamilton's principle can also be deduced in this case from the heuristic consideration that a continuous medium can be regarded as a system composed of a very large but finite number of separate particles.) We shall consider such problems in Section 8. Here we consider only one equilibrium problem, the form of equilibrium of a flexible inextensible string of prescribed length L suspended by its ends. An element of string has mass $\mu(x)\,\mathrm{d}s$, where $\mathrm{d}s$ is the element of

† Cf. for example, G. Ye. Shilov, *Introduction to the Theory of Linear Spaces*, Prentice-Hall Section 76, p. 218.

arc-length and $\mu(x)$ is the density. The gravitational force $\mu(x)\,\mathrm{d}s \cdot g$ acting on this element has a potential function $\mu\, y\, g\, \mathrm{d}s$. The total potential energy of the string is expressed by the integral

$$U = \int_a^b \mu(x)\, y(x)\, g\, \mathrm{d}s = g \int_a^b \mu\, y \sqrt{1 + y'^2}\, \mathrm{d}x .$$

The equilibrium condition is the condition for U to have a minimum. We also know the length of the string

$$\int_a^b \sqrt{1 + y'^2}\, \mathrm{d}x = L .$$

We have arrived at a problem on conditional extrema. In the case of a uniform string $(\mu(x) = \text{const.})$ its solution (Cf. Section 5) is the arc of a catenary. And so a uniform flexible inextensible string hangs in equilibrium along a catenary (whence the name).

6. Functionals with Several Independent Variables

1. We consider functionals of the form

$$F(u) = \iint_G f\left(x, y, u, \frac{\partial u}{\partial x}, \frac{\partial u}{\partial y}\right) \mathrm{d}x\, \mathrm{d}y \tag{1}$$

on the space $D_1(G)$ of functions $\mu(x,y)$, defined on a plane (bounded) region G, continuous, and having continuous first order derivatives with respect to each of the arguments x, y. The norm in this space will be given by

$$\|u\| = \max \left\{ |u(x, y)|, \left|\frac{\partial u(x, y)}{\partial x}\right|, \left|\frac{\partial u(x, y)}{\partial y}\right| \right\}.$$

If the function $f(x, y, u, v, w)$ has continuous derivatives up to the second order with respect to u, v, w, then as we saw in Section 2 (p. 84) the functional (1) is differentiable on the space $D_1(G)$ and its variation has the form

$$\delta F(u, h) = \iint_G \left[\frac{\partial f}{\partial u} h + \frac{\partial f}{\partial u_x} h_x + \frac{\partial f}{\partial u_y} h_y\right] \mathrm{d}x\, \mathrm{d}y .$$

Here $h = h(x, y)$ is the displacement of the function $u(x, y)$. At an extremal point the variation of the functional $F(y)$ vanishes for any displacement $h(x, y)$:

$$\iint_G \left[\frac{\partial f}{\partial u} h + \frac{\partial f}{\partial u_x} h_x + \frac{\partial f}{\partial u_y} h_y\right] \mathrm{d}x\,\mathrm{d}y = 0.$$

We transform this equation by integration by parts, assuming that the values of the function $u(x, y)$ on the boundary Γ of the region G have been fixed, and that the function $h(x, y)$ consequently vanishes on Γ. For the term $\partial f/\partial u_x$ say, we have

$$\int_{A_1}^{B_1} \frac{\partial f}{\partial u_x} h_x\,\mathrm{d}x = -\int_{A_1}^{B_1} h(x, y) \frac{\partial}{\partial x}\left(\frac{\partial f}{\partial u_x}\right)\mathrm{d}x$$

and hence

$$\iint_G \frac{\partial f}{\partial u_x} h_x\,\mathrm{d}x\,\mathrm{d}y = -\iint_G h \frac{\partial}{\partial x} \frac{\partial f}{\partial u_x}\,\mathrm{d}x\,\mathrm{d}y.$$

The next term is transformed similarly and we obtain the equation

$$\delta F(y, h) = \iint_G \left[\frac{\partial f}{\partial y} - \frac{\partial}{\partial x}\left(\frac{\partial f}{\partial u_x}\right) - \right.$$

$$\left. - \frac{\partial}{\partial y}\left(\frac{\partial f}{\partial u_y}\right)\right] h(x, y)\,\mathrm{d}x\,\mathrm{d}y = 0.$$

Since the function $h(x, y)$ is arbitrary we have the result

$$\frac{\partial f}{\partial y} - \frac{\partial}{\partial x}\left(\frac{\partial f}{\partial u_x}\right) - \frac{\partial}{\partial y}\left(\frac{\partial f}{\partial u_y}\right) = 0, \tag{2}$$

which is known as the Euler–Ostrogradsky equation. It is an equation in partial derivatives of the second order; the unknown function $u(x, y)$ must be determined as the solution of this equation which satisfies the prescribed boundary conditions (the function $u(x, y)$ is known on Γ). The problem of determining $u(x, y)$ from equation (2) subject to the given boundary conditions is known as *Dirichlet's problem*. Just as with the corresponding problems in a single variable, Dirichlet's problem may or may not have a solution; for many important equations of the form (2), the existence and uniqueness of a solution are proved in the theory of partial differential equations.

The Euler–Ostrogradsky equation can be formulated quite analogously in the case of three or more independent variables.

2. *Example* 1. For the functional

$$F(u) = \iint\limits_G (u_x^2 + u_y^2)\,\mathrm{d}x\,\mathrm{d}y$$

the Euler–Ostrogradsky equation has the form

$$\frac{\partial^2 u}{\partial x^2} + \frac{\partial^2 u}{\partial y^2} = 0;$$

its solutions are said to be harmonic functions. It is proved in the theory of partial differential equations that in this instance a solution of Dirichlet's problem exists (uniquely) for any region G with a piecewise–smooth boundary Γ and any continuous function $u(x, y)$ assuming prescribed values on Γ. Cf. also Chapter 5, Section 8.

Problem. Find an extremum of the functional

$$F(u) = \int\limits_0^1 \int\limits_0^1 e^{u_y} \sin u_y \,\mathrm{d}x\,\mathrm{d}y$$

subject to the conditions $u(x, 0) = 0$, $u(x, 1) = 1$.
Answer. $u(x, y) = y$.

Note. This problem has a unique solution although the boundary conditions are not prescribed on the whole boundary.

The function $u(x, y)$ can also be made subject to other boundary conditions (in addition to assuming fixed values). In these cases the required solution will again be a solution of Euler's equation (an extremal) satisfying prescribed boundary conditions, but if it is not yet uniquely determined by these conditions, it will be subject to additional conditions on the boundary, obtaining from the requirement that the variation of the functional vanish (as in the free end point problem for a simple functional).

3. *Example* 2. We introduce the equation for small vibrations of a string. A string, placed in a position of equilibrium between points 0 and l on the x-axis, executes small vibrations about this position. It is assumed that each point moves only in a direction perpendicular to the x-axis. We denote by $u(x, t)$ the configuration of the string at an instant t; for definiteness, we suppose that the end-points remain fixed, so that $u(0, t) = u(l, t) = 0$. The kinetic energy of the string, as the sum of the kinetic energies of the

particles that constitute it, can be expressed by the integral

$$T = \int_0^l u_t^2 \tfrac{1}{2} \mu \, dx,$$

where $\mu \, dx$ is the mass of an element of string corresponding to an interval dx on the x-axis. The quantity $\mu = \mu(x)$ is the density of the string at the point x. A most important property of the string is its potential energy; strictly speaking, the expression for the potential energy actually defines the string from the mechanical point of view. *A string is a one-dimensional mechanical system, the potential energy of each part of which is proportional to its extension relative to the position of equilibrium.* Thus we have

$$dU = p(x) [\sqrt{1 + u_x^2} \, dx - dx],$$

the coefficient $\varphi = p(x)$ figuring in this equation is called the modulus of elasticity of the string (Young's modulus). Assuming u_x to be so small that its fourth power can be neglected, we get

$$(\sqrt{1 + u_x^2} - 1) \, dx \cong \frac{u_x^2}{2} \, dx$$

$$U = \int_0^l \frac{p(x)}{2} u_x^2 \, dx.$$

Lagrange's function $L = T - U$ has the form

$$L = \tfrac{1}{2} \int_0^l [\mu u_t^2 - p u_x^2] \, dx.$$

Hamilton's functional $\int L \, dt$ will now be the double integral

$$\tfrac{1}{2} \int_0^l \int_a^b [\mu u_t^2 - p u_x^2] \, dx \, dt.$$

And we can now write the Euler–Ostrogradsky equation as

$$-\frac{\partial}{\partial t} (\mu u_t) + \frac{\partial}{\partial x} (p \, u_x) = 0. \qquad (3)$$

If μ and p are constants (i.e. the string is uniform in respect of density and elasticity, we obtain the equation

$$\mu u_{tt} - p u_{xx} = 0, \tag{4}$$

in which we are interested.

The boundary conditions arise naturally from the physical circumstances and here we can take them as follows: the values of the function $u(x, 0)$ are prescribed for $t = 0$ (i.e. the initial configuration is known), and similarly the values of the function $u_t(x, 0)$ are given (the initial velocity of each point is known). We shall show that they determine just one solution of equation (3). If there were two solutions $u_1(x, t)$, $u_2(x, t)$ for the equation of the string, taking the identical values $u_1(x, 0) = u_2(x, 0)$, $u_{1t}(x, 0) = u_{2t}(x, 0)$, then their difference $u(x, t) = u_1(x, t) - u_2(x, t)$ would also be a solution satisfying the null conditions $u(x, 0) = 0$, $u_t(x, 0) = 0$. We must show that $u(x, t) = 0$. To do this, we make use of the following consideration. The total energy of the string

$$E = T + U = \frac{1}{2}\int_a^b [\mu\, u_t^2 + p\, u_x^2]\,\mathrm{d}x$$

must remain constant throughout, just as in the case of a finite system of point masses (we give a strict proof of this below). But at the initial instant, by hypothesis, $u_t(x, 0) = 0$, $u(x, 0) = 0$, so that $u_x(x, 0) = 0$, and consequently at $t = 0$, we also have $E = 0$; but then $E = 0$ at any instant, and so $u_t = u_x = 0$. It follows that $u(x, t)$ is a constant, and since $u(x, 0) = 0$, $u(x, t) = 0$ for any t.

It remains to verify the law of conservation of energy for a string. The proof preceeds analogously to the case of a finite system of point masses with sums replaced by integrals with respect to x. We multiply the string equation (3) by u_t and integrate with respect to x:

$$\int_0^l u_t\left[\frac{\partial}{\partial x}(p\, u_x) - \frac{\partial}{\partial t}(\mu\, u_t)\right]\mathrm{d}x = 0.$$

We integrate the first term by parts:

$$\int_0^l u_t \frac{\partial}{\partial x}(p\, u_x)\,\mathrm{d}x = u_t\, p\, u_x\Big|_0^l - \int_0^l u_{tx}\, p\, u_x\,\mathrm{d}x.$$

The term under the integral sign vanishes, since $u_t(0, t)$ and $u_t(l, t)$ vanish together with $u(0, t)$ and $u(l, t)$. Thus we have

$$\int_0^l \left[u_t \frac{\partial}{\partial t}(\mu u_t) + p\, u_x u_{tx}\right] \mathrm{d}x = \frac{1}{2}\int_0^l \left[\frac{\partial}{\partial t}(\mu u_t^2) + \right.$$

$$\left. + \frac{\partial}{\partial t}(p\, u_x^2)\right] \mathrm{d}x = 0,$$

and so

$$\frac{\mathrm{d}E}{\mathrm{d}t} = \frac{\mathrm{d}}{\mathrm{d}t}\int_0^l [\mu u_t^2 + p\, u_x^2]\, \mathrm{d}x = 0$$

from which it follows that

$$E = \text{const.}$$

as required.

We show now how to construct the actual solution, at least for sufficiently smooth initial functions $\varphi(x) = u(x, 0)$, $\psi(x) = u_t(x, 0)$. For simplicity we suppose that $p = 1$, $\mu = 1$, $l = \pi$ (the general case is considered in Chapter 5, Section 5). It is easily verified that for any integer n the functions $\sin n x \cos n t$, $\sin n x \sin n t$ satisfy the equation

$$u_{tt} - u_{xx} = 0 \tag{5}$$

and the boundary conditions $u(0, t) = u(l, t) = 0$. We form the series

$$u(x, t) = \sum_1^\infty \sin n x (a_n \cos n t + b_n \sin n t), \tag{6}$$

the coefficients of which are determined from the conditions

$$u(x, 0) = \sum_1^\infty a_n \sin n x = \varphi(x),$$

$$u_t(x, 0) = \sum_1^\infty n\, b_n \sin n x = \varphi(x).$$

If the functions $\varphi(x)$, $\psi(x)$ are sufficiently smooth †, the coefficients a_n, b_n will tend to zero sufficiently rapidly to ensure the absolute convergence of the series (6) together with its first and second

† More precisely, if the odd periodic continuations of $\varphi(x)$, $\psi(x)$ over the whole axis with period 2π are sufficiently smooth.

derivatives with respect to x and t. We can then evaluate u_{xx} and u_{tt} by summing the corresponding derived series; since equation (5) is satisfied for each term that occurs, it will also be satisfied for the sums. A more detailed account of this question and an analysis of cases in which p and μ are not constant would take us beyond the limits of our course.†

Problem. Find the law of vibration of a string fixed at its end-points $x = 0$, $x = \pi$ if $p = \mu = 1$ and $u(x, 0) = 0$, $u_t(x, 0) = \sin 2x \cos x$.

Answer. $u(x, t) = 1/2 \sin x \sin t + 1/6 \sin 3x \sin 3t$.

4. We consider a further case of constrained motion, in which a string is acted on by an external force. Let a force $f(x, t)\, \Delta x$ act on an element Δx of the string. This force possesses a potential function (the work done between 0 and u).

$$U_f = -\int_0^l f(x, t)\, u(x, t)\, dx,$$

and the total potential energy can therefore be expressed by the formula

$$U = \int_0^l \left[\frac{1}{2} u_x^2 - f u\right] dx;$$

Euler's equation will now assume the form

$$p\, u_{xx} - \mu\, u_{tt} = f(x, t).$$

If the external force is actually independent of time, so that $f(x, t) = f(x)$, we can derive the form of the equilibrium position from this equation. In equilibrium $u_{tt} = 0$, and the configuration of the string $u = u(x)$ therefore satisfies the equation

$$p\, u_{xx} = f(x).$$

We determine the solution of this equation that interests us by means of the boundary conditions at the points $x = 0$, $x = l$.

Thus a uniform string (μ, p = const.) hangs under gravity ($f(x) = \mu g$) in the form of a parabola satisfying the equation

$$p\, u_{xx} = \mu g.$$

† Cf. for example, I. G. Petrovsky, *Lectures on Partial Differential Equations*, Interscience, 1955; R. Courant and D. Hilbert, *Methods of Mathematical Physics*, vol. 1, Chapter V, Interscience, N.Y. (1953).

5. *Example* 3. The equation of small vibrations of a membrane. A membrane is the "two dimensional analogue" of a string: in other words, it is a mechanical system in the form of a surface, the potential energy of each part of which is proportional to its increase in area relative to the equilibrium position. Thus if a function $u(x, y, t)$, defined on a region G in the (x, y)-plane for $t \geqq 0$, describes the configuration of a membrane at an instant t, then the expression for its potential energy assumes the form

$$U = \iint_G p(\sqrt{1 + u_x^2 + u_y^2} - 1)\, dx\, dy \cong \tfrac{1}{2} \iint_G p(u_x^2 + u_y^2)\, dx\, dy.$$

The kinetic energy has the form

$$T = \tfrac{1}{2} \iint_G \mu(x, y)\, u_t^2\, dx\, dy.$$

Lagrange's function becomes

$$L = T - U = \tfrac{1}{2} \iint_G [p(u_x^2 + u_y^2) - \mu\, u_t^2]\, dx\, dy.$$

Hence the Euler–Ostrogradsky equation assumes the form

$$\frac{\partial}{\partial x}(p\, u_x) + \frac{\partial}{\partial y}(p\, u_y) - \frac{\partial}{\partial t}(\mu\, u_t) = 0.$$

For μ, k constant, an equation of the form

$$u_{tt} = c^2(u_x^2 + u_y^2)$$

is obtained. As in the case of a string, we can prescribe $u(x, y, 0)$ and $u_t(x, y, 0)$ as initial conditions. The subsequent theory follows a course basically parallel to that of the string; we again recommend to the reader the course in partial differential equations indicated above.

7. Functionals with Higher Derivatives

1. A functional of the form

$$F(y) = \int_a^b f(x, y, y', \ldots, y^{(m)})\, dx \tag{1}$$

is defined on the space $D_m(a, b)$ of functions $y = y(x)$ with m continuous derivatives on the closed interval $[a, b]$. We recall that the norm $D_m(a, b)$ is given by

$$\|y\| = \max_{a \leq x \leq b} \{|y(x)|, |y'(x)|, \ldots, |y^{(m)}(x)|\}.$$

If the function $f(x, y_0, y_1, \ldots, y_m)$ has derivatives up to the second order with respect to the arguments $y_0, \ldots, y_m$ continuous for all $y_0, \ldots, y_m$, then, as was shown in Section 2 (p. 84), the functional (1) is differentiable on $D_m(a, b)$ and its variation has the form

$$\delta F(y, h) = \int_a^b \left[\frac{\partial f}{\partial y} h + \frac{\partial f}{\partial y'} h' + \cdots + \frac{\partial f}{\partial y^{(m)}} h^{(m)}\right] \mathrm{d}x. \qquad (2)$$

The displacement vector $h = h(x)$ is a function in the same space $D_m(a, b)$. We shall solve the extremal problem for the functional $F(y)$ on the manifold of functions $y(x) \in D_m(a, b)$ with prescribed values

$$\left.\begin{aligned} y(a) &= a_0, \quad y'(a) = a_1, \ldots, y^{(m-1)}(a) = a_{m-1}; \\ y(b) &= b_0, \quad y'(b) = b_1, \ldots, y^{(m-1)}(b) = b_{m-1}. \end{aligned}\right\} \qquad (3)$$

The function $h(x)$ then satisfies the conditions

$$\left.\begin{aligned} h(a) &= h'(a) = \cdots = h^{(m-1)}(a) = 0; \\ h(b) &= h'(b) = \cdots = h^{(m-1)}(b) = 0. \end{aligned}\right\} \qquad (4)$$

Let us suppose that the required function $y = y(x)$ has continuous derivatives up to order $2m$. Then all the functions $\dfrac{\partial f}{\partial y^{(k)}}$ that occur in (2) will, as functions of x, have continuous derivatives up to order m. We integrate each of the terms by parts (commencing with the second) as many times as are required to remove all derivatives of the function $h(x)$. In virtue of (4), all the integrated terms vanish, and we have

$$\delta F(y, h) = \int_a^b \left[\frac{\partial f}{\partial y} - \frac{\mathrm{d}}{\mathrm{d}x}\frac{\partial f}{\partial y'} + \frac{\mathrm{d}^2}{\mathrm{d}x^2}\frac{\partial f}{\partial y''} - \cdots + (-1)^m \frac{\mathrm{d}^m}{\mathrm{d}x^m}\frac{\partial f}{\partial y^{(m)}}\right] h(x)\, \mathrm{d}x.$$

Since $h(x)$ is arbitrary, the required function y must satisfy the equation

$$\frac{\partial f}{\partial y} - \frac{\mathrm{d}}{\mathrm{d}x}\frac{\partial f}{\partial y'} + \frac{\mathrm{d}^2}{\mathrm{d}x^2}\frac{\partial f}{\partial y''} - \cdots + (-1)^m \frac{\mathrm{d}^m}{\mathrm{d}x^m}\frac{\partial f}{\partial y^{(m)}} = 0.$$

This is an ordinary differential equation of order $2m$, and it is also known as Euler's equation. The general solution contains $2m$ arbitrary constants, which can be used in conjunction with the conditions (3) to determine the solution that interests us.

Problem. Find the extremals of the functional

$$F(y) = \tfrac{1}{2} \int_a^b (y'')^n \, dx.$$

Answer. $y = \dfrac{(n-1)^2}{\alpha^2 n(2n-1)} (\alpha x + \beta)^{\frac{2n-1}{n-1}} + \gamma x + \delta$. For $n = \frac{1}{2}$ we have $\alpha^2 y = \gamma x + \delta - \log(\alpha x + \beta)$; for $n = 1$ the value of the functional is independent of y.

2. Problems sometimes occur in which not all the boundary conditions (3) are given, but a smaller number only, so that the general solution of Euler's equation retains some free constants though it satisfies the boundary conditions. Such problems are similar to the free end-point problems for simple functionals (Section 4). For their solution we have to transform the variation (2) utilising the existing boundary conditions, and then obtain supplementary conditions on the boundary by equating it to zero.

Example. Find the curve $y = y(x)$ that determines an extremum of the functional

$$F(y) = \tfrac{1}{2} \int_a^b (y'')^2 \, dx$$

subject to the conditions $y(a) = y(b) = 0$.

Solution. The variation of the functional $F(y)$ has the form

$$\delta F(y, h) = \int_a^b y'' h'' \, dx.$$

After one integration by parts the boundary terms remain, since the function $h'(x)$ need not vanish on the boundary, but a second integration produces no new boundary terms ($h(a) = h(b) = 0$). As a result we get

$$\delta F(y, h) = y''(x) \, h'(x) \Big|_a^b - \int_a^b y'''(x) \, h'(x) \, dx$$

$$= [y''(b) \, h'(b) - y''(a) \, h'(a)] + \int_a^b y^{\mathrm{IV}}(x) \, h(x) \, dx.$$

The given expression must vanish for the extremal function $y(x)$ for any function $h(x) \in D_2(a, b)$ with $h(a) = h(b) = 0$. If in addition $h'(a) = h'(b) = 0$, we get

$$\int_a^b y^{\mathrm{IV}}(x)\, h(x)\, \mathrm{d}x = 0,$$

so that $y^{\mathrm{IV}}(x) = 0$, and $y(x)$ is therefore a cubic parabola:

$$y = a_0 + a_1 x + a_2 x^2 + a_3 x^3.$$

Thus the integral term in the expression for the variation vanishes, and we obtain for the general case

$$\delta F(y, h) = y''(b)\, h'(b) - y''(a)\, h'(a) = 0.$$

Since $h'(a)$ and $h'(b)$ are independent, we must have $y''(a) = y''(b) = 0$. These two conditions together with the conditions $y(a)$ $y(b) = 0$ determine the required solution uniquely.

Problem. Find the curve $y = y(x)$ that determines an extremum of the functional

$$F(y) = \tfrac{1}{2} \int_0^1 (y'')^2\, dx$$

subject to the conditions $y(0) = y'(0) = 0$, $y'(1) = 1$.

Answer. $y = 1/2\, x^2$.

3. Completely analogous considerations hold in the case of several independent variables. For simplicity we confine ourselves to a consideration of the functional

$$F(u) = \iint_G f(x, y, u, u_x, u_y, u_{xx}, u_{xy}, u_{yy})\, \mathrm{d}x\, \mathrm{d}y$$

where the integration is over a region G in the plane of the variables x, y. The functional is defined on the space $D_2(G)$ of functions $u(x, y)$ with continuous derivatives up to the second order on the region G. The function f is assumed to have continuous derivatives up to the second order with respect to all its arguments. The first variation of the functional F then has the form

$$\delta F(u, h) = \iint_G [f_u h + f_{u_x} h_x + f_{u_y} h_y + f_{u_{xx}} h_{xx} + f_{u_{xy}} h_{xy} + \\ + f_{u_{yy}} h_{yy}]\, \mathrm{d}x\, \mathrm{d}y.$$

The displacement vector $h = h(x, y)$ is a function in the same space $D_2(G)$. At an extremal point $u = u(x, y)$ of the space $D_2(G)$, the variation $\delta F(u, h)$ vanishes for any $h(x, y)$, so that

$$\iint_G [f_u h + f_{u_x} h_x + f_{u_y} h_y + f_{u_{xx}} h_{xx} + f_{u_{xy}} h_{xy} + \\ + f_{u_{yy}} h_{yy}]\, dx\, dy = 0. \tag{5}$$

We shall consider the values of the functions $u(x, y)$, u_x, u_y on the boundary Γ of the region G fixed, with the consequence that the quantities h, h_x, h_y vanish on Γ; we also suppose that the required function $u(x, y)$ has continuous derivatives up to and including the fourth order. Then each of the terms under the integral sign has derivatives with respect to x and y up to the second order inclusive. We integrate each term, beginning with the second, once or twice so as to remove all derivatives of the function $h(x, y)$; in the process, all integrated terms vanish in virtue of the boundary conditions, and equation (5) becomes

$$\iint_G \left[f_u - \frac{\partial}{\partial x} f_{u_x} - \frac{\partial}{\partial y} f_{u_y} + \frac{\partial^2}{\partial x^2} f_{u_{xx}} + \frac{\partial^2}{\partial x\, \partial y} f_{u_{xy}} + \\ + \frac{\partial^2}{\partial y^2} f_{u_{yy}}\right] h\, dx\, dy = 0.$$

Since $h(x, y)$ is arbitrary, the function u satisfies the 4th-order equation

$$f_u - \frac{\partial}{\partial x} f_{u_x} - \frac{\partial}{\partial y} f_{u_y} + \frac{\partial^2}{\partial x^2} f_{u_{xx}} + \frac{\partial^2}{\partial x\, \partial y} f_{u_{xy}} + \frac{\partial^2}{\partial y^2} f_{u_{yy}} = 0. \tag{6}$$

(the Euler–Ostrogradsky equation). The unknown function $u(x, y)$ must be determined as a solution of this equation satisfying the given boundary conditions. Questions as to the existence and uniqueness of solutions of such equations are considered in the theory of partial differential equations. Just as in the case of functionals with first-order derivatives, the fixed-value conditions can be replaced by others (Section 5).

4. *Example.* The equation of small vibrations of a rod. A rod, placed in a position of equilibrium between points 0 and l on the x-axis, executes transverse vibrations in the (x, u)-plane. We denote by $u(x, t)$ the profile of the rod at an instant t, and we sup-

pose that the ends 0, l are "hermetically sealed", so that

$$u(0, t) = u(l, t) = 0,$$
$$u_x(0, t) = u_x(l, t) = 0.$$

The kinetic energy of the rod, like that of a string, is expressed by the integral

$$T = \tfrac{1}{2}\int_0^l \mu(x)\, u_t^2\, \mathrm{d}x,$$

where $\mu(x)\,\mathrm{d}x$ is the mass of an element corresponding to an interval $\mathrm{d}x$. In contrast with the potential energy of a string, that of the rod is determined, not by the extension, but by the deformation of the profile; more precisely, the rod is determined as a mechanical system by the property that *the potential energy of each element is proportional to the square of the curvature of the profile*:

$$\mathrm{d}U = \frac{k(x)}{2}\,\frac{u_{xx}^2}{(1 + u_x^2)^3}\,\mathrm{d}x.$$

Supposing u_x and u_{xx} so small that we can neglect $u_x^2\, u_{xx}^2$, we can write $\mathrm{d}U$ in the simpler form

$$\mathrm{d}U = \frac{k(x)}{2}\, u_{xx}^2\, \mathrm{d}x,$$

giving for the potential energy

$$U = \tfrac{1}{2}\int_0^l k\, u_{xx}^2\, \mathrm{d}x.$$

Lagrange's function $L = T - U$ has the form

$$L = T - U = \tfrac{1}{2}\int_0^l [\mu\, u_t^2 - k\, u_{xx}^2]\, \mathrm{d}x.$$

Hamilton's functional is therefore represented by the double integral

$$\int_{t_0}^{t_1} L\, \mathrm{d}t = \tfrac{1}{2}\int_{t_0}^{t_1}\int_0^l [\mu\, u_t^2 - k\, u_{xx}^2]\, \mathrm{d}x\, \mathrm{d}t.$$

We write down the Euler–Ostrogradsky equation (6) for the present case:

$$-\frac{\partial}{\partial t}(\mu\, u_t) - \frac{\partial^2}{\partial x^2}(k\, u_{xx}) = 0.$$

For constant μ, k, the equation becomes

$$\mu\, u_{tt} + k\, u_{xxxx} = 0,$$

which is the equation of free vibrations of the rod. For the initial conditions, which are implicit in the physical situation, we can take the following: the values of the functions $u(x, 0)$ (the initial configuration) and $u_t(x, 0)$ (the initial velocities of the points of the rod) are preassigned. As in the case of a string, the uniqueness of the solution to the problem subject to the initial and boundary conditions follows from a consideration of the energy integral. The actual construction of the solution subject to various boundary conditions is elucidated in courses on partial differential equations.†

By the same reasoning as for the string, the equation of forced vibrations of a rod will have the form

$$\frac{\partial}{\partial t}(\mu\, u_t) + \frac{\partial^2}{\partial x^2}(k\, u_{xx}) = f(x, t),$$

where $f(x, t)\,dx$ is the force acting on an element dx. When the external force is independent of time, so that $f(x, t) = f(x)$, the equilibrium configuration of the rod is determined by the condition

$$\frac{\partial^2}{\partial x^2}(k\, u_{xx}) = f(x).$$

In particular a uniform rod (μ, k = const.) hangs under gravity ($f(x) = u\, g$) along some fourth-order curve.

Problem. Find the equilibrium configuration of a uniform rod: (a) hermetically sealed; (b) freely supported at the points $x_{1,\,2} = +l$ (Fig. 8).

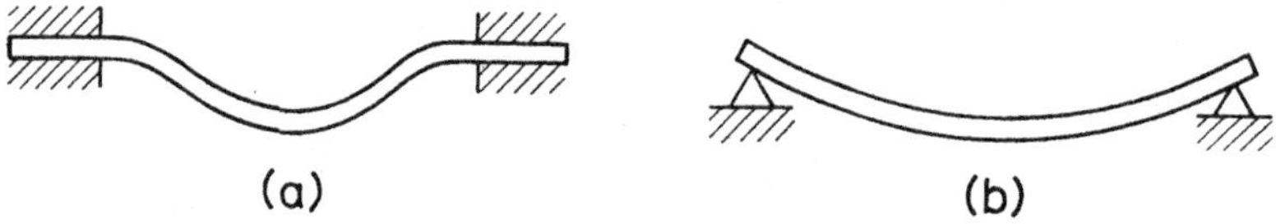

FIG. 8

Hint. In case (b) the boundary conditions are $u(-l) = u(l) = 0$, the derivatives $u'(l)$, $u'(-l)$ remaining free.

Answer. (for $k = \mu = 1$):

(a) $$u(x) = \frac{g}{24}(x^2 - l^2)^2;$$

(b) $$u(x) = \frac{g}{24}(x^2 - l^2)(x^2 - 5l^2).$$

† Cf. for example, A. N. Tikhonov and A. A. Samarskii, *Equations of Mathematical Physics*. Pergamon, London, 1963.

Concluding remark

The calculus of variations emerged as an independent branch of mathematics in 1744 or thereabouts with the discovery of a general method for solving variational problems by Leonard Euler (1707–1783, Swiss by birth, he spent the greater part of his life working in the St. Petersburg Academy of Sciences). Euler's method was the forerunner of current "direct methods" in the calculus of variations. The variational method, of which we have given an account, was first proposed in 1755 by J. Lagrange (French mathematician, 1736–1813) in a letter to Euler. In the development of the classical calculus of variations, the greatest mathematicians of the nineteenth century, Gauss, Poisson, Ostrogradsky, Weierstrass, and others, participated. We have only been able here to sketch the elements of this wide sphere of mathematics, rich as it is in applications. For a deeper acquaintance, we can recommend N. I. Akhiezer, *The Calculus of Variations* Blaisdell, New York, 1962), and *Calculus of variations* by I. M. Gelfand and C. V. Fomin Prentice-Hall, Englewood Cliffs, N. J., 1963).

CHAPTER IV

THEORY OF THE INTEGRAL

WE NOW proceed to extend the concept of the integral. The classical definition of the integral, as given by Cauchy and Riemann, while fully adequate for the case of continuous and piecewise continuous functions, turns out to be indequate from a more general point of view. Thus, we have seen that the space $C_1(a, b)$ of continuous functions $f(x)$ on the closed interval $[a, b]$ with the metric

$$\varrho(f, g) = \int_a^b |f(x) - g(x)|\, dx \qquad (1)$$

is incomplete: there exist fundamental sequences which have no limit in the space. Nothing would be gained by adding discontinuous functions which are Riemann-integrable to the space $C_1(a, b)$. It is only by constructing a new integral, of greater scope than the Riemann integral, that we can obtain a class of functions that will yield a completion of the space $C_1(a, b)$ under the metric (1).

Another problem for the solution of which the old definition of the integral is insufficient is that of describing a class of pairs of functions $\varphi(x)$, $F(x)$, sufficiently wide for the formulae

$$F(x) = F(a) + \int_a^x \varphi(\xi)\, d\xi, \qquad (2)$$

$$F'(x) = \varphi(x) \qquad (3)$$

to be equivalent. The solution of this problem will be given in Chapter VI.

1. Sets of Measure Zero and Measurable Functions

We shall begin the theory of the integral by investigating a class of sets on the closed interval $[a, b]$, which are called sets of measure zero.

We shall see in a moment that they are the sets that can be neglected in evaluating integrals; more precisely, the integral of a function $f(x)$ will be unchanged if the values of the function are changed arbitrarily on a set of measure zero.

Definition. A set A contained in the closed interval $[a, b]$ is said to be a *set of measure zero* if for any $\varepsilon > 0$ it can be covered by a finite or countable system of open intervals, the sum of the lengths of which do not exceed ε.

Sets containing one point, two points, or more generally any finite or countable aggregate of points are examples of sets measure zero. We give a proof of this. Let $A = \{x_1, x_2, \ldots,\}$, be a countable set and let $\varepsilon > 0$ be a prescribed number; then a system of open intervals of lengths $\varepsilon/2, \varepsilon/4, \ldots, \varepsilon/2_n, \ldots$ covering the points $x_1, x_2, \ldots, x_n, \ldots$ respectively will cover the whole set A and the overall sum of the lengths is not greater than $\varepsilon/2 + \varepsilon/4 + \cdots + \varepsilon/2_n + \cdots = \varepsilon$. In particular, the set of rational numbers and the set of algebraic numbers are sets of measure zero.

On the other hand, the whole interval $[a, b]$ is not a set of measure zero. For by a well-known lemma of analysis, if a closed interval is covered by a countable system of open intervals, a finite covering can be extracted from the given covering; the sum of the lengths just of these open intervals certainly exceeds $b - a$, i.e. the length of the whole closed interval.

We can already explain why the values of a function on a set of measure zero are immaterial in the evaluation of its integral. It is enough to show that the integral of a function $f(x)$ equal to 1 on a set A of measure zero and zero on the complement of A must vanish. We cover A with a system of open intervals of total length $< \varepsilon$. It is clear that if the integral of $f(x)$ is well defined, it cannot exceed the sum of the areas of the rectangles of unit height with bases on the given open intervals. This sum is equal to the sum of the lengths of the open intervals themselves and is therefore less than ε, i.e. it can be made as small as desired. It follows that the integral of the function $f(x)$ must vanish.

We observe that in the definition we have given of a set of measure zero we can replace the open covering by a covering of closed intervals or any other covering (with no stipulations as to the inclusion or exclusion of endpoints). For if there exists a covering of a set A by intervals of overall length $< \varepsilon$, then replacing the nth interval by an open interval containing it, of length exceeding that of the nth interval by at most $\varepsilon/2^n$, we obtain a covering of A by open intervals of overall length not exceeding 2ε; hence if A can be covered by some system of intervals of arbitrary small overall length, it also has an open covering of correspondingly small overall length, i.e. the set A has measure zero.

We give a simple construction for closed sets of measure zero. Let us suppose that a closed set F on the closed interval $[a, b]$ is obtained by extracting from the interval an open set comprising a countable aggregate of disjoint open intervals $\Delta_1, \Delta_2, \ldots, \Delta_k, \ldots$ of overall length $b - a$. Then we can certainly assert that the set F has measure zero, since for a given $\varepsilon > 0$ we can find n such that

$$\sum_{k=n+1}^{\infty} |\Delta_k| < \varepsilon.$$

Here and subsequently we denote the length of an open interval Δ by $|\Delta|$. The remaining n intervals $\Delta_1, \ldots, \Delta_n$ are disjoint and together with the intervening closed intervals $\Delta'_1, \ldots \Delta'_m$ (where m can equal $n - 1$, n, or $n + 1$, including the case where Δ_k and Δ_{k+1} have a common end-point so that Δ'_k degenerates into a point) give a finite covering of the whole closed interval $[a, b]$. Since the sum of the lengths of $\Delta_1, \ldots \Delta_n$ exceeds $b - a - \varepsilon$, the system $\Delta'_1, \ldots, \Delta'_m$ has an overall length $< \varepsilon$; and since it evidently covers the whole of F, we find that F is a set of measure zero.

It would seem difficult to anticipate that after extracting from a closed interval of length $b - a$ a system of disjoint open intervals of overall length also $b - a$, any substantial set can remain. Nevertheless it turns out that the set remaining can even be equivalent (in terms of power) to the whole of the original closed interval.

As an example we have the Cantor set on the closed interval [0, 1], with which we are already familiar (Chapter II, Section 4, art. 4). We recall how it is constructed. We first extract from the closed interval [0, 1] the open interval (1/3, 2/3) of length 1/3, which constitutes the middle third of the whole closed interval. We then proceed similarly with each of the two remaining closed intervals [0, 1/3], [2/3, 1], i.e. we extract from each of them the middle third, namely, the open interval (1/9, 2/9) from [0, 1/3] and then open interval (7/9, 8/9) from [2/3, 1]. An analogous procedure is carried out with each of the four remaining closed intervals [0, 1/9], [2/9, 1/3], [2/3, 1/9], [8/9, 1] and the process is continued indefinitely. The set which results is closed and is termed the Cantor set.

It is not hard to calculate the total sum of the lengths of the open intervals extracted: the first was of length 1/3, the next two totalled 2/9, the next four 4/27, and so on; the overall sum is the sum of the series

$$\frac{1}{3} + \frac{2}{9} + \frac{4}{27} + \cdots + \frac{2^n}{3^{n+1}} + \cdots = 1 .$$

By the criteria given above the Cantor set has measure zero.

Further, the Cantor set has no isolated points, since the open intervals extracted in its construction had no common end-points. It follows in virtue of theorem 1, Section 4, Chapter II, that the Cantor set is uncountable; moreover in virtue of theorem 2 of the same paragraph it has the power of the continuum.

We shall subsequently make frequent use of the following property of sets of measure zero:

LEMMA. *The union of a finite or countable aggregate of sets of measure zero is a set of measure zero.*

Proof. We consider at once the case of a countable aggregate $A_1, \ldots, A_n, \ldots$ of sets of measure zero. For a given $\varepsilon > 0$ and for each n, we cover the set A_n with a countable system of open intervals of overall length less than $\varepsilon/2^n$ $(n = 1, 2, \ldots)$. Then the whole set $A = A_1 + \cdots + A_n + \cdots$ will be covered by a countable system of open intervals (the sum of a countable set of countable sets) of overall length less than ε. Consequently A has measure zero, as required.

A set on the closed interval $[a, b]$, complementary to a set of measure zero, is said to be a set of *full measure*. The set of irrational numbers and the set of transcendental numbers are examples.

The intersection of a finite or countable system of sets of full measure is again a set of full measure. For, if $Q_1, Q_2, \ldots$ are sets of full measure and $A_1 = CQ_1, A_2 = CQ_2, \ldots$ are complementary sets of measure zero,

$$C \prod Q_j = \sum C Q_j = \sum A_j$$

has measure zero by the lemma; it follows that $\prod Q_j$ is a set of full measure, as required.

If some property belongs to all the points of a set of full measure on the closed interval $[a, b]$, it is said to hold *for almost all points* of the interval. For example the property that ξ is irrational holds for almost all points $\xi \in [a, b]$. There exist functions which are continuous almost everywhere, i.e. continuous at every point except, perhaps, on a set of measure zero. In the case of functions which are permitted to assume infinite values, a sense can be attributed to the expression "finite almost everywhere"; this means that the set on which the function is infinite is at most a set of measure zero.

We can now describe the class of functions in which our subsequent work on the definition of the integral will originate. The functions which belong to this class are termed *measurable functions.* A function is said to be measurable if it is defined and finite almost everywhere on a closed interval $[a, b]$ and can be represented as the limit of a sequence of *step functions* which converges almost everywhere. A step function in turn is a function which assumes some constant value on each open interval of some partition of the closed interval $[a,b]$ by points $a = x_0 < x_1 < \cdots < x_n = b$. We can disregard the values of a step function at the partitioning points, since these are finite in number and therefore constitute a set of measure zero.

The totality of step functions is a linear space under the usual operations of addition and scalar multiplication; if h, k are step functions, any linear combination $\alpha\, h + \beta\, k$ of them is also a step function. We easily infer that the measurable functions also form a linear space. For if the step functions h_n converge to a function f everywhere except on a set A of measure zero, and the step functions k_n to a function g everywhere except on a set B of measure zero, then the step functions $\alpha\, h_n + \beta\, k_n$ converge to the function $\alpha\, f + \beta\, g$ everywhere except on the set $A + B$, which, as we showed above, is also a set of measure zero; the function $\alpha\, f + \beta\, g$ is therefore also measurable.

Many other properties possessed by the class of step functions can also be carried over by a suitable limiting process to the class of measurable functions. We enumerate some of these.

The product of two step functions is a step function; the product of two measurable functions is accordingly a measurable function.

The quotient of two step functions is a step function, provided the denominator is non-vanishing. The quotient of two measurable functions is accordingly a measurable function provided the denominator is non-zero almost everywhere. For if $h_n \to f$ everywhere except on a set A of measure zero, and $k_n \to g$ everywhere except on a set B of measure zero, then replacing zero values of the functions k_n by the values $1/n$, if necessary, we get a new sequence of non-vanishing step functions k_n which converges to g everywhere except on the set B; but then the step functions h_n/k_n converge to f/g everywhere except on the set $A + B + C$, where C is the set of measure zero on which g vanishes. The set $A + B + C$ has measure zero, and hence the function f/q is also measurable.

The absolute magnitude $|h(x)|$ of a step function $h(x)$ is a step function. It follows easily that the absolute magnitude of a measurable function is a measurable function.

If two step functions h, k are given, then

$$h_1(x) = \max\{h(x), k(x)\}, \quad k_1(x) = \min\{h(x), k(x)\}$$

are also step functions. In the passage to the limit we get that for two measurable functions f, g

$$\max\{f(x), g(x)\}, \quad \mathit{min}\{f(x), g(x)\}$$

are also measurable functions.

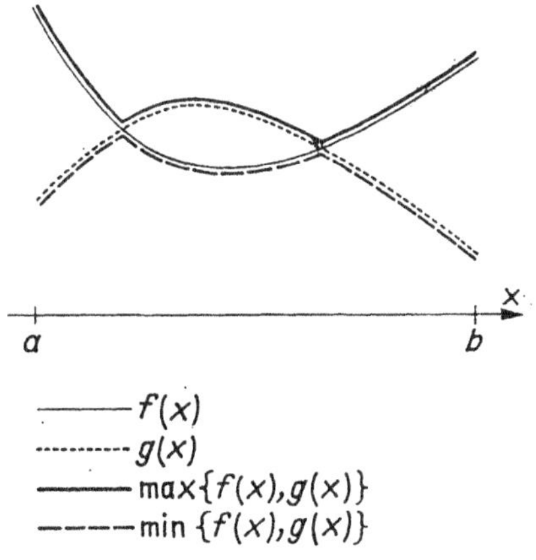

FIG. 9

In particular, the positive part $f^+(x) = \max\{f(x), 0\}$ and the negative part $f^-(x) = \max\{0, -f(x)\}$ of a function $f(x)$ are measurable together with $f(x)$.

We remark the relations

$$f = f^+ - f^-, \quad |f| = f^+ + f^-,$$

which occur frequently and hold for any function $f(x)$.

Problem. It is known that the sum of the lengths of the open intervals contiguous to a closed set $F \subset [a, b]$ is less than $b - a$. Show that F is not a set of measure zero.

2. The Class C^+

We now set about developing the integral concept. We consider first a *step function* $h(x)$, i.e. a function that takes on constant values $b_1, b_2, \ldots, \ldots, b_k$ on each of a finite number of intervals $\Delta_1, \Delta_2, \ldots, \Delta_k$, into which the closed interval $[a, b]$ is partitioned by points $a = x_0 < x_1 < \cdots < x_k = b$. The integral of this function is naturally set equal to

$$\int_a^b h(x)\,dx = \sum_{j=1}^k b_j |\Delta_j|.$$

For the sake of brevity we shall in future denote an expression of the form $\int_a^b h(x)\,dx$ by $I\,h$. It is easily verified that the integrals of step functions possess the following properties:

(a) $I(h_1 + h_2) = I\,h_1 + I\,h_2$ for any two step functions h_1, h_2;

(b) $I(\alpha h) = \alpha\, I(h)$ for any number α;

(c) if $h_1 \leq h_2$, then $I\,h_1 \leqq I\,h_2$; in particular $I\,h \geqq 0$ if $h \geqq 0$.

The next two properties are less obvious; they will be proved shortly:

(d) if a sequence $h_n \geqq 0$ is monotone decreasing (so that, $h_1(x) \geqq h_2(x) \geqq \cdots$) and tends to zero almost everywhere, then $I\,h_n \to 0$;

(e) if a sequence $h_n \geqq 0$ is monotone decreasing and in addition $I\,h_n \to 0$, then the sequence tends to zero almost everywhere.

We shall henceforth denote a monotone decreasing passage to the limit by the sign $\searrow$, so that the notation $f_n \searrow f$, say, denotes that the monotone decreasing sequence of functions $f_n(x)$ tends almost everywhere to the function $f(x)$. The sign $\nearrow$ will have an analogous meaning.

We prove property (d). We are given that $h_n \searrow 0$, and we have to prove that $I\,h_n \searrow 0$. Here we cannot apply the classical theorem on the term-by-term integration of a convergent sequence of functions, since it presupposes the uniform convergence of the sequence to its limit. For the proof we proceed as follows. We denote by A the union of the set on which the sequence h_n does not converge to zero and the countable set of all the points of discontinuity of the h_n; this set is of measure zero. We cover it with a system of open intervals $\{\Delta_k\}$ of overall length less than a prescribed $\varepsilon > 0$. With each of the remaining points x' we associate a number $n = n(x')$ for which the inequality $h_n(x') < \varepsilon$ is satisfied, and an open interval $\Delta(x')$ which contains the point and on which the function h_n is constant. The open intervals $\{\Delta_k\}$ and $\{\Delta(x')\}$ together constitute a covering of the closed interval $[a, b]$, from which we can extract a finite covering. We denote these open intervals by $\Delta_1, \ldots, \Delta_m$, $\Delta'_1, \ldots, \Delta'_p$, primes indicating those constructed from the points x'. If r is the greatest of the numbers mapped onto by the points x', then neither the function h_r nor any of its successors in the sequence h_n exceeds ε on the intervals $\Delta'_1, \ldots, \Delta'_p$. On the intervals $\Delta_1, \ldots, \Delta_m$, the sum of the lengths of which is less than ε by construction, these functions do not exceed a number M, the maximum of the function $h_1(x)$. Now it is clear that for the integral of the function $h_r(x)$ over the closed interval $[a, b]$, and for the integrals of all the succeeding functions, we get a bound of the form

$$I\,h_r < M\,\varepsilon + \varepsilon(b - a).$$

Since ε can be taken arbitrarily small, we have that $I\,h_r \to 0$, as required.

We now prove property (e), the converse of (d). We are given that the non-negative functions h_n decrease monotonely and that $I\,h_n \searrow 0$. It is obvious, since the functions h_n both decrease and remain positive, that they have some limit $g(x) \geqq 0$ as $n \to \infty$. We must show that $g(x)$ vanishes almost everywhere.

For any function $g(x) \geqq 0$, the set F of all the points at which it is non-zero is the countable sum of the sets

$$F_m = \left\{x : g(x) \geqq \frac{1}{m}\right\}.$$

The expression on the right-hand side of the equation indicates that the set F_m is the set of all points for which $g(x) \geqq 1/m$. If we can show that in our case each of the sets F_m has measure zero, then their sum F will also have measure zero. We can therefore restrict ourselves to a study of the set F_m.

It is enough to consider the subset F'_m of F_m on which all the functions h_n are continuous (its complement in F_m is countable and therefore has measure zero). Since $h_n(x) \geqq g(x)$, we have $h_n(x) \geqq 1/m$ at each point of F'_m. We fix the number n; then the intervals on which the function $h_n(x)$ is constant and which correspond to values of $h_n(x)$ greater than or equal to $1/m$ form a covering of the set F'_m. Let δ_n denote the sum of the lengths of these intervals. Since evidently

$$I\,h_n \geqq \delta_n \frac{1}{m}.$$

we get that

$$\delta_n \leqq m\,I\,h_n \to 0 \quad \text{as} \quad n \to \infty.$$

Thus, for sufficiently large n, the set F'_m is covered by a system of open intervals of arbitrarily small overall length. F'_m is therefore a set of measure zero, as required.

We now proceed to extend the definition of the integral from the class of step functions to a wider class.

First of all we recall the scheme by which the Riemann integral is constructed. For a function $f(x)$ the procedure is as follows. The closed interval $[a, b]$ is partitioned by points $a = x_0 < x_1 < \cdots < x_k = b$ into separate open intervals $\Delta_1, \ldots, \Delta_k$; we denote

$$m_j = \inf_{x \in \Delta_j} f(x), \quad M_j = \sup_{x \in \Delta_j} f(x) \quad (j = 1, 2, \ldots, k)$$

and form the two sums (which naturally depend on the aggregate Π of the partitioning points of $[a, b]$):

$$\left.\begin{aligned} s_\Pi &= \sum m_j |\Delta_j|, \\ S_\Pi &= \sum M_j |\Delta_j|. \end{aligned}\right\} \tag{1}$$

The first sum is called the lower, the second the upper. If the partition Π is replaced by a partition Π' by adding further partitioning points, then

$$s_\Pi \leqq s_{\Pi'}, \quad S_{\Pi'} \leqq S_\Pi.$$

It follows in particular that $s_{\Pi_1} < S_{\Pi_2}$ for any subpartitions Π_1, Π_2. We consider further an arbitrary sequence of partitions $\Pi_1, \Pi_2, \ldots, \Pi_n, \ldots$, each of which is obtained by the addition of new partitioning points to its predecessor. The corresponding lower and upper sums form monotone sequences which approach each other:

$$s_1 \leqq s_2 \leqq \cdots \leqq s_n \leqq \cdots \quad \cdots \leqq S_n \leqq \cdots \leqq S_2 \leqq S_1 .$$

Each sequence therefore has a limit: $s_n \nearrow s$, $S_n \searrow S$, with $s \leqq S$. It can be shown that the numbers s and S are independent of the choice of the sequence of partitions $\Pi_1, \Pi_2, \ldots, \Pi_n, \ldots$, provided that as n increases the length of the greatest open interval in the partition Π_n decreases indefinitely. The function $f(x)$ is said to be Riemann-integrable if $s = S$; the value of the integral of $f(x)$ then is taken as the common value of these limits. If $s < S$, the function $f(x)$ is said to be not Riemann-integrable.

We now consider the functioning of this process in respect of step functions. Each partition Π of the closed interval $[a, b]$ by points $a = x_0 \leqq x_1 \leqq \cdots \leqq x_n = b$ determines two step functions $h_\Pi(x)$, $H_\Pi(x)$; the first of these assumes the value m_j on the open interval Δ_j, the second the value M_j. The lower and upper sums (1) represent the integrals of these step functions. The sequence of partitions $\Pi_1, \Pi_2, \ldots, \Pi_n, \ldots$ determines two sequences of functions $h_n(x)$, $H_n(x)$, the first of which is increasing, the second decreasing. Further, let $s(x)$, $S(x)$ be the limiting functions of these sequences, so that

$$h_n(x) \nearrow s(x), \quad H_n(x) \searrow S(x).$$

Since in addition $h_n(x) \leqq f(x) \leqq H_n(x)$, we have $s(x) \leqq f(x) \leqq S(x)$. We claim that if the function $f(x)$ is Riemann-integrable, then these three functions coincide almost everywhere. For the difference $S(x) - s(x)$ is the limit of the sequence of non-negative step functions $H_n(x) - h_n(x)$. This sequence is monotone decreasing and in the case that $f(x)$ is integrable, the integrals of $H_n(x) - h_n(x)$ tend to zero. But then, by property (e), the sequence $H_n(x) - h_n(x)$ tends almost everywhere to zero. Hence if $f(x)$ is integrable the functions $s(x)$, $S(x)$ coincide almost everywhere with each other and with the function $f(x)$. We see that a function which is Riemann-integrable is the limit (almost everywhere) of an increasing sequence of step functions $s_n(x)$ and of a decreasing sequence $S_n(x)$; its integral is the limit of the integrals of the step functions that constitute these sequences.

Conversely if the functions $s_n(x)$, $S_n(x)$ converge almost everywhere to $f(x)$, then the decreasing sequence of differences

$S_n(x) - s_n(x)$ tends almost everywhere to zero. Hence, by property (d) we have $I(S_n(x)) - s_n(x) \searrow 0$. It follows that the numerical sequences $s_n = I(s_n(x)) \nearrow s$, $S_n = I(S_n(x)) \nearrow S$ have a common limit; and this means that the function $f(x)$ is Riemann-integrable.

Thus a function $f(x)$ is Riemann-integrable if and only if it is the limit (in the sense of convergence almost everywhere) of some increasing sequence of step functions $s_n(x) \leqq f(x)$ and simultaneously the limit of some decreasing sequence of step functions $S_n(x) \geqq f(x)$; the integral of f is then the common value of the limits of the integrals of the functions $s_n(x)$, $S_n(x)$.

This observation provides us with a footing on which to extend the definition of the integral to a wider class of functions.

We introduce a class of functions which we shall denote by C^+: a function $f(x)$ belongs, by definition, to the class C^+ if it can be represented as the limit (in the sense of convergence almost everywhere) of a monotone increasing sequence of step functions

$$h_n \nearrow f,$$

where the integrals of these functions are jointly bounded by the inequality

$$I\, h_n \leqq C.$$

We show first that every function f in the class C^+ is finite almost everywhere. Let E be the set of points on which $f(x) = \infty$. We can presuppose that at each point of the set E all the functions $h_n(x)$ are continuous and satisfy the relation $h_n(x) \to \infty$. We choose an arbitrary number N; at each point of E the inequality

$$h_n(x) > N,$$

is satisfied from some n on, so that E is covered by the countable sum of the sets of the form $\{x : h_n(x) > N\}$ $(n = 1, 2, \ldots)$. Each of these sets comprises a finite number of open intervals; hence their union is also a countable sum of open intervals:

$$U = \Delta_1^{(1)} + \cdots + \Delta_{n_1}^{(1)} + \Delta_1^{(2)} + \cdots + \Delta_{n_2}^{(2)} + \cdots + \Delta_1^{(k)} + \cdots + \Delta_{n_k}^{(k)} + \cdots$$

Here $\Delta_1^{(1)}, \ldots, \Delta_n^{(1)}$ denote the component open intervals of the set $\{x : h_1(x) > N\}$; $\Delta_1^{(2)}, \ldots, \Delta_{n_2}^{(2)}$ denote the component open intervals of the set $\{x : h_2(x) > N\} - \{x : h_1(x) > N\}$, so that $\Delta_1^{(1)}, \ldots, \Delta_{n_1}^{(1)}$, $\Delta_1^{(2)}, \ldots, \Delta_{n_2}^{(2)}$ constitute the set $\{x : h_2(x) > N\}$, and so on. We evaluate the sum of the lengths of all these intervals. First we let

$$\delta_k = |\Delta_1^{(1)}| + \cdots + |\Delta_{n_1}^{(1)}| + |\Delta_1^{(2)}| + \cdots + |\Delta_{n_k}^{(k)}|;$$

δ_k is the sum of the lengths of the open intervals which constitute the set $\{h_n(x) > N\}$. We have by hypothesis

$$\delta_k N < I\, h_k \leqq C.$$

It follows that $\delta_k < C/N$; since this is true for any k, we conclude that the overall sum of the lengths of all the open intervals comprising U does not exceed C/N. Since N can be taken arbitrarily large, we see that E can be covered by a countable system of open intervals of arbitrarily small overall length. This means that E is a set of measure zero, as asserted.

In particular, every function $f \in C^+$ is measurable (Section 1). We now define the integral of a function f belonging to the class C^+ by the formula

$$I\, f = \lim_{n \to \infty} I\, h_n, \tag{2}$$

where h_n is the sequence of step functions involved in the definition of the function f. Since the sequence of numbers $I\, h_n$ is monotone increasing and bounded, a limit exists to the right; but we still have to show that it is independent of the choice of the sequence h_n which defines the function f. To do this, we prove the following more general result; if h_n, k_n are step functions with jointly bounded integrals, and if

$$h_n \nearrow f, \quad k_n \nearrow g, \quad f \leqq g,$$

almost everywhere, then

$$\lim_{n \to \infty} I\, h_n \leqq \lim_{n \to \infty} I\, k_n. \tag{3}$$

For the proof, we fix a number m and consider the decreasing sequence of step functions

$$h_m - k_n.$$

Its limit $h_m - g \leqq f - g \leqq 0$; but then $(h_m - k_n)^+ \searrow 0$, and it follows from property (d) that $I(h_m - k_n)^+ \searrow 0$; since $I(h_m - k_n) \leqq I(h_m - k_n)^+$, the decreasing sequence $I(h_m - k_n) = I\, h_m - I\, k_n$ tends to some non-positive limit. We conclude that $I\, h_m \leqq \lim_{n \to \infty} I\, k_n$.

Since this inequality holds for any m, the passage to the limit as $m \to \infty$ gives us (3), as required. Putting $g = f$, we get $I\, f \leqq I\, g$ and $I\, g \leqq I\, f$ by symmetry; it follows that $I\, f = I\, g$. Thus the integral of a function $f \in C^+$ is well-defined by formula (2). If $f \in C^+$, $g \in C^+$, $f \leqq g$, then $I f \leqq I g$.

In particular, every function f which is Riemann-integrable belongs to the class C^+, and its Riemann integral, as the limit of its lower sums, coincides with the integral $I f$, defined by us as the limit of the integrals of the step functions s_n corresponding to those lower sums.

We see that our definition of the integral has at least as broad a compass as Riemann's. In fact it has a considerably broader compass. For example, Dirichlet's function $\chi(x)$, equal to 0 for x irrational and 1 for x rational, is not Riemann-integrable; but seen from our new angle, it vanishes almost everywhere and is therefore integrable, its integral being equal to zero. We could also give more complex examples, where a function integrable in the new sense but not Riemann-integrable cannot be reduced to a Riemann-integrable function by modifying it on a set of measure zero.

In the passage to the limit some, though not all, of the properties of the integrals of step functions carry over to the integrals of functions of the class C^+. It is easily verified that:

(a) The class C^+ contains the function $f + g$ whenever it contains f and g and then

$$I(f + g) = I f + I g.$$

(b) The class C^+ contains together with a function f its scalar product by any number $\alpha \geqq 0$, and

$$I(\alpha f) = \alpha I f.$$

We note that in the class C^+ it is not possible to subtract functions or to multiply them by negative numbers since we are restricted to increasing sequences of step functions.

(c) The class C^+ contains together with functions f, g

$$\min(f, g), \quad \max(f, g).$$

In particular, the function $f^+ = \max(f, 0)$ belongs to C^+ together with f. (This cannot be said of the functions f^-, $|f|$.)

The following property shows that the class C^+ is closed in respect of the limiting passage through increasing sequences of functions with bounded integrals:

THEOREM. *If $f_n \in C^+$ $(n = 1, 2, \dots)$, $f_n \nearrow f$, and $I f_n < C$, then $f \in C^+$ and $I f = \lim I f n$.*

Proof. For each of the functions f_n we construct a definitive sequence of step functions:

$$\begin{aligned} &h_{11} \leqq h_{12} \leqq \cdots \leqq h_{1n} \leqq \cdots, h_{1n} \nearrow f_1, \\ &h_{21} \leqq h_{22} \leqq \cdots \leqq h_{2n} \leqq \cdots, h_{2n} \nearrow f_2, \\ &\cdots\cdots\cdots\cdots\cdots\cdots\cdots\cdots\cdots \\ &h_{k1} \leqq h_{k2} \leqq \cdots \leqq h_{kn} \leqq \cdots, h_{kn} \nearrow f_k, \\ &\cdots\cdots\cdots\cdots\cdots\cdots\cdots\cdots\cdots \end{aligned}$$

Further, we put $h_n = \max(h_{1n}, \ldots, h_{nn})$. It is evident that h_n is also a step function and that the sequence h_n $(n = 1, 2, \ldots)$ is monotone increasing. Moreover $h_n \leqq \max(f_1, \ldots, f_n) = f_n$, whence $I\,h_n \leqq I\,f_n \leqq C$. Let $f^* = \lim h_n$; by the definition of the class C^+, we have $f^* \in C^+$ and $I\,f^* = \lim I\,h_n$. But since $h_{kn} \leqq h_n \leqq f_n$ for any fixed k and $n \geqq k$, we find, passing to the limit as $n \to \infty$, that $f_k \leqq f^* \leqq f$, so that $f^* = f$ (almost everywhere). Thus $f \in C^+$. Further, $I\,h_{kn} \leqq I\,h_n \leqq I\,f_n \leqq I\,f$; since $I\,h_n \nearrow I\,f^* = I\,f$, we have $I\,f_n \nearrow I\,f$, which completes the proof.

COROLLARY. *If for a series* $\sum_{k=1}^{\infty} g_k$, $g_k \in C^+$, $g_k \geqq 0$, *the integrals of the partial sums are jointly bounded, so that*

$$I\left(\sum_{k=1}^{n} g_k\right) \leqq C,$$

then $f = \sum_{k=1}^{\infty} g_k$ *is a function of the class* C^+ *and* $I\,f = \sum_{k=1}^{\infty} I\,g_k$.

For the proof it is sufficient to put $f_n = \sum_{k=1}^{n} g_k$ and apply the preceding theorem.

Problems. 1. If $f(x)$ differs from $f_0(x)$ of class C^+ only on a set of measure zero, $f(x)$ also belongs to C^+.

Hint. The sequence $h_n(x) \nearrow f_0(x)$ is convergent almost everywhere to $f(x)$ also.

2. Show that the Dirichlet function, equal to zero for x rational and 1 for x irrational, belongs to class C^+.

3. A function equal to zero on a closed set F and 1 on its complement, belongs to class C^+.

4. A closed set F exists, such that a function equal to 1 on F and 0 on its complement, does not belong to C^+.

Hint. Take F as nowhere dense and not a set of measure zero. See the problem of Section 1.

5. Show that a function $f(x)$ is Riemann-integrable or differs from such on a set of measure 0 if and only if $f \in C^+$, $-f \in C^+$.

6. Show that a function $f(x)$ is Riemann-integrable if and only if the set of its points of discontinuity has measure 0.

Hint. If $s_n(x) \nearrow f(x)$ and $S_n(x) \searrow f(x)$ and x_0 is a point of continuity of all the step functions $s_n(x)$ and $S_n(x)$, x_0 is a point of continuity of $f(x)$. Conversely, $s_n(x_0) \nearrow f(x_0)$, $S_n(x_0) \searrow f(x_0)$ at every point of continuity x_0.

3. Summable Functions

1. In this paragraph we shall complete the construction of the integral by extending it from the class C^+ to a wider class L in which all the natural functional operations can be carried out.

We shall call *summable (or Lebesgue-integrable)* any function $\varphi(x)$ $(a \leqq x \leqq b)$ that can be represented as the difference

$$\varphi = f - g$$

of two functions of the class C^+. We denote by L the totality of summable functions. In the class of summable functions the following operations can be performed:

(a) *Addition.* If $\varphi = f - g, \varphi_1 = f_1 - g_1$ are summable functions, with f, g, f_1, g_1 functions of the class C^+m, then

$$\varphi + \varphi_1 = (f + f_1) - (g + g_1),$$

and since $f + f_1, g + g_1 \in C^+$, it follows that the function $\varphi + \varphi_1$ belongs to L.

(b) *Multiplication by any real number* α. If $\alpha \geqq 0$, then $\varphi = f - g$, $f \in C^+$, $g \in C^+$ implies $\alpha\varphi = \alpha f - \alpha g$, $\alpha f \in C^+$, $\alpha g \in C^+$, and consequently $\alpha\varphi \in L$; and if $\alpha < 0$, then $-\alpha > 0$ and the equation $\alpha\varphi = (-\alpha) g - (-\alpha) f$ shows that as before $\alpha\varphi \in L$.

It follows from (a) and (b) that any linear combination of functions of the class L is also a function belonging to L.

(c) Taking the modulus of a function. Let $\varphi = f - g$, $f \in C^+$, $g \in C^+$, then $\max(f, g)$, $\min(f, g)$ also belong to the class C^+; it follows that $|\varphi| = \max(f, g) - \min(f, g)$ belongs to the class L. Solving the equations

$$\varphi = \varphi^+ - \varphi^-,$$
$$|\varphi| = \varphi^+ + \varphi^-,$$

we see that the functions φ^+, φ^- belong to the class L together with the function φ.

Further, the equations

$$\max(\varphi, \psi) = (\varphi + \psi)^+ - \psi,$$
$$\min(\varphi, \psi) = -\max(-\varphi, -\psi)$$

show that the class L contains together with functions φ, ψ their maximum and minimum.

2. We now give the definition of the integral in the class L. With the decomposition

$$\left.\begin{aligned} \varphi &= f - g, \\ f &\in C^+, \\ g &\in C^+, \end{aligned}\right\} \tag{1}$$

we put

$$I\varphi = If - Ig.$$

We verify that $I\varphi$ is uniquely defined in this way. Let a second decomposition

$$\varphi = f_1 - g_1, \quad f_1 \in C^+, \quad g_1 \in C^+$$

exist alongside (1). We shall prove that $If - Ig = If_1 - Ig_1$. This equation is equivalent to the equation

$$If + Ig_1 = Ig + If_1. \tag{2}$$

But since $f + g_1 = f_1 + g$, we have in virtue of the uniqueness of the integral in the class C^+

$$I(f + g_1) = I(g + f_1),$$

from which (2) follows.

We show further that the integral obtained possesses the usual linear properties in the class L. Let $\varphi = f - g$, $\varphi_1 = f_1 - g_1$, where f, g, f_1, g_1 belong to the class C^+. Then $\varphi + \varphi_1 = (f + f_1) - (g + g_1)$, and by definition

$$\begin{aligned} I(\varphi + \varphi_1) &= I(f + f_1) - I(g + g_1) = If + If_1 - Ig - Ig_1 \\ &= (If - Ig) + (If_1 - Ig_1) = I\varphi + I\varphi_1. \end{aligned}$$

Thus the integrals of sums are equal to the sums of integrals. Further, for $\alpha > 0$, $I(\alpha\varphi) = I(\alpha f - \alpha g) = I(\alpha f) - I(\alpha g) = \alpha If - \alpha Ig = \alpha(If - Ig) = \alpha I\varphi$; again,

$I(-\varphi) = I(g - f) = Ig - If = -I\varphi$, and hence for $\alpha < 0$ we have $I(\alpha\varphi) = I(-|\alpha|\varphi) = -I(|\alpha|\varphi) = -|\alpha| I(\varphi) = \alpha I\varphi$, so that a scalar α can be carried through the integral sign whatever its sign.

We observe further that if $\varphi \geqq 0$, $\varphi \in L$, then $I\,\varphi \geqq 0$. For if $\varphi = f - g$, $f \in C^+$, $g \in C^+$, $\varphi \geqq 0$, then $f \geqq g$ and $I\,f \geqq I\,g$; hence $I\,\varphi = I\,f - I\,g \geqq 0$. We make the further deduction that $\varphi_1 \leqq \varphi_2$ implies $I\,\varphi_1 \leqq I\,\varphi_2$.

3. We now prove an important theorem on the term-by-term integration of series with positive terms.

THEOREM (Beppo Levi, 1906). *If for a series* $\sum_{k=1}^{\infty} \varphi_k$, $\varphi_k \in L$, $\varphi_k \geqq 0$, *the integrals of the partial sums have a common bound, so that*

$$I\left(\sum_{k=1}^{n} \varphi_k\right) \leqq C,$$

then the function $\varphi = \sum_{k=1}^{\infty} \varphi_k$ *is summable and* $I\,\varphi = \sum_{k=1}^{\infty} I\,\varphi_k$.

Proof. We observe first that in the decomposition of a summable function $\varphi = f - g$, $f \in C^+$, $g \in C^+$, the functions f, g can be made subject to further conditions. For example, g can be chosen so that $g \geqq 0$, $I\,g < \varepsilon$, where ε is a prescribed number. To do this we have to consider a sequence of step functions $h_n \nearrow g$ such that $I\,g = \lim I\,h_n$, and then write

$$\varphi = f - g = (f - h_n) - (g - h_n) = f_n - g_n.$$

It is clear that for sufficiently large n, the required condition is satisfied for the function $g_n = g - h_n$. In addition we observe that if $\varphi \geqq 0$, the function $f_n = f - h_n \geqq f - g = \varphi$ also becomes non-negative.

Now for each of the functions φ_k that occur in the formulation of the theorem we construct a decomposition

$$\varphi_k = f_k - g_k,$$

where $f_k \geqq 0$, $g_k \geqq 0$, $I\,g_k < 1/2^k$ $(k = 1, 2, \ldots)$. The series $\sum_{k=1}^{\infty} g_k$ then satisfies the conditions of the corollary to the theorem of Section 2 $(g_k \geqq 0,\ I(\sum_{k=1}^{n} g_k) < 1)$. Hence $g = \sum_{k=1}^{\infty} g_k$ belongs to the class C^+ and $I\,g = \sum_{k=1}^{\infty} I\,g_k$. We show that the series $\sum_{k=1}^{\infty} f_k$ also satisfies the conditions of the corollary; for we have $f_k \geqq 0$

and
$$I\left(\sum_{k=1}^{n} f_k\right) = I\left(\sum_{k=1}^{n} \varphi_k\right) + I\left(\sum_{k=1}^{n} g_k\right) \leqq C + 1.$$

Hence $f = \sum_{k=1}^{\infty} f_k$ also belongs to the class C^+ and $I f = \sum_{k=1}^{\infty} I f_k$. It follows that $\varphi = \sum_{k=1}^{\infty} \varphi_k = \sum_{k=1}^{\infty} f_k - \sum_{k=1}^{\infty} g_k = f - g$ belongs to the class L and

$$I \varphi = I f - I g = \sum_{k=1}^{\infty} I f_k - \sum_{k=1}^{\infty} I g_k = \sum_{k=1}^{\infty} I (f_k - g_k) = \sum_{k=1}^{\infty} I \varphi_k$$

which completes the proof.

COROLLARY. *If a monotone increasing sequence of summable functions ψ_n tends to a limit ψ and $I \psi_n \leqq C$, then the function ψ is summable and*

$$I \psi = \lim I \psi_n .$$

For the proof it is sufficient to put $\varphi_1 = \psi_2 - \psi_1, \dots, \varphi_n = \psi_{n+1} - \psi_n$ and apply the preceding theorem. An analogous result obviously holds for decreasing sequences $\psi_n \searrow \psi$ provided $I\psi_n \geqq C$.

4. We shall subsequently consider arbitrary (non-monotone) passages to the limit. Classical examples show that theorems of the form $\varphi_n \to \varphi$ implies $I \varphi_n \to I \varphi$ cannot be expected in the absence of additional assumptions about the nature of the convergence of the sequence φ_n to its limit. For example, the functions

$$\varphi_n(x) = \begin{cases} n \sin n x \quad \text{for} \quad 0 \leqq x \leqq \dfrac{\pi}{n}, \\ 0 \text{ for } \dfrac{\pi}{n} \leqq x \leqq \pi \end{cases}$$

converge to zero for every $x \in [0, \pi]$, while their integrals remain constant (with the value 2) and do not tend to the integral of the limiting function.

We consider the aggregate $L(\varphi_0)$ of all summable functions φ that satisfy the inequality

$$-\varphi_0 \leqq \varphi \leqq \varphi_0 ,$$

where φ_0 is a fixed non-negative summable function. Evidently the inequality

$$-I \varphi_0 \leqq I \varphi \leqq I \varphi_0 .$$

is satisfied for each function $\varphi \in L(\varphi_0)$. If there exists a monotone sequence of functions φ_n, increasing or decreasing, that belong to the set $L(\varphi_0)$, the limiting function φ will obviously satisfy this inequality together with the functions φ_n; by the preceding corollary this function is also summable. Consequently the set $L(\varphi_0)$ is closed in respect of monotone passages to the limit. We observe further that we can assert for any sequence $\varphi_n \in (\varphi_0)$ that the functions

$$\sup_n \{\varphi_1(x), \varphi_2(x), \ldots, \varphi_n(x), \ldots\}$$

and

$$\inf_n \{\varphi_1(x), \varphi_2(x), \ldots, \varphi_n(x), \ldots\}$$

also belong to $L(\varphi_0)$: the first of them is the limit as $n \to \infty$ of the increasing sequence of functions

$$\max \{\varphi_1(x), \ldots, \varphi_n(x)\} \in L(\varphi_0),$$

and the second is the limit of the decreasing sequence

$$\min \{\varphi_1(x), \ldots, \varphi_n(x)\} \in L(\varphi_0).$$

Now let $\varphi_n \in L(\varphi_0)$ be any sequence converging almost everywhere to some function ψ; we shall show that ψ also belongs to the class $L(\varphi_0)$. It is enough to prove that ψ can be represented as the limit of some monotone sequence of functions in the class $L(\varphi_0)$. We put

$$\psi_n(x) = \sup \{\varphi_n(x), \varphi_{n+1}(x), \ldots\},$$
$$\psi_n'(x) = \inf \{\varphi_n(x), \varphi_{n+1}(x), \ldots\}.$$

By what we have just proved, these functions are summable and belong to $L(\varphi_0)$. If we consider only those values of x for which the function $\varphi_n(x)$ have the limit $\psi(x)$, it is clear that for each such value

$$\psi_n(x) \geqq \lim_{p\to\infty} \varphi_{n+p}(x) = \psi(x),$$
$$\psi_n'(x) \leqq \lim_{p\to\infty} \varphi_{n+p}(x) = \psi(x).$$

Further, by removing φ_n from the aggregate $\varphi_n, \varphi_{n+1}, \ldots$, we can only diminish its upper bound and increase its lower bound; hence

$$\psi_{n+1}(x) \leqq \psi_n(x),$$
$$\psi_{n+1}'(x) \geqq \psi'_n(x).$$

Consequently the sequence $\psi_n(x)$ is decreasing, and $\psi'_n(x)$ is increasing. Furthermore, it is clear that $\varphi_n(x) \to \psi(x)$ implies

$$\psi_n(x) \searrow \psi(x), \quad \psi'_n(x) \nearrow \psi(x).$$

Thus the function $\psi(x)$ is the limit of an increasing sequence of functions in the class $L(\varphi_0)$ (and at the same time the limit of a decreasing sequence of functions of this class). It follows that $\psi \in L(\varphi_0)$, as asserted. In addition we have $I\,\psi'_n \nearrow I\,\psi$, $I\,\psi_n \nearrow I\,\psi$, $I\,\psi' \leqq I\,\varphi_n \leqq I\,\psi_n$, whence $I\,\varphi_n \to I\,\psi$. We have proved the following theorem:

THEOREM (Lebesgue, 1902). *If a sequence of summable functions φ_n converges almost everywhere to a function φ and satisfies the condition*

$$|\varphi_n(x)| \leqq \varphi_0(x) \in L,$$

then the function φ is summable and $I\,\varphi = \lim I\,\varphi_n$. In particular the equation $I\,\varphi = \lim I\,\varphi_n$ holds if the functions φ_n have common bound.

From this theorem we can obtain an important result in connection with the composition of the class $L(\varphi_0)$. We shall show, namely, that *if some measurable function φ satisfies (almost everywhere) the inequality*

$$-\varphi_0 \leqq \varphi \leqq \varphi_0 \in L,$$

then it is summable (and therefore belongs to the class $L(\varphi_0)$). For let h_n be a sequence of step functions that defines the measurable function φ. Cutting it above by the plane φ_0 and below by the plane $-\varphi_0$, i.e. replacing it by the functions

$$\varphi_n = \max\{-\varphi_0, \min(h_n, \varphi_0)\},$$

we get a sequence of summable functions belonging to the class $L(\varphi_0)$ and converging almost everywhere to the function φ. This means that $\varphi = \lim \varphi_n$ is a summable function, as required.

In particular, *every bounded measurable function is summable.*

Again, the theorem we have proved on summable functions enables us to draw further inferences about measurable functions. We shall show that *if the limit $\varphi(x)$ of an almost everywhere convergent sequence of measurable functions $\varphi_n(x)$ is finite almost everywhere, then it is a measurable function.* It is enough to consider the case $\varphi_n(x) \geqq 0$, since in the contrary event the sequences φ_n^+, φ_n^- can be considered separately. But if the φ_n converge to φ almost everywhere, the sequence of functions $\psi_n = \dfrac{1}{1+\varphi_n}$ will converge almost everywhere to $\dfrac{1}{1+\varphi}$. The functions ψ_n are bounded by 1 and 0, above and below respectively, and are measurable. They are therefore summable and by what we have proved, their limit ψ is also summable and is therefore a measurable function. We note that ψ can vanish only when $\varphi(x) = \infty$, i.e. on a set of measure zero. Hence, inverting the equation we have got, we find that

$$\varphi = \frac{1 - \psi}{\psi}$$

is a measurable function, since both numerator and determinator are measurable and the denominator is non-vanishing almost everywhere.

5. In one instance we can assert the summability of the limiting function of a sequence φ_n when the hypothesis that the functions $|\varphi_n|$ are bounded by a summable function is replaced by certain other assumptions:

LEMMA (Fatou, 1906). *If* $\varphi_n \geqq 0$ *are summable functions,* $\varphi_n \to \varphi$ *almost everywhere, and* $I\,\varphi_n \leqq C$, *then the function* φ *is summable and*

$$0 \leqq I\,\varphi \leqq C.$$

Proof. We put

$$\psi_n = \inf\{\varphi_n, \varphi_{n+1}, \ldots\} \geqq 0.$$

As above, the functions ψ_n form an increasing sequence which converges almost everywhere to the function φ. Further, $\psi_n \leqq \varphi_n$, $I\,\psi_n \leqq I\,\varphi_n \leqq C$; by the corollary to Beppo Levi's theorem the function φ is summable and $I\,\psi_n \nearrow I\,\varphi$. In particular, $0 \leqq I\,\varphi = \lim I\,\psi_n \leqq C$ as required.

Now let the function $\varphi_0 \geqq 0$ be summable and let $I\,\varphi_0 = 0$; we shall show that $\varphi_0 = 0$ almost everywhere. We put $\varphi_n = n\,\varphi_0$; the function φ_n tends to a limit φ, equal to zero for $\varphi_0 = 0$ and infinity for $\varphi_0 > 0$. But since, by Fatou's lemma, the limiting function must be summable, and in particular measurable, the set of those x for which $\varphi_0(x) = \infty$ is of measure zero. At the same time the set of those x for which $\varphi_0(x) > 0$ is a set of measure zero. We get: *if the integral of a non-negative summable function vanishes, the function itself vanishes almost everywhere.*

6. In this article, we shall regard two summable functions as the same if they coincide on a set of full measure.

To be more precise, we pass from summable functions themselves to classes of equivalent functions: two functions are regarded as equivalent if they coincide on a set of full measure. In particular, a function $f(x)$ is equivalent to zero if it differs from zero at most on a set of measure 0. The functions equivalent to zero evidently form a subspace L_0 of the linear space L of all summable functions; in essence, what we are doing is to pass from space L to its factor-space L/L_0 (Chapter II, Section 8, art. 4). In space L we have the quasinorm $\|\varphi\| = I(|\varphi|)$ (we recall that the quasinorm differs from the norm in that there can exist non-zero elements with quasinorm zero). The subspace L_0 consists of those and only those functions whose quasinorm is zero. We can thus introduce a norm into the aggregate of classes of equivalent summable functions by taking it equal to $I(|\varphi|)$, where φ is any function of a class. In accordance with art. 4, Section 8, Chapter II, we get a normed linear space of classes of equivalent summable functions, which we shall term the *Lebesgue space*. We shall avoid unnecessary pedantry in future by speaking simply of a "summable function" rather than of "a class of equivalent summable functions", whilst we shall write L, or more precisely, $L_1(a, b)$, for the Lebesgue space.

THEOREM (E. Fischer and F. Riesz, 1907). *The Lebesgue space is complete: every sequence of functions $\varphi_1, \varphi_2, \ldots$, fundamental with respect to the norm $\|\varphi\| = I(|\varphi|)$, has a limit in the sense of this norm in the space L.*

Before proceeding to the proof of this theorem, we take note of two simple facts. In the first place, the elements of any fundamental sequence are bounded in the norm, since from some point on they are all contained in a sphere of radius ε with centre at some point φ_n. Secondly, to prove the existence of a limit of a fundamental sequence φ_n, it is sufficient to show that some subsequence φ_{n_k} $(k = 1, 2, \ldots)$ has a limit φ; the element φ will also be the limit of the whole sequence φ_n in virtue of the inequality

$$\|\varphi - \varphi_n\| \leqq \|\varphi - \varphi_{n_k}\| + \|\varphi_{n_k} - \varphi_n\|,$$

where the second term on the right tends to zero because the sequence φ_n is fundamental.

We now prove the theorem.

Let $\varphi_1, \varphi_2, \ldots, \varphi_n, \ldots$ be a fundamental sequence in the space L.

We can always find a sequence of indices $n_1 < n_2 < \cdots$ such that for $n > n_k$

$$\|\varphi_n - \varphi_{n_k}\| < \frac{1}{2^k} \quad (k = 1, 2, \ldots).$$

In particular, $\|\varphi_{n_{k+1}} - \varphi_{n_k}\| < \frac{1}{2^k}$; this means that

$$I(|\varphi_{n_{k+1}} - \varphi_{n_k}|) < \frac{1}{2^k}.$$

But then, by Beppo Levi's theorem, the series of summable functions $\sum_{k=1}^{\infty} |\varphi_{n_{k+1}} - \varphi_{n_k}|$ converges almost everywhere. It follows that the series with partial sums

$$\sum_{k=1}^{N} (\varphi_{n_{k+1}} - \varphi_{n_k}) = \varphi_{n_{N+1}} - \varphi_{n_1}.$$

also converges almost everywhere. This means that the function φ_{n_k} has a limit (almost everywhere) as $k \to \infty$. We denote this limit by φ. As a limit of measurable functions, the function φ is measurable. Since the norms of the functions φ_{n_k}, i.e. the numbers $I(|\varphi_{n_k}|)$, are bounded, it follows by Fatou's lemma that the function $|\varphi|$ is summable. The function φ is therefore both measurable and summable. Further, applying the same lemma to the functions $\varphi - \varphi_{n_k}$, we have:

$$\|\varphi - \varphi_{n_k}\| = I(|\varphi - \varphi_{n_k}|) \leqq \sup_{p>k} I(|\varphi_{n_p} - \varphi_{n_k}|)$$
$$= \sup_{p>k} \|\varphi_{n_p} - \varphi_{n_k}\|.$$

But by taking k sufficiently large, the last integral can be made as small as we please, since the sequence φ_n is fundamental. Hence φ_{n_k} converges in the norm of the space L to φ and the theorem is proved.

In conclusion we shall show that the space $L_1(a, b)$ is the completion of the space $C_1(a, b)$ of continuous functions $f(x)$ on the closed interval $[a, b]$ with norm (cf. Chapter II, Section 8)

$$\|f\| = \int_a^b |f(x)| \, dx.$$

The space $L_1(a, b)$ evidently contains the space $C_1(a, b)$ as a subspace with the same metric. It is therefore sufficient for us to prove that the subspace $C_1(a, b)$ is dense in L_p, so that each function $\varphi \in L_1$ can be represented as the limit of a sequence of functions $f_n(x) \in C_1$. We can easily satisfy ourselves that every step function $h(x)$ possesses this property. Again, since each function $\varphi \in L_1$ is the difference of two functions in the class C^+ it is enough to verify our assertion for functions in the class C^+. Let $\varphi \in C^+$ and let $h_n \nearrow \varphi$ be a sequence of step functions. Then

$$\|\varphi - h_n\| = I(\varphi - h_n),$$

and since $\varphi - h_n \searrow 0$, it follows from Beppo Levi's theorem that

$$I(\varphi - h_n) \searrow 0,$$

which proves the theorem.

We have seen (Chapter II, Section 3, art. 3) that the set of polynomials is everywhere dense in space $C(a, b)$, and the set of trigonometric polynomials $T(x) = \sum_{k=0}^{n} a_k \cos k x + b_k \sin k x$ is everywhere dense in space $C_1(-\pi, \pi)$. Since $C(a, b)$ is everywhere dense in its completion $L_1(a, b)$, we can conclude that the set of polynomials is everywhere dense in space $L_1(a, b)$ for any a, b; similarly, the set of trigonometric polynomials is everywhere dense in $L_1(-\pi, \pi)$.

1. If a summable function f vanishes outside a closed interval $[\alpha, \beta]$ which is interior to the original closed interval $[a, b]$, so that it can be "displaced", then it is "continuous in the integral sense" in the space L: for any $\varepsilon > 0$, there can be found $\delta > 0$ such that for $|h| < \delta$:

$$\|f(x + h) - f(x)\| < \varepsilon.$$

Hint. Show that the set of all f for which (*) is satisfied is closed in the space L. Then prove (*) for continuous functions.

4. Measure of Sets and Theory of Lebesgue Integration

1. A set A contained in the closed interval $[a, b]$ is said to be *measurable* if its "characteristic function", i.e. the function $\chi_A(x)$, equal to 1 on A and 0 on the complement of A, is measurable (and consequently summable). The integral of the characteristic function is called the measure of the set A and is denoted by μA. If A is a set of measure zero in the sense of the definition of Section 1, then, as we showed at the time, the integral of the function $\chi_A(x)$ vanishes, so that the set A has measure zero in the new sense too: $\mu A = 0$. Conversely, if the integral of the characteristic function of a set A vanishes, then in agreement with the observation following Fatou's lemma, A has measure zero in the former sense. Thus, for a set of measure zero, the new definition coincides with the earlier one.

The properties of integrals with which we are already familiar enable us to obtain the properties of measurable sets. Thus the union A of a finite or countable aggregate of measurable sets $A_1, A_2, \ldots, A_n, \ldots$ is a measurable set since its characteristic function can be defined by the formula

$$\chi_A(x) = \sup \{\chi_{A_1}(x), \ldots, \chi_{A_n}(x), \ldots\}.$$

In the case that the sets $A_1, A_2, \ldots$, are disjoint, we have

$$\chi_A(x) = \chi_{A_1}(x) + \cdots + \chi_{A_n}(x) + \cdots$$

and by Lebesgue's theorem (Section 3)

$$\mu A = \mu A_1 + \cdots + \mu A_n + \cdots \tag{1}$$

This property is called the *total additivity of measure*. Further, if we subtract a measurable set A_1 from a measurable set A_2 which contains it, we get a new measurable set A_3, since

$$\chi_{A_3} = \chi_{A_2} - \chi_{A_1}.$$

In this case, $\mu A_3 = \mu A_2 - \mu A_1$. In particular, since the closed interval $[a, b]$ is measurable, the complement of any measurable set $A \subset [a, b]$ is measurable. It follows further that the intersection of a sequence of measurable sets $A_1, A_2, \ldots$ is also measurable, since its complement

$$C \prod_{n=1}^{\infty} A_n = \sum_{n=1}^{\infty} C A_n$$

is a measurable set. We observe finally the following limiting relation:

If

$$A_1 \subset A_2 \subset \cdots \subset A_n \subset \cdots \quad \text{and} \quad A = \bigcup_{n=1}^{\infty} A_n,$$

then

$$\mu A = \lim_{n\to\infty} \mu A_n. \tag{2}$$

For we have:

$$\begin{aligned} A &= A_1 + (A_2 - A_1) + (A_3 - A_2) + \cdots + (A_{n+1} - A_n) + \cdots, \\ \mu A &= \mu A_1 + \mu(A_2 - A_1) + \cdots + \mu(A_{n+1} - A_n) + \cdots \\ &= \mu A_1 + (\mu A_2 - \mu A_1) + \cdots + (\mu A_{n+1} - \mu A_n) + \cdots \\ &= \lim_{n\to\infty} \mu A_n. \end{aligned}$$

If $A_1, A_2, \ldots$ is an arbitrary system of measurable sets then the sets $A'_1 = A_1$, $A'_2 = A_1 + A_2$, ..., $A'_n = A_1 + \cdots + A_n$, ... are also measurable and are successively embedded in one another; by what we have proved, we have for $A = \bigcup_{n=1}^{\infty} A_n$

$$\mu A = \lim \mu A'_n = \lim \mu [A_1 + \cdots + A_n].$$

In particular, since it is always the case that

$$\mu(A' + A'') = \mu(A' + [A'' - A' A'']) = \mu(A') + \mu(A'' - A'A'') \leq \mu A' + \mu A'',$$

we have

$$\mu A = \lim \mu A'_n \leq \lim_{n \to \infty} [\mu A_1 + \cdots + \mu A_n] \leq \sum_{n=1}^{\infty} \mu A_n. \quad (3)$$

The analogous results for intersections are obtained by taking complements: for example, if we are given measurable sets $A_1 \supset A_2 \supset \cdots A_n \supset \cdots$ and $A = \prod_{n=1}^{\infty} A_n$, then

$$\mu A = \mu \prod_{k=1}^{\infty} A_k = \lim_{n \to \infty} \mu \prod_{k=1}^{n} A_k. \quad (4)$$

Any interval (closed, open, or semi-open) is evidently a measurable set and its measure equals its length. The foregoing formulae show that any open set, as the (at most) countable sum of disjoint open intervals, is measurable and its measure equals the sum of the lengths of its component open intervals. Further, any closed set is measurable since it is the complement of an open set. In general any set which can be obtained from open intervals by an at most countable number of successive operations of forming unions, intersections, or complements is measurable.

2. It is natural to conjecture whether, in general, there exist unmeasurable sets. The remarks above show that the problem of constructing a concrete unmeasurable set is certainly complex, since the usual methods of construction, which commence with open intervals and employ countable unions and intersections, fail to lead beyond the domain of measurable sets. Up till now no individual instance of an unmeasurable set has been constructed. Nevertheless we cannot doubt the existence of such sets. We give the appropriate reasoning, following N. N. Lusin. It is convenient to imagine the closed interval [0, 1], on which we shall carry out our constructions, rolled into a circle of unit circumference, so that the points 0, 1 coincide. We shall measure all distances along this circumference. Points ξ, η will be called *allied* if the distance between them is rational and *hostile* if it is irrational. It is evident that if ξ is allied to η and η is allied to ζ, then ξ is allied to ξ. We shall call the totality of points situated at rational distances from a given point its *family*; all the members of a family are mutually

allied and are hostile to any point that does not belong to the family. Families A, B which contain prescribed points α, β respectively either coincide (if α, β are allied) or have no element in common (if α, β are hostile). The totality of points of the closed interval [0, 1] is the union of some set of distinct families. The set of all families is certainly uncountable, since the set of all points of the closed interval [0, 1] is uncountable. We imagine now that a representative of each family is somehow chosen. We denote by Z the aggregate of all these representatives and claim that no matter how the representatives are selected, the set Z will be unmeasurable. (We cannot indicate a possible rule for selecting the representatives, and it cannot therefore be said, in effect, that the set Z has been constructed). Let $r_1, r_2, \ldots$ be a sequence containing all the rational numbers; we denote by Z_n the displacement of the entire set Z through the distance r_n, so that Z_n is composed of points of the form $\xi + r_n$, where ξ runs through the whole of Z.

It clearly follows that the sets Z_n, Z_m are disjoint if $m \neq n$; otherwise we should have an equation $\xi + r_n = \eta + r_m$, where ξ, η belong to Z; but then the difference $\xi - \eta = r_m - r_n$ would be rational, i.e. ξ, η would be allied, which is precluded by the construction. Further, every point of the closed interval [0, 1] belongs to some Z_n, since every point λ has the form $\xi + r_n$, where ξ is the representative of the family containing λ. Thus the sets $Z_1, Z_2, \ldots$ are mutually disjoint and their union is the whole closed interval [0, 1]. If the set Z were measurable, all the sets $Z_1, Z_2, \ldots$ would also be measurable and their measures would be equal to that of Z. In virtue of the countable additivity of measure, we should then have to have

$$\mu Z_1 + \mu Z_2 + \cdots + \mu Z_n + \cdots = \mu Z + \mu Z + \cdots + \mu Z + \cdots = 1.$$

But this is not possible for any value of μZ: neither for $\mu Z < 0$ nor for $\mu Z = 0$. Thus the set Z cannot be measurable.

Problems. 1. Given any measurable function f and number c, show that the set $E = \{x : f(x) \geq c\}$ is measurable.

Hint: $E = \prod_{k=1}^{\infty} \bigcup_{n=1}^{\infty} \prod_{m=n}^{\infty} \{x : h_m(x) > c - 1/k\}$, where $h_m \to f$ is a sequence of step functions.

2. If $f_n \to f$ almost everywhere, given any $c > 0$, (5)

$$\lim_{n \to \infty} \mu \{x: |f - f_n| \geq c\} = 0.$$

Hint.

$$\mu \prod_{m=1}^{\infty} \bigcup_{n=m}^{\infty} \{x: f - f_n \geq c\} = 0.$$

3. A sequence of measurable functions $f_n(x)$, satisfying (5) for any $c > 0$, is said to be convergent in measure to the function f. Show that, though the sequence f_n itself may not be convergent almost everywhere to f, it must contain a subsequence convergent almost everywhere to f.

Hint. Given any integers k and m, there exists an $n = n(k, m)$ for which $\mu\{x : |f_{n(k,m)} - f| > 1/k\} < 1/k\, 1/2m$. The functions $f_{n(k,m)}$ converge to f as $k \to \infty$ on a set of measure $> b - a - 1/m$. The required sequence is obtained from $f_{n(k,m)}$ by a diagonal process over m.

4. If every subsequence of a sequence of measurable functions f_n contains a subsequence which converges almost everywhere to a fixed function f, the f_n converge in measure to f.

Hint. Use the result of problem 2 and *reductio ad absurdum.*

5. If $f \geqq 0$ is a summable function and $\mu\{x : f(x) \geqq c\} \geqq b$, then $I\, f \geqq b\, c$.

Hint. The assertion is obvious for step functions. In the general case, use the formula of problem 1 to obtain the inequality

$$b \leqq \mu\{x : f(x) \geqq c\} \leqq \mu\{x : h_m(x) < c - 1/k\} + \varepsilon$$

for any k and ε and sufficiently large m. Hence

$$I\, h_m \leqq (b - \varepsilon)\,(c - 1/k).$$

6. If $f_n \geqq 0$ and $I\, f_n \to 0$, then $f_n \to 0$ in measure. (Convergence almost everywhere does not follow from these conditions.) The condition $f_n \geqq 0$ cannot be dispensed with.

Hint. Use problem 5.

7. A metric is defined on the space of all functions $f(x)$ measurable on the closed interval $a \leqq x \leqq b$, in accordance with the formula

$$\varrho\,(f, g) = I\left(\frac{|f-g|}{1+|f-g|}\right). \tag{6}$$

Show that the metric space axiom is satisfied, and that convergence in the metric (6) is the same as convergence in measure. Prove that the metric space is complete.

8. We denote by $a_k\,(x)$ the function equal at the point $x \in [0, 1]$ to the kth place of the dyadic number x. Show that the functions

$$\sigma_n\,(x) = \frac{1}{n} \sum_{k=1}^{n} a_k\,(x)$$

satisfy the equation $\int\limits_0^1 [\sigma_n(x) - 1/2]^2\, \mathrm{d}x = 1/4n$.

Hint. $\int\limits_0^1 a_j\,(x)\, a_k\,(x)\, \mathrm{d}x = 1/4$ for any $j \neq k$.

9. (continued). The sequence $\sigma_{n^2}(x)$ is convergent almost everywhere to $f(x) = 1/2$.

Hint. $\mu\{x : |\sigma_n(x) - 1/2| > \varepsilon\} \leqq 1/4n\, \varepsilon^2$ for any $\varepsilon > 0$.

10. (continued). The sequence $\sigma_n(x)$ is convergent almost everywhere to $f(x) = 1/2$.

Hint. If $m^2 < n < (m + 1)^2$, then $(1/n) \sum\limits_{m^2+1}^{n} a_k(x) < (2m + 1)/m^2$.

Note. The result of problem 10 shows that, for almost any real number of the closed interval [0, 1], the share of zeros and unities in its dyadic resolution, defined naturally as the limit of the function $\sigma_n(x)$, is equal to 1/2.

3. We shall now examine the structure of a measurable set of positive measure. We shall show that *every measurable set A of positive measure can be regarded, to within a set of arbitrary small measure, as the union of a finite number of closed intervals*. More precisely, *for any given $\varepsilon > 0$, we undertake to find a finite system B*

of closed intervals such that the set A can be obtained from B by adding a set a of measure $< \varepsilon$ and removing a set b of measure also $< \varepsilon$.

For the proof we represent the measurable function $\chi_A(x)$ as the limit of a sequence of step functions $h_n(x)$ which converges almost everywhere. If we replace each function h_n by a function h'_n equal to 0 for $h_n < 1/2$ and 1 for $h_n \geqq 1/2$, we obtain a new sequence of step functions, which also converges almost everywhere to χ_A and which assumes only the values 0 or 1. Each of the functions h'_n is the characteristic function of some set B_n that represents a finite system of closed intervals. We claim that for sufficiently large n the set B_n satisfies the condition stipulated. For consider the functions $(h'_n - \chi_A)^+$, $(h'_n - \chi_A)^-$. The function $(h'_n - \chi_A)^+$ is the characteristic function of the set b_n composed of points which belong to the set B_n but not to the set A; the function $(h'_n - \chi_A)^-$ is the characteristic function of the set a_n of points which belong to A but not to B_n. The set A is obtained from the set B_n by removing b_n and adding a_n.

But

$$\mu(a_n) + \mu(b_n) = I(h'_n - \chi_A)^- + I(h'_n - \chi_A)^+ = I(|h'_n - \chi_A|) \to 0,$$

which completes the proof.

4. The measure of a measurable set as its upper measure. The theorem we have proved, in conjunction with the foregoing observations, enables us to formulate a new, more constructive definition of a measurable set. We cover an arbitrary set A on the closed interval $[a, b]$ with a finite or countable system of intervals and find the sum of their lengths. The exact lower bound of the sums obtained under all possible such coverings is denoted by $\mu^* A$ and is said to be the *upper* or *exterior measure of the set A*. Thus a set of measure zero has exterior measure zero by its very definition. We shall show that *the exterior measure of any measurable set A coincides with the quantity μA.*

If B is a system of non-overlapping intervals covering a set A, then, as above, B is a measurable set with measure equal to the sum of the lengths of the intervals; since $A \subset B$, we have $\mu A \leqq \mu B$ and hence the exterior measure of the set A

$$\mu^* A = \inf_{B \supset A} \mu B \geqq \mu A. \tag{5}$$

On the other hand, using the theorem proved above, we can construct a finite system B_n of intervals for any given $\varepsilon > 0$ and any n, such that

$$A + b_n = B_n + a_n, \tag{6}$$

where each of the measurable sets a_n, b_n has measure $< \varepsilon/2^{n+1}$.

Let $B = \bigcup_{n=1}^{\infty} B_n$, $a = \bigcup_{n=1}^{\infty} a_n$, $b = \bigcup_{n=1}^{\infty} b_n$; obviously,

$$\mu a = \lim \mu a_n = 0, \mu b \leqq \sum_{n=1}^{\infty} \mu b_n < 1/2\, \varepsilon.$$

We have the inclusions

$$A \subset B + a, \tag{7}$$

$$B \subset A + b. \tag{8}$$

For, if $x \in A$ does not belong to B, it does not belong to any B_n, and since $A \subset B_n + a_n$, x belongs to all the a_n, and therefore to a; this proves (7).

Further, we take a point y, belonging to B and hence to some B_n. If $y \in A$, it follows from (6) that y belongs to b_n and therefore to b; this proves (8). Notice that (8) implies $\mu B \leqq \mu A + \mu b < \mu A + + 1/2\, \varepsilon$.

We return to inclusion (7). In the long run the set B is a system of intervals, finite or countable. The set a, as a set of measure zero, can be covered by a system of intervals of overall length $< 1/2\varepsilon$. In short, the whole set A can be covered by a (finite or) countable system of intervals of overall length not exceeding $\mu B + 1/2\varepsilon \leqq \mu A + \varepsilon$. Hence $\mu^* A \leqq \mu A + \varepsilon$. Since ε is arbitrary, $\mu^* A \leqq \mu A$, and combining this with inequality (5), we get that for any measurable set

$$\mu^* A = \mu A.$$

In other words, *the measure of any measurable set can be defined constructively as its exterior measure.*

5. The question naturally arises whether measurable sets themselves can be defined in terms of exterior measure without involving the concept of the integral. The answer to this question is as follows: *a set A on the closed interval $[a, b]$ is measurable if and only if the sum of the exterior measures of A and its complement CA* (in $[a, b]$) *is equal to the length of* $[a, b]$:

$$\mu^* A + \mu^* C A = b - a. \tag{9}$$

The necessity of this condition is immediately apparent; if A is measurable, then so is its complement and the equation (9) follows from $\mu A + \mu C A = b - a$. We prove the sufficiency of the condition. Let (9) hold for a set A. Then for any $\varepsilon > 0$ we can find a covering of the sets A, CA by systems of non-overlapping intervals U, V such that the overall sum of their lengths

$$\mu U + \mu V < b - a + \varepsilon.$$

We denote the characteristic functions of the sets U, V by h_U, h_V. The function $1 - h_V$ is non-zero only within A, and therefore

$$0 \leqq 1 - h_V \leqq \chi_A \leqq h_U,$$

where χ_A is the characteristic function of A.

It follows that

$$0 \leqq I(1 - h_V) \leqq I h_U.$$

At the same time, by hypothesis

$$I h_U - I(1 - h_V) = I h_U + I h_V - (b - a) = \mu U + \mu V - \\ - (b - a) < \varepsilon.$$

As $\varepsilon \to 0$ the sequence $h_U (= h_{U(\varepsilon)})$ can be taken to be monotone decreasing and the sequence $1 - h_V$ monotone increasing. Hence the sequence $h_U - (1 - h_V)$ is also monotone decreasing. Its limit f is a non-negative bounded measurable function, the integral of which vanishes, by the corollary to Beppo Levi's theorem; hence f vanishes almost everywhere. But since

$$h_U - (1 - h_V) \geqq h_U - \chi_A \geqq 0,$$

we have $\chi_A = \lim\limits_{\varepsilon \to 0} h_U$ and it is therefore a measurable and summable function. This proves the sufficiency of the condition formulated.

Note. The theorem proved can be formulated in the following somewhat more general way: *a set A contained in the closed interval $[a, b]$ is measurable if and only if there exist sequences of measurable sets U_n, V_n such that $U_n \supset A$, $V_n \supset CA$,*

$$b - a \leqq \mu U_n + \mu V_n \leqq b - a + \varepsilon_n, \quad \varepsilon_n \to 0.$$

6. In his development of measure theory and the theory of the integral, Lebesgue started from the definition of a measurable set in just the form of the last theorem. That is, he called a set A measurable if it satisfied the relation

$$\mu^* A + \mu^* C A = b - a. \tag{10}$$

This definition is sometimes expressed in another form: the interior (lower) measure of a set A is defined by the formula

$$\mu_* A = \sup_{F \subset A} \mu F,$$

where the upper bound is taken over all closed sets F contained in the given set A; the measure of a closed set F is taken as the number $b - a - \mu\, C F$, where $C F$ is the open set complementary to F. Further, A is said to be a measurable set if

$$\mu_* A = \mu^* A. \tag{11}$$

It is easy to verify that the definitions (6), (7) are equivalent. For

$$\mu_* A = \sup_{F \subset A} \mu F = \sup_{F \subset A} \{b - a - \mu\, C F\} = b - a - \inf_{F \subset A} \mu\, C F.$$

But if a set F is included in a set A, its complement $C F$ covers the set CA; it follows immediately that

$$\mu_* A = b - a - \mu^* C A,$$

which implies that the equations

$$\mu_* A = \mu^* A, \quad \mu^* A + \mu^* C A = b - a$$

are equivalent.

Having established the fundamental properties of measure, of which we were speaking at the beginning of the paragraph, on the basis of the definition (10) [or (11)], Lebesgue proceeds to the definition of measurable functions. He calls a function $\varphi(x)$ *measurable* (and we call it *Lebesgue-measurable*) if any set of the form

$$\{x : \varphi(x) \leqq c\}$$

is measurable, whatever the value of the real number c.

For the moment we shall call functions which are measurable in the sense of Section 1 *measurable in the sense of Riesz*. We shall show that the definitions of functions measurable in the senses of Lebesgue and Riesz are equivalent.

Let $\varphi(x)$ be measurable in the sense of Riesz; we introduce the function

$$\varphi_c(x) = \max\{\varphi(x), c\}.$$

The ratio

$$\frac{\varphi_{c+\varepsilon} - \varphi_c}{\varepsilon}$$

is also a measurable function in the sense of Riesz, vanishing for $\varphi(x) \geqq c + \varepsilon$ and equal to 1 for $\varphi \leqq c$. Its limit as $\varepsilon \to 0$ is the characteristic function of the set $\{x : \varphi(x) \leqq c\}$, which is therefore also measurable in the sense of Riesz. This means that the set itself is measurable in the sense of Riesz and consequently also in the sense of Lebesgue.

Conversely, if a function $\varphi(x)$ is Lebesgue-measurable, i.e. if all sets of the form $\{x : \varphi(x) \leqq c\}$ are measurable, then sets of the form

$$\{x : c_1 < \varphi(x) \leqq c_2\} = \{x : \varphi(x) \leqq c_2\} - \{x : \varphi(x) \leqq c_1\}$$

are also measurable for any c_1, c_2.

Let $\chi_{c_1 c_2}$ be the characteristic function of this set. For a given n, we put a function φ_n equal to m/n at those points where the inequality

$$\left\{x : \frac{m}{n} < \varphi \leqq \frac{m+1}{n}\right\} \quad (m = 1, 2, \ldots).$$

is satisfied. The function φ_n is the sum of an everywhere convergent series of functions measurable in the sense of Riesz

$$\varphi_n(x) = \sum_{-\infty}^{\infty} \frac{m}{n} \chi_{\frac{m}{n}, \frac{m+1}{n}}(x)$$

and is therefore itself measurable in the sense of Riesz. It is evident that the inequality

$$|\varphi_n(x) - \varphi(x)| \leqq \frac{1}{n}$$

holds. Thus $\varphi(x)$ is the limit as $n \to \infty$ of the sequence of Riesz-measurable functions $\varphi_n(x)$ and is therefore measurable in the sense of Riesz, as required.

Lebesgue goes on to construct the definition of an integral for a bounded measurable function $\varphi(x)$. To do this, he partitions the domain of variation $[m, M]$ of the function $\varphi(x)$, into n parts by the points $m = y_0 < y_1 < \cdots < y_n = M$ and forms the "integral sum"

$$s_n(\varphi) = \sum_{j=0}^{n-1} y_j \, \mu\{x : y_j < \varphi(x) \leqq y_{j+1}\}, \tag{8}$$

where the numbers $\mu\{x : y_j < \varphi(x) \leqq y_{j+1}\}$ are meaningful since $\varphi(x)$ is measurable. When the closed interval $[m, M]$ is partitioned indefinitely, the sums $s_\Pi(\varphi)$ tend, as Lebesgue proves, to a (uniquely determined) limit which is called the Lebesgue integral of the function $\varphi(x)$.

We verify that the Lebesgue definition agrees with the Riesz definition assumed in our account. The function $\varphi_\Pi(x)$ equal to y_j on the set $\{x : y_i < \varphi \leqq y_{j+1}\}$ is measurable and its integral in the sense of Riesz is precisely equal to the quantity $s_\Pi(\varphi)$ (8). Since it differs from the function $\varphi(x)$ by at most max $y_{j+1} - y_j)$, the function $\varphi_\Pi(x)$ tends uniformly under an unlimited partitioning of the interval $[m, M]$ to the function $\varphi(x)$. It follows that $I\varphi = \lim I\varphi_\Pi$, which proves the existence of the Lebesgue integral and its equivalence to the Riesz integral. If the function $\varphi(x)$ is unbounded but non-negative, Lebesgue proceeds as follows. Truncating the function φ at the value N, i.e. considering the function

$$\varphi_N(x) = \min(\varphi(x), N),$$

he obtains a non-negative, bounded, measurable function. Its integral $I\varphi_N$, which exists in virtue of the foregoing argument, is non-negative and increases (more precisely, does not decrease) as N increases. If the numbers $I\varphi_N$ have a finite limit as $N \to \infty$, the function φ is said to be *summable in the sense of Lebesgue* and its Lebesgue integral is put equal to $\lim I\varphi_N$. In the general case, where φ can take any sign, it is said to be summable in the sense of Lebesgue if its positive and negative parts φ^+, φ^- are summable in the sense of Lebesgue; its integral is then constructed according to the formula

$$I\varphi = I\varphi^+ - I\varphi^-.$$

We verify now that the class of summable functions arising from our definition (summable in the sense of Riesz) coincides with the class of functions summable in the sense of Lebesgue, and that the values of corresponding integrals coincide. It is sufficient to establish this for non-negative functions. If a non-negative function $\varphi(x)$ is summable in the sense of Lebesgue, it is the limit of an increasing sequence of bounded summable functions φ_N; it follows by Beppo Levi's theorem that φ is summable in the sense of Riesz and we have $I\varphi = \lim I\varphi_N$, i.e. its integral in the sense of Riesz, coincides with its Lebesgue integral. Conversely if $\varphi(x) \geqq 0$ is summable in the sense of Riesz, then all the functions φ_N are bounded, and measurable and their integrals do not exceed $I\varphi$. Since the $I\varphi_N$ are bounded, they have a finite limit and hence $\varphi(x)$ is summable in the sense of Lebesgue; by the foregoing argument, its integral coincides with $I\varphi$. Thus the results of the Lebesgue and Riesz constructions coincide completely. We have preferred to expound the construction originally given by Riesz, because as regards the technique of proofs it is somewhat simpler on the whole than the construction of the Lebesgue method and permits of a direct construction of the theory of the integral with an avoidance of the rather unwieldy Lebesgue theory of measure.

7. Integration over a measurable set. Up till now our domain of integration has been the closed interval $[a, b]$. But an integral can easily be defined over any measurable set $E \subset [a, b]$. Let $\chi_E(x)$ be the characteristic function of the set E; we shall call a function φ summable on the set E if the product $\chi_E \varphi$ is summable on $[a, b]$, and we define

$$\int_E \varphi \, dx = \int_a^b \chi_E \varphi \, dx = I(\chi_E \varphi).$$

The integral over E evidently possesses all the usual properties of the integral. We note some of these specifically.

(a) *If φ is summable on $E = E_1 + E_2 + \cdots$ and the (measurable) sets $E_1, E_2, \ldots$ are mutually disjoint, then φ is summable on each E_n and*

$$\int_E \varphi \, dx = \int_{E_1} \varphi \, dx + \int_{E_2} \varphi \, dx + \cdots \tag{12}$$

For since $\chi_{E_n} \chi_E = \chi_{E_n}$, the function $\chi_{E_n} \varphi = \chi_{E_n}(\chi_E \varphi)$ is summable on E_n. Further, $\chi_{E_1} + \chi_{E_2} + \cdots = \chi_E$, hence

$$\chi_{E_1} \varphi = \chi_{E_2} \varphi + \cdots = \chi_E \varphi,$$

and it follows that the partial sums of this series are bounded by the summable function $\chi_E |\varphi|$. Integrating the series term by term, we get (12).

(b) The converse result, that a function φ is summable on E if it is summable on each E_n and the series (12) converges, can only be substantiated under the additional hypothesis that $\varphi \geqq 0$ on each E_n.

In this case $\chi_E \varphi$ is the limit of the non-decreasing sequence of partial sums of the series $\chi_{E_1} \varphi + \chi_{E_2} \varphi + \cdots$, the integrals of which have a common constant bound; applying Beppo Levi's theorem, we get that $\chi_E \varphi$ is summable and equation (12) holds.

Property (b) is sometimes applied in the following equivalent form: *if a function φ is non-negative and is summable on each of the sets $E_1 \subset E_2 \subset \cdots$* and *the integrals $\int_{E_n} \varphi \, dx$ are bounded by a fixed* *constant, then* φ *is summable on* $E = \sum_{n=1}^{\infty} E_n$ *and*

$$\int_E \varphi\, dx = \lim_{n\to\infty} \int_{E_n} \varphi\, dx.$$

(c) *Absolute continuity of an integral over a set. The integral of a summable* (on $[a, b]$) *function* φ, *taken over a measurable set* E, *tends to zero together with the measure of this set independently of its disposition on* $[a, b]$. In other words, for any $\varepsilon > 0$ there can be found $\delta > 0$ such that $\mu E < \delta$ implies $\int_E |\varphi|\, dx < \varepsilon$. For the proof we find a bounded measurable function $h(x) \geqq 0$, dependent on the prescribed $\varepsilon > 0$, for which

$$\int_a^b \| \varphi(x)| - h(x)|\, dx < \frac{\varepsilon}{2}.$$

For example, let $h(x)$ be bounded by a number M_1 so that $0 \leqq h(x) \leqq M$. Then for any measurable set E of measure less than $\delta = \dfrac{\varepsilon}{2M}$,

$$\int_E |\varphi|\, dx \leqq \int_E \| \varphi(x)| - h(x)|\, dx + \int_E h(x)\, dx < \frac{\varepsilon}{2} + \delta M < \varepsilon,$$

as required.

8. We dwell finally on one further important theorem in the Lebesgue theory of measure:

Theorem (D. F. Egoroff, 1911). *Let there be given a sequence* $\varphi_1(x), \varphi_2(x), \ldots, \varphi_n(x), \ldots$ *of measurable functions which converges almost everywhere on* $[a, b]$ *to some limit* $\varphi(x)$. *For any* $\varepsilon > 0$ *there exists a measurable (even closed) set* A *of measure* $> b - a - \varepsilon$ *on which the sequence converges uniformly to its limit.*

Proof. The limit function $\varphi(x)$ is measurable. By considering the functions $\varphi - \varphi_n$ in place of the φ_n, the problem can be reduced to the case of a sequence that converges to zero; we shall therefore assume at the outset that $\varphi(x) \equiv 0$. Further, the functions φ_n can be supposed to be non-negative and to tend monotonely to zero;

if this is not the case, we can substitute $\sup \{|\varphi_n|, |\varphi_{n+1}|, \ldots\}$ for φ_n.

We denote by A_n^m the set of points on which $0 \leqq \varphi_n(x) \leqq 1/m$. For fixed m the set A_n^m becomes larger as n increases and the system of all these sets covers the whole closed interval $[a, b]$. Hence $\lim\limits_{n\to\infty} \mu A_n^m = b - a$ and we can find a number $n = n(m)$ such that

$$\mu A_{n(m)}^m > b - a - \frac{\varepsilon}{2^m}.$$

We now consider the intersection A of all the sets $A_{n(m)}^m$ $(m = 1, 2, \ldots)$. Since the complement of A has measure

$$\mu C A = \mu C \{\prod A_{n(m)}^m\} = \mu \bigcup C A_{n(m)}^m \leqq \sum_{m=1}^{\infty} \frac{\varepsilon}{2^m} = \varepsilon,$$

the measure of A itself is at least $b - a - \varepsilon$. We claim that on the set A the convergence $\varphi_n \to 0$ is uniform, i.e. for a given $1/m$ there exists a number N such that, for $n \geqq N$, the inequality $\varphi_n(x) < 1/m$ holds everywhere on A. For as N we can take $n(m)$, since $A \subset A_{n(m)}^m$, and hence if $n > n(m)$, then $\varphi_n(x) \leqq \varphi_{n(m)}^{(x)} \leqq 1/m$ at all points of A. Remembering that $\mu A = \sup \mu F$, $F \subset A$, we can replace the measurable set A by a closed set of arbitrary close measure. The theorem is proved.

It is not difficult to see that in the formulation of the theorem the closed interval $[a, b]$ can be replaced by any measurable set. A consequence of Egoroff's theorem is a theorem due to Lusin which establishes a new feature of measurable functions:

THEOREM (N. N. Lusin, 1915†). *A function $\varphi(x)$ defined on the closed interval $[a, b]$ is measurable if and only if for any $\varepsilon > 0$ there exists a closed subset $A \subset [a, b]$ of measure $> b - a - \varepsilon$ on which $\varphi(x)$ is continuous.*

For if $h_n(x) \to \varphi(x)$ is a sequence of step functions that defines the measurable function $\varphi(x)$, we can cover the set of measure zero comprising the points of divergence of the sequence and the points of discontinuity of the functions $h_n(x)$ with a system of open intervals of overall length $< \varepsilon/2$. Further, by Egoroff's theorem, we can remove from the set that remains a subset, also

† D. F. Egoroff (1869–1931), N. N. Lusin (1883–1950) are Muscovite mathematicians, the founders of the Moscow school of the theory of functions of a real variable.

of measure $< \varepsilon/2$, such that the sequence $h_n(x)$ will converge uniformly on the remainder; we can assume also that the subset removed represents a system of open intervals. On the remaining closed subset F of $[a, b]$, the measure of which exceeds $b - a - \varepsilon$, the functions $h_n(x)$ are continuous and converge to their limit $\varphi(x)$ uniformly; it follows that $\varphi(x)$ is continuous on F.

We leave a proof of the converse assertion to the reader.

Problems. 1. Show that to within a set of measure zero every measurable set is the intersection of a sequence of open sets and the union of a sequence of closed sets.

2. Show that an aggregate of measurable sets, every pair of which differ by more than a set of measure zero, has at most the power of the continuum.

Hint. Use problem 1 and problem 5 of Chapter II, Section 2.

3. Construct a measurable set which together with its complement, has positive measure on every open interval.

4. Let A be a set of measure $\geq \varepsilon$ on the closed interval $[0, 1]$ and let $\xi_1, \ldots, \xi_n$ be arbitrary numbers in the interval with $n > 2/\varepsilon$. Show that A contains a pair of points the distance between which coincides with the distance between some pair of the points $\xi_1, \xi_2, \ldots, \xi_n$.

Hint. Show that the sets $\xi_1 + A, \ldots, \xi_n + A$ (which lie in the closed interval $[0, 2]$ cannot be mutually disjoint.

5. Show that every set of positive measure contains a pair of points a rational distance apart.

Hint. Use the preceding problem.

6. A set A on the closed interval $[a, b]$ has measure $> 1/2(b - a)$. Show that A contains a subset of positive measure, symmetrical about the midpoint of the interval.

Hint. Consider the intersection of the set A with its reflection through the mid-point of the interval.

7. Measurable functions f, g are said to be equi-measurable if for any c we have $\mu\{x : f > c\} = \mu\{x : g > c\}$. Show that: (a) any measurable function is equi-measurable with some non-decreasing function; (b) equi-measurable non-decreasing functions coincide almost everywhere.

8. Given an arbitrary sequence $f_1(x), \ldots, f_n(x)$ of measurable functions on the closed interval $[a, b]$, show that the set E of points where $\lim_{n \to \infty} f_n(x)$ exists is measurable.

Hint. $E = \prod_k \bigcup_n \prod_m \{x : |f_n(x) - f_m(x)| < 1/k\}$.

5. Generalisations

1. *The Case of an Infinite Interval*

All the preceding investigations have related to functions defined on a closed interval $[a, b]$ of the x-axis. But the infinite intervals $[a, \infty)$, $(-\infty, b]$, $(-\infty, \infty)$ present no difficulty. In all these cases we shall call a function $h(x)$ *a step function* if it assumes constant values on a finite number of finite open intervals $\Delta_j = (x_j, x_{j+1})$; on the remainder of the infinite interval it is assumed to vanish. A function $\varphi(x)$ is said to be *measurable* if it is the limit of a sequence of step functions which converges almost everywhere on the whole infinite interval. The integral of a step function $h(x)$ that assumes the value b_j on an interval Δ_j of length $|\Delta_j|$ ($j = 1, 2, \ldots, k$) is defined naturally by the formula

$$I\,h = \sum_{j=1}^{k} b_j |\Delta_j| .$$

The class C^+ is defined as before as the aggregate of functions $f(x)$, each of which is the limit of an increasing sequence of step functions $h_n(x)$ with bounded integrals. The class L of summable functions is defined as the class of differences $\varphi = f - g$, $f \in C^+$ $g \in C^+$. All the theorems of Sections 2–4, with a single exception, are preserved without change: in the exceptional case, bounded measurable functions are no longer summable in general. In particular, the equation $I\,\varphi = \lim I\,\varphi_n$ ceases to hold if in converging almost everywhere to a function φ_0 the functions φ_n retain a common bound; the original condition $|\varphi_n| \leqq \varphi_0$, where the function φ_0 is summable, remains sufficient for it to hold. In proving the theorem that the limit of a sequence of measurable functions is a measurable function (Section 3, art. 4) it will be necessary in this connexion to replace the function $\psi_n = \dfrac{1}{1+\varphi_n}$ by $\hat{\psi}_n = \dfrac{h}{h+\varphi_n}$, where h is a strictly positive summable function. The remaining results carry over without change to the case of an infinite interval. In particular, it is possible to define on an infinite interval Δ a normed linear space $L_1(\Delta)$ (the Lebesgue space) of all summable functions with the norm

$$\|\varphi\| = \int_{\Delta} |\varphi(x)|\, dx .$$

It is proved in just the same way as for a finite interval that the space L is complete and that the step functions constitute a set everywhere dense in it.

2. *The Case of Several Independent Variables*

We restrict ourselves to the case of two variables x, y and functions $\varphi(x, y)$ defined on the rectangle $D = \{a_1 \leqq x \leqq b_1, a_2 \leqq y \leqq b_2\}$.

A set on the rectangle D is said to be of measure zero if for any $\varepsilon > 0$ it can be covered by a finite or countable system of rectangles $D_j = \{a_1^{(j)} \leqq x \leqq b_1^{(j)}, a_2^{(j)} \leqq y \leqq b_2^{(j)}\}$, the sum of the areas of which is at most ε. Thus any closed interval or linear set is a set of measure zero.

If D is subdivided into a finite number of rectangles $D_1, \ldots, D_m$, a function which is constant on each of the D_j is said to be a *step function*. The integral of a step function $h(x, y)$ which assumes the value b_j on the rectangle D_j is defined by the formula

$$I\,h = \sum_{j=1}^{m} b_j |D_j|,$$

where $|D_j|$ is the area of D_j.

As before, the class C^+ is the totality of functions, each of which is the limit of an increasing sequence of step functions h_n with bounded integrals $I\,h_n$. The class L—the Lebesgue space—is the class of differences $\varphi = f - g$, $f \in C^+$, $g \in C^+$. The theorems of Sections 2–4 carry over without change to the case of two variables, discounting notation [we write $f(x, y)$ instead of $f(x)$ and use the word "rectangle" instead of "interval" etc.].

But an important new problem arises here, namely that of repeated integration.

In classical analysis a double integral (initially defined as a limit of Riemanian integral sums), say of a continuous function $f(x, y)$, reduces to two ordinary integrals with respect to one variable:

$$\int_{a_1}^{b_1}\int_{a_2}^{b_1} f(x, y)\,\mathrm{d}x\,\mathrm{d}y = \int_{a_1}^{b_1}\left\{\int_{a_2}^{b_2} f(x, y)\,\mathrm{d}y\right\}\mathrm{d}x.$$

In the Lebesgue theory of the integral a similar formula obtains. The corresponding theorem is formulated as follows:

THEOREM (G. Fubini, 1907). *Let $\varphi(x, y)$ be a summable function on the rectangle $D = \{a_1 \leqq x \leqq b_1, a_2 \leqq y \leqq b_1\}$. Then*: (1) *regarded as a function of the argument x with y fixed, it is summable for almost all values of y*; (2) *its integral over the closed interval $a_1 \leqq x \leqq b_1$,* *which we denote by*

$$I_x \varphi(x, y),$$

is summable, as a function of y, *on the closed interval* $a_2 \leq y \leq b_2$;
(3) *we have*

$$I_y\{I_x \varphi(x, y)\} = I\varphi.$$

Interchanging x *and* y, *we get the parallel result*

$$I_x\{I_y \varphi(x, y)\} = I\varphi.$$

Proof. It is sufficient to prove the theorem for a function $\varphi \in C^+$. Let $h_n(x, y)$ be a monotone increasing sequence of step functions which converges to the function $\varphi(x, y)$ on a set of full plane measure. We form the functions

$$g_n(y) = I_x h_n(x, y).$$

The functions $g_n(y)$ are at any rate determined for all y that do not correspond to lines of discontinuity of the $h_n(x, y)$. The sequence $g_n(y)$ increases monotonely as $n \to \infty$, and the integrals of the $g_n(y)$ have a common bound:

$$I_y\, g_n(y) = I_y[I_x h_n(x, y)] = I\, h_n(x, y) \nearrow I\varphi\dagger.$$

By the corollary to Beppo Levi's theorem, the functions $g_n(y)$ converge for almost all y to some summable function $g(y)$, and

$$I_y\, g(y) = \lim_{n\to\infty} I_y\, g_n(y) = I\varphi.$$

Let y be a point of the set E_y of full measure on which the function $g(y)$ is defined and finite. With this value of y, the functions $h_n(x, y)$ as functions of x form a monotone increasing sequence, and their integrals are bounded:

$$I_x\, h_n(x, y) = g_n(y) \nearrow g(y).$$

Hence, in virtue of the same theorem of Beppo Levi, the functions $h_n(x, y)$ tend for almost all x, i.e. on some set E_{yx} of full measure with respect to x, to some function $\varphi_0(x, y)$, so that

$$\lim_{n\to\infty} I_x\, h_n(x, y) = g(y) = I_x\, \varphi_0(x, y).$$

We consequently obtain the result

$$I\varphi = I_y\, g(y) = I_y\{I_x\, \varphi_0(x, y)\}. \tag{1}$$

† The formula for repeated integration is of course valid for step functions.

The function $\varphi_0(x, y)$ coincides with the function $\varphi(x, y)$ on the intersection of the sets E and E_{yx}, since the limit of the sequence $h_n(x, y)$ is $\varphi(x, y)$ on the first of these, and $\varphi_0(x, y)$ on the second.

In particular, if $h(x, y)$ converges *everywhere* to the function $\varphi(x, y)$ then on the set E_{yx}, $\varphi_0(x, y) = \varphi(x, y)$, and the assertion of the theorem holds.

The last condition is realised, for example, if the functions $h_{n+1} - h_n = e_n$ are the characteristic functions of non-intersecting rectangles D_n $(n = 1, 2, \ldots)$. In this case the integral $I\varphi$ is the (countable) sum of the areas of these rectangles. The function $g_n(y) = I_x h_n(x, y)$ is the sum of the lengths of the intervals formed by the intersection of the corresponding horizontal line with the first n-rectangles; the function $g(y) = \lim g_n(y)$ is the n (countable) sum of the lengths of all these intervals.

Formula (1) permits us now to establish an important property for any plane set Z of full measure. We claim that the *set Z is intersected by almost all horizontal lines (i.e. excepting a set of linear measure zero in y) in a set of full linear measure in x.*

For let us form, for a given $\varepsilon > 0$, a covering of the complement CZ of the set Z by a system of non-intersecting rectangles $D_1, \ldots, D_n, \ldots$ of overall area $< \varepsilon$, and let $\varphi_\varepsilon(x, y)$ be the characteristic function of this system. By what we have proved,

$$I\varphi_\varepsilon(x, y) = I_y\{I_x\varphi_\varepsilon(x, y)\} = I_y g_\varepsilon(y).$$

We let ε approach zero; then the functions $\varphi_\varepsilon(x, y)$ can be regarded as forming a decreasing sequence; consequently the functions $g_\varepsilon(y) = I_x \varphi_\varepsilon(x, y)$ also form a decreasing sequence. Since $I\varphi_\varepsilon \to 0$, we have $I_y g_\varepsilon(y) \to 0$, and it follows that for almost all y, $g_\varepsilon(y) \to 0$ too.

We denote by E_0 the set of all y for which $g_\varepsilon(y) \to 0$; this set has full measure. But as we have shown already, each value of the function $g_\varepsilon(y)$ is the sum of the lengths of the intervals formed by the intersection of the corresponding horizontal line $y = y_0$ with the rectangles $D_1, \ldots, D_n, \ldots$, which cover the set CZ. We see that the intersection of the set CZ with the line $y = y_0 \in E_0$ can be covered by a (countable) system of intervals of arbitrarily small overall length. It therefore has measure zero, and the intersection of the set Z with the same line has full measure, as required.

We turn now to the general case.

The sequence h_n converges to φ on a set E of full (plane) measure; on the y-axis, a set E'_y of full (linear) measure can be found such that for $y_0 \in E'_y$ the functions $h_n(x, y_0)$ converge to $\varphi(x, y_0)$ on a set E'_{yx} of full measure in x. To construct the set E_y on which the sequence $g_n(y)$ converges, we can exclude in advance points not contained in E'_y (they form a set of measure zero), and we can therefore suppose that $E_y \subset E'_y$. Further, for a fixed $y_0 \in E_y$, in constructing the set $E_{y_0 x}$ on which the sequence $h_n(x, y_0)$ converges to the function $\varphi_0(x, y_0)$, we can exclude points not contained in $E'_{y_0 x}$ and therefore assume that $E_{y_0 x} \subset E'_{y_0 x}$. But on the set $E'_{y_0 x}$ we have $h_n(x, y_0) \nearrow \varphi(x, y_0)$. Hence $\varphi(x, y_0) = \varphi_0(x, y_0)$ everywhere on $E_{y_0 x}$. Thus we find that for $y \in E_y$ the function $\varphi(x, y)$ is summable with respect to x, its integral $I_x \varphi(x, y) = g(y)$ is a summable function of y, and

$$I_v\{I_x \varphi(x, y)\} = I\varphi.$$

The proof of Fubini's theorem is therefore complete.

Note. It does not, in general, follow from the existence of the integrals

$$I_y\{I_x \varphi(x, y)\}, \quad I_x\{I_y \varphi(x, y)\} \tag{2}$$

either that they are equal or that the function $\varphi(x, y)$ is summable on the set E. But *if the function $\varphi(x, y)$ is measurable and non-negative, then the existence of one of the integrals* (2) *entails its summability on E, and the equation*

$$I\varphi = I_y\{I_x \varphi\} = I_x\{I_y \varphi\}. \tag{3}$$

For let there exist, say, $I_y\{I_x \varphi(x, y)\} = A$ and let $\varphi_n(x, y) = \min\{\varphi(x, y), n\}$. The function $\varphi(x, y)$ is measurable, bounded, and consequently summable on E; by Fubini's theorem

$$I\varphi_n = I_y\{I_x \varphi_n\} \leqq A.$$

As n increases, the functions φ_n increase monotonely and tend to the function φ; since $I\varphi_n \leqq A$, the function φ is summable, by Beppo Levi's theorem. But then we can apply Fubini's theorem, and we get (3), as required.

Example. If $\varphi(x)$ is summable for $a \leqq x \leqq b$ and $\psi(y)$ is summable for $a \leqq y \leqq b$, then $\varphi(x)\psi(y)$ is summable on the set $E = (a \leqq x \leqq b, a \leqq y \leqq b)$ and

$$I(\varphi\psi) = I\varphi\, I\psi. \tag{4}$$

For the function $|\varphi|\,|\psi|$ is evidently measurable and the integral $I_y\{I_x(|\varphi|\,|\psi|)\} = I_y\{(|\psi|\,I_x(|\varphi|)\} = I_y(|\psi|)\,I_x(|\varphi|)$ exists; the function $|\varphi\,\psi|$ is therefore summable on E, and together with it $\varphi\,\psi$. Applying (3), we get (4), as required.

3. The space L_p. Let p be a positive number. L_p is defined as the class of all functions $f(x)$, defined on a region G, for which

$$I(|f|^p) = \int_G |f|^p\,dx < \infty.$$

For any $p > 0$, L_p is a linear space. For clearly, for any α, αf belongs to L_p together with f. Further, if $f \in L_p$, $g \in L_p$, then $f + g \in L_p$, since

$$|f+g|^p \leqq (|f| + |g|)^p \leqq [2 \sup(|f|, |g|)]^p = 2^p[\sup(|f|, |g|)]^p.$$

We shall show that for $p \geqq 1$ this linear space can be converted into a complete normed linear space by introducing the norm

$$\|f\|_p = [I(|f|^p)]^{\frac{1}{p}}. \tag{1'}$$

We have already substantiated this claim for the case $p = 1$ (Section 3). We therefore restrict ourselves to the case $p > 1$. It is evident that the norm (1′) satisfies the condition $\|f\|_p \geqq 0$, and $\|f\|_p = 0$ implies that $f(x) = 0$ almost everywhere; clearly also $\|\alpha f\|_p = |\alpha|\,\|f\|\,p$ for any real α. The triangle inequality presents more difficulty. We consider first the following lemma:

LEMMA. *If $\eta = \omega(\xi)$ is an increasing continuous function, $\omega(0) = 0$, and $\xi = \mu(\eta)$ the inverse function (obviously also continuous and increasing), then for any $x > 0$, $y > 0$,*

$$xy \leqq \int_0^x \omega(\xi)\,d\xi + \int_0^y \mu(\eta)\,d\eta.$$

The proof is obtained quickly by considering Fig. 10.

To be particular, let us put $\omega(\xi) = \xi^{p-1}\,(p > 1)$, $\mu(\eta) = p\,\frac{1}{p-1}$ then we get

$$xy \leqq \frac{x^p}{p} + \frac{y^q}{q}. \tag{1}$$

The number g here is defined by the equation

$$q = \frac{1}{p-1} + 1 = \frac{p}{p-1} = \frac{1}{1 - 1/p}, \tag{2}$$

so that

$$\frac{1}{p} + \frac{1}{q} = 1.$$

Applying inequality (1) to functions $f(x)$, $g(x)$, we get

$$|f(x)|\,|g(x)| \leqq \frac{|f(x)|^p}{p} + \frac{|g(x)|^q}{q}. \tag{3}$$

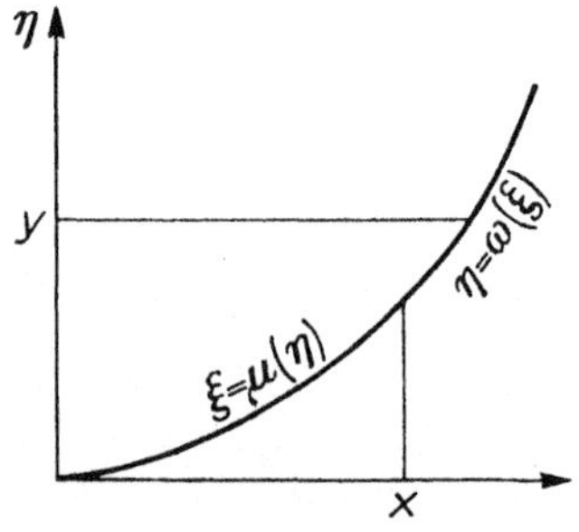

FIG. 10

Let us suppose for the moment that $\|f\|_p = I^{1/p}(|f|^p) = 1$, $\|g\|_q = I^{1/q}(|g|^q) = 1$. Then inequality (3) shows that

$$I(|f\,g|) \leqq \frac{1}{p} + \frac{1}{q} = 1.$$

Now if $f \in L_p$, $g \in L_q$ are arbitrary functions, then for functions $f_0 = \frac{f}{\|f\|_p}$, $g_0 \frac{g}{\|g\|_q}$ the conditions $\|f_0\|_p = \|g_0\|_q = 1$ are satisfied; it follows that

$$I(|f_0\,g_0|) = \frac{I(|f\,g|)}{\|f\|_p\|g\|_q} \leqq 1,$$

and hence for any $f \in L_p$, $g \in L_q$,

$$I(|f\,g|) \leqq \|f\|_p\,\|g\|_q$$

(Hölder's inequality).

We suppose now that $f \in L_p$, $g \in L_p$, and obtain a bound for $\|f + g\|_p$. We have

$$\begin{aligned} |f + g|^p &\leqq (|f| + |g|)^p \\ &= |f|\,(|f| + |g|)^{p-1} + |g|\,(|f| + |g|)^{p-1} \end{aligned} \tag{4}$$

The number $p - 1$ is equal to $\frac{p}{q}$ [as is evident from (2)]; hence

$$(|f| + |g|)^{p-1} = (|f| + |g|)^{p/q} \in L_q,$$

and consequently

$$\|(|f| + |g|)^{p-1}\|_q = I^{1/q}[(|f| + |g|)^p].$$

Integrating (4) and applying Hölder's inequality, we get

$$I(|f| + |g|)^p \leqq \|f\|_p \|(|f| + |g|)^{p-1}\|_q + \|g\|_p \|(|f| + |g|)^{p-1}\|_q$$
$$= (\|f\|_p + \|g\|_p) I^{1/q}[(|f| + |g|)^p].$$

Dividing by $I^{1/q}[(|f| + |g|)^p]$, and remembering $1 - \frac{1}{q} = \frac{1}{p}$, we find:

$$\|f + g\|_p = I^{1/p}(|f + g|)^p \leqq I^{1/p}(|f| + |g|)^p \leqq \|f\|_p + \|g\|_p.$$

Thus, for $p > 1$, the triangle inequality is satisfied.

It remains for us to verify that the space L_p is complete. This is achieved in an analogous way to that for the completeness of the space $L = L_1$. If $\varphi_1, \varphi_2, \ldots$ is a fundamental sequence in the space L_p, a subsequence $\varphi_{n_k} = \psi_k$ can be extracted from it so as to satisfy the conditions

$$\|\psi_{k+1} - \psi_k\|_p < \frac{1}{2^k}.$$

If the region G is bounded, then by Hölder's inequality

$$I(|\psi_{k+1} - \psi|) \leqq \|1\|_q \|\psi_{k+1} - \psi_k\|_p \leqq C \cdot 2^{-k},$$

from which follows, just as for $p = 1$, the convergence (almost everywhere) of the series

$$\sum_{k=1}^{\infty} (\psi_{k+1} - \psi_k)$$

and the existence (almost everywhere) of a limit function $\varphi = \lim_{k \to \infty} \psi_k$. In the case of an infinite region G, integration over an arbitrary bounded subregion also ensures the existence of a limit function almost everywhere on G.

Since $|\varphi|^p = \lim |\psi_k|^p$ and the $I(|\psi_k|^p)$ are bounded by a number M, say, it follows that $\varphi \in L_p$ and therefore $I(|\varphi|_p^k) \leqq M$.

Just as in Section 3, we get a bound

$$I(|\varphi_n - \varphi_n|^p) \leqq \sup_{k \geqq n} I(|\varphi_k - \varphi_n|^p) \leqq \varepsilon$$

for sufficiently large n, whence it follows that $\|\varphi - \varphi_n\|_p \to 0$, completing the proof.

It is proved exactly as in Section 3 that the totality of continuous functions on the closed interval $[a, b]$ is dense in the space L_p with respect to the metric of that space.

Concluding remark

The first work done by A. Lebesgue (French mathematician, 1875–1941) on the theory of measure and of the integral dates back to 1902. Originally designed to solve problems in the theory of functions of a real variable (the original problem concerned the convergence of trigonometric series), Lebesgue theory very soon exceeded these narrow confines. The Fisher-Riesz theorem on the completeness of the space of summable functions (1907) and the introduction by Riesz of the space L_p ultimately rendered the Lebesgue integral indispensable in general contemporary problems in analysis and mathematical physics. F. Riesz (Hungarian mathematician, 1880–1956) also proposed the scheme of construction for integral theory adopted in our exposition. Our recommended literature is: H. L. Lebesgue: *Leçons sur l'integration et la Recherche des functions primitives*, Gauthier-Villars, Paris (1950); F. Riesz and B. Szekefalvi-Nagy, *Functional Analysis*, London (1956).

CHAPTER V

GEOMETRY OF HILBERT SPACE

1. Basic Definitions and Examples

1. A linear space H (under multiplication by real numbers) is said to be a (real) *Hilbert space* if: (1) there is a rule which maps each pair of points (vectors) of H onto a real number, called the scalar product of the vectors x, y and denoted by (x, y); (2) this rule satisfies the following requirements:

(a) $(y, x) = (x, y)$ (transposition law);

(b) $(x, y + z) = (x, y) + (x, z)$ (distributive law);

(c) $(\lambda x, y) = \lambda(x, y)$ for any real λ;

(d) $(x, x) > 0$ for $x \neq 0$ and $(x, x) = 0$ for $x = 0$.

From axioms (a)–(c) we easily obtain the general formula

$$\left(\sum_{i=1}^{k} \alpha_i x_i, \quad \sum_{j=1}^{m} \beta_j y_j\right) = \sum_{i=1}^{k} \sum_{j=1}^{m} \alpha_i \beta_j (x_i, y_j), \tag{1}$$

which holds for arbitrary vectors $x_1, \ldots, x_k, y_1, \ldots, y_m$ and arbitrary real numbers $\alpha_1, \ldots, \alpha_k, \beta_1, \ldots, \beta_m$.

Examples 1. In the n-dimensional space R_n, the elements of which are arrays $x = (\xi_1, \ldots, \xi_n)$ of real numbers, we introduce a scalar product for vectors $x = (\xi_1, \ldots, \xi_n)$, $y = (\eta_1, \ldots, \eta_n)$ by the formula

$$(x, y) = \xi_1 \eta_1 + \cdots + \xi_n \eta_n. \tag{2}$$

(This definition generalizes the well-known expression for the scalar product of vectors in three-dimensional space in terms of their coordinates relative to an orthogonal coordinate system). The reader will easily verify that requirements (a)–(d) are satisfied. A finite-dimensional (real) Hilbert space is usually called a *Euclidean* space.

2. The following example represents the infinite-dimensional analogue of the foregoing.

The space l_2. A sequence of numbers $x = (\xi_1, \ldots, \xi_n, \ldots)$ is an element of this space if the sum of their squares converges:

$$\sum_{n=1}^{\infty} \xi_n^2 < \infty.$$

Linear operations are defined in the natural way:

$$(\xi_1, \ldots, \xi_n, \ldots) + (\eta_1, \ldots, \eta_n, \ldots) = (\xi_1 + \eta_1, \ldots, \xi_n + \eta_n, \ldots),$$
$$\alpha(\xi_1, \ldots, \xi_n, \ldots) = (\alpha\,\xi_1, \ldots, \alpha\,\xi_n, \ldots).$$

The scalar product of vectors $x = (\xi_1, \ldots, \xi n, \ldots)$, $y = (\eta_1, \ldots, \eta_n, \ldots)$ is given by the formula

$$(x, y) = \sum_{n=1}^{\infty} \xi_n \eta_n. \tag{3}$$

We have to verify the correctness of these definitions. In the first place it follows from the elementary inequality

$$\xi_n \eta_n \leqq \tfrac{1}{2}(\xi_n^2 + \eta_n^2)$$

that the series (3) always converges for $x \in l_2$, $y \in l_2$. Further, the equations

$$\sum_{k=n}^{n+m} (\alpha\,\xi_k)^2 = \alpha^2 \sum_{k=n}^{n+m} \xi_k^2,$$

$$\sum_{k=n}^{n+m} (\xi_k + \eta_k)^2 = \sum_{k=n}^{n+m} \xi_k^2 + 2 \sum_{k=n}^{n+m} \xi_k \eta_k + \sum_{k=n}^{n+m} \eta_k^2$$

show that for $x \in l_2$, $y \in l_2$, the series

$$\sum_{k=1}^{\infty} (\alpha\,\xi_k)^2, \quad \sum_{k=1}^{\infty} (\xi_k + \eta_k)^2,$$

converge, and hence the operations we have defined for the addition of vectors and their multiplication by scalars are feasible in the space l_2.

It remains for us to verify that axioms (a)–(d) are satisfied; but a cursory glance suffices to convince us of this.

3. As we have seen (Chapter IV, Section 5), the space $L_2(a, b)$ of functions $\varphi(x)$ with a summable square on the closed interval $a \leqq x \leqq b$ is a linear space; we introduce a scalar product on it by the formula

$$(\varphi, \psi) = \int_a^b \varphi(x)\,\psi(x)\,\mathrm{d}x.$$

The fulfilment of the axioms for a scalar product follows here from the properties of the Lebesgue integral. We observe that the integrability of the product $\varphi \psi$ follows from the inequality

$$|\varphi(x)\,\psi(x)| \leqq \tfrac{1}{2}(|\varphi(x)|^2 + |\psi(x)|^2).$$

2. *Isomorphism of Hilbert spaces*

The definition of an isomorphism between Hilbert spaces is formulated in analogy with the definitions of equivalence of sets, isometry of metric spaces, and isomorphism of linear spaces, familiar to us from the first and second chapters. Two Hilbert spaces H', H'' are said to be *isomorphic* if there exists between them a one-one correspondence with the following properties:

(1) If vectors x'', y'' in the space H'' are the images of x', y' in H', then the vector $x'' + y'' \in H''$ is the image of $x' + y' \in H'$ and the vector $\alpha\, x'' \in H''$ is the image of $\alpha\, x' \in H'$ for any real α.

(2) Under the same hypotheses, the numbers (x', y'), (x'', y'') are equal.

We shall show subsequently (Section 2, art. 3) that any two finite-dimensional Hilbert spaces of equal dimensionality are isomorphic to one another (and therefore isomorphic to the space R_n of example (1). The spaces $L_2(a, b)$ and l_2 are also in fact isomorphic to one another (Section 2, art. 6)).

3. *Length of a Vector and Angle between Vectors*

The scalar product permits us to introduce the concepts of the length (norm) of a vector and of the angle between vectors in Hilbert space by the formulae

$$\|x\| = +\sqrt{(x, x)}\,, \tag{1}$$

$$\cos\left(\widehat{x, y}\right) = \frac{(x, y)}{\|x\|\,\|y\|}\,; \tag{2}$$

these definitions agree with the usual formulae of analytic geometry. We shall often simply write $|x|$ instead of $\|x\|$.

Let us consider these definitions in general Hilbert space. We shall prove that, whatever the vectors x, y, the ratio on the right-hand side of (2) cannot exceed unity in absolute value.

For the proof of this assertion, we consider the vector $\lambda x - y$, where λ is a real number. In virtue of axiom (d) we have for any λ

$$(\lambda x - y, \lambda x - y) \geqq 0. \tag{3}$$

Using formula (1) of art. 1, we can write this inequality in the form

$$\lambda^2(x, x) - 2\lambda(x, y) + (y, y) \geqq 0. \tag{4}$$

The left-hand side of the inequality is a quadratic trinomial in λ with constant coefficients. It cannot have distinct real zeros, since then it would not preserve its sign for all values of λ. Its discriminant $(x, y)^2 - (x, x)(y, y)$ cannot, therefore, be positive. Hence $(x, y)^2 \leqq (x, x)(y, y)$ and taking the square-root, we get

$$|(x, y)| \leqq |x|\,|y|, \tag{5}$$

as required.

We determine when the equality sign in (5) is possible. If

$$|(x, y)| = |x|\,|y|,$$

then the discriminant of the quadratic trinomial (4) vanishes, and the trinomial therefore has one real zero λ_0. Thus we get

$$\lambda_0^2(x, x) - 2\lambda_0(x, y) + (y, y) = (\lambda_0 x - y, \lambda_0 x - y) = 0,$$

and in virtue of axiom (d) we find that $\lambda_0 x - y = 0$, or $y = \lambda_0 x$. Our result is capable of a geometrical formulation: *if the scalar product of two vectors is equal in absolute value to the product of their lengths, then the vectors are collinear.*

The inequality (5) is termed the *Cauchy–Bunyakovsky inequality.*

Examples 1. In the Euclidean space R_n the Cauchy–Bunyakovsky inequality has the form

$$\left|\sum_{j=1}^{n} \xi_j \eta_j\right| \leqq \sqrt{\sum_{j=1}^{n} \xi_j^2}\,\sqrt{\sum_{j=1}^{n} \eta_j}\,; \tag{6}$$

it holds for any pair of vectors $x = (\xi_1, \ldots, \xi_n)$, $y = (\eta_1, \ldots, \eta_n)$, or what is the same thing, for any two systems of real numbers $\xi_1, \xi_2, \ldots, \xi_n$ and $\eta_1, \eta_2, \ldots, \eta_n$. (This inequality was discovered by Cauchy in 1821.)

2. In the space $L_2(a, b)$ the Cauchy–Bunyakovsky inequality becomes

$$\left|\int_a^b \varphi(x)\,\psi(x)\,\mathrm{d}x\right| \leqq \sqrt{\int_a^b \varphi^2(x)\,\mathrm{d}x}\,\sqrt{\int_a^b \psi^2(x)\,\mathrm{d}x}\,; \tag{7}$$

it holds for any pair of square-summable functions $\varphi(x)$, $\psi(x)$. (This inequality for Riemann-integrable functions was discovered by V. Ya. Bunyakovsky in 1859.)

We now proceed to investigate the properties of the norm.

It follows from axiom (d) that each vector x of a Hilbert space H has a norm: any non-zero vector has a positive norm, and the norm of the null vector is zero. The equation

$$|\alpha x| = \sqrt{(\alpha x, \alpha x)} = \sqrt{\alpha^2(x, x)} = |a|\sqrt{(x, x)} = |\alpha|\,|x| \qquad (8)$$

shows that *the modulus of a scalar multiplier of a vector can be carried through the norm sign*. Finally, the norm satisfies the triangle inequality

$$|x + y| \leqq |x| + |y|. \qquad (9)$$

For by the Cauchy-Bunyakovsky inequality,

$$\begin{aligned} |x + y|^2 &= (x + y, x + y) = (x, x) + 2(x, y) + (y, y) \\ &\leqq (x, x) + 2|x|\,|y| + (y, y) = (|x| + |y|)^2, \end{aligned}$$

and (9) follows on taking the square root.

Thus the norm in the space H satisfies the axioms for a normed linear space (Chapter II, Section 8). *All the metric concepts and properties connected with the existence of a norm are valid in Hilbert space.* But since Hilbert space is a very particular instance of a normed space, it is only reasonable to expect that the norm in Hilbert space will possess properties peculiar to Hilbert spaces. One such property is given by the following lemma:

Parallelogramm lemma. *For any two vectors x, y of a Hilbert space H,*

$$|x + y|^2 + |x - y|^2 = 2|x|^2 + 2|y|^2$$

(the sum of the squares of the diagonals of a parallelogram is equal to the sum of the squares of its sides).

The proof follows from the simple computation:

$$\begin{aligned} |x + y|^2 + |x - y|^2 &= (x + y, x + y) + (x - y, x - y) \\ &= 2(x, x) + 2(y, y) = 2|x|^2 + 2|y|^2. \end{aligned}$$

Problem. It is known that the parallelogram lemma holds in some normed space R for any pair of vectors x, y. Consider the function

$$(x, y) = \frac{1}{4}\,(|x + y|^2 - |x - y|^2) \quad (\text{so that } (x, x) = |x|^2)$$

and prove that it satisfies all the axioms of art. 1 for a scalar product.

Hint. To show that axiom (b) is satisfied, apply the parallelogram lemma to parallelograms constructed on the vectors: (1) $x + z, y$; (2) $x - z, y$; (3) $y + z, x$; (4) $y - z, x$. Verify axiom (c) first for integral λ, then for fractions, and then pass to the limit.

4. *The Limiting Process in Hilbert Space*

Together with the metric, there arise in Hilbert space concepts connected with passing to a limit. We say that a sequence $x_1, \ldots, x_n, \ldots$ of elements of a Hilbert space H converges to an element (or has an element x as a limit) if

$$\lim_{n \to \infty} |x - x_n| = 0.$$

A function $f(x)$, defined on the space H, is said to be *continuous* at a point x_0 if $x \to x_0$ implies $f(x) \to f(x_0)$. Continuous functions of two or more variables are defined similarly.

By way of example, we show that the scalar product (x, y) is a continuous function of both variables x, y, i.e. if $x_n \to x, y, y_n \to y$, then $(x_n, y_n) \to (x, y)$.

We put $y - y_n = h_n$, $x - x_n = k_n$; by hypothesis $h_n \to 0$, $k_n \to 0$. By the Cauchy–Bunyakovsky inequality

$$\begin{aligned} |(x, y) - (x_n, y_n)| &= |(x, y) - (x - k_n, y - h_n)| \\ &= |(x, h_n) + (y, k_n) - (k_n, h_n)| \\ &\leqq |x|\,|h_n| + |y|\,|k_n| + |k_n|\,|h_n|; \end{aligned}$$

as η increases, this quantity tends to zero. It follows that $(x, y) = \lim_{n \to \infty} (x_n, y_n)$, as required.

A sequence $\{x_n\} \in H$ is said to be *fundamental* if

$$\lim_{n, m \to \infty} |x_n - x_m| = 0.$$

The space H is said to be *complete* if every fundamental sequence has a limit in it.

All three spaces considered above: R_n, l_2, L_2 are complete.

The proof of the completeness of the space R_n is elementary (cf. Chapter II, Section 4). We proved that the space L_2 was complete in Chapter IV, Section 5, art. 3. We shall show now that the space l_2 is also complete.

Let $x_m = (\xi_1^{(m)}, \ldots, \xi_n^{(m)}, \ldots)$ $(m = 1, 2, \ldots)$ be a fundamental sequence of vectors of the space l_2. Since, by hypothesis, as m and p tend to infinity

$$\|x_m - x_p\|^2 = \sum_{n=1}^{\infty} |\xi_n^{(m)} - \xi_n^{(p)}|^2 \to 0,$$

in particular each term $|\xi_n^{(m)} - \xi_n^{(p)}|^2$ (for fixed n) will tend to zero as m and p increase without bound, and therefore by Cauchy's classical criterion the sequence of coordinates $\xi_n^{(m)}$ $(m = 1, 2, \ldots)$ will converge for each fixed n. We denote $\xi = \lim_{m \to \infty} \xi_n^{(m)}$ and show that the vector $x = (\xi_1, \ldots, \xi_n, \ldots)$ belongs to l_2.

Since the elements of a fundamental sequence have a common bound, we have

$$\|x_m\|^2 = \sum_{n=1}^{\infty} |\xi_n^{(m)}|^2 \leqq K,$$

where K is independent of m. Hence for any fixed N

$$\sum_{n=1}^{N} \xi_n^2 = \lim_{m \to \infty} \sum_{n=1}^{N} |\xi_n^{(m)}|^2 \leqq K;$$

and the convergence of the series $\sum_{n=1}^{\infty} \xi^2$ follows immediately.

It remains to show that $\|x - x_m\|$ tends to zero as $m \to \infty$. To do this, we take the limit as $p \to \infty$ in the inequality

$$\sum_{n=1}^{N} |\xi_n^{(m)} - \xi_n^{(p)}|^2 \leqq \varepsilon,$$

which for a given $\varepsilon > 0$ holds for sufficiently large m, p and any N.

As a result, we obtain the inequality

$$\sum_{n=1}^{N} |\xi_n^{(m)} - \xi_n|^2 \leqq \varepsilon,$$

and by passing to the limit as $N \to \infty$, we get

$$\|x_m - x\|^2 = \sum_{n=1}^{\infty} |\xi_n^{(m)} - \xi_n|^2 \leqq \varepsilon,$$

holding for all sufficiently large m, as required.

If the space H is incomplete, then as we saw in Chapter II, its completion can be constructed. The elements of the completion are all the possible classes of equivalent fundamental sequences.

It was proved in Chapter II that linear operations and a norm can be naturally introduced into the completion of a normed linear space so as to make it again a normed linear space. If H is a Hilbert space, a scalar product can also be naturally introduced into its completion. For let classes, X, Y be given, and let us take any sequences $\{x_n\} \in X$, $\{y_n\} \in Y$. We contend that there exists a limit of the expression (x_n, y_n) as $n \to \infty$.

For we have:

$$|(x_n, y_n) - (x_m, y_m)| = |(x_n, y_n - y_m)| + |(x_n - x_m, y_m)|$$
$$\leqq |x_n|\,|y_n - y_m| + |x_n - x_m|\,|y_m|.$$

The numbers $|x_n|$, $|y_n|$ are bounded since the corresponding sequences $\{x_n\}$, $\{y_n\}$ are fundamental in H. Hence the sequence (x_n, y_n) satisfies the usual Cauchy criterion and therefore has a limit, as asserted. This limit is independent of the choice of sequences $\{x_n\}$, $\{y_n\}$ in the classes X, Y. For if $\{x'_n\} \in X$, $\{y'_n\} \in Y$, then

$$|(x'_n, y'_n) - (x_n, y_n)| = |(x'_n - x_n, y'_n) + (x_n, y'_n - y_n)|$$
$$\leqq |x'_n - x_n|\,|y'_n| + |x_n|\,|y'_n - y_n| \to 0,$$

and since y'_n, x'_n are bounded, $\{x'_n\}$ is equivalent with $\{x_n\}$ and $\{y'_n\}$ with $\{y_n\}$.

It is easily verified that axioms (a)–(d) are satisfied, and we no longer dwell on the details of verification.

Thus the completion of a Hilbert space is again a Hilbert space.

Problems. 1. Let $d_1, d_2, \ldots$ be a fixed sequence of numbers such that $\sum_{n=1}^{\infty} \xi_n d_n$ converges for any sequence $(\xi_n) \in l_2$; show that $(d_n) \in l_2$.

Hint. Show first that $d_n \to 0$. Further, suppose that $\sum_{n=1}^{\infty} d_n^2 = \infty$, and consider the arrays $(d_{n_k}, \ldots, d_{n_{k+1}})$, for which $1 \leqq \sum_{n=n_k}^{n_{k+1}-1} d_n^2 \leqq 2$. Put $\xi_n = \frac{d_n}{k}$ for $n_k \leqq n < n_{k+1}$ and show that $(\xi_n) \in l_2$, but $\sum_{n=1}^{\infty} \xi_n d_n = \infty$.

2. Let $c_1, c_2, \ldots$ be a fixed sequence of positive numbers; show that the set M of all elements $(\xi_1, \ldots, \xi_n, \ldots) \in l_2$ with $|\xi_n| \leqq c_n^2$ $(n = 1, 2, \ldots)$ is locally compact if and only if $\sum_{n=1}^{\infty} c^2 < \infty$.

Hint. If $\sum_{n=N+1}^{\infty} c_n^2 < \varepsilon$, the set of elements $(\xi\ , \ldots, \xi_n, 0, 0, \ldots)$ with $|\xi_n| \leqq c_n$ forms a locally compact ε-net for M. For $\sum_{n=1}^{\infty} c_n^2 = \infty$ consider the arrays $(c_{n_k}, \ldots, c_{n_{k+1}})$ for which $\sum_{n \geq n_k}^{n_{n+1}-1} c_n^2 > 1$; the corresponding elements $(0, \ldots, 0\ C, \ldots, C_{n-1}\ 0, \ldots)$ belong to M and are mutually separated by distances greater than $k + 1$.

3. Prove that a set M of elements $(\xi_1, \ldots, \xi_n, \ldots) \in l_2$ is locally compact if and only if: (1) all the numbers $|\xi_n|$ are bounded by a fixed constant; (2) all the series $\sum_{n=1}^{\infty} \xi_n^2$ converge uniformly on M, i.e. for any $\varepsilon < 0$ a number N can be found such that $\sum_{N}^{\infty} \xi_n^2 < \varepsilon$ for all $(\xi_n) \in M$.

Hint. Use the method of problem 2.

2. Orthogonal Resolutions

1. *Orthogonality*

Vectors x, y are said to be *orthogonal* if $(x, y) = 0$. If $x \neq 0$, $y \neq 0$, then in accordance with the general definition of the angle between two vectors (Section 1, art. 3) this definition means that x and y form an angle of 90^0. The null vector is orthogonal to any vector $x \in H$.

In the space $L_2(a, b)$ the condition for orthogonality of vectors $\varphi(x), \psi(x)$ has the form

$$\int_a^b \varphi(x)\, \psi(x)\, \mathrm{d}x = 0 .$$

The reader will easily verify, on computing the corresponding integrals, that in the space $L_2(-\pi, \pi)$ any two vectors of the "trigonometric system"

$$1, \cos x, \sin x, \cos 2x, \sin 2x, \ldots, \cos nx, \sin nx, \ldots$$

are mutually orthogonal.

We note a few simple properties connected with the concept of orthogonality.

(1) If a vector x is orthogonal to the vectors $y_1, y_2, \ldots, y_k$, it is also orthogonal to any linear combination $\alpha_1 y_1 + \cdots + \alpha_k y_k$ of

these vectors. For we have

$$(\alpha_1 y_1 + \cdots + \alpha_k y_k, x) = \alpha_1 (y_1, x) + \cdots + \alpha_k (y_k, x) = 0.$$

(2) If the vectors $y_1, y_2, \ldots, y_n, \ldots$ are orthogonal to a vector x and $y = \lim_{n\to\infty} y_{n2}$ then the vector y is also orthogonal to x.

For in virtue of the continuity of the scalar product

$$(y, x) = \lim_{n\to\infty} (y_n, x) = 0.$$

It follows from properties (1) and (2) that the totality of vectors orthogonal to a vector x (or to an arbitrary fixed set X of vectors x) constitutes a closed subspace *the orthogonal complement of the vector x (set X)*.

(3) Pythagoras' theorem and its generalisation. Let the vectors x, y be orthogonal; then by analogy with elementary geometry the vector $x + y$ can be called the hypotenuse of the right-angled triangle constructed on the vectors x, y. Forming the scalar product of $x + y$ with itself and using the orthogonality of the vectors x, y, we get:

$$\begin{aligned}|x + y|^2 &= (x + y, x + y) = (x, x) + 2(x, y) + (y, y)\\ &= (x, x) + (y, y) = |x|^2 + |y|^2.\end{aligned}$$

We have thus proved Pythagoras' theorem in general Hilbert space: *the square of the hypotenuse is equal to the sum of the squares of the orthogonal sides*. It is not difficult to generalise this theorem to the case of any finite number of terms. Let the vectors $x_1, x_2, \ldots, x_k$ be mutually orthogonal and let $z = x_1 + x_2 + \cdots + x_k$; then

$$|z|^2 = (x_1 + \cdots + x_k, x_1 + \cdots + x_k) = |x_1|^2 + \cdots + |x_k|^2.$$

2. *Method for Orthogonalisation*

In order to obtain an orthogonal system of vectors, a method of orthogonalising a given non-orthogonal system is often employed. The procedure is as follows. Let there be given a sequence of vectors $x_1, x_2, \ldots, x_n, \ldots$ in which each finite subsystem $x_1, x_2, \ldots, x_n$ is linearly independent. We claim that a special choice of the coefficients a_{jk} in the formulae

$$\left.\begin{aligned} y_1 &= x_1, \\ y_2 &= x_2 + a_{21} x_1, \\ y_3 &= x_3 + a_{32} x_2 + a_{31} x_1, \\ &\dots\dots\dots\dots\dots\dots \\ y_n &= x_n + a_{n,\,n-1} x_{n-1} + \cdots + a_{n1} x_1, \\ &\dots\dots\dots\dots\dots\dots\dots\dots \end{aligned}\right\} \quad (1)$$

enables us to construct a system of non-zero, mutually orthogonal system of vectors $y_1, y_2, \ldots, y_n$. The formulae (1) with the appropriate coefficients a_{jk} are called the *formulae of orthogonalisation.*

The existence of a solution to this system, subject to the required conditions of orthogonality, is easily proved by induction. For let us assume that non-zero, mutually orthogonal vectors $y_1, \ldots, y_{n-1}$, satisfying the first $a - 1$ equations of the system (1) have been constructed; we shall show that a vector y_n can be found, satisfying the nth equation of the system (1) and orthogonal to the vectors $y_1, \ldots, y_{n-1}$. We shall look for the vector y_n as a linear combination of $x_1, x_2, \ldots, x_n$ in the following special form:

$$y_n = b_{n1} y_1 + \cdots + b_{n\ n-1,}\, y_{n-1} + x_n, \quad (2)$$

where $y_1, \ldots, y_{n-1}$ are the vectors already found, and $b_{n1}, \ldots, b_{n,n-1}$ are coefficients which have to be determined.

Multiplying equation (2) scalarly by y_k $(k < n)$ and using the supposed orthogonality of y_k to $y_1, \ldots, y_{k-1}, y_{k+1}, \ldots, y_{n-1}$, we get

$$(y_n, y_k) = (x_n, y_k) + b_{nk}(y_k, y_k).$$

Equating the right-hand side to zero, we get an equation in the coefficient b_{nk}, which is soluble since by hypothesis $(y_k, y_k) \neq 0$. When all the coefficients b_{nk} $(k = 1, 2, \ldots, n^{-1})$ have been found in this way, equation (2) allows the construction of the vector y_n. By construction it will be orthogonal to each of the vectors $y_1, y_2, \ldots, y_{n-1}$; it only remains for us to show that $y_n \neq 0$. For this we substitute the expressions for $y_1, \ldots, y_{n-1}$ of the first $n - 1$ equations (1) in the nth equation; we get a linear expression for y_n in $x_1, \ldots, x_{n-1}, x_n$ in which the coefficient of x_n is equal to 1. If y_n were zero, we should have a linear dependence between $x_1, \ldots, x_n$, which by supposition is impossible. It follows that

$y_n \neq 0$, as required; the correctness of the method of orthogonalisation is thus substantiated.

The orthogonal system of vectors $y_1, y_2, \ldots, y_n, \ldots$ obtained can be further "refined" by dividing each of the vectors y_n by its norm $|y_n|$; the resulting system of vectors $e_n = \frac{y_n}{|y_n|}$ is not only orthogonal, but also normed in such a way that each vector has a unit norm. Such systems of vectors are said to be *orthonormal*.

Problems. **1.** Let a system of vectors $x_1, x_2, \ldots, x_n, \ldots$ be given. Prove that there exists only one (discounting numerical multiples) system of vectors $y_1, \ldots, y_n, \ldots$ satisfying the conditions: (a) $(x_j, y_k) = 0$ for $j < k$; (b) for any n the vector y_n is a linear form in $x_1, x_2, \ldots, x_n, \ldots$

2. The polynomials obtained in orthogonalising the functions $1, x, x^2, \ldots$ over the space $L_2(-1, 1)$ are termed Legendre polynomials. Show that the nth Legendre polynomial has the form

$$p_n(x) = C_n[(x^2 - 1)^n]^{(n)}.$$

Hint. Use problem 1.

3. The functions obtained in orthogonalising the expressions $xe^{-x^2}, xe^{-x^2} \ldots, x^n e^{-x^2}, \ldots$ over the space $L_2(-\infty, \infty)$ are termed Hermite functions. Show that the nth Hermite function has the form

$$\mathcal{E}_n(x) = C_n \, e^{x^2/2}[e^{-x^2}]^{(n)}$$

4. The functions obtained in orthogonalising the expressions $e^{-x}, xe^{-x}, x^2 e^{-x}, \ldots$ over the space $L_2(0, \infty)$ are termed Laguerre functions. Show that the nth Laguerre function has the form

$$L_n(x) = C_n \, e^{-x/2} [x^n e^{-x}]^{(n)}$$

3. *The Isomorphism between Two n-dimensional Euclidean Spaces*

We can now prove a theorem on the isomorphism of any two Euclidean spaces of the same dimension. For the proof of this theorem we construct in each of the spaces R_n', R_n'' an orthonormal basis: $e_1', \ldots, e_n'$ in R_n' and $e_1'', \ldots, e_n''$ in R_n'', orthogonalising by the method of art. 2 an arbitrary linear independent system in each space. Further, we map an arbitrary vector $x' \in R_n'$ with a decomposition, say

$$x' = \sum_{j=1}^{n} \xi_j \, e'_j,$$

relative to the basis $e_1', \ldots, e_n'$ on to the vector $x'' \in R_n''$ which has the decomposition

$$x'' = \sum_{j=1}^{n} \xi_j \, e''_j.$$

with the same coefficients $\xi_1, \ldots, \xi_n$ relative to the basis $e_1'', \ldots, e_n''$. It is clear that this map is one–one and that it preserves linear operations. We shall show that the scalar products of corresponding vectors in R_n', R_n'' coincide. We put

$$y' = \sum_{k=1}^{n} \eta_k e_k',$$

$$y'' = \sum_{k=1}^{n} \eta_k e_k'';$$

then, since the bases $\{e_j'\}$, $\{e_j''\}$ are orthonormal, we have

$$(x', y') = \left(\sum_{j=1}^{n} \xi_j e'_j, \sum_{k=1}^{n} \eta_k e_k'\right) = \sum_{j=1}^{n} \xi_j \eta_j,$$

$$(x'', y'') = \left(\sum_{j=1}^{n} \xi_j e''_j, \sum_{k=1}^{n} \eta_k e_k''\right) = \sum_{j=1}^{n} \xi_j \eta_j = (x', y'),$$

as asserted. This completes the proof.

4. *Orthonormal Systems in an Infinite-dimensional Hilbert Space H*

In an infinite-dimensional space, infinite orthonormal systems are known to exist. We shall call an orthogonal system $e_1, e_2, \ldots, e_n$, *complete* in the space H, if there does not exist a non–zero vector in H orthogonal to each vector of the system. In other words, the system $e_1, e_2, \ldots, e_n, \ldots$ is complete if the conditions

$$x \in H, \quad (e_n, x) = 0 \quad (n = 1, 2, \ldots)$$

imply that $x = 0$.

We shall show that a *complete* orthonormal system in a *complete* Hilbert space is a basis, in the sense that for each vector $f \in H$ there exists a decomposition into a convergent (in the norm) series

$$f = \sum_{j=1}^{\infty} c_j e_j, \tag{1}$$

with

$$|f|^2 = \sum_{j=1}^{\infty} c_j^2. \tag{2}$$

For the proof we first find expressions for the coefficients in the resolution (1), assuming that it exists. To this end we multiply both sides of (1) scalarly by the vector e_k. Since the scalar product is continuous, we get

$$(f, e_k) = \left(\sum_{j=1}^{\infty} c_j e_j, e_k\right) = \left(\lim_{n\to\infty} \sum_{j=1}^{n} c_j e_j, e_k\right)$$

$$= \lim_{n\to\infty} \left(\sum_{j=1}^{n} c_j e_j, e_k\right) = \lim_{n\to\infty} c_k = c_k .$$

We have the formula

$$c_k = (f, e_k). \tag{3}$$

The coefficients c_k defined by (3) are termed the *Fourier coefficients* of the vector f with respect to the system $\{e_k\}$. We observe that these numbers can be constructed directly from the vector f and the system $\{e_k\}$, although we do not yet know whether or not the decomposition (1) exists. They have a simple geometric significance since

$$(f, e_k) = |f|\,|e_k| \cos\left(\widehat{f, e_k}\right) = |f| \cos\left(\widehat{f, e_k}\right),$$

the coefficient c_k is the projection of the vector f on the direction of the vector e_k.

Let f be a prescribed vector and let $e_1, \ldots, e_n$ be a fixed finite orthonormal system (in general incomplete). We construct the vector

$$g = \sum_{k=1}^{n} (f, e_k)\, e_k .$$

This vector belongs to the subspace H_n generated by the vectors $e_1, \ldots, e_n$. We also define a vector h by the condition

$$f = g + h .$$

We claim that the vector h is orthogonal to each of the vectors $e_1, e_2, \ldots, e_n$ (and consequently to the whole subspace H_n). For we have

$$(h, e_j) = (f, e_j) - (g, e_j) = (f, e_j) - \left(\sum_{k=1}^{n} (f, e_k)\, e_k, e_j\right)$$

$$= (f, e_j) - (f, e_j) = 0 \quad (j = 1, 2, \ldots, n).$$

In geometrical language h is the perpendicular dropped from the end of the vector f onto the subspace H_n; and the vector g is the projection of f on this subspace. Hence by Pythagoras' theorem

$$|f|^2 = |g|^2 + |h|^2 = \sum_{k=1}^{n} (f, e_k)^2 + |h|^2 \geq \sum_{k=1}^{n} (f, e_k)^2.$$

The inequality

$$\sum_{k=1}^{n} (f, e_k)^2 \leqq |f|^2, \tag{4}$$

which holds for any vector f and any orthonormal system $e_1, \ldots, e_n$ is known as *Bessel's inequality*. If we are given an infinite orthonormal system $e_1, e_2, \ldots, e_n, \ldots$, then since *Bessel's inequality* holds for any n, we find by passing to the limit that

$$\sum_{k=1}^{\infty} (f, e_k)^2 \leqq |f|^2. \tag{5}$$

In other words, the squares of the Fourier coefficients of any vector f with respect to an orthonormal system $e_1, \ldots, e_n, \ldots$ always form a convergent series.

We shall also refer to inequality (5) as Bessel's inequality.

We now proceed to formulate and prove a fundamental theorem.

THEOREM 1. *Let a complete orthonormal system $e_1, e_2, \ldots, e_n, \ldots$ be chosen in a complete Hilbert space H. Then for any vector f there exists a decomposition*

$$f = \sum_{j=1}^{\infty} (f, e_j)\, e_j, \tag{6}$$

where

$$|f|^2 = \sum_{j=1}^{\infty} (f, e_j)^2. \tag{7}$$

The last equality, which represents an infinite-dimensional generalisation of Pythagoras' theorem, is generally referred to as *Parseval's formula*.

Proof. For the sake of brevity we put $(f, e_n) = a_n$.

Let s_p be the sum of p terms of the series (6) and let $q > p$. Then

$$|s_q - s_p|^2 = \left| \sum_{p+1}^{q} a_j e_j \right|^2 = \sum_{p+1}^{q} a_j^2 .$$

As $p \to \infty$, this quantity tends to zero in consequence of the convergence of the series of numbers a_j^2. The sums s_p therefore constitute a fundamental sequence. Since the space H is supposed to be complete, the sums s_p have some limit $s \in H$ as $p \to \infty$. We shall show that $s = f$. To do this, we observe that for fixed k and $p > k$,

$$(s, e_k) = \lim_{p\to\infty} (s_p, e_k) = \lim_{p\to\infty} \left(\sum_{j=1}^{p} a_j e_j, e_k \right) = \lim_{p\to\infty} a_k = a_k = (f, e_k);$$

it follows that for any k

$$(f - s, e_k) = (f, e_k) - (s, e_k) = 0.$$

Since the system $\{e_k\}$ is supposed to be complete, equation (8) implies that $f = s$.

Thus

$$f = \lim_{p\to\infty} s_p = \sum_{j=1}^{\infty} a_j e_j.$$

Further, in virtue of the continuity of the scalar product,

$$\|f\|^2 = (f, f) = \left(\lim_{p\to\infty} s_p, \lim_{p\to\infty} s_p\right) = \lim_{p\to\infty} (s_p, s_p)$$

$$= \lim_{p\to\infty} \sum_{k=1}^{p} a_k^2 = \sum_{k=1}^{\infty} a_k^2,$$

as required.

Note 1. If g is any other vector of the space, then by a similar evaluation of the scalar product (f, g) we obtain the formula

$$(f, g) = \sum_{k=1}^{\infty} (f, e_k)\,(g, e_k). \tag{9}$$

Note 2. If $a_1, \ldots, a_n, \ldots$ is any sequence of numbers for which $\sum_{k=1}^{\infty} a^2_k$ converges, then the series

$$\sum_{k=1}^{\infty} a_k e_k,$$

converges in H, as is evident from the beginning of our proof. If we denote this sum by f, then as above we shall have $a_n = (f, e_n)$ $(n = 1, 2, \ldots)$. Hence *any numerical sequence with a convergent series of squares is the sequence of Fourier coefficients of some vector of the space H.*

5. *Criterion for the Completeness of a System*

To apply the fundamental theorem of art. 4 we need a complete orthonormal system $e_1, e_2, \ldots, e_n, \ldots$ If it is not immediately apparent whether a given orthonormal system $e_1, e_2, \ldots, e_n, \ldots$ is complete in the space H, the following completeness criterion can be used:

THEOREM 2. *A given orthonormal system* $e_1, \ldots, e_n, \ldots$ *is complete in the complete space* H *if and only if the linear combinations of vectors of the system form a set everywhere dense in* H.

Proof. If the system

$$e_1, e_2, \ldots, e_n, \ldots$$

is complete, then in virtue of theorem 1 each vector $f \in H$ is the limit of linear combinations of vectors of the system $\{e_n\}$, so that the totality of linear combinations of vectors of the system forms a set everywhere dense in the space H. Conversely let it be known that the linear combinations of vectors of the system $\{e_n\}$ form an everywhere dense set in the space H and let the equations $(g, e_k) = 0$ $(k = 1, 2, \ldots)$ hold for some vector g. The orthogonal complement of the vector g contains all the vectors e_k, all their linear conbinations and the closure of the set of their linear combinations, i.e. the whole space H. In particular $(g, g) = 0$ and consequently $g = 0$. Thus the system $\{e_n\}$ is complete, as required.

Let us consider the case when the system $\{e_n\}$ is obtained by orthogonalising some system $\{e_n\}$. In accordance with the formulae for orthogonalisation each vector e_n is a linear combination of the vectors $x_1, \ldots, x_n$ and conversely each vector x_n is a linear combination of the vectors $e_1, \ldots, e_n$ [cf. equation (2) of art. 2]. Hence the totality of linear combinations of the vectors $\{e_n\}$ coincides with that of the vectors $\{x_n\}$. The completeness of the system $e_1, \ldots, e_n, \ldots$ can therefore be established by proving that the totality of linear combinations of the original vectors $\{x_n\}$ is dense in the space H.

Example. As we have seen, the vectors $1, \cos x, \sin x, \ldots$ form an orthogonal system in the space $L_2(-\pi, \pi)$. The linear combinations of vectors of this system—the trigonometric polynomials—form an everywhere dense set in $L_2(-\pi, \pi)$ (Chapter IV, Section 5, art. 3). By theorem 2, the system $1, \cos x, \sin x, \ldots$ is complete in the space $L_2(-\pi, \pi)$ and theorem 1 holds: *every function* $\varphi(x) \in L_2 (-\pi, \pi)$ *resolves into a series in the functions* $1, \cos x, \sin x, \ldots$ *convergent in the metric of the space* $L_2(-\pi, \pi)$. It must be remarked that the functions $1, \cos x, \sin x, \ldots$ are not normalised: it is easily calculated that $\|1\|^2 = 2\pi$, $\|\cos m x\|^2 = \|\sin m x\|^2 = \pi$. The normalised system is composed of the functions

$$\frac{1}{\sqrt{2\pi}}, \quad \frac{\cos x}{\sqrt{\pi}}, \quad \frac{\sin x}{\sqrt{\pi}}, \quad \frac{\cos 2x}{\sqrt{\pi}}, \quad \frac{\sin 2x}{\sqrt{\pi}}, \ldots$$

By formulae (1) and (3) of art. 4 the decomposition we are seeking has the form

$$\varphi(x) = \left(\varphi, \frac{1}{\sqrt{2\pi}}\right)\frac{1}{\sqrt{2\pi}} + \left(\varphi, \frac{\cos x}{\sqrt{\pi}}\right)\frac{\cos x}{\sqrt{\pi}} + \\ + \left(\varphi, \frac{\sin x}{\sqrt{\pi}}\right)\frac{\sin x}{\sqrt{\pi}} + \cdots$$

$$= \frac{1}{2\pi}\int_{-\pi}^{\pi} \varphi(x)\,dx + \frac{\cos x}{\pi}\int_{-\pi}^{\pi} \varphi(x)\cos x\,dx + \\ + \frac{\sin x}{\pi}\int_{-\pi}^{\pi}\varphi(x)\sin x\,dx + \cdots$$

With the notation

$$a_0 = \frac{1}{\pi}\int_{-\pi}^{\pi}\varphi(x)\,dx, \qquad a_n = \frac{1}{\pi}\int_{-\pi}^{\pi}\varphi(x)\cos n x\,dx,$$

$$b_n = \frac{1}{\pi}\int_{-\pi}^{\pi}\varphi(x)\sin n x\,dx,$$

we arrive at the usual Fourier series expansion:

$$\varphi(x) = \frac{a_0}{2} + \sum_{n=1}^{\infty} a_n \cos n x + b_n \sin n x.$$

We now know that this series converges in the metric of $L_2(-\pi, \pi)$ for any function $\varphi(x) \in L_2(-\pi, \pi)$.

Problems. 1. Prove that the system of Legendre polynomials (art. 2, problem 2) is complete in the space $L_2(-1, 1)$.

Hint. Use the Weierstrass theorem on the approximation of continuous functions by polynomials.

Note. The system of Hermite functions (art. 2, problem 3) in space $L_2(-\infty, \infty)$ and the system of Laguerre functions (art. 2, problem 4) in space $L_2(0, \infty)$ are also complete. Cf. Chapter VII, Section 3, art. 5.

2. Construct an orthogonal system of continuous functions $e_1(x), \ldots, e\ (x)$, in the space $L_2(a, b)$ which has the properties:

(a) $\int_a^b f(x)\, e_n(x)\,dx = 0$ $(n = 1, 2, \ldots)$ implies that $f(x) = 0$, for any *continuous* function $f(x)$;

(*b*) the linear combinations of the functions $e_1(x), \ldots, e_n(x), \ldots$ are non-dense in the space $L_2(a, b)$.

Hint. Orthogonalise a sequence of polynomials with rational coefficients which are orthogonal to a fixed discontinuous function, say the characteristic function of the open interval (a, c), $a < c < b$.

Note. The result of this problem shows that in order that the condition for the system given here to be complete should be satisfied, it is essential to presuppose the space H complete.

6. *The Isomorphism of all Countable-dimensional Hilbert Spaces*

We shall call a Hilbert space of *countable dimension* if there exists in it a complete countable orthonormal system. We shall show in this article that any two complete spaces of countable dimension are isomorphic.

By theorem 1 of art. 4 each vector f of a space H of countable dimension with a complete orthonormal system $e_1, e_2, \ldots, e_n, \ldots$ admits a decomposition

$$f = \sum_{k=1}^{\infty} c_k e_k,$$

where $c_k = (f, e_k)$ and the series of squares of the numbers c_k converges. On the other hand, as we observed in art. 4, if $c_1, c_2, \ldots, c_k, \ldots$ is an arbitrary sequence with a convergent series of squares, then there exists a vector $f \in H$ where development in the system $\{e_k\}$ has the numbers $\{c_k\}$ as coefficients.

Using this, we can establish a one–one correspondence between an arbitrary complete space H of countable dimension and the space l_2 (Section 1, art. 1, ex. 2), mapping a vector $f \in H$ on to the sequence $c_n = (f, e_n)$ of its Fourier coefficients with respect to a complete orthonormal system $e_1, e_2, \ldots, e_n, \ldots$ Such a map evidently preserves linear operations. It also preserves scalar products since, if

$$f = \sum_{k=1}^{\infty} a_k e_k,$$

$$g = \sum_{k=1}^{\infty} b_k e_k,$$

then by equation (9) of art. 4

$$(f, g) = \sum_{k=1}^{\infty} a_k b_k,$$

and it was by just such a formula that we defined the scalar product of elements $x = (a_k)$, $y = (b_k)$ of the space l_2. In short, we reaffirm that the space l_2 is complete. We see also that $L_2(a, b)$, *as a complete space f countable diemension, is isomorphic to the space* l_2. *Further, any two complete spaces of countable dimension are isomorphic to the space* l_2 *and are consequently isomorphic to one another.*

7. *Separability*

Finite-dimensional spaces and spaces of countable dimension can be defined as *separable* spaces, i.e. spaces which possess a countable everywhere dense subset.

For if a space H is finite—or countable–dimensional, it must contain a complete finite—or countable—orthonormal system $e_1, e_2, \ldots$ By theorem 1 of art. 4 the linear combinations of the vectors $e_1, e_2, \ldots$ form an everywhere dense set in H. If we confine ourselves to linear combinations with rational coefficients, we get a countable set of elements, dense as before in H; thus H is separable.

Conversely, let H be separable and let $f_1, f_2, \ldots, f_n, \ldots$ be a countable everywhere dense set in H. If we orthogonalise the elements $f_1, f_2, \ldots$, then by art. 5 we get a complete system, orthogonal in H, which will be at most countable by construction; thus H is of finite or countable dimension.

We observe *that in a separable space H each subspace $H' \subset H$ is also separable.* For the proof we fix n and k and select an element $\varphi_{nk} \in H'$, if one exists, which lies in a sphere of radius $1/k$ with centre at a point f_n of a countable set $f_1, f_2, \ldots$ everywhere dense in H. We claim that the resulting countable set of elements φ_{nk} $(k, n = 1, 2, \ldots)$ is dense in H'. For corresponding to any $\varphi \in H'$ and any k we can find some element f_n contained in the sphere of radius $1/k$, centre φ; we are then assured that the sphere of radius $1/k$, centre f_n contains elements of the set H'. Hence there exists an element $\varphi_{nk} \in H'$ for which

$$|\varphi - \varphi_{nk}| < |\varphi - f_n| + |f_n - \varphi_{nk}| < 2/k.$$

In particular, *any closed subspace H' of a separable Hilbert space H contains a complete orthogonal system $e_1, e_2, \ldots$*

Problem. Give a direct proof of the separability of the spaces

$$L_2(-\infty, \infty),\ L_2(0, \infty).$$

8. *Orthogonal Complements*

We turn to the orthogonalisation theorem of art. 2. In this theorem we started from a given system of vectors $x_1, x_2, \dots, x_n, \dots$ and constructed a new system $y_1, y_2, \dots, y_n, \dots$ which was such that the vector was a linear combination of the vectors $x_1, \dots, x_n$ and was orthogonal to the vectors $x_1, \dots, x_{n-1}$. The equation

$$x_n = y_n - a_{n1} x_1 - a_{n2} x_2 - \cdots - a_{n\,n-1} x_{n-1} = y_n + g_n$$

determines a decomposition of the vector x_n into the sum of two vectors g_n, y_n, the first of which lies in the subspace H_{n-1} generated by the vectors $x_1, \dots, x_{n-1}$, while the second is orthogonal to $x_1, \dots, x_{n-1}$ and consequently orthogonal to each vector of the subspace H_{n-1}. It is therefore natural to call the vector g_n the projection of the vector x_n on the subspace H_{n-1}, and y_n the perpendicular dropped from the end-point of x_n on to the subspace H_{n-1}. The process of orthogonalisation consists precisely in replacing the vector x_n by the perpendicular dropped from its end-point onto the subspace generated by the preceding vectors. As we saw in art. 2, the existence of this vector can be demonstrated by analysing the equations of orthoganalisation. The corresponding result essentially uses the finite dimensionality of the subspace H_{n-1}.

Now let L be an arbitrary subspace of a Hilbert space H and let f be a vector not contained in L. We put the question: can we in this case guarantee the existence of a decomposition

$$f = g + h,$$

where $g \in L$ and h is orthogonal to each vector in L (we shall say briefly: orthogonal to L)? It turns out that such a decomposition exists under certain conditions on H and L.

THEOREM. *If H is a complete Hilbert space and $L \subset H$ is a closed subspace, then for any $f \in H$ there exists a decomposition*

$$f = g + h, \tag{1}$$

where $g \in L$ and h is orthogonal to L; moreover g and h are determined uniquely by the vector f.

Proof. We denote $d = \inf |f - g|$. There are two possibilities: either $d = 0$ or $d > 0$. If $d = 0$, we can find a sequence $g_n \in L$ such that $|f - g_n| \to 0$ and therefore f is a limit point for the subspace L; but since L is closed, the point f itself must belong to L. The decomposition (1) is evidently realised for $g = f$, $h = 0$.

Now let $d > 0$. We consider a sequence $g_n \in L$ for which $|f - g_n| \to d$. Applying the parallelogram lemma (Section 1, art. 3) to the vectors $x = f - g_n$, $y = f - g_m$, we get

$$2|f - g_n|^2 + 2|f - g_m|^2 = 4\left|f - \frac{g_n + g_m}{2}\right|^2 + |g_n + g_m|^2.$$

As $n, m \to \infty$ the left-hand side tends to $4d^2$. The first term of the right-hand side is not less than $4d^2$ since $(g_n + g_m)/2 \in L$, $|f - (g_n + g_m)/2| \geqq d$. Hence the final term on the right-hand side tends to zero and it follows that the sequence g_n is fundamental. Since the space H is complete, the sequence g_n has a limit g as $n \to \infty$, which belongs to the subspace L in virtue of its being closed.

We shall show that the vector $h = f - g$ is orthogonal to L. For any $q \in L$ we have for any λ

$$\begin{aligned} \mathrm{d}^2 &\leqq |f - (g - \lambda q)|^2 = |h + \lambda q|^2 \\ &= (h + \lambda q, h + \lambda q) = \mathrm{d}^2 + 2\lambda(h, q) + \lambda^2|q|^2, \end{aligned}$$

and consequently

$$2\lambda(h, q) + \lambda^2|q|^2 \geqq 0,$$

but this can only be the case for arbitrary λ if $(h, q) = 0$.

Thus the decomposition (1) is established. We shall show now that the components g, h are uniquely determined. Let us suppose that

$$f = g + h = g' + h',$$

where g, g' belong to L and h, h' are orthogonal to L. Subtracting, we find:

$$0 = (g - g') + (h - h'),$$

where $g - g' \in L$ and $h - h'$ is orthogonal to L. By Pythagoras' theorem, $g - g' = h - h' = 0$, so that $g = g'$, $h = h'$, as required.

The totality of vectors h orthogonal to the subspace L (including the null vector) constitutes a closed subspace M, which is said to be the *orthogonal complement* of the subspace L.

We have proved that any closed subspace $L \in H$ has an orthogonal complement M and any vector $f \in H$ determines a decomposition

$$f = g + h, \quad g \in L, \quad h \in M.$$

Problem. A system of elements $f_1, f_2, \ldots, f_n, \ldots$ of a Hilbert space H is said to be *minimal* if the vector f_k is not contained in the closed subspace generated by the remaining vectors for any k; it is said to be *complete* if the closed subspace generated by all the vectors $f_1, f_2, \ldots, f_n, \ldots$ coincides with the whole space H.

Systems $f_1, f_2, \ldots, f_n, \ldots$ and $e_1, e_2, \ldots, e_n, \ldots$ are said to be *quadratically proximate* if

$$\sum_{k=1}^{\infty} |f_k - e_k|^2 < \infty.$$

Show that a minimal system $f_1, f_2, \ldots, f_n, \ldots$ which is quadratically proximate to a complete orthonormal system $e_1, e_2, \ldots, e_n, \ldots$ is complete (N. C. Barn).

Hint. We denote by $L_k^r(g)$ the closed subspace generated by vectors $g_k, \ldots, g_r$. Let $\sum_{N+1}^{\infty} |f_k - e_k| < 1$, and show that $L_{N+1}^{\infty}(e)$ does not contain a single vector orthogonal to the whole of $L_{N+1}^{\infty}(f)$. Then deduce that any vector $x \in H$ can be represented in the form $y + z$, where $g \in L_1^N(e)$, $z \in L_{N+1}^{\infty}(f)$.

The factor space $H/L_{N+1}^{\infty}(f)$ (cf. Chapter II, Section 8, art. 4) will then have dimension N at most. At the same time it contains the N linearly independent vector images of $f_1, \ldots, f_N$; they therefore constitute a basis in $H/L_{N+1}^{\infty}(f)$.

The vector z, which is orthogonal to all the f_k ($k = 1, 2, \ldots$), has as its image in $H/L_{N+1}^{\infty}(f)$ the class Z which is orthogonal to the images of the f_k ($k = 1, 2, \ldots, N$); hence $Z = 0$, $z \in L_{N+1}^{\infty}(f)$; consequently $z = 0$.

9. *The General Form of a Linear Functional in Hilbert Space*

We shall now apply the theorem on orthogonal complements to deduce the general form of a bounded linear functional on a complete Hilbert space.

Let x_0 be a fixed vector; for any x we put

$$f(x) = (x, x_0). \tag{1}$$

The functional $f(x)$ is evidently a linear functional on H. It is bounded on the unit sphere in virtue of the Cauchy–Bunyakovsky inequality

$$|f(x)| = |(x, x_0)| \leqq |x|\,|x_0|.$$

We shall show *that formula* (1) *gives the general form of a bounded linear functional on the space H*. We observe that a bounded linear functional is always continuous (Chapter II, Section 9).

Let $f(x)$ be a bounded linear functional on the complete Hilbert space H which does not vanish identically. We consider the subspace $H' \subset H$ defined by the equation $f(x) = 0$. Using the continuity of the functional $f(x)$, it is easy to verify that H' is closed. Let H'' be the orthogonal complement of the subspace H'. We shall show that H'' is *one-dimensional*. Let $z_1, z_2 \in H''$; then $y = f(z_1) z_2 - f(z_2) z_1$ also belongs to H''; but

$$f(y) = f(z_1) f(z_2) - f(z_2) f(z_1) = 0,$$

hence $y \in H'$. But $y \in H'$, $y \in H''$ implies that $(y, y) = 0$ and therefore that $y = 0$. Since $f(z_1) \neq 0$, $f(z_2) \neq 0$, the vectors z_1, z_2 must be linearly dependent. It follows that H'' is one-dimensional.

Now let $e \in H''$ be a normalised vector. Every vector $z \in H''$ is then of the form λe; but as we saw in art. 8 every vector $x \in H$ resolves into the sum $x = z + y$ $(z \in H'', y \in H')$. Since $z = \lambda e$, we have

$$x = \lambda e + y = (x, e) e + y.$$

It follows that

$$f(x) = (x, e) f(e) + f(y) = (x, e) f(e) = (x, f(e) e) = (x, x_0),$$

where $x_0 = f(e) e$ is a fixed vector of the space H.

Thus any *bounded linear functional on the complete Hilbert space H represents the scalar product of the vector x with a fixed vector x_0*. The vector x_0 is determined uniquely, moreover, since the identity $(x, x_0) \equiv (x, x_1)$ would imply that $(x, x_0 - x_1) = 0$ for any x; but then we should have $(x_0 - x_1, x_0 - x_1) = 0$, followed by $x_0 = x_1$.

3. Linear Operators

1. *Definition and Examples*

Let R be a linear space. An operator A, defined on the space R, is a function which maps each element $x \in R$ onto an element $y = A x$ of the same space.

The operator A is said to be linear if the following conditions are satisfied:

(I) $A(x + y) = Ax + Ay$ for any $x, y \in R$;

(II) $A(\alpha x) = \alpha Ax$ for any $x \in R$ and any scalar α.

From formulae (I) and (II) we easily obtain the more general formula

$$A(\alpha_1 x_1 + \cdots + \alpha_k x_k) = \alpha_1 A x_1 + \cdots + \alpha_k A x_k \tag{1}$$

for any $x_1, \ldots, x_k \in R$ and any real numbers $\alpha_1, \ldots, \alpha_k$.

Examples 1. The operator which maps each vector of a space onto the null vector is obviously linear. It is called the *null operator*.

2. The operator E which maps each vector x onto itself is evidently linear; it is called the *unit* or *identity operator*.

3. A linear operator A which maps each vector x onto λx (where λ is a fixed number) is called a *similarity operator*.

4. Let H be a Hilbert space of countable dimension and let $e_1, e_2, \ldots, e_n, \ldots$ be a complete orthonormal system in H. We fix a bounded sequence of real numbers $\lambda_1, \lambda_2, \ldots, \lambda_n, \ldots, |\lambda_n| \leqq C$, and for any vector

$$x = \sum_{j=1}^{\infty} \xi_j e_j \in H$$

we define

$$A x = \sum_{j=1}^{\infty} \lambda_j \xi_j e_j. \tag{2}$$

Since $\sum \lambda_j^2 \xi_j^2 \leqq c^2 \sum \xi_j^2 < \infty$, the operator $A x$ is defined by formula (2) over the whole space H. It is easily verified that this operator satisfies conditions (I), (II). Such an operator will be called an operator of *the normal form*. Each basis vector e_n is mapped by the operator A onto its magnification by the coefficient λ_n:

$$A e_n = \lambda_n e_n.$$

5. We fix a bounded measurable function on the closed interval $[a, b]$. A linear operator can be defined on the space $L_2(a, b)$ as multiplication by $\alpha(x)$:

$$A \varphi(x) = \alpha(x) \varphi(x).$$

6. *Fredholm's integral operator*. In the region

$$G = [a \leqq s \leqq b, a \leqq x \leqq b]$$

we fix a function $K(x, s)$, the square of which is integrable over the region

$$\int_a^b \int_a^b K^2(x, s)\, dx\, ds = K^2 < \infty.$$

We define an operator A on the space $L_2(a, b)$ by the formula

$$y(x) = A\,\varphi(x) = \int_a^b K(x, s)\,\varphi(s)\,\mathrm{d}s. \tag{3}$$

We shall show that formula (3) does indeed define an operator an $L_2(a, b)$. Since by Fubini's theorem (Chapter IV, Section 5, art. 2) the function $K^2(x, s)$ is summable on the region G, it is a summable function of s for almost all x, which means that $K(x, s)$, as a function of s, belongs to $L_2(a, b)$. The integral (3), which represents a scalar product of the functions $K(x, s)$, $\varphi(s)$, exists for any function $\varphi(x) \in L_2(a, b)$. By the same theorem of Fubini, the function

$$k^2(x) = \int_a^b K^2(x, s)\,\mathrm{d}s,$$

is summable with respect to x and

$$\int_a^b k^2(x)\,\mathrm{d}x = \int_a^b\int_a^b K^2(x, s)\,\mathrm{d}x\,\mathrm{d}s = K^2,$$

so that $k(x) \in L_2(a, b)$. Obtaining a bound now for the scalar product (3) with the help of the Cauchy–Bunyakovsky inequality, we find:

$$|y(x)|^2 \leq \int_a^b K^2(x, s)\,\mathrm{d}s \int_a^b \varphi^2(s)\,\mathrm{d}s = k^2(x)\,\|\varphi\|^2, \tag{4}$$

so that the function $y(x)$ belongs to the space $L_2(a, b)$ as required.

It is clear that Fredholm's operator (3) is a linear operator (i.e. conditions (I), (II) are satisfied).

Problems. 1. Show that the boundedness condition on the numbers λ_n cannot be weakened in defining an operator A of the normal form (example 4) if it is desired that the definition should extend over the whole space. In other words, if for some sequence λ_n an operator A is defined by formula (2) for any vector $x \in H$, then the numbers $|\lambda_n|$ have a common bound.

2. The boundedness condition on the function $\alpha(x)$ cannot be weakened in defining the operator A in example 5 if the definition is required to extend over the whole space.

2. *Operations with Linear Operators*

Various operations can be carried out on linear operators defined on a linear space R, resulting in the creation of new linear operators.

(1) *Addition of operators.* If linear operators A, B are given, the operator $C = A + B$ is defined by the formula

$$C\,x = (A + B)\,x = A\,x + B\,x.$$

(2) *Multiplication of an operator by a scalar.* If A is a linear operator and λ is a real number, the operator $B = \lambda\,A$ is defined by the formula

$$B\,x = (\lambda\,A)\,x = \lambda(A\,x).$$

(3) *Multiplication of operators.* If A, B are linear operators, the operator $C = A\,B$ is defined by the formula

$$C\,x = A\,B\,x = A\,(B\,x)$$

(i.e. first operating with B on the vector x, then operating with A on the result).

It is easily verified that new linear operators are obtained as a result of all these operations. The usual algebraic laws hold for the operations specified: the commutativity of addition, associativity, and distributivity (with the exception of commutativity for the multiplication of operators). The powers of an operator A are defined by the natural recurrence formulae

$$A^0 = E, \quad A^n = A \cdot A^{n-1} \quad (n = 1, 2, \ldots).$$

An operator B is said to be the inverse of the operator A if

$$A\,B = B\,A = E;$$

the inverse operator of A is denoted by A^{-1}. If operators C, D have inverses C^{-1}, D^{-1}, then $C\,D$ has the inverse $(C\,D)^{-1} = D^{-1}C^{-1}$.

3. *The Norm of a Linear Operator*

We shall suppose that a linear operator A has as its domain a normed linear space R.

The existence of a metric on the space R yields a mapping of each linear operator A onto a non-negative number $\|A\|$ said to be *the norm of the operator* A. We consider the real-valued function $F(x) = |A\,x|$, defined for vectors $x \in R$. The norm of the operator A is defined as the exact upper bound (possibly ∞) of the values of this function over the unit vectors x:

$$\|A\| = \sup |A\,x|. \qquad (1)$$

An operator A with a finite norm is said to be *bounded.*

In n-dimensional Euclidean space the quantity $\|A\|$ is finite for every linear operator A†. For the length of the vector $A\,x$ is evidently a continuous function of its coordinates $\eta_1, \eta_2, \ldots, \eta_n$; each of these coordinates is a linear function of the coordinates $\xi_1, \xi_2, \ldots, \xi_n$ of the vector x. It follows that $|A\,x|$ is a continuous function of the coordinates $\xi_1, \xi_2, \ldots, \xi_n$ of x. Since the spherical surface $|x| = 1$ is a bounded closed set in n-dimensional space, it follows in virtue of the results of Chapter II, Section 7 that the continuous function $|A\,x|$ is bounded on it. The number $|A|$ therefore exists, since every bounded set has an exact upper bound. Moreover, there exists a point x_0 on the surface $|x| = 1$ at which the function $|A\,x|$ attains its exact upper bound.

Examples 1. The norm of the null operator is evidently equal to zero. Conversely, $\|A\| = 0$ means that the operator A maps each normalised vector x_0 onto zero; but since each vector is collinear with some normalised vector x_0, $A\,x = 0$ for any x. Hence if $\|A\| = 0$, then $A = 0$.

2. The norm of the identity operator E is equal to unity since $|E\,x| = |x|$ for any vector x.

3. The norm of the similarity operator $A\,x = \lambda\,x$ is equal to $|\lambda|$.

4. The norm of an operator of the normal form in Hilbert space (art. 2, example 4)

$$A\,x = A\left(\sum_{j=1}^{\infty} \xi_j\, e_j\right) = \sum_{j=1}^{\infty} \xi_j\, \lambda_j\, e_j$$

is equal to the exact upper bound of the numbers $|\lambda_n|$. For if $C = \sup|\lambda_n|$ and $|x|^2 = \sum_{j=1}^{\infty} \xi_j^2 = 1$, we have

$$|A\,x|^2 = \sum_{j=1}^{\infty} \xi^2_j\, \lambda^2_j \leq C^2 \sum_{j=1}^{\infty} \xi^2_j = C^2,$$

and so $\|A\| \leq C$; on the other hand

$$\|A\| \geq \sup|A\,e_n| = \sup|\lambda_n\, e_n| = \sup|\lambda_n| = C;$$

our assertion follows from these inequalities.

5. The norm of an operator consisting in multiplication by a bounded function $g(x)$ over the space $L_2(a, b)$ (art. 1, example 5)

† The case $\|A\| = \infty$ is possible in infinite-dimensional space.

is equal to the number C determined by the conditions

$$\mu\{x : |g(x)| > C\} = 0,$$
$$\mu\{x : |g(x)| > C - \varepsilon\} > 0 \quad \text{for any} \quad \varepsilon > 0.$$

(For a continuous function, the number C is equal to $\max\limits_{a \leqq x \leqq b} |g(x)|$.) The proof is left to the reader.

6. For the norm of Fredholm's operator (art. 1, example 6) with a square-summable kernel $K(x, s)$ we can obtain the bound

$$\| A \|^2 \leqq \int_a^b \int_a^b K^2(x, s)\, \mathrm{d}x\, \mathrm{d}s, \tag{2}$$

after integrating equation (4) of art. 1 with respect to x.

We consider two simple properties of operators with a finite norm.

(1) *For any vector* $x \in R$ *and any linear operator* A *with finite norm* $\| A \|$,

$$|A x| \leqq \|A\| \, |x|. \tag{3}$$

For inequality (3) holds for any unit vector simply in virtue of the definition of the norm of the operator A. If x is an arbitrary non-zero vector (inequality (3) obviously holds for the null vector), then $|x|$ is a unit vector and consequently

$$\left| A \frac{x}{|x|} \right| \leqq \| A \|. \tag{4}$$

But since A is a linear operator, we have:

$$\left| A \frac{x}{|x|} \right| = \frac{1}{|x|} |A x|;$$

multiplying inequality (4) by $|x|$, we get the required inequality (3).

(2) If A, B are operators with a finite norm, then

$$\|A + B\| \leqq \|A\| + \|B\|, \quad \|AB\| \leqq \|A\| \cdot \|B\|. \tag{5}$$

For if $|x| = 1$, then $|(A + B)x| = |A x + B x| \leqq |A x| + |B x| \leqq \|A\| + \|B\|$, proving the first of the inequalities (5). Further, $|AB x| = |A(B x)| \leqq \|A\| \, |B x| \leqq \|A\| \, \|B\|$, proving the second inequality.

Problems. 1. If A is a bounded linear operator with norm M, then

$$\sup |(A\,x, y)| = M \tag{6}$$

(the upper bound with respect to all normed vectors x and y). Conversely, if the bilinear form $(A\,x, y)$ is bounded on the unit sphere, the operator A is bounded and its norm does not exceed the M in equation (6).

2. If the Hilbert space H contains an orthonormal basis $\{e_n\}$, then every linear operator A can be represented by an infinite matrix $\|a_{jk}\|$, where

$$A\,e_j = \sum_{k=1}^{\infty} a_{jk}\,e_k.$$

For some M and any $\xi_1, \ldots, \xi_m, \eta_1, \ldots, \eta_n$, it is given that

$$\left|\sum_{j=1}^{m} \sum_{k=1}^{n} a_{jk}\,\xi_j\,\eta_k\right|^2 \leq M^2 \sum_{j=1}^{m} \xi_k^2 \sum_{k=1}^{n} \eta_k^2.$$

Prove that A is a bounded operator with norm not exceeding M; conversely, if A is a bounded operator with norm M, the preceding condition is fulfilled.

Hint. The left-hand side is the value of the bilinear form $(A\,x, y)$ on certain vectors x and y.

3. (continued). Obtain the inequalities

$$\sup \sum_{k=1}^{\infty} a_{jk}^2 \leq \|A\|^2 \leq \sum_{j=1}^{\infty} \sum_{k=1}^{\infty} a_{jk}^2. \tag{7}$$

Hint. Firstly, $\|A\,e_j\|^2 = \sum_k (A\,e_j, e_k)^2 = \sum_k a_{jk}^2$. Secondly, if $x_0 = \sum_k \xi_k\,e_k$ is such that $|x_0| = 1$ and $|A\,x_0| > \|A\| - \varepsilon$, then

$$\|A\| - \varepsilon < |A\,x_0| = |\sum \xi_k\,A\,e_k| \leq \sqrt{\sum \xi_k^2} \cdot \sqrt{\sum_k |A\,e_k|^2}$$

$$= \sqrt{\sum_k \sum_j (A\,e_k, e_j)^2} = \sqrt{\sum_k \sum_j a_{kj}^2}.$$

4. An operator of the normal form possesses a bounded inverse if and only if the corresponding numbers λ_n exceed some positive constant in absolute value.

5. An operator of the normal form is said to be positive if all the $\lambda_n > 0$. Show that for a positive operator C of the normal form with a bounded inverse for $0 < \alpha < 2/\|C\|$, the inequality $\|E - \alpha\,C\| < 1$ obtains.

6. Show that every bounded bilinear functional $A(x, y)$ in Hilbert space H can be represented in the form $(A\,x, y)$, where A is a bounded linear operator.

Hint. If we fix the first argument x in the functional $(A\,x, y)$, we get a bounded linear functional of y, which by Section 2, art. 9 can be expressed in the form (x', y). Verify that the operator $x' = A\,x$ is a bounded linear operator.

7. An operator A^* which satisfies the condition $(A\,x, y) = (x, A^*\,y)$ is said to be adjoint to the operator A. Show that every linear operator A in Hilbert space has an adjoint operator.

Hint. Apply the result of problem 6 to the bilinear functional $A(x, y) = (y, A\,x)$.

8. Show that $\|A^*\| = \|A\|$.

Hint. Use problem 1.

9. If the operator A possesses a bounded inverse B, then its adjoint A^* possesses the bounded inverse B^*.

10. Show that the last double sum in inequality (7) does not depend on the choice of orthonormal basis $\{e_n\}$.

Hint. If $\{f_n\}$ is a new basis and $A f_k = \sum b_{jk} f_j$, then

$$\sum \sum a_{jk}^2 = \sum |A e_j|^2 = \sum \sum (A e_j, f_k) = \sum \sum (e_j, A^* f_k) \\ = \sum |A^* f_k|^2 \sum \sum b_{kj}^2 .$$

4. *Characteristic Vectors*

A subspace R' of a linear space R is said to be *invariant under the operator* A if $x \in R'$ implies $A x \in R'$.

In particular, the trivial subspaces—the null space and the whole space—are invariant under any linear operator; we shall naturally concern ourselves only with non-trivial invariant subspaces.

The one-dimensional invariant subspaces of an operator A play a special role. Every (non-zero) vector that belongs to a one-dimensional invariant subspace of the operator A is said to be a characteristic vector of the operator A; in other words, a vector $x \neq 0$ is said to be a *characteristic vector* of the operator A if it is mapped by A onto a collinear vector:

$$A x = \lambda x. \tag{1}$$

The number λ which appears in this equation is said to be *the characteristic value (characteristic number) of the operator A corresponding to the characteristic vector* x.

We consider the examples of linear operators given in art. 1 from this point of view.

(1) For the operators in examples 1–3 every subspace is invariant and every non-zero vector of the space is characteristic with characteristic values 0, 1, λ respectively.

(2) An operator of the normal form (example 4) by definition has characteristic vectors $e_1, e_2, \ldots, e_n, \ldots$ with characteristic values $\lambda_1, \lambda_2, \ldots, \lambda_n, \ldots$ respectively.

(3) The operator of multiplication by $g(x) \equiv x$ (example 5) has no characteristic vectors in the space $L_2(a, b)$ since there is no measurable function $\varphi(x)$, non-zero on a set of positive measure and such that

$$x \varphi(x) = \lambda \varphi(x).$$

(4) The characteristic vectors of Fredholm's operator (example 6) are the solutions of the integral equation

$$\int_a^b K(x, s)\,\varphi(s)\,\mathrm{d}s = \lambda\,\varphi(x).$$

We shall discuss the existence of solutions to this equation later.

The set of *all* characteristic vectors of an operator A with a fixed characteristic value λ evidently constitutes a subspace of the space R. This subspace is said to be the *characteristic subspace corresponding to the characteristic value* λ.

5. *Symmetric and Completely Continuous Operators*

If it is known that a linear operator A in Hilbert space is an operator of the normal form, as in example 4 of art. 1, the investigation of its properties is markedly facilitated. A basis of characteristic vectors of the operator A determines a unique "coordinate system" in which problems connected with the operator A can conveniently be solved.

A *necessary* condition for an operator A to be reducible to the normal form is the equation

$$(A\,x, y) = (x, A\,y),$$

which must be satisfied for any x, y in the space H. For if

$$x = \sum_{j=1}^{\infty} \xi_j\, e_j; \quad y = \sum_{j=1}^{\infty} \eta_j\, e_j, \quad A\,x = \sum_{j=1}^{\infty} \lambda_j\, \xi_j\, e_j, \quad A\,y = \sum_{j=1}^{\infty} \lambda_j\, \eta_j\, e_j,$$

then clearly

$$(A\,x, y) = \sum_{j=1}^{\infty} \lambda_j\, \xi_j\, \eta_j, \quad (y, A\,x) = \sum_{j=1}^{\infty} \eta_j\, \lambda_j\, \xi_j,$$

so that equation (1) holds. Operators which satisfy condition (1) are said to be *symmetric*.

The symmetry condition is not sufficient to ensure that the operator A is reducible to the normal form. For example, the operator of multiplication by x over the space $L_2(a, b)$ is symmetric:

$$(x\,\varphi, \psi) = \int_a^b x\,\varphi(x)\,\psi(x)\,\mathrm{d}x = (\varphi, x\,\psi),$$

but as we saw above this operator has no characteristic vectors and is therefore not reducible to the normal form. In addition to the symmetry requirement, we must impose a further condition on the operator A, which we shall call the condition of *complete continuity*:

From each sequence of vectors $A f_n$, where the numbers $|f_n|$ are bounded, a convergent subsequence can be extracted.

Operators which possess this property are said to be *completely continuous.*

A completely continuous operator is bounded (and consequently continuous): if for some sequence f_n, $|f_n| = 1$, we had $|A f_n| \to \infty$, say $|A f_n| > n$, it would be impossible to extract a convergent subsequence from the sequence $A f_n$, in contradiction to the hypothesis.

Problems. 1. Is the identity operator E completely continuous?

Answer. Not if space H is infinite-dimensional.

2. A and B are completely continuous operators; show that $A + B$ is completely continuous.

3. A is a completely continuous operator, B is bounded; show that AB and BA are completely continuous.

4. If A and A^* are mutually adjoint operators, then $A + A^*$, AA^*, A^*A are symmetric operators, and $\|AA^*\| = \|A^*A\| = \|A\|^2$.

5. Show that the necessary and sufficient condition for an operator of the normal form to be completely continuous is

$$\lim_{n \to \infty} \lambda_n = 0.$$

Hint. Use problem 3, Section 1.

6. We now proceed to the fundamental theorem on symmetric, completely continuous operators.

THEOREM 1. (D. Hilbert) *In a complete separable Hilbert space every symmetric completely continuous operator possesses a complete orthogonal system of characteristic vectors.*

We shall carry out the proof of this theorem in several stages.

LEMMA 1. *If $|e| = 1$ and A is a symmetric operator, then*

$$|A e|^2 \leqq |A^2 e|,$$

with the equality sign possible only if there is a characteristic vector of the operator A^2 with the characteristic value

$$\lambda = |A e|^2.$$

Proof. In virtue of the symmetry of the operator and the Cauchy–Bunyakovsky inequality, we have:

$$|A e|^2 = (A e, A e) = (A^2 e, e) \leqq |A^2 e|\,|e| = |A^2 e|. \quad (1)$$

The Cauchy–Bunyakovsky inequality reduces to equality only if the vectors which figure in it are collinear (Section 1, art. 3), hence in the case of equality we have

$$A^2 e = \lambda e,$$

i.e. e is a characteristic vector of the operator A^2. Substituting this expression in (1), we get for λ:

$$(A^2 e, e) = (\lambda e, e) = \lambda = |A e|^2,$$

as required.

We shall call *the maximal vector of a bounded operator* A the unit vector e, $|e| = 1$, on which the quantity $|A e|$ attains its greatest value $M = \|A\|$. In general, not every bounded operator will have a maximal vector. But we shall show that a symmetric, completely continuous operator always has a maximal vector:

LEMMA 2. *A symmetric, completely continuous operator possesses a maximal vector.*

Proof. We choose a sequence $y_n = A x_n$, where $|x_n| = 1$ ($n = 1, 2, \ldots$) such that $\lim_{n \to \infty} |y_n| = M$. By hypothesis, a convergent subsequence can be extracted from the sequence y_n; deleting the remaining vectors and modifying the numbering, we can assume that the sequence y_n itself converges as $n \to \infty$; let $y = \lim y_n$. In virtue of the continuity of the norm,

$$|y| = \lim_{n \to \infty} |y_n| = M.$$

We claim that the vector $z = \frac{1}{M} y$ is the maximal vector sought.

In the first place, we have in virtue of the continuity of the operator A:

$$A z = \lim_{n \to \infty} A\left(\frac{y_n}{M}\right) = \lim_{n \to \infty} A\left(\frac{A x_n}{M}\right).$$

The vectors $A x_n / M$ belong to the unit spherical surface, and hence the vectors $A (A x_n)/M)$ do not exceed M in length. Applying lemma 1, we get

$$M > \left|A\left(\frac{A x_n}{M}\right)\right| = \frac{1}{M} |A^2 x_n| \geqq \frac{1}{M} |A x_n|^2 \to M,$$

and it follows that

$$|A z| = \lim \left| A\left(\frac{A x_n}{M}\right)\right| = M,$$

i.e. z is the maximal vector of the operator A, as required.

It is not a far cry from maximal vectors to characteristic vectors:

LEMMA 3. *If e_0 is the maximal vector of a symmetric operator A, it is a characteristic vector of the operator A^2 with characteristic value $\|A\|^2$.*

Proof. By LEMMA 1 and the definition of the norm of an operator, we have

$$\|A\|^2 = |A e_0|^2 \leqq |A^2 e_0| \leqq \|A\|^2,$$

and hence

$$|A e_0|^2 = |A^2 e_0| = \|A\|^2.$$

By lemma 1, e_0 is a characteristic vector of the operator A^2 with the characteristic value

$$\lambda = |A e_0|^2 = \|A\|^2,$$

as required.

LEMMA 4. *If the operator A^2 possesses a characteristic vector with characteristic value M^2, the operator A possesses a characteristic vector with characteristic value M or $-M$.*

Proof. The equation $A^2 e_0 = M^2 e_0$ can be written in the form $(A - ME)(A + ME) e_0 = 0$ (where E is the identity operator). Let us suppose that $z_0 = (A + ME) e_0 \neq 0$. Then the condition

$$(A - ME) z_0 = 0$$

or what is the same thing,

$$A z_0 = M z_0$$

implies that z_0 is a characteristic vector of the operator A with the characteristic value $M = \|A\|$. And if $(A + ME) e_0 = 0$, then

$$A e_0 = -M e_0,$$

and we get that e_0 is a characteristic vector of the operator A with the characteristic value $-M = \|A\|$. The lemma is proved.

LEMMAS 1–4 show that *every symmetric, completely continuous operator A possesses a characteristic vector with characteristic value $\pm\|A\|$*. We shall now prove that a complete orthogonal system can be constructed in the space H from the characteristic vectors of such an operator A. We lead up to this construction with the following lemmas:

LEMMA 5. *The characteristic vectors of a symmetric operator corresponding to different characteristic values are mutually orthogonal.*

For let

$$A\,x = \lambda\,x, \quad A\,y = \mu\,y$$

with $\lambda = \mu$. We multiply the first equation scalarly by y, the second by x, and subtract the second from the first:

$$(A\,x, y) - (x, A\,y) = (\lambda - \mu)\,(x, y).$$

The left-hand side of this equation vanishes on account of the symmetry of the operator. Since $\lambda \neq \mu$, we have $(x, y) = 0$, as required.

LEMMA 6. *Every orthonormal system of characteristic vectors of a completely continuous operator A with characteristic values exceeding a positive number δ in modulus is finite.*

Proof. Let us suppose that an infinite system S of such characteristic vectors has been found. Each of them is mapped by the operator onto a scalar multiple of itself under multiplication by a number greater than δ.

Let e_j, e_k be any two of these characteristic vectors:

$$|e_j| = |e_k| = 1, \quad (e_j, e_k) = 0, \quad A\,e_j = \lambda_j\,e_j, \quad A\,e_k = \lambda_k\,e_k.$$

We have

$$|A\,e_j - A\,e_k|^2 = |\lambda_j\,e_j - \lambda_k\,e_k|^2 = \lambda_j^2 + \lambda_k^2 > 2\delta^2.$$

This means that the distances between the vectors obtained by operating with A on the vectors of the system S will exceed δ 2. But then it is impossible to select a convergent sequence from the aggregate of such vectors, which contradicts the complete continuity of the operator A.

In particular, *there exists only a finite number of mutually orthogonal vectors with a given characteristic value $\lambda \neq 0$*; in other words, *every characteristic subspace that corresponds to a non-zero characteristic value of a symmetric, completely continuous operator A is finite-dimensional.*

This lemma allows us to draw definite conclusions in relation to the totality of characteristic vectors and characteristic values of the operator A. We consider the set of all characteristic values of the operator A, which is a subset of the real axis. By lemma 6 there exists only a finite number of characteristic values the moduli of which exceed a given positive number δ, and so if the characteristic values constitute an infinite set, they must form a sequence which converges to zero. We can therefore order them according to their decrease in absolute magnitude. We also agree to repeat each characteristic value a number of times equal to the dimensionality of the corresponding characteristic subspace. We can then map the sequence of all non-zero characteristic values.

$$\lambda_1, \lambda_2, \ldots, \lambda_n, \ldots$$

of the operator onto the sequence of characteristic vectors

$$e_1, e_2, \ldots, e_n, \ldots$$

where $A e_n = \lambda_n e_n$ $(n = 1, 2, \ldots)$. We can assume that the vectors $e_1, e_2, \ldots$ are mutually orthogonal and normalised. In fact if $\lambda_n \neq \lambda_m$, then e_n and e_m are orthogonal by lemma 5; and if $\lambda_n = \lambda_m$, we can always carry out an orthogonalisation in the finite-dimensional characteristic subspace corresponding to the characteristic value $\lambda_n = \lambda_m$. The construction is completed by normalising all the vectors obtained.

We shall show now that *each vector z orthogonal to all the constructed vectors $e_1, e_2, \ldots, e_n, \ldots$ is mapped by the operator A onto zero*. To do this, we use the following lemma:

LEMMA 7. *Let H' be a subspace of the Hilbert space H which is invariant relative to a symmetric operator A. Then the orthogonal complement H'' of the subspace H' is also invariant relative to the operator A.*

Proof. Let x be any vector of the subspace H' and y any vector of the subspace H''. By hypothesis $(A x, y) = 0$. But then in virtue of the symmetry of the operator, $(x, A y) = 0$. This means that the vector $A y$ is orthogonal to any vector $x \in H'$ and consequently $A y \in H''$ for any $y \in H''$, as required.

We now consider the aggregate P of all vectors z orthogonal to each of the constructed vectors $e_1, e_2, \ldots, e_n, \ldots$ It is a closed subspace since it is the orthogonal complement of the linear envelope $L(e_1, e_2, \ldots, e_n, \ldots) = L$†. Since the linear envelope L is evidently invariant relative to A, its orthogonal complement P is also invariant relative to A, by lemma 7. We denote by $M(P)$ the exact upper bound of the values of $|A x|$ on the unit spherical

† The linear envelope of a system of vectors $x_1, x_2, \ldots, x_n, \ldots$ is defined as the subspace comprising all their linear combinations.

surface of the subspace O. By lemma 4 the subspace P contains a characteristic vector e_0 with the characteristic value $\lambda_0 = M(P)$. But by construction P cannot contain a single characteristic vector with non-zero characteristic value. It follows that $\lambda_0 = M(P) = 0$, but this means that $A z = 0$ for any vector $z \in P$, as asserted.

We denote by L' the closure of the linear envelope of the vectors $e_1, e_2, \ldots$; the orthogonal complement of this closure is also the subspace P. Each vector $x \in H$ can be represented in the form of a sum

$$x = x' + x'', \quad x' \in L', \quad x'' \in P.$$

Further, the vector x' can be developed in a Fourier series with respect to the system $e_1, e_2, \ldots, e_n$, which is complete in the space L'; by what we have proved, the vector x'' is mapped by the operator A onto zero. We have obtained the following fundamental theorem:

THEOREM 2. *Each vector x of a complete Hilbert space H in which a symmetric, completely continuous operator A is given can be expressed in the form of an orthogonal sum*

$$x = x' + x'' = \sum_{j=1}^{\infty} \xi_j e_j + x'',$$

where $e_1, e_2, \ldots$ are characteristic vectors of the operator A with non-zero characteristic values and $A x'' = 0$.

Hilbert's theorem is a corollary of this theorem. For in a separable space H the subspace P is also separable and it contains a complete orthogonal system $e_1', e_2', \ldots, e_n', \ldots$; together with the vectors $e_1, e_2, \ldots, e_n, \ldots$ already constructed this yields a complete orthogonal system for the whole space H. Each vector of this system is a characteristic vector of the operator A: the vectors e_n with characteristic values $\lambda \neq 0$ $(n = 1, 2, \ldots)$ and the vectors e_n' with the characteristic value 0. This completes the proof of Hilbert's theorem.

Note. Vectors which belong to the range of an operator A, i.e. vectors of the form

$$\psi = A \varphi,$$

are said to be *source-representable*. We shall explain the significance of this term below (Section 5).

Every source-representable vector ψ admits of a development in the characteristic vectors of the operator A with non-zero characteristic values.

For by Hilbert's theorem

$$\varphi = \sum_{j=1}^{\infty} \xi_j e_j + x'',$$

where $A e_n = \lambda_n e_n$ $(\lambda_n \neq 0)$ and $A x'' = 0$. Operating on this equation with A, we get:

$$\psi = A \varphi = \sum_{j=1}^{\infty} \lambda_j \xi_j e_j,$$

as asserted.

4. Integral Operators With Square-summable Kernels

1. We shall apply the theory expounded in Section 3 to Fredholm's integral operator

$$A \varphi(x) = \int_a^b K(x, s) \varphi(s) \, \mathrm{d}s$$

with a square-summable kernel $K(x, s)$:

$$\int_a^b \int_a^b K^2(x, s) \, \mathrm{d}x \, \mathrm{d}s = K^2 < \infty. \tag{1}$$

As we saw in art. 1 of Section 3 Fredholm's operator is a bounded operator on the Hilbert space $H = L_2(a, b)$ and has a norm equal to at most K.

If the kernel $K(x, s)$ is symmetric, i.e. if we have

$$K(x, s) = K(s, x)$$

almost everywhere in the region $G = \{a \leqq x \leqq b, a \leqq s \leqq b\}$, then Fredholm's operator is also symmetric, i.e. $(A \varphi, \psi) = (\varphi A \psi,)$ for any φ, ψ in $L_2(a, b)$.

For by Fubini's theorem

$$\begin{aligned}
(A \varphi, \psi) &= \int_a^b \left\{ \int_a^b K(x, s) \varphi(s) \, \mathrm{d}s \right\} \psi(x) \, \mathrm{d}x \\
&= \int_a^b \int_a^b K(x, s) \varphi(s) \psi(x) \, \mathrm{d}s \, \mathrm{d}x \\
&= \int_a^b \varphi(s) \left\{ \int_a^b K(x, s) \psi(x) \, \mathrm{d}x \right\} \mathrm{d}s \\
&= \int_a^b \varphi(s) \left\{ \int_a^b K(s, x) \psi(x) \, \mathrm{d}x \right\} \mathrm{d}s = (\varphi, A \psi).
\end{aligned}$$

The existence of the double integral

$$\int_a^b \int_a^b K(x, s)\, \varphi(s)\, \psi(x)\, \mathrm{d}s\, \mathrm{d}x,$$

which is one of the conditions under which Fubini's theorem is applicable, follows from the existence of the integral

$$\int_a^b \int_a^b K^2(x, s)\, \mathrm{d}x\, \mathrm{d}s$$

and

$$\int_a^b \int_a^b \varphi^2(x)\, \psi^2(s)\, \mathrm{d}x\, \mathrm{d}s = \int_a^b \varphi^2(x)\, \mathrm{d}x \int_a^b \psi^2(s)\, \mathrm{d}s.$$

We shall show now that *Fredholm's operator with a square-summable kernel is completely continuous.* We recall that a linear operator A is said to be completely continuous if a convergent subsequence can be extracted from each sequence $A f_n$ for which the $|f_n|$ are bounded. In other words, the operator A is completely continuous if it maps any bounded subset of the space H onto a compact set (Chapter II, Section 7). For instance let A be a bounded operator that maps the space H onto a finite-dimensional subspace Q; we claim that A is then completely continuous. For the vectors $A f_n$ form a bounded set in the finite-dimensional subspace Q, and in accordance with the results of Chapter II, Section 7, such a set is compact, so that A satisfies the condition for complete continuity of an operator.

If the function $K(x, s)$ is of the form

$$K(x, s) = \sum_{k=1}^{m} \varphi_k(x)\, \psi_k(s),$$

where $\varphi_k(x)$, $\psi_k(S)$ $(k = 1, 2, \ldots, m)$ are functions the squares of which are integrable (such a kernel $K(x, s)$ is said to be *degenerate*), then $K(x, s)$ is bounded and

$$A\varphi(x) = \int_a^b \sum_{k=1}^{m} \varphi_k(x)\, \psi_k(s)\, \varphi(s)\, \mathrm{d}s = \sum_{k=1}^{m} \left\{ \int_a^b \psi_k(s)\, \varphi(s)\, \mathrm{d}s \right\} \varphi_k(s),$$

i.e. the operator A maps the whole space $L_2(a, b)$ onto the finite-dimensional subspace generated by the functions $\varphi_1(x), \ldots, \varphi_m(x)$. Hence a *Fredholm operator with a degenerate kernel is completely continuous.*

To deal with the general case we use the following lemma:

LEMMA. *Let there be given in a Hilbert space H a sequence $A_1, A_2, \ldots, A_n, \ldots$, of linear operators which converges to an operator A in the sense that $\| A - A_n \| \to 0$ as $n \to \infty$. If the operators A_n $(n = 1, 2, \ldots)$ are completely continuous, then so is the limit operator A.*

Proof. Let the vector f run through a bounded set B, say a sphere of radius r with centre at O; we have to show that the vector $A f$ then runs through a compact set. It is sufficient to establish that for any $\varepsilon > 0$ the set $\{A f\}$ possesses a compact ε-net. We find an operator A_n in the sequence $A_n \to A$ such that $\| A_n - A \| < \varepsilon / r$. By hypothesis the set of elements $\{A_n f\}$ $(f \in B)$ is compact, and for any f we have $|A_n f - A f| \leqq \| A_n - A \| \, |f| < < \varepsilon / r \cdot r = \varepsilon$. It follows that the set $\{A_n f\}$ is a compact ε-net for $\{A f\}$, as required.

Now let $K(x, s)$ be an arbitrary function the square of which is integrable over the region $G = \{a \leqq x \leqq b, a \leqq s \leqq b\}$. We claim that this function can be developed [in the metric of $L_2(G)$] in a series of the form

$$K(x, s) = \sum_{m, n = 1}^{\infty} a_{mn} e_m(x) e_n(s). \tag{2}$$

We take as the functions $e_n(x)$ any complete orthogonal system in the space $L_2(a, b)$. Then the *products* $e_m(x) e_n(s)$ $(m, n = 1, 2, \ldots)$ *form a complete orthogonal system in the space* $L_2(G)$. That the system is orthogonal is obvious; we verify that it is complete. If the space $L_2(G)$ contained a function $f(x, s)$ orthogonal to all the products $e_m(x) e_n(s)$, so that

$$\int\int_G f(x), s) e_m(x) e_n(s) \, \mathrm{d}x \, \mathrm{d}s = 0 \quad (m, n = 1, 2, \ldots),$$

then by Fubini's theorem we should have for any fixed n

$$\int_a^b e_m(x) \left\{ \int_a^b f(x, s) e_n(s) \, \mathrm{d}s \right\} \mathrm{d}x = 0.$$

It would then follow, since the system $e_m(x)$ is complete, that for any $n = 1, 2, \ldots$ and for almost all x

$$\int_a^b f(x, s) e_n(s) \, \mathrm{d}s = 0.$$

This would imply, since the system $e_n(s)$ is complete, that for almost all x, s

$$f(x, s) = 0$$

and $f(x, s)$ would therefore be the zero element of the space $L_2(G)$.

Thus the functions $e_m(x)\, e_n(s)$ really do form a complete orthogonal system in the space $L_2(G)$. But then by the fundamental theorem of Section 2 each element of the space $L_2(G)$ admits of a development (2), as asserted.

The degenerate kernel, formed by the partial sums $K_{pq}(x, s)$ of the series (2),

$$K_{pq}(x, s) = \sum_{m=1}^{p} \sum_{n=1}^{q} a_{mn}\, e_m(x)\, e_n(s),$$

determines a sequence of completely continuous operators

$$A_{pq}\, \varphi(x) = \int_a^b K_{pq}(x, s)\, \varphi(s)\, \mathrm{d}s.$$

Using the bound for the norm of Fredholm's operator (cf. p. 207), we get the inequality

$$\|A - A_{pq}\|^2 \leqq \int_a^b \int_a^b [K(x, s) - K_{pq}(x, s)]^2\, \mathrm{d}x\, \mathrm{d}s,$$

from which it follows that the operators A_{pq} converge in the norm to the operator A as $p \to \infty$, $q \to \infty$.

Applying lemma 1, we conclude that the operator A is completely continuous together with the operators A_{pq}, as asserted.

2. Thus a Fredholm operator with a symmetric square-summable kernel $K(x, s)$ is symmetric and completely continuous. We can therefore apply Hilbert's theorem (Section 3); in virtue of this theorem, *there exists in the space $L_2(a, b)$ a complete orthonormal system composed of characteristic functions of Fredholm's operator.* We shall show *that the squares of the characteristic values of Fredholm's operator form a convergent series.*

Let us consider the equation which determines the normalized characteristic functions of Fredholm's operator:

$$\int_a^b K(x, s)\, e_n(s)\, \mathrm{d}s = \lambda_n\, e_n(x); \tag{1}$$

it shows that the quantity $\lambda_n e_n(x)$ is a Fourier coefficient of the function $K(x, s)$ (for a constant value of x). Hence, applying Bessel's inequality (Section 2, art. 4) to (1), we get

$$\int_a^b K^2(x, s)\, \mathrm{d}s \geqq \sum_{n=0}^{N}{}' \lambda_n^2 e_n^2(x) \tag{2}$$

for every value of N. Integrating this inequality with respect to x, we get

$$\int_a^b \int_a^b K^2(x, s)\, \mathrm{d}s\, \mathrm{d}x \geqq \sum_{n=0}^{N}{}' \lambda_n^2$$

for any natural number N. It follows that this series converges, as required.

3. We consider Fredholm's operator with a kernel which satisfies *the Hilbert–Schmidt condition*:

$$\int_a^b K^2(x, s)\, \mathrm{d}s \leqq C. \tag{1}$$

When this condition is satisfied, every function $\varphi(x) \in L_2(a, b)$ is mapped by the operator A onto a bounded function, since

$$|A\,\varphi(x)|^2 = \left[\int_a^b K(x, s)\,\varphi(s)\, \mathrm{d}s\right]^2$$

$$\leqq \int_a^b K^2(x, s)\, \mathrm{d}s \int_a^b \varphi^2(s)\, \mathrm{d}s \leqq C \int_a^b \varphi^2(s)\, \mathrm{d}s. \tag{2}$$

In particular, *each characteristic function of the operator A with non-zero characteristic value is bounded.*

We shall call a function $g(x)$ of the form

$$g(x) = \int_a^b K(x, s)\,\varphi(s)\, \mathrm{d}s = A\,\varphi$$

with an arbitrary function $\varphi(x) \in L_2(a, b)$ *source-representable in the kernel* $K(x, s)$, (in accordance with the general definition given at the end of Section 3).

At the end of Section 3 it was shown that, when A is a symmetric operator, every source-representable vector $\psi = A\,\varphi$ admits of a development of the following form in the characteristic vectors of A:

$$\psi = \sum_{j=1}^{\infty} \lambda_j (\varphi, e_j)\, e_j. \tag{3}$$

We shall show that when condition (1) is satisfied the series (3) converges for any source-representable function not only in the norm, but also absolutely and uniformly. For by Cauchy's inequality (cf. p. 192) we have for any m, n

$$\left\{\sum_{j=1}^{n+m} |(f, e_j)\, \lambda_j\, e_j(x)|\right\}^2 \leq \sum_{j=1}^{n+m} (f, e_j)^2 \sum_{j=1}^{n+m} \lambda^2_j\, e^2_j(x). \tag{4}$$

Since the sum $\lambda_m^2\, e_n^2(x) + \cdots$ on the right–hand side of (4) is always bounded [in virtue of inequality (2) of art. 2] and the sum $(f, e_n) + \cdots$ can be made arbitrarily small by taking n sufficiently large [by Bessel's inequality (cf. p. 203)], the sum on the left-hand side of (4) can be made arbitrarily small similarly; hence the Fourier series under consideration converges absolutely and uniformly, as required.

The property established here is sometimes called the Hilbert–Schmidt theorem.

Note. If the kernel $K(x, s)$ is continuous over the region $a \leqq x \leqq b$, $a \leqq s \leqq b$, the corresponding Fredholm operator maps every function $\varphi(x) \in L_2(a, b)$ *onto a continuous function.* For if we put

$$\psi(x) = \int_a^b K(x, s)\, \varphi(s)\, ds;$$

then for any x', x'',

$$|\psi(x') - \psi(x'')|^2 \leqq \left\{\int_a^b |K(x', s) - K(x'', s)|\, |\varphi(s)|\, ds\right\}^2$$
$$\leqq \int_a^b [K(x', s) - K(x'', s)]^2\, ds \int_a^b [\varphi(s)]^2\, ds,$$

and the continuity of the function $\psi(x)$ follows immediately. Of course all the characteristic functions with non–zero characteristic values are then also continuous.

4. *Determination of Characteristic Functions and Characteristic Values*

The practical application of our results requires a knowledge of the system of characteristic functions corresponding to a Fredholm operator.

If the kernel $K(x, s)$ of a Fredholm operator A is degenerate, so that

$$K(x, s) = \sum_{j=1}^{m} p_j(x)\, q_j(s),$$

then, as we have already seen, the operator A maps the whole space L_2 onto the finite-dimensional subspace generated by the functions $p_j(x)$ $(j = 1, 2, \ldots, m)$. The characteristic functions with non-zero characteristic values should therefore be sought only in this subspace; they must be of the form

$$e(x) = \sum_{j=1}^{m} c_j\, p_j(x). \tag{1}$$

To determine the coefficients c_j we substitute the function (1) in the defining equation of the characteristic functions

$$\int_a^b K(x, s)\, e(s)\, \mathrm{d}s = \lambda\, e(x).$$

Then we get:

$$\lambda\, e(x) = \sum_{j=1}^{m} \lambda\, c_j\, p_j(x) = \int_a^b K(x, s) \sum_{j=1}^{m} c_j\, q_j(s)\, \mathrm{d}s$$

$$= \sum_{i=1}^{m} \sum_{j=1}^{m} p_i(x)\, c_j \int_a^b q_i(s)\, q_j(s)\, \mathrm{d}s = \sum_{j=1}^{m} \sum_{i=1}^{m} c_j\, q_{ij}\, p_i(x),$$

where

$$q_{ij} = \int_a^b q_i(s)\, q_j(s)\, \mathrm{d}s.$$

Consequently

$$\lambda\, c_j = \sum_{i=1}^{m} q_{ij}\, c_i \quad (j = 1, 2, \ldots, m). \tag{2}$$

This system of equations allows λ and the constants c_j to be found in the usual way. It assumes a particularly simple form if $p_i(x) \equiv q_i(x)$ and $(p_i, q_j) = 0$ for $i = j$.

In this case the q_{ij} vanish for $i \neq j$ and the system (2) has the obvious solution: $c_j = 1$ for some j, $c_i = 0$ for $i \neq j$, $\lambda = q_{jj}$. By formula (1] the characteristic function corresponding to this solution coincides with the function $p_j(x) = q_j(x)$. Thus the characteristic functions are the functions $p_j(x)$ ($j = 1, 2, \ldots, m$) themselves and the characteristic values are the numbers q_{jj}, i.e. the squares of the norms of these functions.

If the kernel $K(x, s)$ of the Fredholm operator A is non-degenerate, the following method of approximation is frequently useful in determining its characteristic functions and characteristic values. We replace the given kernel $K(x, s)$ by a degenerate kernel $K_n(x, s)$ which approximates it (for example, by a partial sum of its Fourier development) and find the characteristic functions and characteristic values of the corresponding operator A_n by the method described above. Under certain assumptions regarding the smoothness of the kernel $K(x, s)$ the characteristic values of the operator A_n obtained approximate those of the operator A. We cannot deal at length with these questions here, but we recommend the reader to consult the special literature†.

Problems. 1. What are the characteristic functions of Fredholm's integral operator with the kernel $K(x, s) = \cos(x + s)$ on the intervals (a) $[0, \pi]$, (b) $[0, \pi/2]$?

Answer. (a) $\cos x$, $\sin x$; (b) $\cos x + \sin x$, $\cos x - \sin x$.

2. Show that a square-summable symmetric function $K(x, s)$ can be developed in a bilinear series which converges in the metric of $L_2(G)$:

$$K(x, s) = \sum \lambda_k \varphi_k(x) \varphi_k(s), \qquad (1)$$

where the $\varphi_k(x)$ are the normalised characteristic functions and the λ_k the corresponding characteristic values of the integral operator K (with kernel $K(x, s)$).

Hint. The products $\varphi_k(x) \varphi_k(s)$ form a complete orthogonal system in $L_2(G)$.

3. If with the conditions of problem 2 the function $K(x, s)$ is developed in a series which converges in the metric of $L_2(G)$:

$$K(x, s) = \sum \mu_k u_k(x) u_k(s),$$

and the functions $u_k(s)$ are mutually orthogonal (in $L_2(a, b)$) and normalised, then $u_k(x)$ is a characterised function of the operator k and μ_k is the corresponding characteristic value.

† Cf. R. Courant and D. Hilbert, *Methods of Mathematical Physics*, Vol. 1, Chapter 3, Section 9; L. V. Kantorovich and V. I. Krilov, *Methods of Approximation in Higher Analysis*, State Technical Publishing Dept. 1949, Chapter 2, Section 4.

4. Analogous to the quadratic form $\sum_{j=1}^{n} \sum_{k=1}^{n} a_{jk} \xi_j \xi_k$ of finite rank is the *integral quadratic form*

$$(K\varphi, \varphi) = \int_a^b \int_a^b K(x, s)\, \varphi(x)\, \varphi(s)\, \mathrm{d}x\, \mathrm{d}s. \tag{2}$$

The quadratic form (2) is said to be *positive-definite* if $(K\varphi, \varphi) > 0$ for any function $\varphi \in L_2(a, b)$ not identically zero. Show that all the characteristic values of the Fredholm operator K corresponding to a given positive-definite quadratic form are positive.

5. If the kernel $K(x, s)$ of a positive-definite quadratic form (1) is symmetric and continuous, then $K(s, s) \geqq 0$.

Hint. Suppose that $K(s_0, s_0) < 0$ and construct a function $\varphi_0(x)$ for which $(K\varphi_0, \varphi_0)$ is < 0.

6. If the kernel $K(x, s)$ corresponding to a positive-definite form $(K\varphi, \varphi)$ is continuous and symmetric, the development (1) converges absolutely and uniformly (Mercer's theorem).

Hint. Applying the result of problem 5 to the kernel

$$K(x, s) - \sum_{k=1}^{n} \lambda_k \varphi_k(x)\, \varphi_k(s)$$

we get the convergence of the series $\sum_{k=1}^{\infty} \lambda_k \varphi_k^2(s)$; use Cauchy's inequality to deduce that the series $\sum_{k=1}^{\infty} \lambda_k \varphi_k(x)\, \varphi_k(s)$ converges uniformly in each coordinate (with the other held fixed). Then use this result in conjunction with that of problem 3 to deduce that the sum of this series is $K(x, s)$. Applying Dini's theorem (Chapter II, Section 7, problem 4) to $K(s, s) = \sum_{k=1}^{\infty} \lambda_k \varphi_k^2(s)$ and again using Cauchy's inequality, obtain the uniform convergence of the development (1) in the region G.

7. Show that the operator A, given in the orthonormal basis $\{e_j\}$ by the matrix $\|a_{jk}\|$ in accordance with

$$A e_j = \sum_{k=1}^{\infty} a_{jk} e_k,$$

is completely continuous if $\sum \sum a_{jk}^2 < \infty$.

Hint. Express A as the limit of operators mapping the entire space on to finite-dimensional subspaces.

8. If A is an integral operator in $L_2(a, b)$ with a square-summable kernel and $\{e_k(x)\}$ is an orthonormal system in $L_2(a, b)$, then

$$\sum |A e_k|^2 = \sum\sum (A e_k, e_n)^2 < \infty.$$

Hint. Use the method of art. 2.

9. If an operator A is given in $L_2(a, b)$ in the orthonormal basis $\{e_n(x)\}$ by the matrix $\|a_{jk}\|$ with $\sum\sum a_{jk}^2 < \infty$, a square-summable kernel $K(x, s)$ exists such that $A\varphi = \int K(x, s)\varphi(s)\,ds$.

Hint. Put $K(x, s) = \sum\sum (A e_j, e_k)\, e_j(x)\, e_k(s)$.

Note. The results of problems 8 and 9 show that, among the completely continuous operators that act in space $L_2(a, b)$, the Fredhold integral operators are distinguished by the condition that the sum of the squares of all the matrix elements in any orthogonal basis of the space is finite.

5. The Sturm–Liouville Problem

1. The general theorem on integral operators with symmetric kernels has many applications in mathematical physics. One of the most important of these is *the solution of the Sturm–Liouville problem.*

Consider the differential operator

$$S[u] = (p(x)\,u'(x))' - q(x)\,u(x) \qquad (1)$$
$$(p(x) \in D_1(a, b),\ q(x) \in C(a, b)),$$

defined on the closed interval $[a, b]$ for twice differentiable functions $u(x)$ subject to certain homogeneous boundary conditions, for example $u(a) = u(b) = 0$.

The operator S is essentially different from the operators we have so far considered; it is not a bounded operator, nor is it defined on the whole space $L_2(a, b)$. Nevertheless, it is symmetric on its domain of definition, i.e. for any two twice differentiable functions u, v which satisfy the prescribed boundary conditions, the equation

$$(S\,u, v) = (u, S\,v) \qquad (2)$$

holds. For

$$\left.\begin{aligned} (S\,u, v) &= \int_a^b [(p\,u')' - q\,u]\,v\,dx \\ &= p\,u'\,v\Big|_a^b - \int_a^b (p\,u'\,v' + q\,u\,v)\,dx, \\ (u, S\,v) &= \int_a^b u[(p\,v')' - q\,v]\,dx \\ &= u\,p\,v'\Big|_a^b - \int_a^b (p\,u'\,v' + q\,u\,v)\,dx. \end{aligned}\right\} \qquad (3)$$

and in virtue of the boundary conditions,

$$p(u' v - u v')\Big|_a^b = 0,$$

and consequently (2) holds.

A function $e(x)$ is said to be a characteristic function of the operator S if it is contained in the domain of S (i.e. if it is twice differentiable and satisfies the prescribed boundary conditions) and satisfies the equation

$$S e = \lambda e.$$

By the symmetry of S, just as in Section 3, its characteristic functions corresponding to distinct values of λ are mutually orthogonal in the space $L_2(a, b)$.

We wish to prove *that the characteristic functions of the operator* S *form a complete system in* $L_2(a, b)$. This problem, which incorporates first the question of the existence of an infinite set of characteristic functions, and secondly the question of the completeness of such a system, is called the Sturm–Liouville problem.

2. There is an important problem in which the necessity of solving the Sturm–Liouville problem becomes apparent.

Let us consider the vibrations of a non-uniform string fixed at the points $x = a$, $x = b$. As we know from Chapter III, these vibrations are described by the equation

$$(p\, u_{xt})_x = \mu\, u_t\,, \tag{1}$$

where $p(x)$ is the modulus of elasticity and $\mu(x)$ is the density of the string; the required solution $u(x, t)$ must satisfy the boundary conditions

$$u(a, t) = u(b, t) = 0 \tag{2}$$

and the initial conditions

$$u(x, 0) = \varphi(x), \quad u_t(x, 0) = \psi(x). \tag{3}$$

We shall look for a solution of (1) subject to the condition (2) in the form

$$u(x, t) = X(x)\, T(t). \tag{4}$$

Substituting (4) in (1) and separating the variables, we find:

$$\frac{(p\, X')'}{\mu\, X} = \frac{T''}{T}\,, \tag{5}$$

where primes denote differentiation with respect to the corresponding variable. The right-hand side of (5) is independent of x and the left independent of t, which means that the ratio (5) is constant. Denoting its value by λ, we get the two equations

$$(p X')' = \lambda \mu X, \tag{6}$$
$$T'' = \lambda T.$$

If $\mu(x) \equiv 1$, the function $X(x)$ must be a characteristic function of the Sturm–Liouville equation (6).

If $\mu(x) \not\equiv 1$, we can make the substitution

$$z = X\sqrt{\mu},$$

when equation (6) will reduce to the form

$$(p_1 z')' - q z = \lambda z, \tag{7}$$

where

$$p_1 = \frac{p}{\mu}, \quad q = -\frac{1}{\sqrt{\mu}}\left[p\left(\frac{1}{\sqrt{\mu}}\right)'\right]',$$

and the function $z(x)$ again turns out to be a characteristic function of the Sturm–Liouville equation (7). For the sake of simplicity we shall assume in what follows that $\mu(x) \equiv 1$.

Let us suppose that the corresponding Sturm–Liouville problem has a positive solution: there exists a complete orthogonal system of functions $e_1(x), \ldots, e_n(x), \ldots$ satisfying the equations

$$(p(x) e_n'(x))' = \lambda_n e_n(x)$$

and the boundary conditions $e_n(a) = e_n(b) = 0$. If $p(x) > 0$, the λ_n are clearly negative, since

$$\lambda_n \int_a^b e_n^2(x)\,\mathrm{d}x = \int_a^b (p e_n')' e_n\,\mathrm{d}x = p e_n' e_n \Big|_a^b - \int_a^b p[e_n'(x)]^2\,\mathrm{d}x < 0.$$

The equation

$$T'' = \lambda_n T$$

evidently has a solution of the form

$$T_n = A_n \cos \nu_n t + B_n \sin \nu_n t,$$

where $\nu_n^2 = -\lambda_n$ and A_n, B_n are arbitrary constants. Equation (1) has a set of solutions of the form

$$u_n(x, t) = e_n(x)(A_n \cos \nu_n t + B_n \sin \nu_n t),$$

which represent pure vibrations of frequencies $\nu_1, \nu_2, \ldots$ The numbers $\nu_1, \nu_2, \ldots$ are said to be *natural frequencies* of the string given by conditions (1)–(3). A solution $u(x, t)$ which will also satisfy the initial conditions (3) can now be obtained by combining the solutions $u_n(x, t)$ already found:

$$u(x, t) = \sum_{n=1}^{\infty} u_n(x, t).$$

To determine the coefficients A_n, B_n we have the conditions

$$u(x, 0) = \varphi(x) = \sum_{n=1}^{\infty} A_n e_n(x),$$

$$u_t(x, 0) = \psi(x) = \sum_{n=1}^{\infty} \nu_n B_n e_n(x).$$

The functions $\varphi(x)$, $\psi(x)$ have developments in the $e_n(x)$ since the $e_n(x)$ are assumed to form a complete system and the coefficients A_n, B_n can therefore be found. They are evidently determined directly by the coefficients of the developments of φ, ψ. Thus in principle the problem of the vibrations of a non-uniform string has been solved†.

3. To tackle the solution of the general Sturm–Liouville problem, let us suppose that the equation

$$S u \equiv (p u')' - q u = 0$$

has no solution in the domain of the operator S (i.e. among the twice differentiable functions with the fixed boundary conditions). The operator S is then said to be *non-singular.*

We shall show that a *non-singular operator S has an inverse A, which is a Fredholm integral operator with a continuous symmetric kernel.*

The phrase "A is the inverse operator of the operator S" means the following:

(1) For any function $u(x)$ in the domain of S we have

$$A(S u) = u.$$

(2) For any continuous function $\psi(x)$ the function $A\,\psi(x)$ belongs to the domain of S and

$$S(A\,\psi) = \psi.$$

† We leave aside questions as to the convergence of the series obtained and the sense in which the function $u(x, t)$ is a solution of equation (1).

The existence of an operator A with the specified properties solves the problem. For as a symmetric Fredholm operator, the operator A possesses an orthogonal system of characteristic functions $e_1(x), \ldots, e_n(x), \ldots$ with non-zero characteristic values $\lambda_1, \ldots, \lambda_n, \ldots, \ldots$ The functions $e_n(x)$ are all continuous in virtue of the concluding results of Section 4. Operating with S on the equation

$$e_n = \frac{1}{\lambda_n} A e_n$$

we get

$$S e_n = \frac{1}{\lambda_n} S A e_n = \frac{1}{\lambda_n} e_n,$$

so that $e_n(x)$ is a characteristic function of S with the characteristic value $1/\lambda_n$. We shall show that the functions $e_n(x)$ $(n = 1, 2, \ldots)$ form a complete system in the space $L_2(a, b)$ (and that consequently the operator A has no characteristic functions corresponding to the characteristic value zero). For any function u in the domain of S, we have

$$A(S u) = u,$$

which shows that u belongs to the range of the operator A. But then u can be developed in the characteristic functions of A with non-zero characteristic values, i.e. in the functions $e_1(x), e_2(x), \ldots, e_n(x), \ldots$

The functions $u(x)$ evidently constitute a set which is dense in the space $L_2(a, b)$. It follows that the functions $e_n(x)$ form a complete systems in $L_2(a, b)$, as required.

4. Thus the solution of the Sturm–Liouville problem reduces to the construction of an inverse operator A for the operator S in the form of a Fredholm integral operator with a symmetric kernel.

Let $u(x)$ be an arbitrary function in the domain of S, i.e. a function which satisfies the boundary conditions and has two continuous derivatives, and let $S u = \psi$. We have to reconstruct u from ψ. We observe that the domain of operator S can contain only one function u which satisfies the equation $S u = \psi$. For if there were two possible solutions to such an equation, their difference v would satisfy the homogeneous equation $S v = 0$ and would also lie in the domain of S, when it would follow by hypothesis that $v \equiv 0$.

So we must solve the equation

$$S\,u = \psi, \tag{1}$$

where $\psi(x)$ is a given continuous function. We apply the usual method of undetermined coefficients. Let $u_1(x)$, $u_2(x)$ be two linearly independent solutions of the equation

$$S\,u = (p\,u')' - q\,u = 0, \tag{2}$$

and for definiteness let $u_1(x)$ vanish at $x = b$, and $u_2(x)$ at $x = a$. We shall seek a solution of equation (2) in the form

$$u(x) = C_1(x)\,u_1(x) + C_2(x)\,u_2(x), \tag{3}$$

where $C_1(x)$, $C_2(x)$ are some undetermined (once) differentiable functions. Differentiating (3), we find

$$u'(x) = C_1'\,u_1 + C_2'\,u_2 + C_1\,u_1' + C_2\,u_2'.$$

As usual we impose on the functions $C_1(x)$, $C_2(x)$ the condition

$$C_1'\,u_1 + C_2'\,u_2 = 0, \tag{4}$$

so that

$$u' = C_1\,u_1' + C_2\,u_2'. \tag{5}$$

Differentiating again and substituting in (1), we get

$$C_1'\,p\,u_1' + C_2'\,p\,u_2' = \psi. \tag{6}$$

C_1' and C_2' can now be found from equations (4) and (6). Their discriminant

$$p(u_1'\,u_2 - u_2'\,u_1)$$

is actually independent of x. In fact, by a well-known theorem of Liouville the Wronskian of the equation $S\,u = 0$ can be expressed in the form

$$W(x) = W(a)\,e^{-\int_a^x \frac{p'(t)}{p(t)}\,dt} = W(a)\,e^{-\log\frac{p(x)}{p(a)}} = \frac{W(a)\,p(a)}{p(x)},$$

so that

$$p(u_1'\,u_2 - u_2'\,u_1) = p(x)\,W(x) = W(a)\,p(a) = c_0 = \text{const},$$

as required.

Solving the system (4), (6), we get

$$C_1' = \frac{u_2\,\psi}{c_0}, \quad C_2' = -\frac{u_1\,\psi}{c_0}.$$

We then take the primitives $C_1(x)$, $C_2(x)$ in the following form:

$$C_1(x) = \frac{1}{c_0}\int_a^x u_2(\xi)\,\psi(\xi)\,d\xi, \quad C_2(x) = \frac{1}{c_0}\int_x^b u_1(\xi)\,\psi(\xi)\,d\xi;$$

this choice of the primitives entails that the required solution

$$u = C_1 u_1 + C_2 u_2 = \frac{u_1(x)}{c_0}\int_a^x u_2(\xi)\,\psi(\xi)\,d\xi + \frac{u_2(x)}{c_0}\int_x^b u_1(\xi)\,\psi(\xi)\,d\xi \tag{7}$$

satisfies both boundary conditions $u(a) = u(b) = 0$.

Thus we have obtained the following result: if we have $S\,u = \psi$ for some function $u(x)$ in the domain of the operator S, then

$$u(x) = \frac{u_1(x)}{c_0}\int_a^x u_2(\xi)\,\psi(\xi)\,d\xi + \frac{u_2(x)}{c_0}\int_x^b u_1(\xi)\,\psi(\xi)\,d\xi,$$

or, in other words,

$$u(x) = \int_a^x K(x, \xi)\,\psi(\xi)\,d\xi,$$

where

$$K(x, \xi) = \begin{cases} \dfrac{1}{c_0}\,u_1(x)\,u_2(\xi) & \text{for} \quad \xi < x, \\ \dfrac{1}{c_0}\,u_1(\xi)\,u_2(x) & \text{for} \quad \xi > x. \end{cases} \tag{8}$$

The function $K(x, \xi)$ is clearly symmetric and continuous for $a \leqq \xi \leqq b$, $a \leqq x \leqq b$. We denote

$$A\,\psi = \int_a^b K(x, \xi)\,\psi(\xi)\,d\xi. \tag{9}$$

Then for any function $u(x)$ in the domain of S we have $A(S\,u) = u$, so that the first of the two conditions implicit in "A is the inverse operator of S" is satisfied. We verify the second condition. Let $\psi(x)$ be an arbitrary continuous function on the closed interval $[a, b]$; we shall show that $u = A\,\psi$ belongs to the domain of S and that the equation

$$S\,A\,\psi = \psi \tag{10}$$

is satisfied.

On substituting the expression for $K(x, s)$ from (8) into (9), we get

$$u(x) = \frac{u_1(x)}{c_0} \int_a^x u_2(\xi)\, \psi(\xi)\, d\xi + \frac{u_2(x)}{c_0} \int_x^b u_1(\xi)\, \psi(\xi)\, d\xi. \quad (11)$$

This function is obviously continuous and vanishes at $x = a$, $x = b$; in addition, it has a continuous derivative:

$$u'(x) = \frac{u_1'(x)}{c_0} \int_a^x u_2(\xi)\, \psi(\xi)\, d\xi + \frac{u_1(x)}{c_0} u_2(x)\, \psi(x) +$$

$$+ \frac{u_2'(x)}{c_0} \int_x^b u_1(\xi)\, \psi(\xi)\, d\xi - \frac{u_2(x)}{c_0} u_1(x)\, \psi(x) \quad (12)$$

$$= \frac{u_1'(x)}{c_0} \int_a^x u_2(\xi)\, \psi(\xi)\, d\xi + \frac{u_2'(x)}{c_0} \int_x^b u_1(\xi)\, \psi(\xi)\, d\xi.$$

It is clear from (12) that $u(x)$ also has a second derivative:

$$u''(x) = \frac{u_1''(x)}{c_0} \int_a^x u_2(\xi)\, \psi(\xi)\, d\xi + \frac{u_1'(x)}{c_0} u_2(x)\, \psi(x) +$$

$$+ \frac{u_2''(x)}{c_0} \int_x^b u_1(\xi)\, \psi(\xi)\, d\xi - \frac{u_2'(x)}{c_0} u_1(x)\, \psi(x) \quad (13)$$

$$= \frac{u_1''(x)}{c_0} \int_a^x u_2(\xi) \psi(\xi)\, d\xi + \frac{u_2''(x)}{c_0} \int_b^b u_1(\xi) \psi(\xi)\, d\xi + \frac{1}{c_0} W(x) \psi(x).$$

On multiplying (13) by $p(x)$, (12) by $p'(x)$, (11) by $-q(x)$ and adding, we get

$$p\, u'' + p'\, u' - q\, u = \frac{p\, u'' + p'\, u_1 - q\, u_1}{c_0} \int_a^x u_2(\xi)\, \psi(\xi)\, d\xi +$$

$$+ \frac{p\, u'' + p'\, u_2 - q\, u_2}{c_0} \int_x^b u_1(\xi)\, \psi(\xi)\, d\xi + \frac{1}{c_0} p\, W\, \psi(x) = \psi(x),$$

or

$$S u = S A \psi = \psi,$$

as required.

We now have a full solution of the Sturm–Liouville problem in the non-singular case.

The function $K(x, s)$ which we have constructed is called *Green's function* for the boundary problem under consideration.

Note. A definite physical significance can be attached to Green's function. We interpret the equation

$$(p u')' - q u = \psi(x)$$

as the condition for equilibrium of a non-uniform string under the action of a steady force of line-density $\psi(x)$. As we have shown, the required equilibrium configuration can be described in the form of an integral

$$u(x) = \int_a^b K(x, \xi) \psi(\xi) \, d\xi .$$

Now let $\psi(x)$ vanish everywhere except on an interval of length $2h$, centre ξ_0, where it takes on the constant value $1/2h$. Such a function $\psi(x)$ is the line-density of a unit force distributed uniformly along the interval $[\xi_0 - h, \xi_0 + h]$. The equilibrium configuration of the string is given in this case by the function

$$\frac{1}{2h} \int_{\xi_0 - h}^{\xi_0 + h} K(x, \xi) \, d\xi .$$

In the limit as $h \to 0$ this expression tends to $K(x, \xi_0)$. We can therefore say that the function $K(x, \xi_0)$ *describes the equilibrium configuration of a string to which is applied a unit force concentrated at the single point* ξ_0.

Conversely, if we know for any ξ the equilibrium configuration $K(x, \xi)$ of the string under the action of a unit force concentrated at the point ξ, we naturally assume that the corresponding configuration under the action of a continuously distributed force of line-density $\psi(x)$ is given by the integral

$$\int_b^a K(x, \varepsilon) \, \psi(\varepsilon) \, d\varepsilon,$$

This consideration could equally well be taken as a starting-point for the solution of the Sturm–Liouville problem †.

A similar situation often obtains with other physical models. For instance let $K(x, \xi)$ denote the steady-state temperature of a diathermal bar occupying the closed interval $[a, b]$ under the agency of a heat source of unit power situated at the point ξ; then the steady-state temperature when the source is continuously distributed with line-density $\psi(\xi)$ will be given by the integral

$$u(x) = \int_a^b K(x, \xi)\, \psi(\xi)\, d\xi\,. \tag{9}$$

It is for this reason that functions of the form (9) are called "source-representable".

5. Let us consider the special case when there exists a function $u(x) \not\equiv 0$ which satisfies the boundary conditions and the equation $S u = 0$.

This means that the operator S possesses a characteristic value $\lambda = 0$. Since S can have only a countable set of characteristic values (we recall that the characteristic functions of a symmetric operator S which correspond to distinct characteristic values are orthogonal), there exists a number λ_0 which is not a characteristic value of S. Then all our arguments can be applied to the operator $S_1 = S - \lambda_0 E$, which has the same structure as S but is non-singular. The operator S_1 has an inverse operator A_1—an integral operator with a symmetric kernel—and therefore possesses a complete system of characteristic functions $e_1(x), \ldots, e_n(x), \ldots$ with characteristic values $\lambda_1, \ldots, \lambda_n, \ldots$, it follows that S also has a complete system of characteristic functions, viz. the same ones $e_1(x), \ldots, e_n(x), \ldots$ with characteristic values $\lambda_1 + \lambda_0, \ldots, \lambda_n + \lambda_0, \ldots$ Thus the Sturm–Liouville problem is also soluble in the singular case.

Problem. Construct Green's function $K(x, s)$ for the differential operators

$$S u = u_{xx}$$

with the boundary conditions: (1) $u(0) = u(\pi) = 0$; (2) $u(0) = u'(\pi) = 0$. Write down the developments of these functions in bilinear series (Section 4, problem 2).

† Cf. R. Courant and D. Hilbert, *Methods of Mathematical Physics*, Vol. 1, Chapter 5, Section 14.

Answer.

$$K_1(x, \xi) = \begin{cases} x(\pi - \xi) \text{ for } x \leqslant \xi, \\ (\pi - x)\xi \text{ for } x \geqslant \xi, \end{cases} \qquad K_2(x, s) = \min(x, s),$$

$$K_1(x, s) = \frac{2}{\pi} \sum_{n=1}^{\infty} \frac{\sin nx \sin ns}{n^2},$$

$$K_2(x, s) = \frac{2}{\pi} \sum_{n=1}^{\infty} \frac{\sin\left(n + \frac{1}{2}\right) x \sin\left(n + \frac{1}{2}\right) s}{\left(n + \frac{1}{2}\right)^2}.$$

6. Non-homogeneous Integral Equations With Symmetric Kernels

1. In this section we shall consider the equation

$$\varphi(x) = f(x) + \int_a^b K(x, s)\varphi(s)\,ds, \tag{1}$$

where $f(x)$ and $K(x, s)$ are given, and $\varphi(x)$ is the function required. The kernel $K(x, s)$ is assumed to be symmetric and square-summable.

In abstract space the analogous equation is of the form

$$\varphi = f + A\varphi, \tag{2}$$

where A is a completely continuous symmetric operator.

Let us suppose to begin with that a solution φ of equation (2) exists.

Projecting both sides of (2) onto the line determined by the characteristic vector e_k (where $A e_k = \lambda_k e_k$), we get

$$\begin{aligned}(\varphi, e_k) &= (f, e_k) + (A\varphi, e_k) = (f, e_k) + (\varphi, A e_k) \\ &= (f, e_k) + (\varphi, \lambda_k e_k) = (f, e_k) + \lambda_k(\varphi, e_k),\end{aligned} \tag{3}$$

and hence for $\lambda_k \neq 1$ we find

$$(\varphi, e_k) = \frac{(f, e_k)}{1 - \lambda_k}. \tag{4}$$

Thus, provided the number 1 is not a characteristic value of the operator, all the Fourier coefficients, of the solution φ are well-defined. In this case there can only be one solution, namely

$$\sum_{k=1}^{\infty} \frac{(f, e_k)}{1 - \lambda_k} e_k. \tag{5}$$

Let us see that the series (5) does actually constitute a solution of (2). We observe first that the series is convergent in the norm since the squares of its coefficients evidently form a convergent series†. We denote its sum by φ. Then

$$A\varphi = \sum (f, e_k)\frac{\lambda_k e_k}{1-\lambda_k} = -\sum (f, e_k)\, e_k + \sum \frac{(f, e_k)}{1-\lambda_k} e_k = \varphi - f,$$

so that the vector φ really does satisfy equation (2).

Let us consider now the case when the set of characteristic values of the operator A contains the number 1. If $\lambda_k = 1$ and $(f, e_k) \neq 0$, equation (3) yields a contradiction and (2) has no solution. But if when $\lambda_k = 1$ we also have $(f, e_k) = 0$, then there is no contradiction, but equation (3) imposes no condition at all on the unknown coefficient (φ, e_k).

We claim that in this case the series (5) gives a solution of (2), in which for $\lambda_k = 1$ we can take any value for $(f, e_k)/1 - \lambda_k$. For let us put $\varphi = \varphi' + \varphi''$, where

$$\varphi' = \sum \frac{(f, e_k)}{1-\lambda_k} e_k \qquad (\lambda_k \neq 1),$$

$$\varphi'' = \sum \xi_k e_k \qquad (\lambda_k = 1).$$

By hypothesis, the vector f is orthogonal to the vectors e_k for which $\lambda_k = 1$ and therefore has a development in the vectors e_k of the first group. Applying the theorem proved above to the subspace generated by these vectors, we get

$$A\varphi' = \varphi' - f.$$

Using the condition $\lambda_k = 1$, we find further:

$$A\varphi'' = \sum \xi_k \lambda_k e_k = \sum \xi_k e_k = \varphi''.$$

If follows that

$$A\varphi = A\varphi' + A\varphi'' = \varphi' - f + \varphi'' = \varphi - f,$$

as required.

We have obtained the following result:

THEOREM: *If the number 1 is not included amongst the λ_k, then the equation*

$$\varphi = f + A\varphi \tag{6}$$

† The quantities $\frac{1}{|1-\lambda_k|}$ have a common bound since $\lambda_k \to 0$ (Section 3).

has a unique solution for any f. If the number 1 does occur in the λ_k, then a solution of (6) exists only for vectors f which are orthogonal to the corresponding characteristic subspace of the operator A; moreover the solution is determined only up to an arbitrary term of this subspace.

Problems. 1. Solve the equations:

(a) $$\varphi_1(x) = 3\int_0^2 xs\varphi_1(s)\,ds + 3x - 2.$$

Answer $$\varphi_1(x) = \frac{9}{7}x - 2,$$

(b) $$\varphi_2(x) = 3\int_0^1 xs\varphi_2(s)\,ds + 3x - 2.$$

Answer $\varphi_2(x) = Cx - 2$ (C arbitrary),

(c) $$\varphi_3(x) = \int_0^1 (x+s)\,\varphi_3(s)\,ds + 18x^2 - 9x - 4.$$

Answer $$\varphi_3(x) = 18x^2 + 12x + 9,$$

(d) $$\varphi_4(x) = \int_0^\pi \cos(x+s)\,\varphi_4(s)\,ds + 1.$$

Answer $$\varphi_4(x) = 1 - \frac{2\sin x}{1 - \frac{\pi}{2}}.$$

Hint. Use the method of Section 4, art. 4.

2. An integral equation in which the unknown function $\varphi(x)$ enters only under the integral sign

$$\int_a^b K(x, s)\,\varphi(s)\,ds = f(x),$$

is sometimes termed *an equation of the first kind* (in distinction to equations of the second kind such as were considered above where the unknown function figures separately). Show that an equation of the first kind (in the case of a square-summable symmetric kernel) has a solution in the space $L_2(a, b)$ if and only if the series $\sum_{n=2}^{\infty} \frac{c_n^2}{\lambda_n^2}$ converges, where the c_n are the coefficients of the development of $f(x)$ in the characteristic functions of the kernel $K(x, s)$ and the λ_n are the corresponding characteristic values.

2. By way of example, an equation of the form (1) arises in solving the problem of forced vibrations of a non-uniform string

fixed at its end-points under the action of a periodic force $g(x, t) = g(x) \cos \omega t$.

These vibrations are described by the equation

$$(p(x) u_x)_x = \mu(x) u_{tt} + g(x) \cos \omega t, \qquad (1)$$

where $p(x)$, $\mu(x)$ are physical characteristics of the string and $g(x)$ is the external force per unit length of string.

We shall look for a particular solution of equation (1) in the form of the product

$$v(x, t) = \varphi(x) \cos \omega t, \qquad (2)$$

where $\varphi(x)$ is some twice differentiable function that vanishes for $x = a$ and $x = b$. Substituting (2) in (1), we get for the function $\varphi(x)$

$$(p \varphi')' + \omega^2 \mu \varphi = g(x). \qquad (3)$$

Operating with A on both sides of the equation, A being the inverse operator of $S = (p \varphi')'$, we get

$$\varphi + \omega^2 A(\mu \varphi) = A g.$$

Recalling that A is Fredholm's operator with a symmetric kernel $K(x, s)$, we arrive at the equation

$$\varphi(x) = f(x) - \omega^2 \int_a^b K(x, s) \mu(s) \varphi(s) \, ds, \qquad (4)$$

where $f(x) = A g(x) = \int_a^b K(x, s) g(s) \, ds$ is a known function. If $\mu(s) \not\equiv 1$, the Fredholm operator has a non-symmetric kernel. In this case, the substitution $\psi(x) \equiv \varphi(x) \sqrt{\mu(x)}$ reduces the equation to one with the symmetric kernel $\omega^2 K(x, s) \sqrt{\mu(x) \mu(s)}$. For simplicity we shall assume that $\mu(s) \equiv 1$.

The condition for equation (4) to have a solution for any $f(x)$ is that the kernel $-\omega^2 K(x, s)$ should not be associated with the characteristic value 1, or what is the same thing, that the Sturm–Liouville kernel of S, the inverse operator of A, should lack the characteristic value $\lambda = -\omega^2$. We observe that the frequencies of the characteristic vibrations of the string were determined by the condition $\lambda_n = -\nu_n^2$ (Section 5, art. 2). Thus the condition for equation (4) to have a solution for any function $f(x)$ is that the frequency ω should not coincide with any natural frequency ν_n of the string. The external force must not be in resonance with the natural frequencies of the string.

If the frequency ω of the forced vibration does coincide with one of the natural frequencies of the string, the condition for the problem to be soluble is that $f(x)$ should be orthogonal to the corresponding characteristic function $e_n(x)$, i.e. $(f, e_k) = (A\,g, e_k) = 0$. Since the operator A is symmetric, this condition quickly reduces to one of orthogonality between the function $g(x)$ and the characteristic function $e_k(x)$:

$$(A\,g, e_k) = (g, A\,e_k) = (g, \lambda_k\, e_k) = \lambda_k (g, e_k) = 0.$$

The solution obtained for the integral equation (4) allows us to construct a particular solution of (3). Any solution of (3) can be obtained by adding to the solution found some solution of the homogeneous equation

$$(p\, v_x)_x = \mu\, v_{tt}.$$

These considerations enable us also to obtain a solution of equation (1) in the case when we have initial conditions.

7. Non-homogeneous Integral Equations With Arbitrary Kernels

1. Let us consider the integral equation

$$\varphi(x) - \int_a^b K(x, s)\, \varphi(s)\, \mathrm{d}s = f(x), \tag{1_1}$$

where the kernel $K(x, s)$ is square-summable but not in general symmetric. The function $(f(x)$ is assumed to belong to the space $L_2(a, b)$ and it is in this space that we seek the unknown function $\varphi(x)$.

For $f(x) \equiv 0$ we get a homogeneous equation in which we denote the unknown function by $\varphi_0(x)$:

$$\varphi_0(x) - \int_a^b K(x, s)\, \varphi_0(s)\, \mathrm{d}s = 0. \tag{1_2}$$

It seems natural to consider alongside equations (1_1), (1_2) the "allied" equations with kernel $K(s, x)$ distinguished from the original kernel $K(x, s)$ by a transposition of the arguments:

$$\psi(x) - \int_a^b K(s, x)\, \psi(s)\, \mathrm{d}s = g(x), \tag{1_3}$$

$$\psi_0(x) - \int_a^b K(s, x)\, \psi_0(s)\, \mathrm{d}s = 0. \tag{1_4}$$

The following fundamental theorem establishes a connection between the solution of equations (1_1), (1_2), (1_3), (1_4).

We observe first that only two cases are logically possible:

(a) *equation* (1_2) *has the unique solution* $\varphi_0(x) \equiv 0$;

(b) *equation* (1_2) *has a solution* $\varphi_0(x) \not\equiv 0$.

THEOREM 1. (E. Fredholm, 1903). *In case* (a) *equation* (1_1) *has a solution for any* $f(x) \in L_2$ *which is, moreover, unique; equation* (1_4) *has the unique solution* $\psi_0(x) \equiv 0$; *equation* (1_3) *has a unique solution for any* $g(x) \in L_2$.

In case (b) *the number of linearly independent solutions of equation* (1_2) *is finite; we denote it by* ν. *Equation* (1_4) *has the same number of linearly independent solutions. Equation* (1_1) *has a solution if and only if the function* $f(x)$ *is orthogonal to all the* ν *solutions of* (1_4); *this solution is determined not uniquely, but up to a term which is a solution of* (1_2); *among the solutions of equation* (1_1) *there exists one and only one which is orthogonal to all the solutions of* (1_2). *Similar assertions hold for the solutions of equation* (1_3).

2. Let us consider first of all the analogue of theorem 1 for the case of a linear system of algebraic equations.

$$\sum_{j=1}^{m} a_{ij}\,\xi_j = b_i \quad (i = 1, 2, \dots, m). \tag{2_1}$$

We write down the corresponding homogeneous system

$$\sum_{j=1}^{m} a_{ij}\,\xi_j^0 = 0 \tag{2_2}$$

and the allied systems

$$\sum_{j=1}^{m} a_{ji}\,\eta_j = c_i, \tag{2_3}$$

$$\sum_{j=1}^{m} a_{ji}\,\eta_0^j = 0, \tag{2_4}$$

for which the matrix is obtained by transposing the matrix of the systems (2_1) and (2_2). We examine the assertions that parallel those of Fredholm's theorem.

(a) Let us suppose that the system (2_2) has only the null solution. As we know from algebra, this means that the rank of the matrix $A = \|a_{ij}\|$ is equal to the number m, i.e. $\det A \neq 0$. Hence system (2_1) has a solution for any b_i. The system (2_4) has the determinant

$\det \|a_{ji}\| = \det \|a_{ij}\|$ which is therefore also non-zero; hence system (2_3) possesses a solution for any c_i, which is moreover unique; in particular, for $c_i = 0$ it has the unique solution $\eta_j^0 = 0$. Thus in case (a) all the assertions that constitute the analogue to Fredholm's theorem are verified.

(b) We suppose now that the system (2_2) has a non-trivial solution ξ^0. This means that the rank r of the matrix A is less than m. The number ν of linearly independent solutions of the system (2_2) is equal to $m - r$. Since the rank of a matrix is invariant under transposition, the number of linearly independent solutions of the system (2_4) is also equal to $m - r = \nu$. The system (2_1) no longer has a solution for arbitrary b_i. To ascertain what conditions must be imposed on the b_i to ensure that the system (2_1) should have a solution, we interpret the system geometrically, regarding the aggregate of numbers $(\xi_1, \ldots, \xi_m)$ as a vector in the m-dimensional Euclidean space R_m. The existence of a solution of the system (2_1) is equivalent to asserting that the vector $b = (b_1, b_2, \ldots, b_m)$ is contained in the linear envelope L of the vectors $a_1 = (a_{11}, a_{21}, \ldots, a_{m1})$, $a_2 = (a_{12}, a_{22}, \ldots, a_{m2})$, $\ldots$, $a_m = (a_{1m}, a_{2m}, \ldots, a_{mm})$. If Z is the orthogonal complement of this envelope, the interpretation can be expressed as follows: the system (2_1) has a solution if and only if the vector b is orthogonal to the subspace Z. The condition that a vector η^0 should belong to the subspace Z is formulated in system (2_4). It follows that system (2_1) has a solution if and only if the vector b is orthogonal to any solution of system (2_4). Further, in the case considered the system (2_1) has a whole complex of solutions which in geometrical terms constitute a hyperplane parallel to the solution subspace of system (2_2). The perpendicular dropped from the origin of coordinates onto this hyperplane identifies uniquely that solution which is orthogonal to all the solutions of the system (2_2). Thus we have also verified for case (b) the analogues of the assertions of Fredholm's theorem.

3. Returning now to the integral equations, let us begin by considering the equations with degenerate kernels:

$$K(x, s) = \sum_{k=1}^{m} p_k(x)\, q_k(s),$$

$$K(s, x) = \sum_{k=1}^{m} p_k(s)\, q_k(x).$$

We can assume that the functions $p_k(x)$, and likewise the $q_k(s)$, are linearly independent. Equations (1_1)–(1_4) acquire the form

$$\varphi(x) - \sum_{k=1}^{m} p_k(x) \int_a^b q_k(s)\, \varphi(s)\, \mathrm{d}s = f(x), \tag{3_1}$$

$$\varphi_0(x) - \sum_{k=1}^{m} p_k(x) \int_a^b q_k(s)\, \varphi_0(s)\, \mathrm{d}s = 0, \tag{3_2}$$

$$\psi(x) - \sum_{k=1}^{m} q_k(x) \int_a^b p_k(s)\, \psi(s)\, \mathrm{d}s = g(x), \tag{3_3}$$

$$\psi_0(x) - \sum_{k=1}^{m} q_k(x) \int_a^b p_k(s)\, \psi_0(s)\, \mathrm{d}s = 0. \tag{3_4}$$

These equations can be expressed in the abstract form

$$\varphi - \sum p_k(q_k, \varphi) = f, \tag{4_1}$$

$$\varphi_0 - \sum p_k(q_k, \varphi_0) = 0, \tag{4_2}$$

$$\psi - \sum q_k(p_k, \psi) = g, \tag{4_3}$$

$$\psi_0 - \sum q_k(p_k, \psi_0) = 0, \tag{4_4}$$

where the vectors $\varphi, f, p_k, q_k, \ldots$ belong to some Euclidean space E.

The operators involved in these equations belong to the class of *degenerate operators*; an operator is said to be degenerate if it is defined by an equation of the form

$$B\varphi = \sum_{k=1}^{m} p_k(q_k, \varphi).$$

It is evident that the degenerate operator B maps the whole space onto the finite-dimensional subspace generated by the vectors $p_1, p_2, \ldots, p_m$. It is clear from equation (4_1) that if it possesses a solution it will be of the form

$$\varphi = f + \sum \xi_k p_k, \tag{5_1}$$

where the ξ_k are certain unknown coefficients. Similarly the solutions of the remaining equations will be of the form

$$\varphi_0 = \sum \xi_k^0 p_k, \tag{5_2}$$

$$\psi = g + \sum \eta_k q_k, \tag{5_3}$$

$$\psi_0 = \sum \eta_k^0 q_k. \tag{5_4}$$

Substituting (5_1) in (4_1) we find that the numbers ξ_k must satisfy the equation

$$\sum_{k=1}^{m} \xi_k p_k - \sum_{k=1}^{m} p_k (q_k, f) - \sum_{k=1}^{m} p_k \left(q_k, \sum_{i=1}^{m} \xi_i p_i \right) = 0$$

or, (since the vectors p_k are linearly independent), the system

$$\xi_k - \sum_{i=1}^{m} \xi_i (p_i, q_k) = (f, q_k) \quad (k = 1, 2, \dots, m).$$

With the notation $(p_i, q_k) = a_{ik} (i \neq k)$, $1 - (p_i, q_i) = a_{ii}$, $(f, q_k) = b_k$, we reduce this system to the form

$$\sum_{j=1}^{m} a_{ij} \xi_j = b_i \quad (i = 1, 2, \dots, m). \tag{6_1}$$

Equations (4_2)–(4_4) reduce similarly to systems of the form

$$\sum_{j=1}^{m} a_{ij} \xi_j^0 = 0, \tag{6_2}$$

$$\sum_{j=1}^{m} a_{ji} \eta_j = c_i \tag{6_3}$$

$$\sum_{j=1}^{m} a_{ji} \eta_j^0 = 0, \tag{6_4}$$

respectively, where $c_i = (g, p_i)$. If a solution exists for any one of the systems (6_1)–(6_4), a solution of the corresponding member of the set (4_1)–(4_4), or what is the same thing, a solution of the corresponding equation in (3_1)–(3_4) can be constructed using the corresponding formula in the group (5_1)–(5_4).

But as we have seen, all the assertions paralleling the Fredholm theorem hold for the systems (6_1)–(6_4). They will therefore be valid for equations (3_1)–(3_4) also. We need only verify that the scalar product, say of the vector f and the solution ψ_0 of equation (4_4), coincides in the sense of the metric of the abstract Euclidean space E with the scalar product of the vector b with coordinates $b_k = (f, q_k)$ and the vector $\eta^0 = (\eta^0, \eta^0, \dots, \eta^0)$ of the finite-dimensional Euclidean space R_m. This verification is accomplished by the simple computation:

$$(f, \psi_0) = \left(f, \sum_{k=1}^{m} \eta_k^0 q_k \right) = \sum_{k=1}^{m} \eta_k^0 (f, q_k) = \sum_{k=1}^{m} b_k \eta_k^0 = (b, \eta^0).$$

We have thus established the Fredholm theorem in the case of a degenerate kernel $K(x, s)$.

4. We now consider the general case. Let $K(x, s)$ be an arbitrary function, square-summable over the region $a \leqq x, s \leqq b$. As was shown in Section 4, the integral operator

$$A\varphi = \int_a^b K(x, s)\,\varphi(s)\,\mathrm{d}s$$

can be represented as the limit (in the sense of the norm) of integral operators

$$A_n\varphi = \int_a^b K_n(x, s)\,\varphi(s)\,\mathrm{d}s$$

with degenerate kernels $K_n(x, s)$. It is evident that the adjoint integral operator

$$A^*\psi = \int_a^b K(s, x)\,\psi(s)\,\mathrm{d}s$$

is then the limit of the integral operators

$$A_n^*\psi = \int_a^b K_n(s, x)\,\psi(s)\,\mathrm{d}s,$$

the kernels of which are also degenerate.

The adjoint operator A^* is related to A by the equation

$$(A^* p, q) = (p, A q) \tag{7}$$

for any vectors p, q. To see this, we observe that

$$(A^* p, q) = \int_a^b \left\{ \int_a^b (K(s, x)\,p(s)\,\mathrm{d}s \right\} q(x)\,\mathrm{d}x,$$

$$(p, A q) = \int_a^b p(x) \left\{ \int_a^b K(x, s)\,q(s)\,\mathrm{d}s \right\} \mathrm{d}x.$$

The first of these integrals is transformed into the second by interchanging the variables x, s and reversing the order of integration, a procedure which is valid in the general case in virtue of Fubini's theorem.

Equations (1_1)–(1_4) can be written in abstract form, regarding the vectors $\varphi, f, \ldots$ as elements of some Euclidean space E:

$$\varphi - A\varphi = f, \tag{8_1}$$

$$\varphi_0 - A\varphi_0 = 0, \tag{8_2}$$

$$\psi - A^*\psi = g, \tag{8_3}$$

$$\psi_0 - A^*\psi_0 = 0. \tag{8_4}$$

The solutions of equations of the form (8_2) are the characteristic vectors of the corresponding operators with characteristic value 1. For brevity we shall simply call them the characteristic vectors. We denote by A_n^*, A_n the operators corresponding to the degenerate kernels $K_n(x, s)$, $K_n(s, x)$, and consider the homogeneous equation

$$\varphi_n^0 - A_n \varphi_n^0 = 0. \qquad (9_2)$$

LEMMA 1. *If for each* η *equation* (9_2) *has a solution* $\varphi_n^0 \neq 0$, *equation* (8_2) *will also have a non-zero solution.*

Proof. We can always take the solution φ_n^0 of (9_2) to be normalised, so that $\|\varphi_n^0\| = 1$. Since the operator A is completely continuous, the sequence $A\varphi_n^0$ contains a convergent subsequence; discarding unwanted terms and renumbering, we can regard the sequence $A\varphi_n^0$ itself as convergent. Then $A_n \varphi_n^0$ also converges, since

$$A_n \varphi_n^0 = (A_n - A)\varphi_n^0 + A\varphi_n^0$$

and

$$\|(A_n - A)\varphi_n^0\| \leqslant \|A_n - A\| \, \|\varphi_n^0\| \to 0.$$

The sequence $\varphi_n^0 = A_n \varphi_n^0$ converges together with $A_n \varphi_n^0$; we put $\varphi_0 = \lim\limits_{n\to\infty} \varphi_n^0$. The vector φ_0, like the φ_n^0, has the norm 1 and

$$A\varphi_0 = \lim_{n\to\infty} A\varphi_n^0 = \lim_{n\to\infty} A_n \varphi_n^0 = \lim_{n\to\infty} \varphi_n^0 = \varphi_0;$$

thus equation (8_2) does in fact possess a non-zero solution φ_n.

LEMMA 2. *If for each n equation* (9_2) *has some number k of linearly independent solutions* $\varphi_n^0, \varphi_2^0, \ldots, \varphi_{nk}^0$, *then equation* ($8_2$) *also has k linearly independent solutions* $\varphi_1^0, \varphi_2^0, \ldots, \varphi_k^0$.

Proof. We can take the solutions $\varphi_{1n}^0, \varphi_{2n}^0, \ldots, \varphi_{kn}^0$ of (9_2) to be orthogonal and normalised. We form the sequences

$$\varphi_{11}^0, \varphi_{12}^0, \ldots, \varphi_{1n}^0, \ldots,$$
$$\varphi_{21}^0, \varphi_{22}^0, \ldots, \varphi_{2n}^0, \ldots,$$
$$\cdots\cdots\cdots\cdots\cdots\cdots$$
$$\varphi_{k1}^0, \varphi_{k2}^0, \ldots, \varphi_{kn}^0, \ldots$$

As was shown in Lemma 1, each of them contains a convergent subsequence, and with the same convention as before we can assume the sequences themselves to converge and their limits $\varphi_1^0, \varphi_2^0, \ldots, \varphi_k^0$ to be non-zero solutions of (8_2). Moreover, since the functions $\varphi_{1n}, \varphi_{2n}, \ldots, \varphi_{kn}$ are orthogonal for each n, their limits $\varphi_1^0, \varphi_2^0, \ldots, \varphi_k^0$

are also orthogonal and are therefore linearly independent as required.

The totality of solutions of equation (8_2) constitutes *a* subspace, which we denote by Φ_0. By lemma 6, Section 3, the subspace Φ_0 is finite-dimensional. We denote its dimension by ν.

LEMMA 3. *If the completely continuous operator A can be represented as the limit of a sequence of degenerate operators $\tilde{A}_n$, it can also be represented as the limit of a sequence of degenerate operators A_n for each of which the space formed by its characteristic vectors coincides with the subspace Φ_0.*

Proof. Let $\varphi_1^0, \ldots, \varphi_\nu^0$ be an orthonormal system in the subspace Φ_0. We denote

$$h_i^n = \varphi_i^0 - A_n \varphi_i^0 \quad (i = 1, 2, \ldots, \nu).$$

We define an operator $\tilde{A}_n$ by the formula

$$\tilde{A}_n \varphi = A_n \varphi + \sum_{i=1}^{\nu} h_i^n (\varphi, \varphi_i^0).$$

Evidently $\tilde{A}_n$ is degenerate together with the operator A_n. It is clear that $\|\tilde{A}_n - A_n\| \to 0$, since $h_i^n \to 0$, and it follows that $\|\tilde{A}_n - A_n\| \to 0$. We shall show that the vectors φ_j^0 are characteristic vectors of the operator $\tilde{A}_n$, for

$$\begin{aligned} \tilde{A}_n \varphi_i^0 &= A_n \varphi_j^0 + \sum_{i=1}^{\nu} (\varphi_i^0 - A_n \varphi_i^0)(\varphi_j^0, \varphi_i^0) \\ &= A_n \varphi_j^0 + \varphi_j^0 - A_n \varphi_j^0 = \varphi_j^0, \end{aligned}$$

as required.

Generally speaking, the operators $\tilde{A}_n$ can have more than ν linearly independent characteristic vectors, but this cannot be the case for infinitely many of them. For if it were, then considering only those values of n for which the operator $\tilde{A}_n$ has at least $\nu + 1$ linearly independent characteristic vectors and applying lemma 2, we should obtain the contradiction that the operator A has at least $\nu + 1$ linearly independent characteristic vectors.

Thus only finitely many of the operators $\tilde{A}_n$ can have more than ν linearly independent characteristic vectors. Discarding these and reindexing the sequence $\tilde{A}_n$, we get a sequence which satisfies the conditions of lemma 3.

We can now prove that *the solution subspace* Ψ_0 of *equation* (8_4) *has the same dimensionality as the subspace* Φ_0.

We consider the sequence of degenerate operators $\tilde{A}_n$ defined in lemma 3 and the sequence of adjoint operators $\tilde{A}_n^*$. Since the Fredholm alternative holds for degenerate operators, each of the operators $\tilde{A}_n^*$ has a space of characteristic vectors of dimension exactly ν. Since $\tilde{A}_n \to A$, we have also $\tilde{A}_n^* \to A^*$.

By lemma 2 there exists a system of ν orthonormal solutions of the equation $A^* \psi_0 = \psi_0$. There cannot be more than ν such solutions since, arguing in the reverse order from the operator A^* to the operator A with $\nu + 1$ such solutions, we should get $\nu + 1$ linearly independent solutions for the equation $A \varphi_0 = \varphi_0$, which is excluded by hypothesis.

Thus the number of linearly independent solutions of the equation $A^* \psi_0 = \psi_0$ is always identical with the corresponding number for $A \varphi_0 = \varphi_0$.

We shall now discuss the question of the existence of solutions to equation (8_1). Let us suppose that the vector φ is a solution of this equation and that ψ_0 is a solution of (8_4). Multiplying (8_1) by ψ_0 and using (7), we get

$$(\varphi, \psi_0) - (A \varphi, \psi_0) = (f, \psi_0).$$

But using (8_4), we have

$$(\varphi, \psi_0) - (\varphi, A^* \psi_0) = (\varphi, \psi_0) - (\varphi, \psi_0) = 0,$$

so that for any solution ψ_0 of (8_4)

$$(f, \psi_0) = 0.$$

Thus equation (8_1) can only have a solution on condition that the vector f is orthogonal to all the solutions of (8_4). We shall show that when this condition is satisfied a solution of (8_1) always exists.

We consider a sequence of degenerate operators $\tilde{A}_n \to A$ with the same subspace Φ_0 of characteristic vectors as the operator A, its dimension being ν. The existence of such a sequence was established in lemma 3.

Each of the operators $\tilde{A}_n^*$ possesses the same number ν of orthonormal characteristic vectors ψ_i^n, since the alternative holds for degenerate operators. By lemma 2, we can assume that as $n \to \infty$ these vectors tend respectively to the orthonormal characteristic vectors ψ_i^0 ($i = 1, 2, \ldots, \nu$) of the operator A^*. We consider equation (8_1) with the right-hand side $f_n = f - \sum_{i=1}^{v} (f, \psi_i^n) \psi_i^n$. The vector f_n is orthogonal to all the vectors ψ_i^n (f_n is the perpendicular dropped from the end-point of the vector f onto the subspace $L\{\varphi_1^n, \varphi_2^n, \ldots, \psi_r^n\}$), hence, applying the alternative for the degenerate operator A_n, we establish the existence of a vector φ_n satisfying the equation

$$\varphi_n - \tilde{A}_n \varphi_n = f_n .$$

As $n \to \infty$ the vectors f_n converge to f, since $\psi_i^n \to \psi_i^0 . (f, \psi_i^0) = 0$ ($i = 1, 2, \ldots, \nu$). The vector φ_n can be chosen for any n so as to be orthogonal to the subspace of characteristic vectors of the operator $\tilde{A}_n$, i.e. to the subspace Φ_0.

We shall show that the vectors φ_n so obtained are bounded in the norm. Let us suppose the contrary: let the norms of the φ_n be unbounded; then discarding superfluous vectors, we can suppose that $\|\varphi_n\| \to \infty$. Putting $\tilde{\varphi}_n = \frac{\varphi_n}{\|\varphi_n\|}$, we get a sequence of normalised vectors which satisfy the equation

$$\tilde{\varphi}_n - \tilde{A}_n \tilde{\varphi}_n = \frac{f}{\|\varphi_n\|}. \tag{10}$$

As $n \to \infty$ the right-hand side of this equation tends to zero. By the same considerations as above, the sequence of vectors $\tilde{A}_n \tilde{\varphi}_n$ can be assumed convergent, and with it will converge the sequence $\tilde{\varphi}_n$. Let $\varphi_0 = \lim \tilde{\varphi}_n$; since $\|\tilde{\varphi}_n\| = 1$, we have also $\|\varphi_0\| = 1$. Passing to the limit in equation (10), we find that the vector φ_0 satisfies the equation

$$\varphi_0 - A \varphi_0 = 0,$$

and it follows that φ_0 is contained in the subspace Φ_0. At the same time, since all the vectors $\tilde{\varphi}_n$ are orthogonal to Φ_0, so also is the limiting vector φ_0. The contradiction obtained shows that the vectors φ_n are in fact bounded in the norm.

Since the φ_n are bounded, the sequence $A\varphi_n$ can, as before, be supposed convergent; with it will converge the sequence $\tilde{A}_n \varphi_n = (\tilde{A}_n - A) \varphi_n + A \varphi_n$ and also the sequence $\varphi_n = f_n + \tilde{A}_n \varphi_n$. We denote the limit of the sequence φ_n by φ; passing to the limit in the equation

$$\varphi_n = f_n + (\tilde{A}_n - A_n)\,\varphi_n + A\,\varphi_n,$$

we get

$$\varphi = f + A\,\varphi,$$

i.e. the vector φ is a solution of equation (8_1).

Thus *if the right-hand side f of equation* (8_1) *is orthogonal to every solution of* (8_4), *then* (8_1) *is soluble.* The solution φ is then determined up to an arbitrary solution φ_0 of the homogeneous equation (8_2) and since the solution space of (8_2) is of finite dimension, it can be chosen orthogonal to the whole of this space; such a choice then determines it uniquely.

We have verified all the assertions of the alternative in case (b).

In case (a) equation (8_2) has no non-zero solutions and the subspace Φ_0 contains only the null vector. By what we have proved the solution subspace Ψ_0 of equation (8_4) also contains only the null vector in this case. As we have seen, equation (8_1) has a solution for any f orthogonal to the subspace Ψ_0; in the given case this condition is satisfied by any vector f and (8_1) therefore has a solution for any f. This solution is unique; for the difference of any two solutions of (8_1) is a solution of (8_2) and is therefore zero by hypothesis.

The proof of the theorem is thus complete.

We observe that if both the kernel $K(x, s)$ and the free term $f(x)$ are continuous, then in view of the concluding note to art. 3 of Section 4 the solution of the integral equation (1_1) is a continuous function.

An important corollary of Fredholm's theorem must be mentioned.

Fredholm's alternative. *Given the conditions of art.* 1, *one of the two cases is possible: either the complete integral equation* (1_1) *has a solution for any right-hand side* $f(x) \in L_2(a, b)$, *or the homogeneous equation* (1_2) *has a non-zero solution.*

Problems. 1. If A is an operator in Hilbert space H, mapping H on to a finite-dimensional subspace L, A^* also maps H on to a finite-dimensional subspace, of the same dimensions as L.

Hint. The orthogonal complement of L is mapped on to zero by A^*. Hence $A^*H = A^*L$ and has a number of dimensions not exceeding that of L. It follows from the symmetry of the construction that the number of dimensions cannot be diminished.

2. Show that for any completely continuous operator A in Hilbert space H an orthogonal resolution $H = H_1 + H_2$ can be found such that H_1 is finite- or countable-dimensional, $AH_1 \subset H_1$, $AH_2 = 0$.

Hint. Show that the closure of the subspace AH cannot contain a non-countable set of orthogonal vectors.

3. Show that every completely continuous operator in Hilbert space H is a limit (in the norm) of degenerate operators.

Hint. By problem 2, we can assume that $H = l_2$. If $Ax = (\xi_1, \ldots, \xi_n, \ldots) \in l_2$, put $A_n x = (\xi_1, \ldots, \xi_n, O, \ldots)$. Use problem 3 of Section 1, art. 1.

4. Show that the adjoint to a completely continuous operator is also completely continuous.

Hint. Use problems 1 and 3.

5. Show that Fredholm's theorem still holds if the Fredholm integral operator figuring in it is replaced by any completely continuous operator in Hilbert space.

Hint. Use problems 1–4.

8. Applications to Potential Theory

1. We shall assume a knowledge of the following results from the theory of differential equations†:

(a) A function $u(x, y)$ which satisfies the equation

$$\Delta u = \frac{\partial^2 u}{\partial x^2} + \frac{\partial^2 u}{\partial y^2} = 0 \tag{1}$$

on a region G of the xy-plane is said to be harmonic. For example the function

$$\log \frac{1}{\sqrt{(x - \xi)^2 + (y - \eta)^2}}, \tag{2}$$

which depends on the parameters ξ, η, is harmonic wherever the expression under the square-root sign is non-zero (i.e. everywhere except at the point $x = \xi$, $y = \eta$). Putting $(x, y) = P$, $(\xi, \eta) = Q$, we shall denote the function (2) briefly by

$$\log \frac{1}{r(P, Q)}. \tag{3}$$

The partial derivatives of the function (1) are also harmonic functions for $P \neq Q$.

† Cf., for example, I. G. Petrovsky, *Lectures on Partial Differential Equations*, Interscience, 1955, Chapter 3.

(b) Let C be some simple closed smooth contour dividing the plane into two regions: an interior G_i and an exterior G_e. We consider a function v which is continuous and differentiable in both regions (with a possible discontinuity across the contour C), denoting by v_i its limiting value as it approaches the contour from within and by v_e its limiting value as it approaches from without.

The normal derivatives $\partial v_i/\partial n$, $\partial v\,/\partial n$ have a similar sense (the normal is taken to be positive inward).

If v is a harmonic function, $v_i = 0$ implies $v(P) \equiv 0$ in G_i and $\partial v_i/\partial n \equiv 0$ implies $v(P) \equiv$ const. in G_i; if, in addition, v is bounded as $P \to \infty$, $v_e = 0$ implies $v(P) \equiv 0$ in G_e and $\partial v_e/\partial n \equiv 0$ implies $v(P) =$ const. in G_e.

(c) The point Q is always situated on the contour C in what follows. We write $l = l_Q$ for the coordinate of a point of the contour, say the arc measured from a fixed initial point Q_0 to the point Q, and ω for the angle between the ray PQ and PQ_0. Let a continuous function $\varrho(l)$ be given. Then the function

$$v(P) = \int\limits_C \varrho(l) \frac{\partial}{\partial n_Q} \log \frac{1}{r(P,Q)}\, dl \tag{4}$$

[the potential of a double layer density $\varrho(l)$] is harmonic in both regions G_i and G_e.

Since

$$\frac{\partial}{\partial n_Q} \log \frac{1}{r(P,Q)} = -\frac{1}{r(P,Q)} \frac{\partial r(P,Q)}{\partial n_Q} = -\frac{1}{r(P,Q)} \cos\left(\widehat{\overline{PQ}, \overline{n}}\right) \tag{4'}$$

and at the same time

$$dl_Q = r(P,Q)\, d\omega \frac{1}{\cos\left(\overline{PQ}, \overline{n}\right)},$$

we get a new expression for the potential as

$$v(P) = -\int\limits_C \varrho(l)\, d\omega. \tag{4''}$$

In particular, integral (4) is seen to exist at points P of the contour C itself. If $\varrho(l) = 1$, the function $v(P)$ acquires a simple geometric meaning; it gives the total increment of the angle described by

PQ when Q runs over the contour in the negative direction, so that

$$\int_C \frac{\partial}{\partial n} \log \frac{1}{r(P,Q)} \, \mathrm{d}l = -2\pi, \quad \text{if} \quad P \in G_i, \tag{5}$$

$$\int_C \frac{\partial}{\partial n} \log \frac{1}{r(P,Q)} \, \mathrm{d}l = -\pi, \quad \text{if} \quad P \in C, \tag{6}$$

$$\int_C \frac{\partial}{\partial n} \log \frac{1}{r(P,Q)} \, \mathrm{d}l = 0, \quad \text{if} \quad P \in G_e. \tag{7}$$

Thus in this case the value of $v(P)$ in the region G_i is π less than on the boundary and in the region G_e it exceeds the boundary value by π. In the general case with an arbitrary continuous density $\varrho(Q)$ we have

$$v_i(Q) = v(Q) - \pi\varrho(Q), \tag{8}$$

$$v_e(Q) = v(Q) + \pi\varrho(Q), \tag{9}$$

$$\frac{\partial v_i(Q)}{\partial n} = \frac{\partial v_e(Q)}{\partial n}. \tag{10}$$

The function

$$u(P) = \int_C \varrho(l) \log \frac{1}{r(P,Q)} \, \mathrm{d}l \tag{11}$$

[the potential of a simple layer density $\varrho(l)$] is also harmonic in both the regions G_i, G_e and we have

$$u_i(Q) = u_e(Q) = u(Q), \tag{12}$$

$$\frac{\partial u_i(P)}{\partial n} = \int_C \varrho(l) \frac{\partial}{\partial n_P} \log \frac{1}{r(P,Q)} \, \mathrm{d}l + \pi\varrho(P) \quad (P \in C), \tag{13}$$

$$\frac{\partial u_e(P)}{\partial n} = \int_C \varrho(l) \frac{\partial}{\partial n_P} \log \frac{1}{r(P,Q)} \, \mathrm{d}l - \pi\varrho(P) \quad (P \in C). \tag{14}$$

We note that the potentials $v(P)$ and $u(P)$, which possess continuous ($v(P)$) or piecewise-continuous ($u(P)$) normal derivatives in the vicinity of the boundary, cannot have any well-behaved (even, say, square-summable) derivatives in other directions.

2. The following problems arise:

(1) The first boundary problem (Dirichlet's problem): to find a function $v(P)$ which is harmonic on the region G_i (the interior problem) or on G_e (the exterior problem) and reduces to a prescribed function $f(Q)$ on the contour C.

(2) The second boundary problem (Neumann's problem): to find a function $u(P)$ which is harmonic on the region G_i (the interior problem) or on G_e (the exterior problem) and the normal derivative of which reduces to a prescribed function $g(Q)$ on the contour C.

The last assertion of section (b) can be interpreted as a uniqueness theorem for the solutions of these problems, viz, *Dirichlet's interior problem can only have one solution, and Neumann's interior problem a single solution, discounting an added constant; Dirichlet's exterior problem can only have one solution in the class of bounded functions, and Neumann's exterior problem one solution in the same class, discounting an added constant.*

Fredholm gives a solution to these problems in the case of a contour C with continuous curvature.

3. The solution of Dirichlet's interior problem is sought in the form of a double layer potential

$$v(P) = \int_C \varrho(l) \frac{\partial}{\partial n} \log \frac{1}{r(P,Q)} \, dl \tag{15}$$

where the continuous function $\varrho(l)$ is unknown. In virtue of equation (8) and the conditions of the problem (we denote the coordinate of a point P on the contour C by x)

$$v_i(P) \equiv v_i(x) = \int_C \varrho(l) \frac{\partial}{\partial n} \log \frac{1}{r(P,Q)} \, dl - \pi \varrho(x) = f(x),$$

so that the function $\varrho(x)$ must be determined from Fredholm's integral equation of the second kind with kernel

$$K(x,l) = \frac{\partial}{\partial n} \log \frac{1}{r(P,Q)} = -\frac{d\omega}{\partial l}.$$

This function is continuous for all P, Q on the contour C; it follows from (4″) that as $P \to Q$ it has as its limit the curvature of C at the point Q, taken with the opposite sign.

In virtue of the Fredholm alternative established in Section 7, it is sufficient to show that the corresponding homogeneous equation

$$\int_C \varrho(l)\, K(x, l)\, \mathrm{d}l - \pi\varrho(x) = 0$$

has only a zero solution. Let us suppose the contrary; let $\varrho_0(l)$ be a non-zero solution of this equation. Then for the harmonic function

$$v_0(P) = \int_C K(x, l)\varrho_0(l)\, \mathrm{d}l = -\int_C \varrho_0(l)\, \mathrm{d}\omega$$

we get:

$$v_{i0}(P) = v_0(P) - \pi\varrho_0(P) = 0,$$

and hence by note (c) $v_0(P) = 0$ on the region G_i. But then $\partial v_{i0}(P)/\partial n \equiv 0$ as well, and by (10) $\partial v_{e0}(P)/\partial n \equiv 0$. Since the function $v_0(P)$ evidently tends to zero as P tends to infinity, we see by note (b) that $v_0(P) \equiv 0$ in G_e. Hence $v_{e0}(P) = 0$. We can now infer from (8) and (9) that $\varrho_0(P) = 0$, as required.

Applying the results of Section 7, we get that the integral equation (15) has a solution for any function $f(P)$. If $f(P)$ is continuous, then in virtue of the continuity of the kernel $K(x, l)$ and the concluding note of Section 4, art. 3, the solution $\varrho(P)$ will also be a continuous function. Formulae (8)–(10) will then hold for this solution and the reduction of Dirichlet's problem to the potential (15) will be valid.

Thus *Dirichlet's interior problem has a solution for any continuous boundary function $f(P)$*

The solution of Dirichlet's exterior problem is sought in the same form (15). Here we get the equation

$$v_e(P) = \int_C \varrho(l) \frac{\partial}{\partial n} \log \frac{1}{r(P, Q)}\, \mathrm{d}l + \pi\varrho(P) = f(P). \tag{16}$$

But this time the corresponding homogeneous equation

$$\int \varrho(l)\, K(x, l)\, \mathrm{d}l + \pi\varrho(x) = 0$$

already has a non-zero solution $\varrho(x) \equiv 1$ (cf. equation (6)). This solution is unique (up to scalar multiples). For reproducing the arguments given above with the indexes i, e interchanged, we reach the conclusion $\partial v_i(P)/\partial n = 0$ and then by (b), $v(P) = \text{const.}$ on G_i. From (8) and (9) we deduce that $\varrho(P) = \text{const.}$

It follows by the Fredholm alternative that equation (16) has a solution, not for all f, but only for those that are orthogonal to a certain fixed function $\varrho_0(P)$—the solution of the adjoint equation, which is unique up to scalar multiples.

But we can also make a solution feasible for any boundary function f if, instead of restricting ourselves to solutions determined by (15), which evidently tend to zero at infinity, we also consider the ones obtained from these by the addition of a constant. In fact if f is any function defined on the boundary, we can always find a constant c such that the difference $f - c$ is orthogonal to the function ϱ_0. Then by what we have proved there exists a solution $v(P)$ of the exterior problem with boundary values $f - c$. On the other hand $v_0(P) \equiv c$ is a solution of the exterior problem with the constant boundary value c. It follows that $v(P) + v_0(P)$ is a solution of the problem with boundary values f. Thus *Dirichlet's exterior problem is soluble for any continuous boundary function f.*

4. We look for a solution of Neumann's interior problem in the form of a simple layer potential

$$u(P) = \int\limits_C \varrho(l) \log \frac{1}{r(P,Q)} \, dl \tag{17}$$

with the function $\varrho(l)$ unknown. From (13) and the conditions of the problem

$$\frac{\partial u_i(P)}{\partial n} = \int \varrho(l) \frac{\partial}{\partial n_P} \log \frac{1}{r(P,Q)} \, dl + \pi \varrho(P) = g(P), \tag{18}$$

so that for the function $\varrho(P)$ we again have Fredholm's integral equation, with the kernel

$$K_1(x, l) = \frac{\partial}{\partial n_P} \log \frac{1}{r(P,Q)},$$

the transpose of the kernel $K(x, l)$ which figures in Dirichlet's problem. By what we have proved, the homogeneous adjoint equation

$$\int \varrho(l) \frac{\partial}{\partial n} \log \frac{1}{r(P,Q)} \, dl + \pi \varrho(x) = 0$$

has only a constant solution. By the Fredholm alternative equation (18) has a solution if and only if the function $\varrho(x)$ is orthogonal to 1, i.e.

$$\int_C \varrho(l)\,\mathrm{d}l = 0. \tag{19}$$

But we know that the equation

$$\int_C \frac{\partial u(l)}{\partial n}\,\mathrm{d}l = 0,$$

holds for any harmonic function $u(P)$ on the region G_i, whether of the form (11) or not, provided that the function $\partial u_i/\partial n$ exists and is continuous.

We see that the condition (19) is necessary and sufficient for the solubility of Neumann's interior problem.

Finally the solution of Neumann's exterior problem, which we seek in the same form (17), leads to the integral equation

$$\frac{\partial u_e(P)}{\partial n} = \int_C \varrho(l)\,\frac{\partial}{\partial n_P}\log\frac{1}{r(P,Q)}\,\mathrm{d}l - \pi\varrho(P) = g(P).$$

By what we have proved, the homogeneous adjoint equation

$$\int_C \varrho(l)\,K(x,l)\,\mathrm{d}l - \pi\varrho(x) = 0$$

has no non-zero solutions. Hence *Neumann's exterior problem has a solution for continuous boundary function* $g(P)$.

But this solution, specified by potential (17), has a logarithmic growth at infinity in the general case and therefore falls outside the uniqueness class mentioned in Section (b). It can be shown that potential (17) is bounded at infinity (and moreover tends to zero) if and only if condition (19) is fulfilled. Thus (19) *is the necessary and sufficient condition for solubility of the Neumann exterior problem in the class of bounded functions.*

9. Integral Equations With Complex Parameters

1. *Complex Hilbert Space*

It is often necessary in analysis to consider functions which assume complex values; it is natural to try to construct from them a space with scalar products. But axioms (a)–(d) of Section 1, art. 1 cannot be simply carried over to the new scalar product, for by axiom (d) the expression $(i\,x, i\,x)$ must be positive, while by (a) and (c) it must be equal to—(x, x), i.e. negative.

We reformulate axiom (a) in the complex space as follows:

(a′) $(y, x) = \overline{(x, y)}$, where the bar denotes complex conjugation.

We can then preserve the remaining axioms:

(b) $(x, y + z) = (x, y) + (x, z)$;

(c) $(\lambda x, y) = \lambda (x, y)$ for any complex λ;

(d) $(x, x) > 0$ for $x \neq 0$, and $(x, x) = 0$ for $x = 0$.

From axioms (a′) and (c) we get a new rule for carrying a complex multiplier in the second member of a scalar product through the product sign:

$$(x, \lambda y) = \overline{(\lambda y, x)} = \overline{\lambda (y, x)} = \bar{\lambda} \overline{(y, x)} = \bar{\lambda} (x, y),$$

i.e. a *complex multiplier in the second member of a scalar product can be carried through the product sign under conjugation.*

An example of a complex Hilbert space is the space $L_2(a, b)$ of complex functions $\varphi(x)$ with a square-summable modulus; the scalar product is defined by the formula

$$(\varphi, \psi) = \int_a^b \varphi(x) \overline{\psi(x)} \, dx.$$

Another example is the space l_2 of *complex* sequences $x = (\xi_1, \xi_2, \ldots)$ in which the squares of the moduli form a convergent series:

$$\sum_{n=1}^{\infty} |\xi_n|^2 < \infty.$$

The scalar product of elements $x = (\xi_1, \xi_2, \ldots)$, $y = (\eta_1, \eta_2, \ldots)$ is given by the formula

$$(x, y) = \sum_{n=1}^{\infty} \xi_n \bar{\eta}_n.$$

All the basic results of this chapter carry over to the case of complex Hilbert spaces with more or less obvious modifications to formulations and results. Thus for the Cauchy–Bunyakovsky inequality (Section 1, art. 3), we proceed as before from the inequality

$$(\lambda x - y, \lambda x - y) \geqq 0,$$

which holds for any complex λ. Expanding the left-hand side, we get

$$\lambda \bar{\lambda} (x, x) - \lambda (x, y) - \bar{\lambda} (y, x) + (y, y) \geqq 0.$$

We put $\lambda = t\, e^{-i \arg (x, y)}$ (t real), and then this inequality transforms to the form

$$t^2 (x, x) - 2t |(x, y)| + (y, y) \geqq 0,$$

so that, as before,

$$|(x, y)| \leqq |x|\,|y|.$$

If we have a real Hilbert space H, we can always construct a "complex extension" $\bar{H}$ of it from the formal sums $x + i\, y$, where $x \in H$, $y \in H$. In the space $\bar{H}$ linear operations of addition and complex multiplication are introduced in the natural way; we also define a scalar product by means of the formula

$$(x_1 + i\, y_1, x_2 + i\, y_2) = [(x_1, x_2) + (y_1, y_2)] + i[(y_1, x_2) - (x_1, y_2)].$$

It is easily verified that it satisfies conditions (a'), (b), (c), (d). In particular,

$$(x + i\, y, x + i\, y) = (x, x) + (y, y).$$

The space $\bar{H}$ contains the space H as a subspace (under real multiplication only!) with the same scalar product.

The complex spaces we have introduced, $L_2(a, b)$ and l_2, are evidently the complex extensions of the real spaces $L_2(a, b)$, l_2 considered earlier.

Every complete orthogonal system $e_1, \ldots, e_n, \ldots$ in H will also be a complete orthogonal system in $\bar{H}$; if $(x + i\, y, e_n) = 0$ for all n, then for all n

$$(x, e_n) = 0, \quad (y, e_n) = 0,$$

and therefore $x = 0$, $y = 0$, $x + i\, y = 0$.

Of course there exist new orthonormal systems. Thus in the space $L_2(-\pi, \pi)$ the system of functions $1/\sqrt{(2\pi)}\, e^{inx}$ ($n = 0, \pm 1, \pm 2, \ldots$) is an example of a complete orthonormal system. Its completeness follows from the fact that each function of the complete system 1, $\cos x$, $\sin x$, ... is a linear combination of the functions e^{inx}.

In complex Hilbert space the development of a vector f in an orthonormal system $e_1, e_2, \ldots, e_n, \ldots$ has the form

$$f = \sum_{n=1}^{\infty} c_n e_n,$$

where $c_n = (f, e_n) = \overline{(e_n, f)}$. In particular the Fourier coefficients in the space $L_2(-\pi, \pi)$ for the development

$$f(x) = \sum_{-\infty}^{\infty} c_n e^{inx}$$

are given by the formula

$$c_n = \frac{1}{2\pi} \int_{-\pi}^{\pi} f(x)\, e^{-inx}\, dx.$$

The fundamental theorem of Section 2, art. 4 carries over to the complex case with the single modification that in Parseval's equation (7) and Bessel's inequality (5) $|(f, e_k)|^2$ replaces $(f, e_k)^2$.

A linear operator A which has as its domain a complex space H is said to be *symmetric if*, as before, it satisfies the equation

$$(A\,x, y) = (x, A\,y)$$

for any x, y in H. In general, linear operators in complex space can have characteristic vectors with complex characteristic values. *But a symmetric operator A cannot have non-real characteristic values.* For let $A\,x = \lambda\,x$; then

$$(A\,x, x) = (\lambda\,x, x) = \lambda(x, x),$$
$$(x, A\,x) = (x, \lambda\,x) = \bar{\lambda}(x, x)$$

and it follows from the equation $(A\,x, x) = (x, A\,x)$ that $\lambda = \bar{\lambda}$ is real.

It is easily verified that Fredholm's integral operator

$$A\,\varphi = \int_a^b K(x, s)\,\varphi(s)\,ds$$

will be a completely continuous symmetric operator on the complex space $L_2(a, b)$ if the kernel $K(x, s)$ satisfies the conditions

$$\int_a^b \int_a^b |K(x, s)|^2\,dx\,ds < \infty, \quad K(x, s) = \overline{K(s, x)}.$$

Every linear operator A which has as its domain a real space H can be extended to operate over the complex extension $\bar{H}$ by meas of the formula

$$\bar{A}(x + i\,y) = A\,x + i\,A\,y.$$

If in addition the operator A is symmetric on H, then the operator $\bar{A}$ will be symmetric on $\bar{H}$:

$$\begin{aligned}(A(x_1 + i\,y_1),\, x_2 + i\,y_2)& \\ &= (A\,x_1, x_2) + i(A\,y_1, x_2) - i(A\,x_1, y_2) + (A\,y_1, y_2)\\ &= (x_1, A\,x_2) + i(y_1, A\,x_2) - i(x_1, A\,y_2) + (y_1, A\,y_2)\\ &\qquad = (x_1 + i\,y_1, A(x_2 + i\,y_2))\end{aligned}$$

The theorems on the existence of characteristic vectors (Sections 3 and 4) and the solution of integral equations (Section 6) carry over to the complex case without change.

The Fredholm theorem (Section 8) remains valid as it stands for an integral operator with a square-summable kernel on a complex space $L_2(a, b)$:

$$A\,\varphi = \int_a^b K(x, s)\,\varphi(s)\,\mathrm{d}s;$$

the adjoint operator A^* is now defined by the formula

$$A^*\varphi = \int_a^b \overline{K(s, x)}\,\varphi(s)\,\mathrm{d}s.$$

2. We can obtain some new information on the general properties of operators if in place of the single equation

$$\varphi = A\,\varphi + f$$

we consider a family of equations with a complex parameter μ:

$$\varphi = \mu\,A\,\varphi + f.$$

This family of equations can be written in the form

$$(E - \mu\,A)\,\varphi = f. \tag{1}$$

The operator A is supposed completely continuous as before, e.g. the Fredholm operator, but not in general symmetric. Applying the Fredholm alternative to equation (1), we get: for a fixed value of μ either equation (1) has a unique solution for any f (the value of μ is then said to be *regular*), or the homogeneous equation

$$(E - \mu\,A)\,\varphi = 0$$

has a non-zero solution φ_0 which will evidently be a characteristic vector of the operator A with characteristic value $\lambda = 1/\mu$ (the value of μ is then said to be singular). As μ varies both these possibilities can be realised, but it turns out that the first is the rule and the second the exception: more precisely, a completely continuous operator *A can possesses only a finite number of distinct characteristic values the moduli of which exceed a given positive number.* We have already established this for a completely continuous symmetric operator in Section 3; we shall now show that it is true for any completely continuous operator.

Let us suppose that the completely continuous operator A has an infinite set of characteristic values $\lambda_1, \lambda_2, \ldots$ the moduli of which exceed the positive number δ; let $g_1, g_2, \ldots$ be the corresponding characteristic vectors. Orthonormalising the sequence $g_1, g_2, \ldots,$ we get a new sequence $e_1, e_2, \ldots, e_n, \ldots$ The vectors e_n are in general no longer characteristic for the operator A; for if

$$e_n = a_{nn} g_n + a_{n, n-1} g_{n-1} + \cdots + a_{n1} g_1, \tag{2}$$

then

$$A e_n = a_{nn} \lambda_n g_n + a_{n, n-1} \lambda_{n-1} g_{n-1} + \cdots + a_{n1} \lambda_1 g_1. \tag{3}$$

The equation

$$A e_n = \lambda_n e_n + \sum_{k=1}^{n-1} a_{nk} (\lambda_k - \lambda_n) g_k,$$

which derives from (2) and (3), determines a resolution of the vector $A e_n$ into a component lying in the subspace $(g_1, \ldots g_{n-1})$ and a component orthogonal to this subspace. The length of the latter component is equal to $|\lambda_n|$. Thus the distance of the vector $A e_n$ from any vector in the subspace $(g_1, \ldots, g_{n-1})$, in particular from the vector $A e_m$ for $m < n$, exceeds $|\lambda_n| \geqq \delta$. But then it is impossible to extract a fundamental sequence from the sequence $A e_1, A e_2, \ldots,$ which contradicts the complete continuity of the operator A.

Thus the characteristic values λ_n of any completely continuous operator form at most a countable sequence which converges to zero. Hence the singular values $\mu = 1/\lambda_n$, for which the equation $(E - \mu A) \varphi = 0$ has a non-zero solution, form at most a countable sequence which diverges to ∞.

Let us consider a regular value μ for which the equation $(E - \mu A) \varphi = f$ has a unique solution φ for any $f \in H$. We denote the solution φ, as a function of f, by $R_\mu f$. The operator R_μ is evidently linear. We shall show that it is bounded. Let us suppose the contrary: for some bounded sequence f_n the vectors $\varphi_n = R_\mu f_n$

become unbounded in the norm. We put $\varphi_n/|\varphi_n| = e_n, f_n/|\varphi_n| = g_n$; the vectors e_n have the norm 1, the vectors g_n tend to zero. The sequence $A\,e_n$ contains a convergent subsequence; by renumbering we can assume that the sequence $A\,e_n$ itself converges. But then the sequence $e_n = g_n + \mu\,A\,e_n$ converges to some vector e, $|e| = 1$. Since $g_n \to 0$, we have $e = \mu\,A\,e$ and μ is a singular value, in contradiction to the hypothesis. Thus the operator R_μ is bounded.

Since the relations $(E - \mu\,A)\,\varphi = f$ and $\varphi = R_\mu\,f$ are equivalent, the operator R_μ is the inverse of $E - \mu\,A$. It is said to be *the resolvent operator* or the *resolvent* of the operator A.

3. The question arises as to how an explicit expression for the resolvent can be constructed.

We apply the fixed point principle (Chapter II, Section 5) to the operator $B\,\varphi = \mu\,A\,\varphi + f$. We recall that in order for an operator B which has as its domain a complete space to have a fixed point it is sufficient that it should be compressive, i.e. that for some constant $0 < 1$ it should satisfy the inequality

$$|B\,\varphi - B\psi| \leqq \theta\,|\varphi - \psi|. \tag{1}$$

In the given instance

$$|B\,\varphi - B\,\psi| = |\mu\,A\,\varphi - \mu\,A\,\psi| \leqq |\mu|\,\|A\|\,|\varphi - \psi|$$

and the sufficiency condition reduces to $|\mu| < 1/\|A\|$. *Equation* (1) of art. 2 *is then soluble and for sufficiently small* μ *the solution is unique.*

As we showed in the same section, Chapter II, Section 5, the actual solution is the limit of the sequence $\varphi_0, B\,\varphi_0, B^2\,\varphi_0, \ldots, B_n\varphi_0, \ldots$, where the initial vector φ_0 is arbitrary. Putting $\varphi_0 = 0$, we get:

$$B\,\varphi_0 = f, B^2\,\varphi_0 = \mu\,A\,f + f,$$
$$B^3\,\varphi_0 = \mu\,A\,(\mu\,A\,f + f) + f = \mu^2\,A^2\,f + \mu\,A\,f + f, \ldots$$
$$B^n\,\varphi_0 = f + \mu\,A\,f + \mu^2\,A^2\,f + \cdots + q^{n-1}\,A^{n-1}\,f,$$
$$\cdots\cdots\cdots\cdots\cdots\cdots\cdots\cdots\cdots\cdots$$

The convergence of this process is equivalent to the convergence of the series

$$f + \mu\,A\,f + \mu^2\,A^2\,f + \cdots \tag{2}$$

Thus for $|\mu| < 1/\|A\|$ *equation* (1) of art. 2 *has a unique solution in the form of the series* (2).

It follows that the operator R_μ can be expressed for $|\mu| < 1/\|A\|$ in the form

$$R_\mu = E + \mu A + q^2 A^2 + \cdots \tag{3}$$

Note. The solution could be obtained by means of the formal expansion of the expression $(E - \mu A)^{-1}$ in a power series in μ:

$$(E - \mu A)^{-1} = E + \mu A + \mu^2 A^2 + \cdots \tag{4}$$

The series on the right-hand side of (4) converges for all μ with $|\mu| < 1/\|A\|$, since $\|\mu^n A^n\| \leq |\mu|^n \|A\|^n$. (Left or right) multiplication by $E - \mu A$ gives E, so that the operator represented by it really is the inverse of $E - \mu A$.

If A is Fredholm's integral operator

$$A\varphi = \int_a^b K(x, s)\, \varphi(s)\, ds, \quad \int_a^b\int_a^b |K^2(x, s)|\, dx\, ds = K^2, < \infty,$$

then for $|\mu| < 1/K$, as we shall now show, the operator R_μ is of the form $E + \Gamma_\mu$, where Γ_μ is a Fredholm integral operator the kernel of which depends on the parameter μ.

For the proof we reason as follows. Let there be given two integral operators:

$$A\varphi = \int_a^b K(x, s)\, \varphi(s)\, ds, \quad \int_a^b\int_a^b |K^2(x, s)|\, dx\, ds = K^2,$$

$$B\varphi = \int_a^b L(x, s)\, \varphi(s)\, ds, \quad \int_a^b\int_a^b |L^2(x, s)|\, dx\, ds = L^2.$$

We shall construct the operator AB. We have:

$$AB\varphi = \int_a^b K(x, s)\left\{\int_a^b L(s, t)\, \varphi(t)\, dt\right\} ds$$

$$= \int_a^b \left\{\int_a^b K(x, s)\, L(s, t)\, ds\right\} \varphi(t)\, dt.$$

Reversing the order of integration is permissible in virtue of Fubini's theorem, applied to the summable function of s and t

$$K(x, s)\, L(s, t)\, \varphi(t),$$

which is the product of two square-summable functions $L(s, t)$ and $K(x, s)\,\varphi(t)$. We further denote

$$M(x, t) = \int_a^b K(x, s)\, L(s, t)\, \mathrm{d}s.$$

By the Cauchy–Bunyakowsky inequality

$$|M^2(x, t)| \leq \int_a^b |K^2(x, s)|\, \mathrm{d}s \int_a^b |L^2(s,t)|\mathrm{d}s,$$

so that $M(x, t)$ is square-summable and

$$M^2 = \int_a^b \int_a^b |M^2(x, t)|\, \mathrm{d}t\, \mathrm{d}x$$
$$\leq \int_a^b \int_a^b |K^2(x, s)|\, \mathrm{d}s\, \mathrm{d}x \int_a^b \int_a^b |L^2(s, t)|\, \mathrm{d}s\, \mathrm{d}t = K^2 L^2.$$

Thus the operator AB is a Fredholm integral operator the kernel $M(x, t)$ of which satisfies the inequality

$$M \leq K L.$$

It follows that each of the operators $A^2, A^3, \ldots, A^n, \ldots$ is an integral operator together with A, and the kernel $K_n(x, s)$ of the operator A^n will satisfy the inequality

$$K_n^2 = \int_a^b \int_a^b |K_n^2(x, s)|\, \mathrm{d}x\, \mathrm{d}s \leq K^{2n},$$

$$\text{where} \quad K^2 = \int_a^b \int_a^b |K^2(x, s)|\, \mathrm{d}x\, \mathrm{d}s.$$

On the space $L_2(G)$, $G = \{a \leq x \leq b,\ a \leq s \leq b\}$, the series

$$\mu K(x, s) + \mu^2 K^2(x, s) + \cdots + \mu^n K^n(x, s) + \cdots \tag{5}$$

is majorized in the norm for $|\mu| < 1/K$ by the convergent series

$$|\mu|\, K + |\mu|^2 K^2 + \cdots + |\mu|^n K^n + \cdots$$

and therefore converges in the mean to some function $\Gamma(x, s, \mu)$ which is square-summable for each μ, $|\mu| < 1/K$. But since for integral operators we have

$$\| A \|^2 \leq \int_a^b \int_a^b K^2(x, s)\, dx\, ds = K^2,$$

the convergence in the mean of the series (5) implies the convergence of the series of operators $\mu A + \mu^2 A^2 + \cdots$; with the addition of the unit operator the sum of this series is the resolvent R_μ. Thus for $|\mu| < 1/K$ the operator R_μ is the sum of the unit operator and an integral operator with the kernel $\Gamma(x, s, \mu)$, as required.

The condition $|\mu| < 1/K$ is evidently not necessary for the solubility of equation (1). If the series (5) converges, it always determines a solution, but it can converge over a wider range of values of μ. For example, if some iterate of the kernel $K(x, s)$ vanishes, i.e.

$$K_m(x, s) = 0$$

for some m, then (5) converges for all μ. An example of such an operator is an operator with the kernel $K(x, s) = p(x)\, q(s)$, where the functions $p(x)$, $q(x)$ are orthogonal. For then we need look no further than the second iterated kernel

$$K_2(x, s) = \int_a^b p(x)\, q(t)\, p(t)\, q(s)\, dt = 0.$$

More generally the series (5) will converge for all μ if the iterated kernels satisfy an inequality of the form

$$|K_n(x, s)| \leq \frac{C_n}{n!}. \tag{6}$$

In all these cases the resolvent R_μ exists for all values of μ and all values of μ are regular. By way of example we consider the Volterra operator

$$A\, \varphi(x) = \int_a^x K(x, s)\, \varphi(s)\, ds \quad (a \leq x \leq b)$$

with a bounded kernel $K(x, s)$, $|K(x, s)| \leq M$. The Volterra operator can be regarded as a particular case of Fredholm's integral operator, when the kernel $K(x, s)$ vanishes on the triangle $s \geq x$; we therefore have an integral between the limits a, x instead of

a, x instead of a, b for s. We have:

$$|K(x, s)| \leqq M,$$

$$|K_2(x, s)| = \left|\int_a^b K(x, t)\, K(t, s)\, \mathrm{d}t\right|$$

$$= \begin{cases} \left|\int_s^x K(x, t)\, K(t, s)\, \mathrm{d}t\right| \leqq M^2(x - s) & (x \geqq s), \\ 0 & (x \leqq s), \end{cases}$$

$$|K_3(x, s)| = \left|\int_a^b K_2(x, t)\, K(t, s)\, \mathrm{d}t\right|$$

$$= \begin{cases} \left|\int_s^x K_2(x, t)\, K(t, s)\, \mathrm{d}t\right| \leqq \frac{M^3}{2}(x - s)^2 & (x \geqq s), \\ 0 & (x \leqq s), \end{cases}$$

and so on, so that for any n

$$K_n(x, s) \begin{cases} \leqq \frac{M^n}{(n-1)!}(x - s)^{n-1} & (x \geqq s), \\ = 0 & (x \leqq s). \end{cases}$$

Hence

$$\int_a^b\int_a^b |K_n^2(x, s)|\, \mathrm{d}x\, \mathrm{d}s \leqq \frac{M_1^n}{(n-1)!}$$

and inequality (6) is satisfied.

In the general case, the series which determines the resolvent

$$R_\mu = E + \mu A + \mu^2 A^2 + \cdots,$$

has a finite circle of convergence, outside which it does not allow the resolvent to be directly calculated. We give (without derivation) Fredholm's formulae, which give an expression for the resolvent for any non-singular value μ in the case of a bounded continuous kernel $K(x, s)$†. We fix values $x_1, \ldots, x_n$ and $s_1, \ldots, s_n$

† See e.g. I. I. Privalov, *Integral equations,* Gost (1937), Chapter II. T. Carlemann (1921) showed that these formulae remain in force for a square summable kernel. A simple proof (with a generalisation to an infinite interval) may be found in S. G. Mikhlin, *Dokl. Akad. Nauk SSSR,* **92**, No. 9, 387–90 (1944).

and introduce the notation

$$K\begin{pmatrix} x_1, & \dots, & x_n \\ s_1, & \dots, & s_n \end{pmatrix} = \begin{vmatrix} K(x_1, s_1) & \cdots & K(x_1, s_n) \\ \dots & \dots & \dots \\ \dots & \dots & \dots \\ K(x_n, s_1) & \cdots & K(x_n, s_n) \end{vmatrix}.$$

We define two functions $D(\mu)$, $D(x, s, \mu)$ by the formulae

$$D(\mu) = 1 - \mu \int_a^b K\begin{pmatrix} \xi_1 \\ \xi_1 \end{pmatrix} d\xi + \frac{\mu^2}{2} \int_a^b \int_a^b K\begin{pmatrix} \xi_1 & \xi_2 \\ \xi_1 & \xi_2 \end{pmatrix} d\xi_1\, d\xi_2 + \cdots$$

$$\cdots + (-1)^n \frac{\mu^n}{n!} \int_a^b \cdots \int_a^b K\begin{pmatrix} \xi_1 & \cdots & \xi_n \\ \xi_1 & \cdots & \xi_n \end{pmatrix} d\xi_1 \dots d\xi_n + \cdots$$

$$D(x, s, \mu) = K\left(\frac{x}{s}\right) - \mu \int_a^b K\begin{pmatrix} x & \xi_1 \\ s & \xi_1 \end{pmatrix} d\xi +$$

$$+ \frac{\mu^2}{2} \int_a^b \int_a^b K\begin{pmatrix} x & \xi_1 & \xi_2 \\ s & \xi_1 & \xi_2 \end{pmatrix} d\xi_1\, d\xi_2 + \cdots$$

$$\cdots + (-1) \frac{\mu^n}{n!} \int_a^b \cdots \int_a^b K\begin{pmatrix} x & \xi_1 & \cdots & \xi_n \\ s & \xi_1 & \cdots & \xi_n \end{pmatrix} d\xi_1 \dots d\xi_n + \cdots$$

Both of these functions are entire analytic functions of μ; moreover the function $D(\mu)$ vanishes for precisely those μ that are singular values of the operator A (when the resolvent R_μ does not exist). For any regular value of μ the solution of the integral equation

$$\varphi(x) + \mu \int_a^b K(x, s)\, \varphi(s)\, ds = f(x)$$

can be expressed in the form

$$\varphi(x) = f(x) + \int_a^b \frac{D(x, s, \mu)}{D(\mu)} f(s)\, ds.$$

Thus the resolvent R_μ of the operator A is, whenever it exists, the sum of the unit operator and Fredholm's integral operator with the ratio of the Fredholm functions $D(x, s, \mu)$, $D(\mu)$ as kernel. The characteristic functions of A can also be expressed in terms of $D(\mu)$ and $D(x, s, \mu)$; however these formulae are somewhat intractable and we shall not give them here.

Problems. 1. Show that the set of solutions of the equation

$$P(A)\,\varphi = (A - \lambda_1 E) \ldots (A - \lambda_m E)\,\varphi = 0$$

($\lambda_1, \lambda_2, \ldots, \lambda_m$ distinct complex numbers) coincides with the set of linear combinations of the characteristic vectors of the operator A which correspond to the characteristic values $\lambda_1, \lambda_2, \ldots, \lambda_m$.

Hint. The development in partial fractions

$$\frac{1}{(\lambda - \lambda_1) \ldots (\lambda - \lambda_m)} = \frac{a_1}{\lambda - \lambda_1} + \cdots + \frac{a_m}{\lambda - \lambda_m}$$

has an operator analogue

$$E = a_1 (A - \lambda_2 E) \ldots (A - \lambda_m E) + \cdots + a_m (A - \lambda_1 E) \ldots (A - \lambda_{m-1} E).$$

2. If the operator A^m has a characteristic value λ, then the operator A has a characteristic value μ equal to one of the mth roots of λ.

Hint. Use problem 1.

3. If the operator A is symmetric and A^m is completely continuous, then A is completely continuous.

Hint. The operator A has a complete orthogonal system of vectors with characteristic values which tend to zero. Use problem 5 of Section 3, art. 5.

4. Show that an operator A defined on an orthonormal basis $e_1, e_2, \ldots, e_n, \ldots$ by the formulae

$$A\,e_{2k-1} = e_{2k}, \quad A\,e_{2k} = 0 \quad (k = 1, 2, \ldots),$$

is not completely continuous, while A^2 is completely continuous.

In problems 5–8 the operator A^m is assumed to be completely continuous (A itself may not be).

5. Prove that the characteristic values of A form at most a countable sequence which converges to zero.

Hint. Use problem 2.

6. For sufficiently large prime p the characteristic vectors of A^p-corresponding to the characteristic value 1 are precisely those of A corresponding to the characteristic value 1.

Hint. Use problems 2 and 5.

7. The dimension of the subspace of characteristic vectors of A corresponding to the characteristic value 1 is the same as that of the corresponding subspace for A^*.

Hint. Use problem 6 and the complete continuity of A^p for $p < m$.

8. The equation $\varphi - A\,\varphi = f$ has a solution if and only if $(f, \psi_0) = 0$ where ψ_0 is a characteristic vector of A^* with the characteristic value 1.

Hint. consider the equation

$$(E - A^p)\,\varphi = (E + A + \cdots + A^{p-1})\,f$$

where p^m is such that none of the complete p^{th} roots of 1 is a characteristic value of the operators A or A^*. Verify that it has a solution. If φ_0 is a solution, then $(E - A)\varphi_0 - f$ is mapped onto zero by the operators $E + A + \cdots + A^{p-1}$. Applying the result of problem 1, obtain the relation $(E - A)\varphi_0 - f = 0$ (S. L. Soboliev).

Note. Problems 7 and 8 show that the Fredholm alternative is valid for operators A for which some power A^m is completely continuous.

9. Let μ be a non-singular value of a completely continuous operator A (i.e. $E - \mu A = B$ has an inverse) and let $Q = E - \alpha B^* B$, where $\alpha > 0$ can be taken arbitrarily small. Then the sequence $\varphi_{n+1} = Q\varphi_n + \alpha B^* f$ with φ_0 arbitrary converges to a vector φ which satisfies the equation $(E - \mu A)\varphi = f$.

Hint. The equation $(E - \mu A)\varphi = f$ is equivalent to $\varphi = Q\varphi + x B^* f$. Use the result of Section 3, art. 3, problem 5 (with $C = B^*B$) and the fixed point method (I. P. Natanson, 1948).

Note. The result of this problem suggests an iterative procedure for solving the equation $(E - \mu A)\varphi = f$ for any non-singular value of μ.

10. Show that if an equation of the first kind, $A\varphi = f$, has a solution, it is the limit of a sequence

$$\varphi_n = \varphi_{n-1}(E - \mu A A^*) + \mu f, \tag{1}$$

where $0 < \mu < 2/\|A\|^2$ (B. M. Friedmann).

Hint. Substitute $\varphi_n = \varphi + \mu_n$ in (1) and obtain the formula $u_n = (E - \mu A A^*) u_{n-1}$. If e_j is an orthonormal system of characteristic vectors of the operator AA^* with characteristic values λ_j, then $(u_n, e_j) = (1 - \mu\lambda_j)(u_{n-1}, e_j) = (1 - \mu\lambda_j)(u_0, e_j)$. Choose p so that $|(u_0, e_p)|^2 + |(u_0, e_{p+1})| + \cdots < \varepsilon$ and obtain the result that, for sufficiently large n,

$$|u_n|^2 = \sum_{j=p}^{\infty} |(u_n, e_j)|^2 \leq \sum_{j=1}^{p-1} (1 - \mu\lambda_j)^{2n} |u_0|^2 + \sum_{j=p}^{\infty} |(u_0, e_j|^2 < 2\varepsilon.$$

Concluding Remark

The theory of integral equations with a variable upper limit was developed in 1887 by V. Volterra (Italian mathematician, 1860–1940). In the years 1900–1903 a series of fundamental papers by E. Fredholm (Swedish mathematician, 1866–1927) appeared and in these he introduced the entire functions $D(\mu)$, $D(x, s, \mu)$ in terms of which he expressed the solution of the general equation ("Fredholm's equation") and the characteristic functions. In papers of the years 1904–1910 D. Hilbert (German mathematician, 1862–1943) first related geometrical concepts to integral equations, regarding the problem of characteristic functions as the problem of reducing a quadratic form of infinite rank to principle axes. "Hilbert space" is one of the most important mathematical concepts of the twentieth century. Hilbert determined the canonical development of a bounded symmetric operator (the theorem proved in Section 3 is a special case relating to Hilbert space), which became the starting-point of the contemporary spectral theory of linear operators, of wide application in mathematics and physics. The class of completely continuous operators (in Banach space) was first distinguished by F. Riesz in 1919. For further applications of integral equations in mathematical physics, cf.: S. G. Mikhlin, *Integral Equations*, Pergamon, London, 1951; for spectral theory and its applications cf.: N. I. Akhiezer and I. M. Glazman, *Theory of Linear Operators in Hilbert Space*, Ungar, New York, 1961, and M. A. Naimark, *Differential Operators*, State Tech. Pub. Dept., 1952.

CHAPTER VI

DIFFERENTIATION AND INTEGRATION

IT IS well known that in the classical analysis of continuous functions differentiation and integration are inverse operations. The precise sense to be attached to this assertion is the following:

A. Given a continuous function $\varphi(x)$, its indefinite integral

$$F(x) = \int_a^x \varphi(\xi)\, d\xi + C \tag{1}$$

where C is an arbitrary constant, is a differentiable function and for each x

$$F'(x) = \varphi(x). \tag{2}$$

B. If $F(x)$ is a function which possesses a continuous derivative then

$$\varphi(x) = F'(x),$$

$$\int_a^x \varphi(\xi)\, d\xi = F(x) + C, \tag{3}$$

where C is a certain constant (viz. $-F(a)$).

We now propose a much more general concept of the integral, which can be applied to a wide class of discontinuous functions. The continuity clause for $\varphi(x)$ will no longer hamper us in constructing the indefinite integral (1).

The following problems arise:

Problem I. Does equation (2) hold when $\varphi(x)$ in (1) is an arbitrary summable function?

Since $F(x)$ remains unaltered under an arbitrary modification of the function $\varphi(x)$ by a set of measure zero, it is natural to require in answering problem I that equation (2) hold not everywhere, but only almost everywhere.

Problem II. Under what conditions does a given $F(x)$ possess a summable derivative $\varphi(x)$ (even if defined only almost everywhere) and does formula (3) hold?

In this chapter we shall consider the answers to problems I and II and also certain closely related topics, chief among which is the construction of the Stieltjes integral.

1. Derivative of a Non-decreasing Function

1. We first consider the problem: is the existence of a derivative an elementary property of a function $f(x)$ or can it be deduced from more elementary properties? It is natural to regard e.g. continuity as a possible more elementary property. A function $f(x)$, having a derivative at a point x_0, is well known to be continuous at x_0. Is the converse true, in other words, must a continuous simple function $f(x)$ have a derivative? Of course the answer is no; very examples of the type $f(x) = |x|$ show that a function can be continuous without having a derivative everywhere. It still seems conceivable that points where there is no derivative may be exceptional for a continuous function, and that any such function must always have a substantial set of points at which it has a derivative. This was the opinion of many mathematicians at the beginning of the last century. In the end, the answer proved to be in the negative: Weierstrass' famous example (of 1860) of a continuous function† with no derivative at any point surprised the mathematical world and put an end to the attempts to find the points at which a continuous function must be differentiable. (Van der Waerden's simpler example is given in one of the problems on this article.) Thus the continuity of a function does not imply its differentiability.

Let us try to approach the problem from another direction. A function $f(x)$, having a derivative $f'(x_0) > 0$ at $x = x_0$, is not decreasing in the neighbourhood of x_0, in the sense that we have $f(x) > f(x_0)$ if $x > x_0$, and $f(x) < f(x_0)$ if $x < x_0$, for x sufficiently

† The Czech mathematician, B. Bolzano, constructed a similar example in 1830. The manuscript was only discovered in 1920, and published in 1930—a hundred years after it was written.

close to x_0. Is it possible that the monotoneness of a function implies the existence of a derivative? If we are speaking of individual points, the answer is again no: for instance, $f(x) = |x| + 2x$ is increasing everywhere, but has no derivative at $x = 0$. However, the points where a monotonic function has no derivative are in fact the exception, not the rule. In fact, as shown by Lebesgue (1902), a non-decreasing function can only lack a derivative on a set of measure zero.

THEOREM 1 (A. Lebesgue). *A non-decreasing function $F(x)$ defined on the closed interval $a \leqq x \leqq b$ has a finite derivative almost everywhere on this interval.*

We begin by giving the proof for the case when the function $F(x)$ is continuous.

Let $G(x)$ be a function on the closed interval $a \leqq x \leqq b$. We call a point $x \in [a, b]$ a point of ascent (Fig. 11) if there exists a point ξ to the right of x on $[a, b]$ at which G takes a greater value than at x:

$$G(x) < G(\xi) \quad (x < \xi). \tag{1}$$

We establish the following lemma:

LEMMA (F. Riesz). *The set of all points of ascent of a continuous function $G(x)$ is open on the closed interval $[a, b]$ and at the end-points of each component open interval (a_k, b_k) of this set the inequality*

$$G(a_k) \leqq G(b_k). \tag{2}$$

holds.

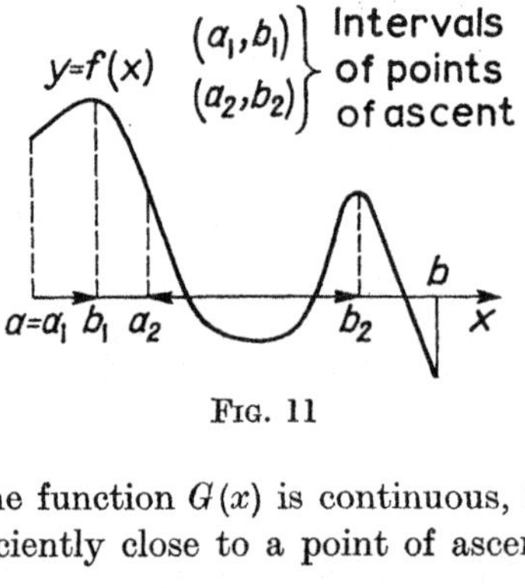

FIG. 11

Proof. Since the function $G(x)$ is continuous, it is evident that all points x sufficiently close to a point of ascent are also points of ascent; thus the set Z of all points of ascent is open. Let (a_k, b_k) be some component open interval of this set. We shall show that for any $x \in (a_k, b_k)$ we have $G(x) \leq G(b_k)$; then proceeding to the limit as $x \to a_k$ we get the required inequality (2).

Let us suppose that $G(x_0) > G(b_k)$; we find the point x_0' furthest to the right in (a_k, b_k) for which $G(x_0') = G(x_0)$. Then since $G(x_0') = G(x_0) > G(b_k)$; it follows that $G(x_0) > G(x)$ everywhere in the interval (x_0', b_k). Further, the point b_k does not belong to the set of points of ascent and so everywhere to the right of b_k we have $G(x_0) \leqq G(b_k)$. Consequently we get $G(x_0') > G(x)$ everywhere to the right of x_0'. But then x_0' cannot be a point of ascent, in contradiction to the construction.

Note. We shall call a point x with the property (1) more precisely a point of ascent to the right. A point of ascent to the left can be defined analogously as a point to the left of which exists a point ξ such that

$$\xi < x, \quad G(\xi) > G(x).$$

Just as above, it can be proved that the set δ points of ascent to the left is open and that at the end-points of its component open intervals (a_k, b_k) we have

$$G(a_k) \geqq G(b_k).$$

Before proceeding to the proof of the Lebesgue theorem itself, we shall make a few further observations of a general nature.

The derivative of a function $F(x)$ is defined as the limit of the ratio

$$\frac{F(\xi) - F(x)}{\xi - x}, \tag{3}$$

as ξ approaches x in accordance with any rule. Of course this limit may not exist. But in every case the following quantities (for which we admit infinite values) always exist:

Λ_r—the limit superior of the ratio (3) as ξ approaches x from the right (the upper right-derived number);

λ_r—the corresponding limit inferior of (3) (the lower right-derived number);

Λ_l—the limit superior (3) as ξ approaches x from the left (the upper left-derived number);

λ_l—the corresponding limit inferior of (3) (the lower left-derived number).

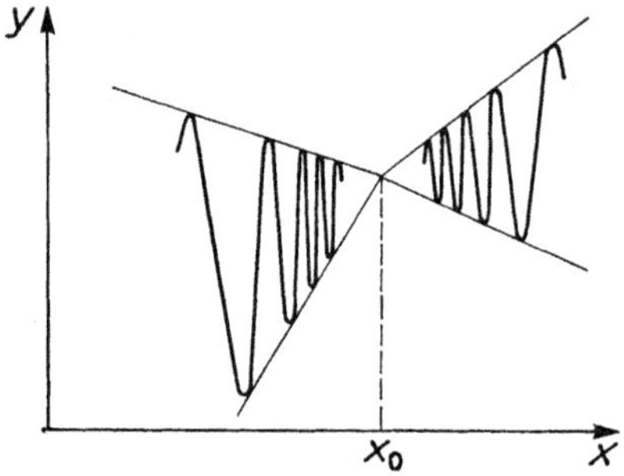

Fig. 12

Figure 12 illustrates a case in which all for derived numbers are finite and distinct at the given point x_0. It is left to the reader to render graphically the case in which all four values are prescribed at the given point with $-\infty \leqq \lambda_r \leqq \Lambda_r \leqq +\infty$, $-\infty \leqq \lambda_l \leqq \Lambda_l \leqq +\infty$.

If $\Lambda_r = \lambda_r \neq \pm\infty$, the function $F(x)$ possesses a derivative to the right at the given point; if $\Lambda_l = \lambda_l \neq \pm\infty$, it has a derivative to the left. Finally if all four derived values are finite and equal, it has a derivative at the point in question.

Bearing all this in mind, we set about proving Lebesgue's theorem.

We recall that the theorem stipulates a non-decreasing function: it is always the case for $x < \xi$ that $F(x) \leq F(\xi)$. Hence the ratio (3) is non-negative, and together with it all four derived numbers $\Lambda_r, \lambda_r, \Lambda_l, \lambda_l$ are non-negative at every point.

We shall show that almost everywhere

$$\Lambda_r < +\infty.$$

If at some point x we have $\Lambda_r = +\infty$, then for any C we can find a point ξ to the right of x at which

$$\frac{F(\xi) - F(x)}{\xi - x} > C,$$

or, what is the same thing,

$$F(x) - Cx < F(\xi) - C\xi.$$

Thus every point x at which $\Lambda_r = +\infty$ is a point of ascent to the right for the function $G(x) = F(x) - Cx$. By Riesz' lemma the set of all points of ascent to the right is open and at the end-points of its component open intervals

$$F(a_k) - C a_k \leqq F(b_k) - C b_k,$$

or, what is the same thing,

$$C(b_k - a_k) \leqq F(b_k) - F(a_k). \tag{4}$$

Summing the inequalities (4) over all component open intervals, we get

$$C \sum (b_k - a_k) \leqq \sum [F(b_k) - F(a_k)] \leqq F(b) - F(a).$$

We see that the set Z of all points x at which $\Lambda_r = +\infty$ can be covered by a system of open intervals of overall length

$$\sum (b_k - a_k) \leqq \frac{1}{C} [F(b) - F(a)].$$

Since C can be chosen arbitrarily large, we conclude that the set Z has measure zero, as asserted.

The next step in the proof is to verify that almost everywhere

$$\Lambda_r \leqq \lambda_l.$$

The set of points where $\Lambda_r > \lambda_l$ can be represented in the form of a countable sum of sets Z_{cC}, determined by the inequalities

$$\lambda_l < c < C < \Lambda_r$$

where c, C range over all possible pairs of rational constants ($c < C$). It is therefore enough for us to prove that each of the sets Z_{cC} has measure zero.

Let $x \in Z_{cC}$. Then, since $\lambda_l < c$, there exists a point ξ lying to the left of x at which

$$\frac{F(\xi) - F(x)}{\xi - x} < c. \tag{5}$$

We observe that here $\xi - x < 0$; hence (5) is equivalent to the inequality

$$F(\xi) - c\xi > F(x) - cx.$$

Thus x is a point of ascent to the left for the function $G(x) = F(x) - cx$. Applying Riesz' lemma (cf. the note on p. 285), we get for the component open intervals of the set of all points of ascent to the left inequalities of the form

$$F(a_k) - c\,a_k \geqq F(b_k) - c\,b_k,$$

or, what is the same thing,

$$F(b_k) - F(a_k) \leqq c(b_k - a_k). \tag{6}$$

The point x in question lies in one of the given open intervals (a_k, b_k). Since at this point $\Lambda_r > C$, we can find a point $\xi > k$ in (a_k, b_k) at which

$$\frac{F(\xi) - F(x)}{\xi - x} > C. \tag{7}$$

We shall carry out the construction that follows within the interval (a_k, b_k). As above, inequality (7) shows that x is a point of ascent to the right for the function $F(x) - C\,x$. The set of all points of ascent to the right for this function in the interval (a_k, b_k) is open and has a decomposition into the sum of component open intervals (a_{kj}, b_{kj}) ($j = 1, 2, \ldots$), at the end-points of which we have

$$F(a_{kj}) - C\,a_{kj} \leqq F(b_{kj}) - C\,b_{kj}.$$

In other words

$$F(b_{kj}) - F(a_{kj}) \geqq C(b_{kj} - a_{kj}).$$

Summing over the index j, we get

$$\sum_j (b_{kj} - a_{kj}) \leqq \frac{1}{C} \sum [F(b_{kj}) - F(a_{kj})] \leqq \frac{1}{C} [F(b_k) - F(a_k)].$$

Using inequality (6), we get

$$\sum_j (b_{kj} - a_{kj}) \leqq \frac{c}{C} \sum (b_k - a_k).$$

Summing over k, we find

$$\sum_k \sum_j (b_{kj} - a_{kj}) \leqq \frac{c}{C} \sum_k (b_k - a_k). \tag{8}$$

The system of intervals (a_{kj}, b_{kj}), like the system (a_k, b_k), covers the set Z_{cC} and we see that the first system covers it more "economically" than the second.

This construction can be repeated for each point $x \in Z_{cC}$ in the corresponding open interval (a_{kj}, b_{kj}). We obtain a new "third-order" system (a_{kjm}, b_{kjm}) $(m = 1, 2, \ldots)$ with and a fourth-order system (a_{kjmn}, b_{kjmn}) $(m, n = 1, 2, \ldots)$

$$\sum_m \sum_n (k_{kjmn} - a_{kjmn}) \leqq \frac{c}{C} \sum_m (b_{kjm} - a_{kjm}) \leqq \frac{c}{C} (b_{kj} - a_{kj}).$$

Summing over k, j and using (8), we get:

$$\sum_k \sum_j \sum_m \sum_n (b_{kjmn} - a_{kjmn}) \leqq \left(\frac{c}{C}\right)^2 \sum_k (b_k - a_k).$$

Continuing this process, we can cover the set Z_{cC} with still finer systems of open intervals, the overall length of the $2p^{\text{th}}$ covering not exceeding $(c/C)^p (b - a)$ and this quantity can be made arbitrarily small by taking p sufficiently large. The set Z_{cC} therefore has measure zero, as asserted. Thus *for any non-decreasing function the inequalities* $\Lambda_r < +\infty$, $\Lambda_r \leqq \lambda_l$ *hold almost everywhere*. Let us replace the function $F(x)$ by $F^*(x) = -F(-x)$; the function $F^*(x)$ is also non-decreasing, and again the inequality $\Lambda_r^* \leqq \lambda_l^*$ holds almost everywhere. But it is easy to see that at corresponding points $\Lambda_r^* = \Lambda_l$, $\lambda_l^* = \lambda_r$; hence the inequality $\Lambda_l \leqq \lambda_r$ holds almost everywhere. Thus we get the chain of inequalities

$$0 \leqq \Lambda_r \leqq \lambda_l \leqq \Lambda_l \leqq \lambda_r \leqq \Lambda_r < +\infty,$$

which hold simultaneously on a set of full measure; we see that on this set

$$0 \leqq \Lambda_r = \lambda_l = \Lambda_l = \lambda_r < +\infty,$$

i.e. the function $F(x)$ has a finite derivative.

Lebesgue's theorem is therefore proved for a continuous non-decreasing function.

We now return to the case of discontinuous non-decreasing functions.

We observe that an arbitrary non-decreasing function $F(x)$ can only have discontinuities of the first kind, so that $F(x)$ has limiting values on both right and left at every point:

$$F(x + 0) = \lim_{\substack{\xi \to x \\ \xi > x}} F(\xi), \quad F(x - 0) = \lim_{\substack{\xi \to x \\ \xi < x}} F(\xi).$$

For a multiplicity of limiting values on either side would contradict the monotoneness of the function. The interval $(F(x-0), F(x+0))$ is said to be the interval of discontinuity and its length $F(x+0) - F(x-0)$ the saltus of the function $F(x)$ at the point x. Since $F(x)$ is non-decreasing, the intervals $(F(x-0), F(x+0))$ corresponding to distinct points of discontinuity are non-overlapping (at most they can have an end-point in common); hence the set of such intervals is at most countable. It follows that *the set of discontinuities of a non-decreasing function is at most countable.*

To verify the existence of a derivative for a discontinuous non-decreasing function we generalise Riesz' lemma accordingly. Let $G(x)$ be a function having at worst discontinuities of the first kind. We shall call a point x a point of ascent to the right if there exists to the right of x a point ξ at which

$$\max\,[G(x), G(x-0), G(x+0)] < G(\xi).$$

Repeating the arguments adduced above in support of Riesz' lemma we get that the set of all points of ascent to the right is open and for each of its component open intervals (a_k, b_k) that

$$G(a_k + 0) \leqq \max\,\{G(b_k), G(b_k - 0), G(b_k + 0)\}.$$

But this is already enough to carry through unaltered the proof of the theorem itself.

Thus *every non-decreasing function has a finite derivative almost everywhere.*

Problems. Prove the following propositions:

1. If one of the right-derived numbers of a continuous function $F(x)$ is non-negative in the open interval (a, b) then $F(a) \leqq F(b)$.

Hint. Each point is a point of ascent to the right.

2. If one of the right-derived numbers is restricted to the range $[\alpha, \beta]$ in the interval $a < x \leqq b$, then for any x_1, x_2, in (a, b) we have

$$\alpha \leqq \frac{F(x_2) - F(x_1)}{x_2 - x_1} \leqq \beta.$$

Hint. Apply the result of problem 1 to the function $F(x) - \alpha x$.

3. If one of the derived numbers of the function $F(x)$ is continuous at the point x_0, then $F'(x_0)$ exists.

Hint. Use problem 2.

4. (Van der Waerdens' example). Let

$$\varphi_0(x) = \begin{cases} x & \text{for } 0 \leqq x \leqq \frac{1}{2}, \\ 1 - x & \text{for } \frac{1}{2} \leqq x \leqq 1 \end{cases}$$

and let us continue this function throughout the axis with period 1. Further, let

$$\varphi_n(x) = \frac{1}{4^n} \varphi_0(4^n x).$$

The function $\varphi_n(x)$ has period 4^{-n} and a derivative (everywhere except at angular points with abscissae $p/2^n$) equal to $+1$ or -1. Finally, let

$$f(x) = \sum_{n=1}^{\infty} \varphi_n(x).$$

Show that $f(x)$ is continuous but lacks a derivative at every point.

Hint. Fix m for a given x_0 and take the increment $h = \pm 1/4^m$. The increments of all the $\varphi_n(x)$, as from the mth, will now vanish. The function $\varphi_{m-1}(x)$ has intervals without angular points of length $2/4^m$; the one that contains x_0 will also contain one of the intervals $(x_0, x_0 + 1/4^m)$ or $(x_0, x_0 - 1/4^m)$. But all the preceding functions $\varphi_k(x)$, $k < m - 1$, have no angular points in this interval; their increments will be equal in modulus to the increment of the argument. All in all, we have

$$\frac{\Delta f(x)}{\Delta x} = \sum_{k=0}^{m-1} \frac{\Delta \varphi_k(x)}{\Delta x} = \begin{cases} \text{an even number if } m \text{ is even,} \\ \text{an odd number if } m \text{ is odd.} \end{cases}$$

Thus $\Delta f(x)/\Delta x$ has no limit as $\Delta x \to 0$.

5. At a point x, $0 \leqq x < 1$, having the dyadic resolution $0, a_1, a_2 \ldots a_n \ldots$ (as usual, we exclude resolutions of period 1), we put $f(x) = 0, a_1\, 0\, a_3\, 0 \ldots$, replacing the dyadic digits at even positions by zeros. Show that $f(x)$ is continuous from the right and nowhere has a right-hand derivative.

Hint. Consider the increment of the function on changing the nth dyadic digit in x from 0 to 1 or the group 01 to 10.

2. To the usual rules for differentiating sums and products, we add here a theorem on the term-by-term differentiation of a series of monotone functions:

THEOREM 2. ("Fubini's little theorem"). *A series of monotone (non-decreasing) functions*

$$\sum_{n=1}^{\infty} F_n(x) = F(x) \tag{1}$$

which converges everywhere admits term-by-term differentiation almost everywhere:

$$\sum_{n=1}^{\infty} F'_n(x) = F'(x).$$

Proof. We can suppose without loss of generality that all the functions $F_n(x)$ are non-negative and vanish at $x = a$: in the contrary event we could replace $F_n(x)$ by $F_n(x) - F_n(a)$.

The sum of a series of non-decreasing functions is of course a non-decreasing function. Let us consider the set E of full measure on which all the $F'_n(x)$ and $F'(x)$ exist. For $x \in E$ and any ξ we have

$$\frac{\sum_{n=1}^{\infty} [F_n(\xi) - F_n(x)]}{\xi - x} = \frac{F(\xi) - F(x)}{\xi - x}.$$

Since the terms on the left are non-negative, we have for any N

$$\frac{\sum_{n=1}^{N} [F_n(\xi) - F_n(x)]}{\xi - x} \leqq \frac{F(\xi) - F(x)}{\xi - x}.$$

Proceeding to the limit as $\xi \to x$, we get

$$\sum_{n=1}^{N} F'_n(x) \leqq F'(x),$$

so that, letting N tend to ∞ and remembering that all the $F'_n(x)$ are non-negative, we find

$$\sum_{n=1}^{\infty} F'_n(x) \leqq F'(x). \tag{2}$$

We shall show that in fact the equality sign holds her for almost all x. For a given k we find a partial sum $S_{n_k}(x)$ of the series (1) for which

$$0 \leqq F(b) - S_{n_k}(b) < \frac{1}{2^k} \quad (k = 1, 2, \ldots).$$

It follows, since the difference $F(x) - S_{n_k}(x) = \sum_{j > n_k} F_j(x)$ is a non-decreasing function, that for all x

$$0 \leqq F(x) - S_{n_k}(x) < \frac{1}{2^k},$$

and hence the series of non-decreasing functions

$$\sum_{k=1}^{\infty} [F(x) - S_{n_k}(x)]$$

converges (uniformly, what is more) on the whole closed interval $a \leqq x \leqq b$. But then by what we have proved the series of derivatives converges almost everywhere. The general term $F'(x) - S'_{n_k}(x)$ of this series tends almost everywhere to zero, which means that $S'_{n_k}(x) \to F'(x)$ almost everywhere. But if we had the $<$ sign in inequality (2), no sequence of partial sums could have the limit $F'(x)$. We must therefore have the equality sign for almost all x, as asserted.

3. The decomposition of a non-decreasing function into a saltus function and a continuous function. Let $A = \{x_1, x_2, \ldots\}$ be an arbitrary finite or countable subset of the closed interval $[a, b]$ and let $B = \{h_1, h_2, \ldots\}$ be an equivalent set of positive numbers with a finite sum $\sum_n h_n$. We set up a one–one correspondence between A and B so that the point x_n corresponds to the number h_n bearing the same index n, and define the function $H(x)$ as the sum of all the h_n which correspond to points x_n lying not to the right of x:

$$H(x) = \sum_{x_n \leqq x} h_n.$$

The function $H(x)$ so constructed is said to be a *saltus function*. Since it is non-decreasing, it has an at most countable set of discontinuities. We shall show that it is continuous to the right that all its discontinuities lie at the points $x_0, x_1, \ldots, x_n, \ldots$ and that the corresponding salti $H(x_n) - H(x_n - 0)$ are equal to just the numbers h_n. For a given $x = x_0$, we have

$$H(x_0 + 0) = \lim_{x \to x_0 + 0} \sum_{x_n \leqq x} h_n = \sum_{x_n \leqq x_0} h_n = H(x_0),$$

$$H(x_0 - 0) = \lim_{x \to x_0 - 0} \sum_{x_n \leqq x} h_n = \sum_{x_n < x_0} h_n.$$

If x_0 does not coincide with one of the x_n, then $H(x_0 + 0) = H(x_0) = H(x_0 - 0)$ and $H(x)$ is continuous at x_0. If x_0 coincides with one of the x_n, then the difference between $H(x_0 + 0) = H(x_0)$ and $H(x_0 - 0)$, i.e. the saltus at x_n, is equal to h_n. Our assertion is therefore proved. By Fubini's little theorem the function $H(x)$, as the sum of a convergent series of non-decreasing functions

$$H_n(x) = \begin{cases} 0 & \text{for} \quad x < x_n, \\ h_n & \text{for} \quad x \geqq x_n, \end{cases}$$

the derivatives of which vanish almost everywhere, also has a derivative which vanishes almost everywhere.

THEOREM. *Any non-decreasing function $F(x)$, continuous to the right, can be represented as the sum of two non decreasing functions:*

$$F(x) = H(x) + G(x),$$

where $H(x)$ is a saltus function.and $G(x)$ a continuous function.

Proof. Let $x_1, x_2, \ldots$ be all the discontinuities of the function $F(x)$ and $h_1, h_2, \ldots$ the corresponding salti. We construct a saltus function by means of the formula

$$H(x) = \sum_{x_n \leqq x} h_n .$$

We shall show that the difference $G(x) = F(x) - H(x)$ is continuous non-decreasing. For $x' < x''$ we have

$$G(x'') - G(x') = [F(x'') - F(x')] - [H(x'') - H(x')],$$

where the difference on the right is non-negative [it represents the measure of the range of $F(x)$ over the set of points of continuity of the interval x',x'']. Further, at each point x there exist $G(x + 0) = G(x)$ and $G(x - 0)$, and

$$G(x + 0) - G(x - 0)$$
$$= [F(x + 0) - F(x - 0)] - [H(x + 0) - H(x - 0)];$$

in virtue of the properties of the function $H(x)$ indicated above, this difference vanishes for all x, so that the function $G(x)$ is continuous. The theorem is proved.

Note. The condition of continuity to the right, which figures in all our constructions, can be discarded if we generalize the concept of a saltus function suitably. For instance let points x_n and non-negative numbers h_n, h'_n be given; the function defined by the formula

$$H(x) = \sum_{x_n < x} h'_n + \sum_{x_n \leqq x} h_n,$$

has at each point x_n a saltus h_n to the left (so that $H(x_n) - H(x_n - 0) = h_n$), a saltus h'_n to the right (so that $H(x_n + 0) - H(x_n) = h'_n$), and a total saltus $h'_n + h_n$. If $F(x)$ is a non-decreasing function which has salti h_n to the left and h'_n to the right at x_n then subtracting $H(x)$ we get a continuous non-decreasing difference $G(x) = F(x) - H(x)$.

2. Functions of Bounded Variation

1. Starting with non-decreasing functions, we can construct a wider class of functions which have a derivative almost everywhere. Together with non-decreasing functions their differences evidently have a derivative almost everywhere. We shall give an intrinsic definition of functions which are differences of non-decreasing functions. We start with the following definition:

Definition: A function $F(x)$, defined on the closed interval $a \leqq x \leqq b$, is said to be a *function of bounded variation* if for any partition of the interval

$$a = x_0 < x_1 < \cdots < x_n = b$$

the sum

$$\sum_{k=0}^{n+1} |F(x_{k+1}) - F(x_k)| \tag{1}$$

remains bounded above by a fixed constant.

Every *non-decreasing* function $F(x)$ is of bounded variation since the sum (1) is then independent of the partition and is equal to $F(b) - F(a)$. The sum and difference of two functions of bounded variation are obviously themselves of bounded variation. In particular, the difference of two non-decreasing functions is of bounded variation. We shall now see that the converse is true; every function of bounded variation can be represented as the difference of two non-decreasing functions.

Let $F(x)$ be a function of bounded variation; we shall call the exact upper bound of the sums (1) under all possible partitions of the integral $[a, b]$ the *total variation* of $F(x)$ on $[a, b]$ and denote it by $\mathsf{V}_a^b[F]$. We shall show that for $a < c < b$

$$\mathsf{V}_a^b\,[F] = \mathsf{V}_a^c\,[F] + \mathsf{V}_c^b\,[F]. \tag{2}$$

If c is one of the partitioning points of $[a, b]$, so that $x_m = c$, say, then

$$\sum_{k=0}^{n-1} |F(x_{k+1}) - F(x_k)|$$
$$= \sum_{k=0}^{m-1} |F(x_{k+1}) - F(x_k)| + \sum_{k=m}^{n-1} |F(x_{k+1}) - F(x_k)|. \tag{3}$$

The sums on the right can be made arbitrarily close to $\mathsf{V}_a^c[F] + \mathsf{V}_c^b[F]$ by a sufficient refinement of the subpartitions. We can therefore assert that

$$\mathsf{V}_a^b\,[F] = \sup \sum_{k=0}^{n-1} |F(x_{k+1}) - F(x_k)| \geqq \mathsf{V}_a^c\,[F] + \mathsf{V}_c^b\,[F]. \quad (4)$$

On the other hand the addition of a further partitioning point c to an arbitrary partition of the interval can only increase the sum (1). Hence for any partition, whether it contains the point c or not, we have in virtue of (3):

$$\sum_{k=0}^{n-1} |F(x_{k+1}) - F(x_k)| \leqq \mathsf{V}_a^c[F] + \mathsf{V}_c^b[F].$$

Taking the upper bound on the left-hand side, we get

$$\mathsf{V}_a^b[F] \leqq \mathsf{V}_a^c[F] + \mathsf{V}_c^b[F]. \quad (5)$$

The result (2) follows, as required, on combining (3) and (5). In particular, $V(x) = \mathsf{V}_a^x[F]$ is a non-decreasing function. The difference $G(x) = V(x) - F(x)$ is also a non-decreasing function, since evidently for $x' < x''$ we have $V(x'') - V(x') = \mathsf{V}_{x'}^{x''}\,[F] \geqq F(x'') - F(x')$ and therefore

$$V(x'') - F(x'') \geqq V(x') - F(x').$$

Thus a function of bounded variation can be represented as the difference of two non-decreasing functions

$$F(x) = V(x) - G(x).$$

2. Many additional properties which may be possessed by the function $F(x)$ carry over to its non-decreasing components $V(x)$, $G(x)$. Such a property, for example, is continuity, two-sided or one-sided. We shall show that if $F(x)$ is continuous, say to the right, at $x = x_0$, then so also are $V(x)$ and $G(x)$. It is enough to show this for $V(x)$. Since $F(x)$ is continuous to the right at x_0, we can find for a given $\varepsilon > 0$ a corresponding $\delta > 0$, such that for any $x_1 > x_0$ which exceeds x_0 by less than δ we have

$$|F(x_1) - F(x_0)| < \frac{\varepsilon}{2}.$$

We construct a partition $x_0 < x_1 < \cdots < x_n = b$ of the closed interval $[x_0, b]$ such that

$$\sum_{k=0}^{n-1} |F(x_{k+1}) - F(x_k)| > \mathsf{V}_{x_0}^{b}[F] - \frac{\varepsilon}{2}.$$

The point x_1 here can always be required to satisfy the condition $x_0 < x_1 < x_0 + \delta$. We get

$$\mathsf{V}_{x_0}^{b}[F] < \sum_{k=0}^{n-1} |F(x_{k+1}) - F(x_k)| + \frac{\varepsilon}{2}$$

$$< \sum_{k=1}^{n-1} |F(x_{k+1}) - F(x_k)| + \varepsilon < \mathsf{V}_{x_1}^{b}[F] + \varepsilon$$

and therefore

$$V(x_1) - V(x_0) = \mathsf{V}_{x_0}^{x_1}[F] = \mathsf{V}_{x_0}^{b}[F] - \mathsf{V}_{x_1}^{b}[F] < \varepsilon,$$

from which it follows that $V(x)$ is continuous to the right at $x = x_0$. Similarly $V(x)$ can be proved continuous to the left if the original function $F(x)$ is supposed continuous to the left. We therefore get: *if a function $F(x)$ of bounded variation is continuous on the closed interval $[a, b]$, then so is the function $V(x) = \mathsf{V}_a^x[F]$ and likewise $G(x) = V(x) - F(x)$.*

Conversely many properties of the function $F(x)$ can be foreseen from the properties of its non-decreasing components. Thus by Lebesgue's theorem (Section 1) every non-decreasing function has a derivative almost everywhere. Hence *any function of bounded variation has a derivative almost everywhere, since it is the difference of two non-decreasing functions.*

Problems. 1. Show that the product of two functions $F_1(x)$, $F_2(x)$ of bounded variation is again a function of bounded variation, with

$$\mathsf{V}_a^b[F_1 F_2] \leq \max_x |F_1(x)| \mathsf{V}_a^b[F_2] + \max_x |F_2(x)| \mathsf{V}[F_1].$$

2. Let $F(x) \geq \alpha > 0$ be a function of bounded variation. Show that $1/F(x)$ is also a function of bounded variation, with

$$\mathsf{V}_a^b\left[\frac{1}{F(x)}\right] \leq \frac{1}{\alpha^2}\mathsf{V}_a^b[F].$$

3. A curve $y = F(x)$ $(a \leq x \leq b)$ is said to be *rectifiable* if the length of the polygonal lines with successive vertices at the points $(x_1, F(x_1))$, ... $(x_n, F(x_n))$, where $a = x_1 < x_2 < \cdots < x_n = b$, is bounded by a fixed con-

stant which is independent of the number n and the choice of points $x_2, \ldots, x_{n-1}$. Show that the curve $y = F(x)$ is rectifiable if and only if the function $F(x)$ is of bounded variation.

Hint. Use the inequality

$$|\Delta y_j| < \sqrt{|\Delta x_j|^2 + |\Delta y_j|^2} < |\Delta x_j| + |\Delta y_j|.$$

4. Prove that the continuous function $x^\alpha \sin \frac{1}{x^\beta}$ on the closed interval $[0, 1]$ is of bounded variation for $\alpha > \beta$ but not for $\alpha \leqq \beta$.

5. Does there exist a continuous function $F(x)$ which is not of bounded variation over any interval?

Answer. For instance, a function such as Weierstrass' function, which has no derivative anywhere.

6. Define a norm

$$||F|| = \mathsf{V}_a^b[F]$$

on the space of all functions $F(x)$ of bounded variation on the closed interval $[a, b]$, (functions which differ by a constant are considered equivalent). Show that a complete normed space results.

7. Put $V(x) = \mathsf{V}_a^x[F]$, where $F(x)$ is a function of bounded variation; show that almost everywhere

$$V'(x) = |F'(x)|.$$

Hint. For a given n construct a partition $a = x_0 < x_1 < \cdots < x_n = b$ such that the sum $\sum_{k=0}^{n-1} |F(x_{k+1}) - F(x_k)|$ differs from $\mathsf{V}_a^b[F]$ by less than $\frac{1}{2^n}$. Put $F_n(x) = \pm F(x) + C_{kn}$ on the interval (x_k, x_{k+1}), choosing the signs $\pm$ and the constants C_{kn} so that $F_n(x_{k+1}) - F_n(x_k) = |F(x_{k+1}) - F(x_k)|$. Show that $V(x) - F_n(x)$ is non-decreasing, and apply Fubini's little theorem to the convergent series $\sum_{n=1}^{\infty} [V(x) - F_n(x)]$.

8. Show that the functions $F(x)$, $V(x)$ have the same points of discontinuity and that their salti at these points coincide, disregarding sign.

3. Let us consider the indefinite integral of a summable function $\varphi(x)$:

$$F(x) = \int_a^x \varphi(\xi)\, d\xi.$$

The function $F(x)$ is of bounded variation since

$$\sum_{k=0}^{n-1} |F(x_{k+1}) - F(x_k)| = \sum_{k=0}^{n-1} \left| \int_{x_k}^{x_{k+1}} \varphi(\xi)\, d\xi \right|$$

$$\leqq \sum_{k=0}^{n-1} \int_{x_k}^{x_{k+1}} |\varphi(\xi)|\, d\xi = \int_a^b |\varphi(\xi)|\, d\xi.$$

(Alternatively $F(x)$ can be represented immediately as the difference of two non-decreasing functions, using the decomposition of $\varphi(x)$ into positive and negative parts). By what we have proved, the function $F(x)$ *has a derivative* $F'(x)$ *almost everywhere.*

We shall show that this derivative coincides almost everywhere with the original function $\varphi(x)$. It is sufficient if we restrict ourselves to a summable function $\varphi(x)$, in the class C^+ (Chapter IV, Section 2). We have $\varphi(x) = \lim_{n\to\infty} h_n(x)$, where $h_n(x)$ is a step function and $h_n(x) \leqq h_{n+1}(x)$ $(n = 1, 2, \ldots)$. For integrals of the functions $h_n(x)$ our assertion is immediately evident: if

$$H_n(x) = \int_a^x h_n(\xi)\,\mathrm{d}\xi,$$

then $H'_n(x) = h_n(x)$ everywhere save at the discontinuities of $h_n(x)$. Since the monotone increasing sequence $h_n(x)$ converges to $\varphi(x)$, the sequence $H_n(x)$ has the function $F(x)$ as its limit for every x. Moreover $F(x)$ can be written in the form of a series of non-decreasing functions

$$F(x) = H_1(x) + \sum_{n=1}^{\infty} [H_{n+1}(x) - H_n(x)].$$

Applying Fubini's little theorem, we get

$$F'(x) = h_1(x) + \sum_{n=1}^{\infty} [h_{n+1}(x) - h_n(x)] = \varphi(x),$$

almost everywhere, the result required.

We have thus solved problem 1, proposed in the introduction to this chapter. We formulate the result in the form of a theorem:

Theorem 1. (A. Lebesgue). *If* $\varphi(x)$ *is a summable function, its indefinite integral*

$$F(x) = \int_a^x \varphi(\xi)\,\mathrm{d}\xi$$

is a continuous function of bounded variation and it has almost everywhere a derivative equal to $\varphi(x)$.

4. *Lebesgue Points*

We shall say that a point $x_0 \in [a, b]$ is a *Lebesgue point* of the summable function $\varphi(x)$ if

$$\lim_{x \to x_0} \frac{1}{x - x_0} \int_{x_0}^{x} |\varphi(\xi) - \varphi(x_0)| \, d\xi = 0 \tag{1}$$

If the function $\varphi(x)$ is continuous, it is easily seen that any point $x_0 \in [a, b]$ is a Lebesgue point of $\varphi(x)$. We shall show that in the general case when $\varphi(x)$ is summable, *almost every point of the interval $[a, b]$ is a Lebesgue point $\varphi(x)$.*

Let r be a fixed number. Then by theorem 1 the limiting relation

$$\lim \frac{1}{x - x_0} \int_{x_0}^{x} |\varphi(\xi) - r| \, d\xi = |\varphi(x_0) - r|$$

holds on some set E_r of full measure. We consider a countable everywhere dense set of values of r, say the set of rational r. The intersection E of all the E_r is also a set of full measure. Let $x_0 \in E$ be such that $\varphi(x_0)$ is finite; we show that x_0 is a Lebesgue point. For, given $\varepsilon > 0$, we can find r such that $|\varphi(x_0) - r| < \varepsilon/3$; further, we can write:

$$\frac{1}{|x - x_0|} \int_{x_0}^{x} |\varphi(\xi) - \varphi(x_0)| \, d\xi$$

$$\leqq \frac{1}{|x - x_0|} \int_{x_0}^{x} |\varphi(\xi) - r| \, d\xi + \frac{1}{|x - x_0|} \int_{x_0}^{x} |r - \varphi(x_0)| \, d\xi$$

$$= \left\{ \frac{1}{|x - x_0|} \int_{x_0}^{x} |\varphi(\xi) - r| \, d\xi - |\varphi(x_0) - r| \right\} + 2|\varphi(x_0) - r|.$$

For sufficiently small $|x - x_0|$ the expression in curly brackets becomes smaller than $\varepsilon/3$; hence for such $|x - x_0|$ the total sum on the right-hand side will be less than ε, and the required equation (1) follows.

A summable function $\varphi(x)$ has many of the properties of a continuous function at its Lebesgue points.

We shall subsequently require a property of $\varphi(x)$ which is expressed in the following lemma:

LEMMA. *If x_0 is a Lebesgue point of the function $\varphi(x)$ and E_n is a sequence of measurable sets which contract to the point x_0, in the sense that E_n is disposed along an interval Δ_n which contains x_0 and is of length $\delta_n \to 0$, and are such that $\mu E_n \geqq \alpha \delta_n$* (where $\alpha > 0$ is a fixed constant), *then*

$$\lim \frac{1}{\mu E_n} \int_{E_n} \varphi(\xi)\, d\xi = \varphi(x_0). \tag{2}$$

Proof. We have, clearly,

$$\left|\varphi(x_0) - \frac{1}{\mu E_n} \int_{E_n} \varphi(\xi)\, d\xi\right|$$

$$= \left|\frac{1}{\mu E_n} \int_{E_n} [\varphi(x_0) - \varphi(\xi)]\, d\xi\right| \leqq \frac{1}{\mu E_n} \int_{E_n} |\varphi(x_0) - \varphi(\xi)|\, d\xi$$

$$\leqq \frac{\alpha^{-1}}{\mu \Delta_n} \int_{\Delta_n} |\varphi(x_0) - \varphi(\xi)|\, d\xi.$$

If x_0 is the end-point of an interval Δ_n, this expression has the limit 0 by hypothesis (x_0 is a Lebesgue point). If x_0 is an interior point of $\Delta_n = (\alpha_n, \beta_n)$, the ratio obtained lies between

$$\frac{\alpha^{-1}}{x_0 - \alpha_n} \int_{\alpha_n}^{x_0} |\varphi(x_0) - \varphi(\xi)|\, d\xi, \text{ and } \frac{\alpha^{-1}}{\beta_n - x_0} \int_{x_0}^{\beta_n} |\varphi(x_0) - \varphi(\xi)|\, d\xi\,†$$

and therefore tends to zero, together with them, as required.

As a corollary, we get that *at each Lebesgue point the function $\varphi(x)$ is equal to the derivative of its indefinite integral.*

Problems. 1. *A* point x is said to be a *density point* of a measurable set E if

$$\lim_{\Delta \to \xi} \frac{\mu E(\Delta)}{\mu \Delta} = 1,$$

where $E(\Delta)$ denotes the part of E which is contained in the interval Δ. Show that almost all points of a set E are density points of that set.

† By a well-known arithmetical inequality the fraction $(a + c)/(b + d)$ lies between a/b and c/d.

Hint. Apply the lemma to the characteristic function of E, putting $E_n = \Delta_n$ (the interval which contracts to x).

2. Given that for any interval Δ, $\mu E(\Delta) \geqq \alpha \mu \Delta$, where $\alpha > 0$ is fixed, show that E has full measure.

Hint. The complementary set has no density points.

COROLLARY. There exists no measurable set such that the measure of the part of it falling in an arbitrary interval is precisely equal to half the length of this interval.

3. A point ξ is said to be a *point of asymptotic continuity* of a function $F(x)$, measurable over $[a, b]$, if there exists a measurable set E on which $F(x)$ is continuous and for which ξ is a density point. Prove that almost every point $\xi \in [a, b]$ is a point of asymptotic continuity of $F(x)$.

Hint. Every density point of a set E_ε, $\mu E_\varepsilon < b - a - \varepsilon$, on which $F(x)$ is continuous (Chapter IV, Section 4, art. 7) satisfies the condition.

4. Verify that a summable function $F(x)$ is asymptotically continuous at every Lebesgue point; if $F(x)$ is bounded, the converse is also true.

5. Construct an example of a summable function which has a Lebesgue point at which (2) is not satisfied if for the E_n are chosen certain sets for which $\mu E_n/\mu \Delta_n \to 0$.

3. Determination of a Function from its Derivative

We proceed now to the solution of problem II of the introduction to the present chapter.

1. A function $F(x)$ is given on the closed interval $a \leqq x \leqq b$, having almost everywhere a derivative $F'(x) = \varphi(x)$.

It is asked: is the function $\varphi(x)$ summable and does the formula

$$F(x) = \int_a^x \varphi(\xi)\, d\xi + C \tag{1}$$

hold?

As we know, a necessary condition for (1) to hold is that $F(x)$ should be of bounded variation. We shall assume that this condition is satisfied and initially we shall suppose that $F(x)$ is non-decreasing. We begin by answering the first question: we shall show that *the derivative of a non-decreasing function is always summable.* The derivative of the function $F(x)$ is the limit of the ratio

$$\Phi_h(x) = \frac{F(x+h) - F(x)}{h}.$$

The functions $\Phi_h(x)$ are non-negative and as $h \to 0$ converge almost everywhere on the interval $[a, b]$ to the limit $F'(x)$†. To prove $F'(x)$ summable we apply Fatou's theorem, Chapter IV, Section 3; this guarantees that $F'(x)$ is summable if the integrals of the $\Phi_h(x)$ over $[a, b]$ remain bounded. Assuming α, β to be points of continuity for all the $\Phi_h(x)$, we have:

$$\int_\alpha^\beta \Phi_h(x)\,\mathrm{d}x = \frac{1}{h}\int_{\alpha+h}^{\beta+h} F(x)\,\mathrm{d}x - \frac{1}{h}\int_\alpha^\beta F(x)\,\mathrm{d}x$$

$$= \frac{1}{h}\int_\beta^{\beta+h} F(x)\,\mathrm{d}x - \frac{1}{h}\int_\alpha^{\alpha+h} F(x)\,\mathrm{d}x.$$

This quantity has the limit $F(\beta) - F(\alpha)$ as $h \to 0$ and is therefore bounded. In consequence we are able to apply Fatou's theorem, and we get in addition to the summability of $F'(x)$ the inequality

$$\int_\alpha^\beta F'(x)\,\mathrm{d}x \leqq F(\beta) - F(\alpha) \leqq F(b-0) - F(a+0).$$

$$\int_a^b F'(x)\,\mathrm{d}x \leqq F(b-0) - F(a+0).$$

In particular, if a, b are points of continuity of $F(x)$, then

$$\int_a^b F'(x)\,\mathrm{d}x \leqq F(b) - F(a). \tag{2}$$

The $<$ sign can in fact obtain here, for example in the case of a step function $F(x)$ the derivative of which vanishes almost everywhere.

2. Indeed it turns out that the $<$ sign in inequality (2) of art. 1 can arise in practice even for a *continuous* non-decreasing function $F(x)$.

As an example let us consider some closed nowhere dense set Z (e.g. the Cantor set).

We showed previously (Chapter II, Section 4) that such a set has the power of the continuum. We recall the construction. We began by establishing a one–one correspondence between the set

† Letting $\alpha \to a$, $\beta \to b$, we get that $F'(x)$ is summable over $[a, b]$ and a. If $x + h$ goes outside $[a, b]$ we continue $F(x)$ as a constant.

of all contiguous open intervals of Z and the set of dyadic rationals of the interval $[0, 1]$ with order preserved, i.e. such that if the interval Δ' lies to the left of Δ'', then the corresponding dyadic rationals r', r'' are connected by the inequality $r' < r''$. We then extended it, on the one hand to all points of Z of the second type, on the other to all the dyadic irrationals of $[0, 1]$, still preserving order. This correspondence determines a non-decreasing function $F(x)$ of a variable x which runs over the whole closed interval $[a, b]$ along which the set Z is disposed; the function $F(x)$ varies between 0 and 1; it is constant on the contiguous open intervals of Z, taking on the corresponding dyadic rational value, while at points of Z of the second type, it assumes the corresponding dyadic irrational value. Since it is non-decreasing and assumes all values in the interval $[0, 1]$, it is continuous.

Its derivative exists in every case and vanishes at all points of contiguous open intervals; hence if the set Z is of measure zero, the derivative of $F(x)$ vanishes almost everywhere and the $<$ sign holds in (2).

3. Thus we have to impose stronger conditions than continuity on the non-decreasing function $F(x)$ to ensure equality in (2) of art. 1.

Definition. A function $F(x)$, defined on the closed interval $[a, b]$, is said to be *absolutely continuous* if for any $\varepsilon > 0$ we can find $\delta > 0$ such that, for any (finite) system of non-overlapping intervals $(a_1, b_1), \ldots, (a_n, b_n)$ of overall length

$$\sum_{k=1}^{n} (b_k - a_k) < \delta,$$

the corresponding sum of absolute increments of the function does not exceed ε:

$$\sum_{k=1}^{n} |F(b_k) - F(a_k)| < \varepsilon. \tag{1}$$

For example every function that satisfies the Lipschitz condition

$$|F(x'') - F(x')| \leq C|x'' - x'|,$$

is absolutely continuous, since for any system of intervals $(a_1, b_1), \ldots (a_n, b_n)$

$$\sum_{k=1}^{n} |F(b_k) - F(a_k)| \leq C \sum_{k=1}^{n} (b_k - a_k),$$

and to ensure that the sum on the left is less than a given $\varepsilon > 0$, it is only necessary that the overall length of the system chosen should not exceed $\delta < \varepsilon/C$.

On the other hand the continuous non-decreasing function $F(x)$ of art. 2 with a derivative vanishing almost everywhere is not absolutely continuous. For the set Z can be covered with a countable system of non-overlapping intervals of arbitrarily small overall length, on which the sum of the increments of $F(x)$ is equal to 1; on a sufficiently large finite subsystem this sum will exceed 1/2, which is incompatible with the definition of absolute continuity.

A function $F(x)$ which is the indefinite integral of a summable function $\varphi(x)$, so that

$$F(x) = \int_a^x \varphi(\xi)\, \mathrm{d}\xi,$$

is always absolutely continuous.

For we have:

$$\sum_{k=1}^{n} |F(b_k) - F(a_k)| = \sum_{k=1}^{n} \left| \int_{a_k}^{b_k} \varphi(\xi)\, \mathrm{d}\xi \right|$$
$$\leq \sum_{k=1}^{n} \int_{a_k}^{b_k} |\varphi(\xi)|\, \mathrm{d}\xi = \int_{\Sigma(a_k,\, b_k)} |\varphi(\xi)|\, \mathrm{d}\xi,$$

and in virtue of the absolute continuity of the integral over the set (cf. p. 177), the result tends to zero together with the measure of the system of intervals (a_k, b_k).

We observe a few simple properties of absolutely continuous functions.

As a matter of fact, instead of inequality (1) in the definition of absolutely continuous function, we can write

$$\left| \sum_{k=1}^{n} [F(b_k) - F(a_k)] \right| < \varepsilon, \tag{2}$$

which seems rather surprising at first sight. But suppose (2) is satisfied for any system of intervals of overall length $< \delta$. Having fixed such a system, we distinguish two subsystems in it, so that the increments of $F(x)$ are positive in intervals of the first subsystem, and negative in intervals of the second. On writing (2) for each of these subsystems and adding, we in fact obtain (1), with the unimportant replacement of ε by 2ε.

Every absolutely continuous function $F(x)$ is of bounded variation. For let $\delta > 0$ correspond to a given $\varepsilon > 0$ in the condition for absolute continuity of $F(x)$. Then on any interval of length $\leqq \delta$ the variation of $F(x)$ is bounded and does not exceed ε. Hence on the whole interval $[a, b]$, which we can represent as the union of a finite number of closed intervals of length $\leqq \delta$, the variation of $F(x)$ is again finite.

An absolutely continuous function, like any function of bounded variation, can be represented as the difference of two non-decreasing functions

$$F(x) = V(x) - G(x), \quad V(x) = \bigvee_a^x[F]. \tag{3}$$

We claim that the minuend and subtrahend in the decomposition (3) are also absolutely continuous. It is sufficient to confirm this just for the minuend $V(x)$.

Let $\delta > 0$ corresponding to a given $\varepsilon > 0$ in the condition for absolute continuity of $F(x)$ be found and let us consider a system of intervals $(a_1, b_1), \ldots, (a_n, b_n)$ of overall length $< \delta$. The sum

$$\sum_{k=1}^{n} \{V(b_k) - V(a_k)\} = \sum_{k=1}^{n} \bigvee_{a_k}^{b_k} [F]. \tag{4}$$

is the exact upper bound of the quantities

$$\sum_{k=1}^{n} \sum_{j=1}^{n_k} |F(x_{k\,j+1}) - F(x_{kj})|, \tag{5}$$

where $a_k = x_{k0} < x_{k1} < \cdots < x_{knk} = b_k$ is an arbitrary subpartition of the interval (a_k, b_k). Since the sum of the lengths of the subpartition intervals for (a_k, b_k) is equal to the length of the interval (a_k, b_k) itself and the total sum of the lengths of the intervals (a_k, b_k) is less than δ, none of the expressions (5) can exceed ε in virtue of the absolute continuity of $F(x)$; the required absolute continuity of $V(x)$ follows.

Like every function of bounded variation, an absolutely continuous function has derivative almost everywhere. We shall require the following lemma:

LEMMA: *If an absolutely continuous non-decreasing function $F(x)$ has a derivative which vanishes almost everywhere, then $F(x)$ is constant.*

Proof. The domain of variation of $F(x)$ is the closed interval $S = [F(a), F(b)]$; to prove the lemma, we shall show that it is of measure zero. We denote by Z the set (of measure zero) of points x at which the derivative is either non-existent or non-zero, and by E the set (of full measure) on which the derivative exists and vanishes. The set Z is mapped by the function $F(x)$ onto some set $F(Z)$, and the set E onto $F(E)$; clearly $S = F(E) + F(Z)$. Given $\varepsilon > 0$, we find $\delta > 0$ in the condition for absolute continuity of $F(x)$ and cover Z with a (possibly countable) system of non-overlapping intervals of overall length $< \delta$†; then the set $F(Z)$ will be covered by a system of intervals of overall length $\leqq \varepsilon$. Since ε is arbitrarily small, the measure of $F(Z)$ is equal to zero. Further, since the derivative of $F(x)$ vanishes at each point $x \in E$, we can find for each such point a point $\xi > x$ for which

$$F(\xi) - F(x) \leqq \varepsilon(\xi - x),$$

or, what is the same thing,

$$\varepsilon \xi - F(\xi) \geqq \varepsilon x - F(x).$$

Thus x is a point of ascent to the right for the function $G(x) = \varepsilon x - F(x)$ (Section 1). By Riesz' lemma the set of points of ascent to the right is open and for its component open intervals (a_k, b_k) we have

$$\varepsilon a_k - F(a_k) \leqq \varepsilon b_k - F(b_k)$$

or

$$F(b_k) - F(a_k) \leqq \varepsilon(b_k - a_k),$$

and therefore

$$\sum [F(b_k) - F(a_k)] \leqq \varepsilon \sum (b_k - a_k) \leqq \varepsilon(b - a).$$

Thus the set $F(E)$ is covered by the system of intervals $(F(a_k), F(b_k))$, which have arbitrarily small overall length. It follows that $F(E)$ is of measure zero. Hence the closed interval $S = F(Z) + F(E)$ is also of measure zero, i.e. it reduces to a single point, as required.

Note. Analysing the proof given, we observe that it incorporates some more general results, viz:

† A system of non-overlapping intervals can be obtained from an arbitrary covering $\Delta_1 + \Delta_2 + \cdots \supset Z$ by discarding from each Δ_n any points that belong to $\Delta_1 + \cdots + \Delta_{n-1}$.

(a) if $F(x)$ is an absolutely continuous non-decreasing function, then the image $F(Z)$ of any set Z of measure zero is also a set of measure zero.

(b) if the function $F(x)$ is non-decreasing and its derivative vanishes on a set E, then the image $F(E)$ of E is a set of measure zero.

We can now prove the fundamental theorem of this paragraph, which yields the answer to problem II.

THEOREM (A. Lebesgue). *The derivative $\varphi(x)$ of an absolutely continuous function $F(x)$ defined on the closed interval $[a, b]$ is summable, and for every x*

$$\int_a^x \varphi(x)\,dx = F(x) - F(a).$$

Proof. An absolutely continuous function $F(x)$ can be represented as the difference of two absolutely continuous non-decreasing functions; we can therefore assume without loss of generality that the function $F(x)$ is non-decreasing. Let $\varphi(x)$ be its derivative; by what has been proved, $\varphi(x)$ is summable. We put

$$G(x) = \int_a^x \varphi(\xi)\,d\xi.$$

The function $G(x)$ is absolutely continuous and, as we saw in Section 2, its derivative coincides almost everywhere with $\varphi(x)$. But the derivative of the absolutely continuous function $F(x)$ also coincides with $\varphi(x)$; hence the derivative of the difference $H(x) = F(x) - G(x)$ vanishes almost everywhere. We observe that the function $H(x)$ is also non-decreasing, since in virtue of inequality (2) of art. 1

$$H(\beta) - H(\alpha) = F(\beta) - F(\alpha) - \int_\alpha^\beta \varphi(x)\,dx \geqq 0.$$

Hence in accordance with the lemma $H(x)$ is constant, equal to C_0 say. But then we have:

$$F(x) = G(x) + H(x) = \int_a^x \varphi(\xi) + C_0.$$

Putting $x = a$, we get $C_0 = F(a)$, and the theorem is proved.

Let us consider a non-decreasing continuous, but not absolutely continuous function $F(x)$. Repeating the foregoing argument, we find that the difference

$$Z(x) = F(x) - G(x)$$

is a continuous non-decreasing function and that its derivative vanishes almost everywhere. Since $Z(x)$, together with $F(x)$, is not absolutely continuous, it is not constant. We get a representation of the continuous non-decreasing function $F(x)$ in the form of a sum of two continuous non-decreasing functions

$$F(x) = G(x) + Z(x),$$

of which the first is absolutely continuous and the second has a derivative which vanishes almost everywhere.

For an arbitrary non-decreasing function (possibly discontinuous) this decomposition is completed in accordance with the last article of Section 1 by the addition of one further term—a saltus function $H(x)$.

$$F(x) = G(x) + Z(x) + H(x).$$

From non-decreasing functions it is easy to pass to functions of bounded variation. Equation (6) holds good for a continuous function $F(x)$ of bounded variation; $G(x)$ is then an absolutely continuous function, and $Z(x)$ a so-called singular function, i.e. a continuous function of bounded variation the derivative of which vanishes almost everywhere.

We observe incidentally that the decomposition of a given continuous function $F(x)$ into absolutely continuous and singular components is unique up to an additive constant. For let

$$F(x) = G(x) + Z(x) = G_1(x) + Z_1(x),$$

where G, G_1 are absolutely continuous and Z, Z_1 are singular; then

$$G - G_1 = Z_1 - Z$$

and the function $G - G_1$, which is absolutely continuous, has together with the right-hand side a derivative which vanishes almost everywhere and must therefore be constant.

4. *Integration by Parts*

Let $\varphi(x)$, $\psi(x)$ be summable functions and $\Phi(x)$, $\Psi(x)$ their indefinite integrals. We then have the formula

$$\int_a^b \Phi(x)\,\psi(x)\,\mathrm{d}x + \int_a^b \Psi(x)\,\varphi(x)\,\mathrm{d}x = \Phi(b)\,\Psi(b) - \Phi(a)\,\Psi(a). \quad (1)$$

For the proof it is sufficient to observe that the function $\Phi(x)\,\Psi(x)$, together with $\Phi(x)$ and $\Psi(x)$, is absolutely continuous and its derivative is given by the usual formula

$$(\Phi(x)\,\Psi(x))' = \Phi(x)\,\psi(x) + \varphi(x)\,\Psi(x),$$

from which (1) follows on integration between the limits a, b.

Problems. 1. Construct a continuous function, the derivative of which exists everywhere but is not summable.

Answer. For example, $y = x^2 \sin 1/x^2$.

2. It is given that a function $F(x)$ satisfies the following condition: for any $\varepsilon > 0$ there exists $\delta > 0$ such that the inequality $\Sigma\, b_k - a_k) < \delta$ implies the inequality $\Sigma\, |F(b_k) - F(a_k)| < \varepsilon$ (i.e. in the condition for absolute continuity the requirement that the intervals (a_k, b_k) be non-overlapping is dropped). Show that $F(x)$ satisfies the Lipschitz condition

$$|F(x'') - F(x')| \leqq C|x'' - x'|.$$

3. Prove the following converse of the note on the lemma (p. 308): if the function $F(x)$ is non-decreasing and the image $F(Z)$ of the set Z on which the derivative of $F(x)$ is non-existent of infinite is of measure zero, then $F(x)$ is absolutely continuous.

4. If $F(x)$ is absolutely continuous, then for $p \geqq 1$, but not necessarily for $p < 1$, $|F(x)|^p$ is also absolutely continuous.

5. Prove that the total variation of the function $F(x) = \int_a^x \varphi(\xi)\,\mathrm{d}\xi$, where $\varphi(x)$ is a summable function, over the closed interval $[a, b]$ is equal to $\int_a^b |\varphi(\xi)|\,\mathrm{d}\xi$.

Hint. A corollary to problem 7, art. 2, Section 2. An independent construction: approximate $\varphi(x)$ in the metric of L_1 by step functions.

6. Show that absolutely continuous functions form a closed subspace in the space of functions of bounded variation (see problem 6, art. 2, Section 2).

4. Functions of Several Variables

1. We now undertake the task of generalising the results of Sections 1–3, which relate to the case of a single variable, to the case of several variables. For simplicity we shall consider the case of two variables.

We begin by formulating the facts familiar to us from classical analysis. If $\varphi(x, y)$ is a continuous function, we can form its integral over any (closed) region G:

$$\Phi(G) = \iint_G \varphi(\xi, \eta)\, d\xi\, d\eta .$$

This naturally replaces the indefinite integral $\int_a^x \varphi(\xi)\, d\xi$ which we considered when we were dealing with a single variable.

The function $\varphi(x, y)$ of a point can be obtained from the function $\Phi(G)$ of a region by means of a limiting operation in place of differentiation

$$\varphi(x, y) = \lim_{G \to (x, y)} \frac{1}{|G|} \iint_G \varphi(\xi, \eta)\, d\xi\, d\eta ,$$

where $|G|$ denotes the area of the region G, and $G \to (x, y)$ denotes that G shrinks to the point (x, y). If now $\varphi(\xi, \eta)$ is an arbitrary summable function defined on the rectangle

$$\Delta = \{a_1 \leqq x \leqq b_1, a_2 \leqq y \leqq b_2\},$$

we can take any measurable set $\mathscr{E}$ in place of the region G. The following problems naturally arise:

Problem III. *Does the limiting relation*

$$\varphi(x, y) = \lim_{\mathscr{E} \to (x, y)} \frac{1}{\mu \mathscr{E}} \iint_{\mathscr{E}} \varphi(\xi, \eta)\, d\xi\, d\eta$$

hold everywhere in the region Δ if $\varphi(x, y)$ is an arbitrary summable function?

The converse problem can be as follows:

Problem IV: *Let a function $\Phi(\xi)$ of a measurable set be given; when can we assert that its "density"*

$$\varphi(x, y) = \lim_{\mathscr{E} \to (x, y)} \frac{\Phi(\mathscr{E})}{\mu \mathscr{E}} ,$$

exists (if only almost everywhere) and under what conditions is the function $\Phi(\mathscr{E})$ recoverable from its density in the form of an integral

$$\Phi(\mathscr{E}) = \iint_{\mathscr{E}} \varphi(\xi, \eta)\, d\xi\, d\eta\, ?$$

We shall reduce the solution of both problems to the solution of the analogous problems for a single variable, which were considered in Sections 1–3, by Riesz' so-called principle of transition.

2. THEOREM 1 (F. Riesz' principle of transition). *It is possible to establish between the set of points of the plane and the set of points of the line an "almost" one–one correspondence such that measurable subsets correspond to measurable subsets and measure is preserved.*

Proof. We partition the ξ-axis into intervals $(m, m + 1)$ of unit length and the $x\,y$-plane into a square grill of unit mesh with vertices at the integral points, then set up an arbitrary one–one correspondence between the intervals of the axis and the squares of the plane; the map is possible since each set is countable. We further divide each square into four equal squares of area $1/4$ and the corresponding interval into four equal intervals of length $1/4$, and once more map each of the smaller squares arbitrarily onto one of the smaller intervals to give a one–one correspondence. We continue this construction indefinitely, taking care at each step that intervals obtained by quadrisection of an antecedent interval Δ correspond to squares obtained by quadrisection of the square corresponding to Δ. We shall call the aggregate of all the squares obtained in this way a *plane net* and the aggregate of all the intervals on the line a *linear net.*

By construction every sequence of nested squares of a plane net is mapped onto a sequence of nested intervals of the corresponding linear net, and conversely. We extend this correspondence as follows into a correspondence between the points of the plane and the points of the line.

For definiteness we shall consider only "proper" sequences of squares and intervals, i.e. those in which the nth term has measure (area or length) exactly $\frac{1}{4^n}$.

If both coordinates of a point of the plane are dyadic-irrational, there exists a unique proper sequence of squares of the plane nes all of which contain it. The corresponding sequence of intervalt on the line determines uniquely a point on the line, which we make the image of the chosen point in the plane. If the point in the plane has one coordinate (or both) dyadic-rational, there exist two (or four) proper sequences of squares of the plane net which contain it; we can therefore map it onto two (or four) points on the line. Since the set A of all points of the plane with at least one rational coordinate is of measure zero and can therefore be covered by a system of squares of the net of arbitrarily small total area, the set of corresponding doubletons and quadrupletons on the line can be covered by a system of intervals of the same total length and therefore has linear measure zero. (The set of quadrupletons is evidently at most countable). Conversely, if the coordinate of a point on the line is dyadic–irrational, it can be mapped onto a uniquely determined point of the plane, using the unique proper sequence of intervals containing it. And if its coordinate is dyadic–rational, it can be mapped in general onto a pair of points of the plane; but the set of all points of the plane included in such pairs is at most countable.

We can thus establish a one–one correspondence between two sets of full measure: the set of points of the plane with both coordinates dyadic-irrational excepting those included in the pairs indicated, and the set of dyadic-irrational points of the line, excepting the zero-measure set of points included in the specified doubletons and quadrupletons. We shall show that under this correspondence *every measurable set*, say on the interval [0, 1], *is mapped onto a measurable set of equal measure* in the corresponding square. We begin by observing that every open set on the line can be expressed in the form of a union of some (countable) set of intervals of the net without common interior points. Let E be an arbitrary measurable set on the closed interval [0, 1]. There exist sequences O_n, O'_n $(n = 1, 2, \dots)$ of open sets with the property

$$O_n \supset E, \quad O'_n \supset CE,$$
$$1 \leqq \mu O_n + \mu O'_n \leqq 1 + \varepsilon_n, \quad \varepsilon_n \to 0.$$

Let $\mathscr{E}$ be the plane image of the set E. Considering the aggregate of sets G_n, G'_n in the plane, composed of the squares which correspond to dyadic-rational intervals contained in O_n, O'_n, we get the relations

$$G_n \supset \mathscr{E}, \quad G'_n \supset C\mathscr{E},$$
$$1 \leqq \mu G_n + \mu G'_n \leqq 1 + \varepsilon_n, \quad \varepsilon_n \to 0.$$

In follows from these relations that $\mathscr{E}$ is measurable (Chapter IV, Section 4, art. 5) and has measure

$$\mu \mathscr{E} = \lim \mu G_n = \lim \mu O_n = \mu E,$$

as required.

We can further extend the correspondence to measurable functions on the line and on the plane, mapping a function $f(\xi)$ onto the function $\varphi(x, y)$ for which $\varphi(x, y) = f(\xi)$ if (x, y) is the image of ξ. The functions $f(\xi)$, $\varphi(x, y)$ are then not only measurable, but even "equi-measurable", in the sense that for any C the sets $\{\xi: f(\xi) < C\}$, $\{(x, y): \varphi(x, y) < C\}$ are of equal measure. It follows further that the integrals of the functions $f(\xi)$, $\varphi(x, y)$ over corresponding measurable sets E, $\mathscr{E}$ are equal.

3. We are now in a position to solve problem III. We shall say that a sequence $\mathscr{E}_n$ of plane measurable sets *converges properly* to the point $P = (x, y)$ if there exists a sequence of squares Q_n of the plane net, containing P, and such that $\mu Q_n \to 0$, $\mathscr{E}_n \subset Q_n$, $\mu \mathscr{E}_n \geqq \alpha \mu Q_n$, where $\alpha > 0$ is a fixed constant.

THEOREM 2. *If $\varphi(x, y)$ is summable over a rectangle Δ, then for almost all points of this rectangle*

$$\varphi(x, y) = \lim_{\mathscr{E}_n \to (x, y)} \frac{1}{\mu \mathscr{E}_n} \iint\limits_{\mathscr{E}_n} \varphi(\xi, \eta) \, d\xi \, d\eta, \tag{1}$$

for any sequence of measurable sets $\mathscr{E}_n$ that converges properly to the point (x, y).

Proof. By the principle of transition the summable function $\varphi(x, y)$ maps onto a summable function $f(\xi)$, the point (x, y) in the plane onto a point ξ on the line, the sequence of squares Q_n onto a sequence of intervals Δ_n which contain ξ, and the sequence of sets $\mathscr{E}_n$ onto a sequence of measurable sets $E_n \subset \Delta_n$ with $\mu E_n \geqq \alpha \mu \Delta_n$; furthermore

$$\frac{1}{\mu \mathscr{E}_n} \iint\limits_{\mathscr{E}_n} \varphi(x, y) \, dx \, dy = \frac{1}{\mu E_n} \int\limits_{E_n} f(\xi) \, d\xi.$$

By the lemma of Section 2 (art. 4) the last expression converges to $f(x)$ almost everywhere, to be precise, at all the Lebesgue points of the function $f(\xi)$. Reverting to the function $\varphi(x, y)$, we get that the limiting relation (1) holds almost everywhere in the rectangle Δ, as required.

4. We proceed now to the solution of problem IV.

We are given a function $\Phi(\mathscr{E})$ of a measurable set $\mathscr{E}$; it is asked: when can we assert that $\Phi(\mathscr{E})$ is expressible in the form

$$\Phi(\mathscr{E}) = \iint\limits_{\mathscr{E}} \varphi(\xi, \eta) \, d\xi \, d\eta, \tag{1}$$

where $\varphi(x, y)$ is a summable function?

We shall begin by giving necessary conditions for such a representation to be possible. If $\Phi(\mathscr{E})$ has a representation in the form (1), it obviously possesses the following properties:

(a) additivity: $\Phi(\mathscr{E}_1 + \mathscr{E}_2) = \Phi(\mathscr{E}_1) + \Phi(\mathscr{E}_2)$ if the measurable sets $\mathscr{E}_1$, $\mathscr{E}_2$ are disjoint;

(b) absolute continuity: given $\varepsilon > 0$, we can find $\delta > 0$ such that $|\Phi(\mathscr{E})| < \varepsilon$ whenever $\mu\,\mathscr{E} < \delta$ (cf. p. 177).

We shall show that if these properties of the function $\Phi(\mathscr{E})$ obtain, even if not for all measurable sets, then for those composed of squares of the grill, we are assured that $\Phi(\mathscr{E})$ is expressible in the form (1).

By the principle of transition, we can map the function $\Phi(\mathscr{E})$ onto a function $\Psi(E)$ defined on measurable subsets of the x-axis. We can pass from the set function $\Psi(E)$ to a point function $F(x)$ by putting $F(x)$ equal to the value of $\Psi(E)$ on the interval $E = (a, x)$. The function $F(x)$ obtained is absolutely continuous in the usual sense, for given a system of intervals $\Delta_k = (a_k, b_k)$ of total length $< \delta$ we have

$$\sum [F(b_k) - F(a_k)] = \sum \Psi(\Delta_k) = \sum \Phi(\mathscr{E}_k) = \Phi\left(\sum \mathscr{E}_k\right), \quad (2)$$

where $\mathscr{E}_k$ is the plane set corresponding to the interval Δ_k. Since together with the sum of the lengths of the Δ_k the sum of the measures of the $\mathscr{E}_k$ is less than δ, each of the sums (2) is less than ε in absolute value, as required. In accordance with the fundamental theorem of Section 3 the function $F(x)$ can be expressed as the integral of its derivative $\varphi(x) = F'(x)$:

$$F(x) = \int_a^x \varphi(\xi)\,\mathrm{d}\xi + F(a).$$

Under the principle of transition the summable function $\varphi(\xi)$ is mapped onto a summable function $\varphi(x, y)$ on the plane. We shall show that the set function

$$G(\mathscr{E}) = \int\int_{\mathscr{E}} \varphi(x, y)\,\mathrm{d}x\,\mathrm{d}y$$

coincides with the original function $\Phi(\mathscr{E})$. Since both $\Phi(\mathscr{E})$ and $G(\mathscr{E})$ are additive and absolutely continuous, it is sufficient to prove coincidence for sets $\mathscr{E}$ which are squares of the grill, in view of the fact that by the theorem of Chapter IV,

Section 4 art. 3 any measurable set can be approximated by a finite sum of such sets to within a set of arbitrarily small measure. But if $\mathscr{E}$ is a square of the grill, it corresponds to an interval, say $\Delta = (\alpha, \beta)$, of the linear net; hence

$$\Phi(\mathscr{E}) = \Psi(\Delta) = F(\beta) - F(\alpha) = \int_\alpha^\beta \varphi(x)\, dx,$$

$$G(\mathscr{E}) = \iint_{\mathscr{E}} \varphi(\xi, \eta)\, d\xi\, d\eta = \int_\alpha^\beta \varphi(x)\, dx,$$

so that $\Phi(\mathscr{E}) = G(\mathscr{E})$, as required.

By what we have proved above the function $\varphi(x, y)$ can be obtained directly from $\Phi(\mathscr{E})$ by means of a limiting operation

$$\varphi(x, y) = \lim_{\mathscr{E}_n \to (x, y)} \frac{1}{\mu \mathscr{E}_n} \iint_{\mathscr{E}_n} \varphi(\xi, \eta)\, d\xi\, d\eta = \lim_{\mathscr{E}_n \to (x, y)} \frac{\Phi(\mathscr{E}_n)}{\mu \mathscr{E}_n},$$

where $\mathscr{E}_n$ is a sequence of measurable sets which converges properly to the point (x, y); the limit exists almost everywhere. We thus have the following theorem:

THEOREM 3. *A necessary and sufficient condition for a set function $\Phi(\mathscr{E})$ to have density $\varphi(x, y)$ at almost every point and for the formula*

$$\Phi(\mathscr{E}) = \iint_{\mathscr{E}} \varphi(\xi, \eta)\, d\xi\, d\eta,$$

to hold for every measurable set $\mathscr{E}$ is that $\Phi(\mathscr{E})$ be additive and absolutely continuous.

5. As an application of the theorem just proved we shall find the general form of a continuous linear functional $f[\varphi]$ on the space $L_1(D)$ of all functions integrable over a given region D.

The condition of continuity of the functional f can be written in the form of the inequality

$$|f[\varphi]| \leq C\|\varphi\| = C \iint_D |\varphi(x, y)|\, dx\, dy$$

with fixed constant C.

We map the functional f onto the function $F(\mathscr{E})$ of a measurable set $\mathscr{E}$ which has as value the value of the functional at the characteristic function $\chi_{\mathscr{E}}$ of $\mathscr{E}$. The function $F(\mathscr{E})$ is additive together with the functional f and satisfies the inequality

$$|F(\mathscr{E})| = |f(\chi_{\mathscr{E}})| \leq C \iint_{\mathscr{E}} \chi_{\mathscr{E}}(x, y)\, dx\, dy = C \mu \mathscr{E},$$

from which it follows that $F(\mathscr{E})$ is absolutely continuous. By theorem 3 it has at almost every point the density

$$g(x, y) = \lim_{\mathscr{E}_n \to (x, y)} \frac{F(\mathscr{E}_n)}{\mu \mathscr{E}_n}.$$

Here evidently the absolute value of the function $g(x, y)$ does not exceed C. By the same theorem the function $F(\mathscr{E})$ is recoverable in terms of $g(x, y)$ by the formula

$$F(\mathscr{E}) = \int\int_{\mathscr{E}} g(\xi, \eta)\,\mathrm{d}\xi\,\mathrm{d}\eta = \int\int_{D} \chi_{\mathscr{E}}(\xi, \eta)\, g(\xi, \eta)\,\mathrm{d}\xi\,\mathrm{d}\eta.$$

Thus for the characteristic functions of measurable sets we have

$$f[\chi_{\mathscr{E}}] = F(\mathscr{E}) = \int\int_{D} \chi_{\mathscr{E}}(x, y)\, g(x, y)\,\mathrm{d}x\,\mathrm{d}y.$$

The functional

$$f_1[\varphi] = \int\int_{D} \varphi(x, y)\, g(x, y)\,\mathrm{d}x\,\mathrm{d}y,$$

is evidently a continuous linear functional on the space $L_1(D)$. It coincides with the functional $f[\varphi]$ at the characteristic functions of measurable sets; but then it must coincide with f at all step functions and in the limit in general at all functions $\varphi \in L_1(D)$.

Thus $f = f_1$ and we have proved the following theorem:

THEOREM: *Every continuous linear functional $f[\varphi]$ on the space $L_1(D)$ is of the form*

$$f[\varphi] = \int\int_{D} \varphi(x, y)\, g(x, y)\,\mathrm{d}x\,\mathrm{d}y,$$

where $g(x, y)$ is a bounded measurable function.

An analogous theorem obviously holds for the case of functions of any number of independent variables.

6. THEOREM 3, proved in art. 4, is the analogue of the theorem (Section 3) on the differentiability and integral representation of an absolutely continuous function of one variable.

We consider now what analogue exists for the theorem on the decomposition of an arbitrary monotone—say non-decreasing—continuous function $F(x)$ into absolutely continuous and singular components.

A continuous non-decreasing function $F(x)$ determines a non-negative additive function of an interval (α, β) of the linear net, equal to $F(\beta) - F(\alpha)$, which can be extended over all finite systems of intervals of the net (using additivity). By the principle of transition a non-negative additive function $\Phi(\mathscr{E})$ will be given in the plane on the system of squares of the grill. The function $\Phi(\mathscr{E})$ is continuous in the sense that if a square $\mathscr{E}_n$ of the grill converges to a point (possibly a boundary point), then $\lim_{n\to\infty} \Phi(\mathscr{E}_n) = 0$. Conversely, any non-negative additive continuous function $\Phi(\mathscr{E})$ defined on the finite system of grill squares maps onto a non-negative continuous function of the intervals (α, β) of the linear net, and the function $F(x)$ is consequently continuous non-decreasing.

We decompose $F(x)$ into absolutely continuous and singular components:

$$F(x) = G(x) + Z(x).$$

To this decomposition corresponds a decomposition of the set function, also into two components:

$$\Phi(\mathscr{E}) = \mathfrak{g}(\mathscr{E}) + \mathfrak{Z}(\mathscr{E}).$$

The first component $g(\mathscr{E})$ is absolutely continuous and is expressible as the integral of its density. Let us consider what property is possessed by the second component $\mathfrak{Z}(\mathscr{E})$. Let ξ_0 be any point at which the derivative of the function $Z(x)$ vanishes, and let (x_0, y_0) be the corresponding point in the plane; further, let $\mathscr{E}_n$ be a sequence of measurable sets of a special kind, viz. finite systems of grill squares, which converge properly to the point (x_0, y_0). The set $\mathscr{E}_n$ is included in the grill square Q_n and $\mu\mathscr{E}_n \geqq \alpha\mu Q_n$, $\alpha > 0$. The grill square Q_n corresponds to the interval (α_n, β_n) which converges to the point ξ_0. Hence

$$\frac{\mathfrak{Z}(\mathscr{E}_n)}{\mu\mathscr{E}_n} \leqq \frac{\alpha^{-1}}{\mu Q_n}\mathfrak{Z}(\mathscr{E}_n) \leqq \frac{\alpha^{-1}}{\beta_n - \alpha_n}[Z(\beta_n) - Z(\alpha_n)] \to 0,$$

since the derivative of $Z(x)$ vanishes at ξ_0. Thus if we define density, not in terms of arbitrary measurable subsets, but in terms of grill squares and their properly convergent finite sums only, the density of the function $\mathfrak{Z}(\mathscr{E})$ also vanishes almost everywhere.

Note. Instead of considering the values of non-absolutely continuous set functions $\Phi(\mathscr{E})$ on all sets measurable in the sense of Lebesgue, we have been forced to confine ourselves to their values on square nets and finite unions of such. This has not been a matter of chance. By using a process analogous to the construction of a system of Lebesgue-measurable sets, we can start from intervals (Chapter IV, Section 4) and define for any non-negative countably-additive function $\Phi(\mathscr{E})$ of rectangles a system $\sigma(\Phi)$ of sets measurable with respect to the function Φ, or more briefly, Φ-measurable. However, *if $\Phi(\mathscr{E})$ is continuous but not absolutely continuous, the system $\sigma(\Phi)$ certainly does not contain all Lebesgue-measurable sets.* For there always exists in this case a countable set $\mathscr{E}_0$, for which $\mu\,\mathscr{E}_0 = 0$, $\Phi(\mathscr{E}_0) > 0$; further we can find a Φ-measurable $\mathscr{E}_1 \subset \mathscr{E}_0$, so that $\mathscr{E}_1 \notin \sigma(\Phi)$; at the same time $\mathscr{E}_1$ is Lebesgue-measurable and $\mu\,\mathscr{E}_1 = \mu\,\mathscr{E}_0 = 0$.

5. The Stieltjes Integral

1. In constructing a theory of the integral we started from the known values of the integrals of step functions. The integral of an "elementary step"—a function equal to 1 on an interval Δ and 0 outside it—was put equal to the length of the interval. Intervals of equal length gave rise to integrals having the same value, so that under integration the line became a perfectly homogeneous manifold, of uniform construction throughout. But many problems by their nature prohibit us from regarding the line as uniform. In some cases, e.g. a non-uniform string or rod, we discounted non-uniformity, by introducing a density variable. Unfortunately the introduction of density does not always overcome the difficulty (for example, a string loaded with a point bead). The most expeditious approach to the study of such problems lies in introducing an interval measure which allows for non-uniformity of the line.

By an interval on the line we shall understand one of the following five types of set:

(1) the closed interval $[\alpha, \beta]$ (both end-points included);

(2) the open interval (α, β) (both end-points excluded);

(3) the semi-interval $(\alpha, \beta]$ (the right end-point included);

(4) the semi-interval $[\alpha, \beta)$ (the left end-point included);

(5) the isolated point $[\alpha]$.

Let each interval Δ on the closed interval $[a, b]$ be mapped onto some non-negative number $\varrho\,\Delta$ so as to satisfy the condition of total additivity: if an interval Δ is the union of intervals

$\Delta_1, \Delta_2, \ldots \Delta_n, \ldots$ without common points, then

$$\varrho \Delta = \varrho \Delta_1 + \varrho \Delta_2 + \cdots + \varrho \Delta_n + \cdots$$

We say in such a case that we are given a *Stieltjes measure.*

Cases can occur where isolated points have a positive Stieltjes measure. However, this can happen comparatively seldom, at most for a countable set of points, since the sum of the measures of any finite number of points cannot exceed a fixed constant $c > 0$.

Examples 1. $\varrho\Delta$ is the length of the interval Δ (Lebesgue measure).

2. $\varrho \Delta = \int\limits_{\Delta} \varphi(x)\, dx,$

where $\varphi(x)$ is a fixed non-negative summable function.

3. $\varrho \Delta$ is equal to 1 for every Δ containing the point c, and 0 for every Δ not containing c.

The condition for total additivity can be replaced by two conditions: for (finite) additivity and continuity.

(a) The additivity condition: if the intervals Δ_1, Δ_2 are disjoint,

$$\varrho(\Delta_1 + \Delta_2) = \varrho \Delta_1 + \varrho \Delta_2.$$

Naturally, if the additivity property holds for any pair of non-overlapping terms, it also holds for any finite number of such terms $\Delta_1, \ldots, \Delta_n$:

$$\varrho(\Delta_1 + \cdots + \Delta_n) = \varrho \Delta_1 + \cdots + \varrho \Delta_n.$$

(b) The continuity condition: if a sequence of nested intervals $\Delta_1 \supset \Delta_2 \supset \ldots$ has Δ as its intersection, $\varrho \Delta = \lim \varrho \Delta_n$.

Properties (a) and (b) are easily seen to follow from the total additivity condition. The converse, that total additivity is a consequence of properties (a), (b), is more difficult; it will be proved below by using the theory of the integral (p. 323).

The measure $\varrho \Delta$ can be considered not only on a closed interval $[a, b]$, but also on the entire line $-\infty < x < \infty$. If the measure is finite for the entire line, there is no difference between this case and the case of measure specified on a finite closed interval, so that $[a, b]$ can be replaced throughout what follows by the entire line, on condition that its measure be finite.

We shall now define the integral corresponding to the Stieltjes measure. We begin as usual with the integration of step functions.

We partition $[a, b]$ into a finite set of non-overlapping intervals $\Delta_1, \Delta_2, \ldots, \Delta_n$. A function $h(x)$ which assumes a constant value h_j on the interval Δ_j $(j = 1, 2, \ldots, n)$ is said to be a step function. We define the Stieltjes integral of $h(x)$ by the formula

$$I_\varrho h = \sum_{j=1}^{n} h_j \, \varrho \Delta_j .$$

We shall extend this concept of the integral from step functions to ϱ-measurable functions—the limits of sequences of step functions, as we did in Chapter IV for the Lebesgue integral. An important stage in the course of this construction was reached with the concept of a set of measure zero. In our new scheme a set $Z \subset [a, b]$ is said to be a set of (Stieltjes) measure zero if for any $\varepsilon > 0$ it can be covered by a finite or countable system of intervals of total Stieltjes measure $< \varepsilon$. We observe that now isolated points can have positive measure while entire closed intervals may be of measure zero (as in example 3). A sequence of functions $f_n(x)$ is said to converge almost everywhere if it converges at all points of the closed interval, except perhaps on a set of (Stieltjes) measure zero.

In particular, a sequence that converges almost everywhere must actually converge at points bearing positive measure, whereas it can behave quite arbitrarily on closed intervals of measure zero.

We must define an integral for functions of the class C_ϱ^+ composed of functions $f(x)$ which are the limits (almost everywhere) of increasing sequences of step functions. In Chapter IV, Section 2 this construction was based on two lemmas relating to decreasing sequences of step functions and asserting the equivalence of the relations $h_n \searrow 0$, $I h_n \searrow 0$. In proving these lemmas it was essential that the discontinuities of the functions h_n $(n = 1, 2, \ldots)$ should constitute a set of measure zero. In our case, if we start from arbitrary step functions, this condition will not be satisfied, since there is nothing to stop a step function indulging in a jump just at a point of positive measure. We extricate ourselves from this difficulty very simply: we shall require that the initial step functions do not have salti at points of positive measure. (As we saw above, there is at most a countable set of such points).

With the fulfilment of this condition we can reiterate the scheme of Chapter IV, Sections 2–3 in its entirety. The (Stieltjes) class

C_ϱ^+ is composed of the functions $f(x)$ which are limits almost everywhere of convergent increasing sequences of step functions $h_n(x)$ with bounded integrals $I_\varrho h_n$ and without salti at points of the axis which have positive measure. The Stieltjes integral $I_\varrho f$ is defined by the formula

$$I_\varrho f = \lim I_\varrho h_n .$$

The differences of functions of the class C_ϱ^+ form a class of functions L_ϱ which are said to be *summable in the Lebesgue–Stieltjes sense.*

If $\varphi = f_1 - f_2$, where $f_1, f_2 \in C_\varrho^+$, the Stieltjes integral of the function φ is defined by the formula

$$I_\varrho \varphi = I_\varrho f_1 - I_\varrho f_2 .$$

The correctness of all these definitions carries over from the scheme given in Chapter IV.

The integral $I_\varrho \varphi$ is termed the *Lebesgue–Stieltjes integral* with respect to the measure ϱ and has the more pedantic denotation

$$I_\varrho \varphi = \int_a^b \varphi(x)\, \varrho(\mathrm{d}x) .$$

We observe that a step function $h(x)$ with a discontinuity at a point of positive measure can always be expressed as the limit of an increasing or decreasing sequence of permissible step functions and will therefore belong to the class C_ϱ^+ or to L_ϱ, so that our restriction on the class of step functions used in constructing the integral will not lead to any diminution of the totality of summable functions. The class L_ϱ is a complete normed linear space with the norm

$$\|\varphi\| = I_\varrho\, |\varphi|) .$$

The spaces $L_\varrho^p\, (p \geqq 1)$ are defined similarly. ϱ-measurable functions are defined in the natural way as the limits of sequences of step functions which converge almost everywhere (in the measure ϱ). Further, ϱ-measurable sets can be defined as sets the characteristic functions of which are ϱ-measurable; the ϱ-measure of a ϱ-measurable set E is defined by means of the formula

$$\varrho\, E = I_\varrho(\chi_E(x)), \quad \chi_E(x) = \begin{cases} 1 & \text{for} \quad x \in E, \\ 0 & \text{for} \quad x \notin E . \end{cases}$$

The ϱ-measure is countably additive on a system of ϱ-measurable sets, i.e.

$$\varrho(E_1 + E_2 + \cdots) = \varrho\, E_1 + \varrho\, E_2 + \cdots$$

if the sets $E_1, E_2, \ldots$ are ϱ-measurable and have no common points.

A ϱ-measurable function can now be defined as a function $f(x)$ for which the set $\{x: f(x) < C\}$ is ϱ-measurable for any C. This enables us to extend the whole of the Lebesgue theory (Chapter IV, Section 4, art. 6) to ϱ-integrals.

An essential role is played in all these constructions by the condition for countable additivity of the ϱ-measure on intervals, or the formally weaker conditions for additivity and continuity (p. 320). Firstly, these conditions are essential in the definition of sets of ϱ-measure zero. On the one hand, an interval Δ is of measure zero if it can be ascribed the ϱ-measure $\varrho\,\Delta = 0$. On the other hand, in accordance with the above conditions, an interval Δ of measure zero can be covered by a system of intervals—or even by one interval here—of arbitrarily small measure. The continuity condition ensures the equivalence of these definitions. Further, by our construction, the measure of an interval Δ, one or both ends of which are of positive measure, is defined by the integral $I_\varrho(\Delta)$, which is the limit of the measures $\varrho\,\Delta_n$ of intervals, the ends of which are of measure zero; the continuity condition ensures that $I_\varrho(\Delta) = \varrho\,\Delta$. We note finally that a proof has now been obtained via the theory of the integral of the total additivity of a finitely-additive, continuous Stieltjes measure.

2. The measure ϱ can be used to determine a non-decreasing function

$$F(x) = \varrho[a, x].$$

Moreover ϱ is uniquely recoverable from the function $F(x)$ for any interval; in fact we have:

$$\varrho[a, x] = F(x), \tag{1}$$

$$\varrho[x, x'] = \varrho[a, x'] - \varrho[a, x] = F(x') - F(x). \tag{2}$$

For the semi-interval $[a, x)$ we can write the following:

$$\varrho[a, x] = \sum_{n=1}^{\infty} \varrho(x_n, x_{n+1}) + \varrho[a, x_1]$$

$$\left(a < x_1 < x_2 < \cdots < x,\ \lim_{n\to\infty} x_n = x\right).$$

We note that the total additivity of the function ϱ is used here. With the aid of the function F the last relation can be written:

$$\begin{aligned}\varrho[a, x] &= \sum_{n=1}^{\infty} [F(x_{n+1}) - F(x_n)] + F(x_1) \\ &= \lim_{n\to\infty} F(x_n) = F(x - 0).\end{aligned}$$

It is now easy to find $\varrho\Delta$ for all the remaining types of interval:

$$\varrho[x, x') = \varrho[a, x') - \varrho[a, x) = F(x' - 0) - F(x - 0), \quad (3)$$

$$\varrho(x, x') = \varrho[a, x') - \varrho[a, x] = F(x' - 0) - F(x), \quad (4)$$

$$\varrho[x, x'] = \varrho[a, x'] - \varrho[a, x) = F(x') - F(x - 0), \quad (5)$$

$$\varrho[x] = \varrho[a, x] - \varrho[a, x) = F(x) - F(x - 0). \quad (6)$$

The function $F(x)$ defined by (1) is continuous to the right at every point, since

$$\begin{aligned}F(x + 0) &= \lim_{x_n \to x+0} F(x_n) = F(x_1) - [F(x_1) - F(x_2)] - \cdots \\ &= \varrho[a, x_1] - \varrho[x_2, x_1] - \cdots = \varrho[a, x] = F(x).\end{aligned}$$

As shown by equation (6), at each point x of non-zero ϱ-measure, the function $F(x)$ has a saltus equal to the ϱ-measure of the point; elsewhere $F(x)$ is continuous.

Conversely, an arbitrary non-decreasing function $F(x)$, continuous to the right, determines a non-negative interval function in accordance with formulae (1)–(6); it is easily seen that this interval function is additive and continuous, and hence countably additive. The function $F(x)$ is said to be the *generating function* of the Stieltjes measure ϱ.

In the case of finite measure on an infinite interval extending to $-\infty$ on the left, the corresponding generating function is given by the equation

$$F(x) = \varrho[-\infty, x],$$

which is formally obtained from (1) by substituting $-\infty$ for a. We observe that the point $-\infty$ (and $+\infty$) can also carry a positive measure.

In the general case the Lebesgue–Stieltjes integral constructed from the function $F(x)$ is denoted by the symbol

$$I_\varrho \varphi = \int_a^b \varphi(x)\, \mathrm{d}F(x). \quad (7)$$

This notation is very convenient and, as we shall see below, it conforms with the other universally adopted notations. We draw attention to one distinction. Putting $\varphi(x) \equiv 1$, we get in accordance with the definition

$$\int_a^b \mathrm{d}F(x) = \varrho[a, b] = F(b),$$

and not $F(b) - F(a)$, as we might expect. Of course no importance attaches to this exception unless $F(a) \neq 0$, i.e. unless the point a carries positive measure. In relation to the integral (7), $F(x)$ is said to be the *integrating function.*

3. Let us consider some different types of integrating function.

(a) The integrating function $F(x)$ is a saltus function: then there exist points x_n and numbers p_n, with $\sum p_n < \infty$, such that $F(x)$ is given by an equation of the form

$$F(x) = \sum_{x_n \leqq x} p_n.$$

As was shown in Section 1, art. 3, $F(x)$ is continuous to the right.

In this case the measure $\varrho\Delta$ of an interval Δ is the sum of all p_n which correspond to points x_n contained in Δ. The integral of a step function $h(x)$ which takes the value h_j on an interval Δ_j is equal to

$$\sum_j h_j \varrho \Delta_j = \sum_j h_j \sum_{x_n \in \Delta_j} p_n = \sum_{n=1}^{\infty} h(x_n)\, p_n.$$

If the increasing sequence of step functions $h_\nu(x)$ converges almost everywhere in the measure ϱ to $h(x)$ with the $I_\varrho h_\nu$ bounded, then as $\nu \to \infty$ the $h_\nu(x)$ tend to $f(x)$ at each of the points x_n $(n = 1, 2, \ldots)$ and

$$I_\varrho f = \lim_{\nu \to \infty} I_\varrho h_\nu = \sum_{n=1}^{\infty} f(x_n)\, p_n.$$

Taking differences, we find that the class L_ϱ is formed by functions $\varphi(x)$ which are defined (uniquely) only at the point x_n $(n = 1, 2, \ldots)$, with

$$I_\varrho \varphi = \sum_n \varphi(x_n)\, p_n \quad \text{and} \quad \sum_n p_n |\varphi(x_n)| < \infty.$$

It is easily verified that any function $\varphi(x)$ with $\sum_n p_n |\varphi(x_n)| < \infty$ falls in the class L_ϱ. This completely determines L_ϱ.

(b) The integrating function $F(x)$ is absolutely continuous.

Then the Stieltjes measure on any interval is of the form

$$\varrho[\alpha, \beta] = \varrho[\alpha, \beta) = \varrho(\alpha, \beta) = \varrho(\alpha, \beta] = F(\beta) - F(\alpha) = \int_{\alpha}^{\beta} g(\xi)\,\mathrm{d}\xi,$$

where $g(\xi) \geqq 0$ is the derivative of $F(x)$. There are no points with positive ϱ-measure. The integral of a step function $h(x)$ taking the value h_j on Δ_j is equal to

$$I_\varrho(h) = \sum h_j\, \varrho\, \Delta_j = \sum h_j \int_{\Delta_j} g(\xi)\,\mathrm{d}\xi = \int_a^b h(\xi)\, g(\xi)\,\mathrm{d}\xi = I(h\,g),$$

where I denotes the Lebesgue integral.

If the sequence h_n is increasing and converges almost everywhere in the measure ϱ to a function f, with the integrals $I_\varrho h_n$ bounded, then the Lebesgue integrals $I(h_n\, g)$ are bounded and the limit $f\,g$ of the sequence $h_n\, g$ is consequently summable in the usual sense for a function; we then have

$$I_\varrho f = \lim I_\varrho h_n = \lim I(h_n\, g) = I(f\,g).$$

Taking differences, we see that the class L_ϱ consists of the functions φ for which the products $\varphi\, g$ are summable in the usual sense; we then have for any $\varphi \in L_\varrho$

$$\int_a^b \varphi\,\mathrm{d}F = I_\varrho \varphi = I(g\,\varphi) = \int_a^b \varphi\, g\,\mathrm{d}x \qquad (1)$$

(so that in this case we can replace dF by $F'(x)\,dx = g\,\mathrm{d}x$).

Every set of Lebesgue measure zero (we shall use the abbreviation: L-measure 0) can be covered by a finite or countable system S of intervals of arbitrarily small overall length; the ϱ-measure of this system is equal to the integral over S of the function $g(x)$ and therefore tends to zero together with the ordinary measure of S. It follows that every set of L-measure 0 is a set of ϱ-measure 0. We shall consider an arbitrary L-measurable set E. We know that there exist closed sets F and open sets G, $F \subset E \subset G$, such that the difference $G - F$ has arbitrarily small L-measure. The sets G and F are ϱ-measurable, and by what has been proved, the ϱ-measure of $G - F$ is also arbitrarily small; it follows that the set E is ϱ-measurable. Thus *any L-measurable set is also ϱ-measurable*. Moreover by formula (1) the ϱ-measure of E is equal to

$$I_\varrho \chi_E = I(\chi_E\, g) = \int_E g(x)\,\mathrm{d}x.$$

Let us consider now the structure of ϱ-measurable sets. Let E be a ϱ-measurable set and let $\chi(x)$ be its characteristic function; then by what we have proved χg is summable in the usual sense and

$$\varrho E = \int_a^b \chi(x) \, \mathrm{d} F(x) = \int_a^b \chi g \, \mathrm{d} x .$$

Let E_1 be the set on which $\chi g > 0$; it is contained in E and is L-measurable. The set E itself may be unmeasurable in the ordinary sense, but, representing it in the form

$$E = E_1 + (E - E_1),$$

we see that *it can be expressed as the sum of two sets such that one is L-measurable and $g(x)$ vanishes on the other*. The converse is also true: *if some set E can be expressed as the sum of two disjoint sets E_1, E_2 such that one is L-measurable and $g(x)$ vanishes on the other, then E is ϱ-measurable.* Since an L-measurable set is ϱ-measurable, it is enough to show that the set E_2 on which $g(x)$ vanishes is ϱ-measurable (and has ϱ-measure zero). But the set E_0 of *all* points at which $g(x)$ vanishes is measurable in the ordinary sense. Hence it is ϱ-measurable and its ϱ-measure is

$$\varrho E = \int_{E_0} g \, \mathrm{d} x = 0 .$$

It follows that any subset $E \subset E_0$ is ϱ-measurable and has ϱ-measure zero, which completes the proof.

It was shown above that if the function φ is ϱ-summable, the product φg is L-summable. We shall show that the converse is also true: if for some function φ the product φg is L-summable, then φ is ϱ-summable. We show first that φ is ϱ-measurable. For a given C let us consider the set E on which $\varphi(x) \leqq C$. This set coincides with the set E' on which $\varphi(x) \, g(x) \leqq C \, g(x)$, discounting some set E'' on which $g(x)$ vanishes. The set E' is L-measurable and consequently ϱ-measurable, while E'' is of ϱ-measure zero; hence E is ϱ-measurable. Since C is arbitrary, we infer that the function φ is ϱ-measurable. To prove that φ is ϱ-summable it is enough to show that the integrals $I_\varrho \varphi_N$ are bounded, where $\varphi_N = \min(|\varphi|, N)$. But the function φ_N is bounded and ϱ-measurable and consequently ϱ-summable, so that in accordance with formula (1)

$$I_\varrho \varphi_N = I(\varphi_N g) \leqq I(|\varphi| \, g),$$

as required. We have thus obtained a complete description of ϱ-summable functions: *they are those and only those functions* $\varphi(x)$ *for which the product* $\varphi\, g$ *is summable in the ordinary sense.*

4. It is also possible in the general case when the integrating function $F(x)$ is an arbitrary non-decreasing function to give some definitive though less effective property of the corresponding space L_ϱ. For this we consider the map of the x-axis into the y-axis determined by the function $y = F(x)$. At points of continuity this map is well-defined; we shall map a point of discontinuity x_0 onto the entire interval $[F(x_0 - 0), F(x_0)]$. The inverse function $x = G(y)$ has the same properties; each point y can be mapped by it onto a single point x or onto an entire interval, the latter only in the case that there exists an interval of the x-axis on which $F(x)$ maintains a constant value y_0 (the set of such values y_0 is at most countable). Each set E on the x-axis is mapped onto some set $\mathscr{E}$ on the y-axis; moreover an interval Δ of the x-axis maps onto an interval of the y-axis with length exactly equal to the ϱ-measure of Δ. To each function $f(x)$ corresponds a function $g(y) = f(G(y))$, defined for those y for which $G(y)$ is single-valued. Thus the function $g(y)$ can be indeterminate at most on a fixed countable set. A step function $h(x)$, equal to h_j on the interval Δ_j, determines a corresponding step function $k(y)$ equal to h_j on the interval $F(\Delta_j)$; moreover

$$I_\varrho h = \sum h_j\, \varrho\, \Delta_j = \sum h_j |F(\Delta_j)| = I\, k,$$

where $(F(\Delta_j))$ denotes the length of the interval $F(\Delta_j)$. We see that under the given map a step function h on the x-axis corresponds to a step function k on the y-axis with the ordinary integral $I\,k$ equal to the ϱ-integral of h.

It is now clear that the limiting process which produces a ϱ-summable function $\varphi(x)$ on the x-axis has an image on the y-axis which leads to the function $\psi(y) = \varphi(G(y))$, which is summable in the ordinary sense. Thus if $\varphi(x)$ is ϱ-summable, the corresponding function $\psi(y) = \varphi(G(y))$ is summable in the sense of Lebesgue and $I_\varrho \varphi = I\,\psi$. In addition $\psi(y)$ is constant on every interval that corresponds to a point x of positive ϱ-measure. Conversely if the function $\psi(y)$ is summable in the sense of Lebesgue and constant on all the intervals that correspond to points x of positive ϱ-measure, then it can be approximated with respect to the metric of L_ϱ by step functions which are also constant on these intervals;

it follows that the function $\varphi(x) = \psi(F(x))$ is ϱ-summable and $I_\varrho \varphi = I \psi$. *Thus a function* $\varphi(x)$ *is* ϱ*-summable if and only if the function* $\psi(y) = \varphi(G(y))$ *is summable in the ordinary sense.* In particular, a set E on the x-axis is ϱ-measurable if and only if the corresponding set $F(E)$ is measurable in the sense of Lebesgue.

A particular consequence of these results is that we can obtain a generalisation of formula (1] of art. 3, which relates to the case of an absolutely continuous integrating function, for the general case.

We shall call a function $\Phi(x)$ *absolutely continuous with respect to the non-decreasing function* $F(x)$ if for any $\varepsilon > 0$ there exists $\delta > 0$ such that for any system of non-intersecting intervals (a_k, b_k) $(k = 1, 2, \ldots m)$ the condition

$$\sum [F(b_k) - F(a_k)] < \delta$$

entails the inequality

$$\sum |\Phi(b_k) - \Phi(a_k)| < \varepsilon.$$

We can say that Φ is absolutely continuous relative to F say if Φ satisfies a "Lipschitz condition" relative to F, i.e. for any α and β,

$$|\Phi(\beta) - \Phi(\alpha)| \leqq [F(\beta) - F(\alpha)]\, C$$

with a fixed constant C.

By what we have proved above the function $\Phi(G(y)) = \Psi(y)$ is absolutely continuous with respect to the ordinary Lebesgue measure on the y-axis and can therefore be expressed as the integral of its derivative $\psi(y)$:

$$\Psi(y) = \int_{a_1}^{y} \psi(\eta)\, d\eta \quad (a_1 = F(a)).$$

This in turn signifies that the function $\Phi(x)$ is the integral of the function $g(x) = \psi(G(y))$ with respect to the measure dF:

$$\Phi(x) = \int_{a}^{x} g(\xi)\, dF(\xi).$$

We note that $g(x)$ is bounded in absolute value by the constant C when the Lipschitz condition (2) is satisfied.

Let us suppose further that Φ is also non-decreasing. Then it is easily verified in the same way that every function $f(x)$ summable in the measure $d\Phi$ is also summable in the measure dF, and

$$\int_a^b f(x)\,d\Phi(x) = \int_a^b f(x)\,g(x)\,dF(x).$$

The considerations which relate the Stieltjes and Lebesgue integrals will be developed at the end of the next paragraph in application to the case of functions of several variables.

6. The Stieltjes Integral (continued)

1. In the last paragraph the measure which defined the Stieltjes integral was supposed non-negative. But in fact our procedure can be extended without difficulty to certain measures to which non-positive values can be assigned.

We shall find it more convenient now to speak, not of actual measures, but of their generating functions. Let $\Phi(x)$ be some function of bounded variation on the closed interval $[a, b]$ $(-\infty \leqq a < b \leqq \infty)$, continuous to the right. Like every such function it can be expressed as the difference of two non-decreasing functions, also continuous to the right, one of which is the total variation of $\Phi(x)$ (Section 2):

$$\Phi(x) = V(x) - G(x).$$

The function Φ satisfies a Lipschitz condition with constant 1 relative to V:

$$|\Phi(\beta) - \Phi(\alpha)| \leqq V_\beta^\alpha[\Phi] = V(\beta) - V(\alpha).$$

Hence, by (6),

$$\Phi(x) = \int_a^x h(\xi)\,dV(\xi).$$

Going by analogy with (7), we put by definition, for any function $\varphi(x)$, summable with respect to $V(x)$:

$$\int_a^b \varphi(x)\,d\Phi = \int_a^b \varphi(x)\,h(x)\,dV(x). \tag{1}$$

This defines the integral with respect to $\Phi(x)$. We recall our assumption that $\Phi(x)$ is of bounded variation, and not necessarily a non-decreasing function.

Integral (1) can also be expressed directly in terms of integrals with respect to non-decreasing functions; we have

$$G(x) = V(x) - \Phi(x) = \int_a^b [1 - h(\xi)]\, \mathrm{d}\, V(\xi),$$

or

$$\int_a^b \varphi\, \mathrm{d} G = \int_a^b (1 - h)\, \varphi\, \mathrm{d}\, V = \int_a^b \varphi\, \mathrm{d}\, V - \int_a^b \varphi\, h\, \mathrm{d}\, V$$

so that

$$\int_a^b \varphi\, h\, \mathrm{d}\, V = \int_a^b \varphi\, \mathrm{d}\, V - \int_a^b \varphi\, \mathrm{d} G. \tag{2}$$

Finally, the integral defined by equation (1) can be obtained by a direct construction, analogous to the construction of an integral with respect to a non-negative measure. Let $\varrho\, \Delta$ be the measure corresponding to the generating function $V(x)$ and $\sigma\, \Delta$ the measure corresponding to $G(x)$. In accordance with formulae (2)–(6) Section 5, art. 2, we construct a measure $\tau\, \Delta$ derived from $\Phi(x)$:

$$\tau(x, x'] = \Phi(x') - \Phi(x), \qquad \tau[x, x') = \Phi(x' - 0) - \Phi(x - 0),$$
$$\tau(x, x') = \Phi(x' - 0) - \Phi(x), \quad \tau[x, x'] = \Phi(x') - \Phi(x - 0).$$

The interval function $\tau\, \Delta$ is additive and continuous; this can be seen immediately from the definition or from the equation $\tau\, \Delta = \varrho\, \Delta - \sigma\, \Delta$. It can also assume negative values; but for any partition of the closed interval $[a, b]$

$$[a, b] = \Delta_1 + \Delta_2 + \cdots + \Delta_m$$

the condition

$$|\tau\, \Delta_1| + |\tau\, \Delta_2| + \cdots + |\tau\, \Delta_m| \leqq \sum_{j=1}^{m} \varrho\, \Delta_j + \sum_{j=1}^{m} \sigma\, \Delta_j < C \tag{3}$$

is satisfied, where C is a fixed constant. The inequality (3) expresses the fact that the original function $\Phi(x)$ is of bounded variation. Further, notwithstanding that $\tau\, \Delta$ can take negative values, we can define the integrals of step functions with respect to the measure $\tau\, \Delta$ and then obtain the space L_τ by the usual limiting process. In all evaluations of bounds the constant C in (3) will have to take the place of $\varrho[a, b]$. Using the equation $\varrho\Delta - \sigma\Delta = \tau\Delta$ at each step, we arrive at formula (2), and together with this formula (1).

2. The Riemann-Stieltjes integral. Given an arbitrary function $f(x)$ and some generating function $\Phi(x)$ (of bounded variation), let us consider the sums analogous to the Riemann sums

$$\sum_{j=0}^{m-1} f(\xi_j)\,[\Phi(x_{j+1}) - \Phi(x_j)] \tag{1}$$

$$(a = x_0 < x_1 < \cdots < x_m = b,\ x_j \leqq \xi_j \leqq x_{j+1}).$$

The limit of such sums when the partitioning intervals $\Delta x_j = x_{j+1} - x_j$ are indefinitely refined is said, if it exists, to be the *Riemann–Stieltjes integral of the function $f(x)$ with respect to the function $\Phi(x)$*. We shall show that it exists and coincides with the already defined Lebesgue–Stieltjes integral of f with respect to Φ over the interval $(a, b]$ if $f(x)$ is continuous. The integral sum expressed is the Lebesgue–Stieltjes integral of the step function $h_m(x)$, defined on the interval $(a, b]$ and equal to $f(\xi_j)$ on the interval $\Delta_j = (x_j, x_{j+1}]$. When the Δ_j are refined indefinitely the step function $h_m(x)$ tends uniformly to $f(x)$; hence in virtue of the fundamental theorems of integral theory

$$\sum f(\xi_j)\,[\Phi(x_{j+1}) - \Phi(x_j)] = \int\limits_{(a,b]} h_m(x)\,\mathrm{d}\Phi \to \int\limits_{(a,b]} f(x)\,\mathrm{d}\Phi,$$

as required.

The actual definition of the Riemann–Stieltjes integral gives rise to the inequality

$$\left|\int_a^b f(x)\,\mathrm{d}\Phi(x)\right| \leqq \sup_{x\in(a,b]} |f(x)|\, \mathrm{V}_a^b[\Phi],$$

which replaces the bound

$$\left|\int_a^b f(x)\,\mathrm{d}x\right| \leqq \sup_{b\leqq x\leqq b} |f(x)|\,(b-a)$$

for the ordinary Riemann integral.

We observe that in defining the Riemann–Stieltjes integral there is no need to ensure that the function $\Phi(x)$ is continuous to the right. In fact if $f(x)$ is continuous the integral sums (1) have a limit for any function $\Phi(x)$ of bounded variation and this limit is independent of the values of Φ at its discontinuities. For the proof, given some function $\Phi(x)$ of bounded variation, we put

$$\Phi(x) = \Phi_0(x) + D(x),$$

where $\Phi_0(x)$ coincides with $\Phi(x)$ at all the points of continuity of $\Phi(x)$ and is equal to $\Phi(x+0)$ at its discontinuities, being therefore continuous to the right; the function $D(x)$ is evidently distinct from zero on an at most countable set $z_1, z_2, \ldots$ of discontinuities of $\Phi(x)$.

By what we have proved, the integral sums corresponding to the function $\Phi_0(x)$ have the limit

$$\int_a^b f(x)\, \mathrm{d}\Phi_0(x).$$

We shall show that the integral sums corresponding to $D(x)$ tend to zero. Since, together with $\Phi(x)$ and $\Phi_0(x)$, $D(x)$ is of bounded variation, the sum of the moduli of all its values is finite. For a given $\varepsilon > 0$ we find a number N such that $\sum\limits_{n>N} |D(z_n)| < \varepsilon$. Further, we put

$$D(x) = D_1(x) + \cdots + D_N(x) + \bar{D}_N(x),$$

where $D_j(x)$ is non-zero only at the point z_j $(j = 1, 2, \ldots, N)$ and $\bar{D}_N(x)$ is non-zero only at the points $z_n (n > N)$. The integral sum corresponding to the function $D_j(x)$ is equal either to 0 or to

$$[f(\xi_j') - f(\xi_j'')]\, D_j(z_j),$$

where $\xi_j' \leqq z_j \leqq \xi_j''$ and ξ_j', ξ_j'', lie in adjacent elements of the partition of $[a, b]$. Since $f(x)$ is continuous, this quantity can be made arbitrarily small by a suitable refinement of the partition. The integral sum corresponding to the function $\bar{D}(x)$ admits the bound

$$\left| \sum_{j=1}^{m} f(\xi_j) [\bar{D}_N(x_{j+1}) - \bar{D}_N(x_j)] \right| \leqq 2M \sum_{n \geqq N} |D(z_n)| \leqq 2M\varepsilon,$$

where $M = \max |f(x)|$; as we see, this sum can also be made arbitrarily small. Thus the integral sums corresponding to the function $D(x)$ do in fact tend to zero as the partition is increasingly refined, and our assertion is proved.

Problems. 1. Find the values of the Stieltjes integrals:

$$I_1 = \int_{-1}^{3} x \, \mathrm{d}F(x), \quad F(x) = \begin{cases} 0 & \text{for} \quad x = -1, \\ 1 & \text{for} \quad -1 < x < 2, \\ -1 & \text{for} \quad 2 \leqq x \leqq 3; \end{cases}$$

$$I_2 = \int_{0}^{2} x^2 \, \mathrm{d}F(x), \quad F(x) = \begin{cases} -1 & \text{for} \quad 0 \leqq x < \frac{1}{2}, \\ 0 & \text{for} \quad \frac{1}{2} \leqq x < \frac{3}{2}, \\ 2 & \text{for} \quad x = \frac{3}{2}, \\ -2 & \text{for} \quad \frac{3}{2} < x \leqq 2. \end{cases}$$

Answer. $I_1 = -5$; $I_2 = -17/4$.

2. Write down an expression for the moment about the origin of a mass distributed over the closed interval $[a, b]$ in such a way that the mass of the closed interval $[a, x]$ is equal to $F(x)$.

Answer. $M = \int_a^b x \, \mathrm{d}F(x)$.

3. *The Limiting Passage under the Riemann–Stieltjes Integral Sign*

THEOREM 1. (E. Helly). *If the functions $F_n(x)$ of bounded variation converge at each point of the closed interval $a \leqq x \leqq b$ to some function $F(x)$ and the total variations of all the $F_n(x)$ have a common bound:*

$$\mathsf{V}_a^b [F_n] \leqq C,$$

then the limiting function $F(x)$ is also bounded variation and for any continuous function $\varphi(x)$

$$\lim_{n \to \infty} \int_a^b \varphi(x) \, \mathrm{d}F_n(x) = \int_a^b \varphi(x) \, \mathrm{d}F(x). \tag{1}$$

Proof. We begin by showing that $F(x)$ has a bounded variation not exceeding C. For given any partition $a = x_0 < x_1 < \cdots < x_m = b$ of $[a, b]$ we have:

$$\sum_{j=0}^{m-1} |F(x_{j+1}) - F(x_j)| = \lim_{n \to \infty} \sum_{j=0}^{m-1} |F_n(x_{j+1}) - F_n(x_j)| \leqq C,$$

and it follows that $\mathsf{V}_a^b[F] \leqq C$.

We now proceed to the proof of relation (1). First let $\varphi(x)$ be a step function, equal to h_j on the interval

$$\Delta_j = (x_j, x_{j+1}).$$

Then

$$\int_a^b \varphi(x)\,\mathrm{d}F_n(x) = \sum h_j[F_n(x_{j+1}) - F_n(x_j)],$$

$$\int_a^b \varphi(x)\,\mathrm{d}F(x) = \sum h_j[F(x_{j+1}) - F(x_j)].$$

It is obvious that for sufficiently large $n \geqq N$ these expressions differ by less than a given $\varepsilon > 0$. In the general case for a given $\varepsilon > 0$ we construct a step function $\varphi_\varepsilon(x)$ such that $|\varphi(x) - \varphi_\varepsilon(x)| < \varepsilon/C$. Then

$$\left|\int_a^b \varphi(x)\,\mathrm{d}F_n(x) - \int_a^b \varphi_\varepsilon(x)\,\mathrm{d}F_n(x)\right|$$

$$= \left|\int_a^b [\varphi(x) - \varphi_\varepsilon(x)]\,\mathrm{d}F_n(x)\right| \leqq \frac{\varepsilon}{C}\,\mathsf{V}_a^b[F_n] \leqq \varepsilon,$$

$$\left|\int_a^b \varphi(x)\,\mathrm{d}F(x) - \int_a^b \varphi_\varepsilon(x)\,\mathrm{d}F(x)\right|$$

$$= \left|\int_a^b [\varphi(x) - \varphi_\varepsilon(x)]\,\mathrm{d}F(x)\right| \leqq \frac{\varepsilon}{C}\,\mathsf{V}_a^b[F] \leqq \varepsilon,$$

and hence for $n > N$

$$\left|\int_a^b \varphi(x)\,\mathrm{d}F_n(x) - \int_a^b \varphi(x)\,\mathrm{d}F(x)\right| \leqq \left|\int_a^b [\varphi(x) - \varphi_\varepsilon(x)]\,\mathrm{d}F_n(x)\right| +$$

$$+ \left|\int_a^b [\varphi(x) - \varphi_\varepsilon(x)]\,\mathrm{d}F(x)\right| + \left|\int_a^b \varphi_\varepsilon(x)\,\mathrm{d}F_n(x) - \int_a^b \varphi_\varepsilon(x)\,\mathrm{d}F(x)\right|$$

$$\leqq 3\varepsilon,$$

which proves the theorem.

Note 1. This theorem can be generalised somewhat by allowing the integrand $\varphi(x)$ to depend on n. We claim that the relation

$$\int_a^b \varphi(x)\,\mathrm{d}F(x) = \lim_{n\to\infty} \int_a^b \varphi_n(x)\,\mathrm{d}F_n(x),$$

holds if the following conditions are satisfied:

(a) the functions $F_n(x)$ are of uniformly bounded variation and converge to $F(x)$ at every point of $[a, b]$;

(b) the continuous functions $\varphi_n(x)$ converge uniformly to their limit $\varphi(x)$.

The proof follows quickly from the bounds

$$\left|\int_a^b [\varphi(x) - \varphi_n(x)]\,\mathrm{d}F_n(x)\right| \leq \max|\varphi(x) - \varphi_n(x)|\, \mathrm{V}_a^b[F_n],$$

$$\left|\int_a^b [\varphi(x) - \varphi_n(x)]\,\mathrm{d}F(x)\right| \leq \max|\varphi(x) - \varphi_n(x)|\, \mathrm{V}_a^b[F]$$

in conjunction with the theorem just proved.

Note 2. Theorem 2 (along with note 1) can be generalised in an obvious way to an infinite interval of integration, say $[0, \infty]$, if the functions $F_n(x)$ are of uniformly bounded variation throughout the interval and $\varphi(x)$ (or $\varphi_n(x)$), if we are speaking of note (1) is continuous, including at infinity; this latter property enables $\varphi(x)$ to be uniformly approximated by step functions, which forms an essential part of the proof.

Note 3. If $\varphi(x)$ is merely bounded instead of continuous at infinity, so that $|\varphi(x)| \leq M$, Helly's theorem remains valid if the $F_n(x)$ satisfy the following supplementary condition (the role of which amounts to the fact that the mass carried by the distribution $F_n(x)$ does not become infinite as n increases):

(*) *Given any* $\varepsilon > 0$, *an* $N = N(\varepsilon)$ *can be found such that, for all* n,

$$\operatorname*{Var}_{|x| \geqq N} F_n(x) \leq \varepsilon. \tag{2}$$

For, on passing to the limit in (2) as $n \to \infty$, we find that

$$\operatorname*{Var}_{|x| \geqq N} F(x) \leq \varepsilon.$$

or the limit function $F(x)$ also.

Further, having found the N from condition (*) for a given $\varepsilon \geqq 0$, we have

$$\int_{-\infty}^{\infty} \varphi(x)\,\mathrm{d}F_n(x) - \int_{-\infty}^{\infty} \varphi(x)\,\mathrm{d}F(x) = \int_{-N}^{N} \varphi(x)\,\mathrm{d}[F_n(x) - F(x)] +$$

$$+ \int_{|x| \geqq N} \varphi(x)\,\mathrm{d}F_n(x) - \int_{|x| \geqq N} \varphi(x)\,\mathrm{d}F(x). \tag{3}$$

Knowing N, and using Helly's theorem for the finite interval $[-N, N]$, we can find an n_0 such that, for $n > n_0$,

$$\left| \int_{-N}^{N} \varphi(x) \, d[F_n(x) - F(x)] \right| < \varepsilon .$$

The remaining two integrals do not exceed $2M\varepsilon$ in absolute value, by construction; we see that the entire left-hand side of (3) does not exceed $(2M + 1)\varepsilon$, which justifies the passage to the limit under the Stieltjes integral sign in the case in question.

4. The application of theorem 1 is much facilitated by the following theorem, which allows the extraction of a convergent sequence from a given set of functions of (uniformly) bounded variation.

THEOREM 2. (E. Helly). *From any infinite set K of functions $f(x)$ which are defined on the closed interval $a \leqq x \leqq b$ and have the properties*

$$\max |f(x)| \leqq C, \tag{1}$$

$$\mathrm{V}_a^b [f] \leqq \mathrm{V} \tag{2}$$

(*where C, V are constants independent of the choice of $f \in K$*), *it is possible to choose a sequence $f_n(x)$ which converges at every point of the interval $a \leqq x \leqq b$.*

Proof. Let us suppose first that the functions $f(x)$ are non-decreasing. Let $r_1, \ldots, r_n, \ldots$ be a sequence containing all the rational points of $[a, b]$. Since the numbers $f(r_1)$ are bounded, there exists a sequence of functions $f_{n1} \in K$ for which the numbers $f_{n1}(r_1)$ tend to a limit. From this sequence $f_{n1}(x)$ we can extract a subsequence $f_{n2}(x)$ for which the numbers $f_{n2}(r_2)$ converge (and also of course the numbers $f_{n2}(r_1)$); continuing in this way we get for each k a sequence $f_{nk}(x)$ which converges at the points $r_1, r_2, \ldots, r_k$ as $n \to \infty$. The diagonal sequence $f_{nn}(x)$, which we shall denote simply by $f_n(x)$, converges at each $x = r_1, r_2, \ldots$ The limit $f(x)$ of the sequence $f_n(x)$, defined as yet only at the rational points, represents a non-decreasing function. We complete its definition by putting for each irrational x

$$f(x) = \lim_{r \to x - 0} f(r) \quad (r \text{ rational}).$$

As a result we get a non-decreasing function defined at all points of the interval $[a, b]$. We shall show, that it remains the limit of the sequence $f_n(x)$ wherever it is continuous. Let x_0 be a point at which $f(x)$ is continuous; given $\varepsilon > 0$, we find $\delta > 0$ such that $|f(x) - f(x_0)| < \varepsilon$ whenever $|x - x_0| < \delta$, and choose rational points $r' < x_0 < r''$ such that $r'' < x_0 + \delta$, $r' > x_0 - \delta$. We also find a number N such that for $n \geqq N$ we have $|f_n(r') - f(r')| < \varepsilon$, $|f_n(r'') - f(r'')| < \varepsilon$. It follows that $|f_n(r') - f_n(r'')| < 4\varepsilon$. Since the function $f_n(x)$ is non-decreasing, the number $f_n(x_0)$ lies between $f_n(r')$ and $f_n(r'')$; hence

$$\begin{aligned} |f(x_0) - f_n(x_0)| & \\ &\leqq |f(x_0) - f(r')| + |f(r') - f_n(r')| + |f_n(r') - f_n(x_0)| \\ &\leqq \varepsilon + \varepsilon + |f_n(r') - f_n(r'')| \leqq 6\varepsilon, \end{aligned}$$

and it follows that $f(x_0) = \lim f_n(x_0)$.

The sequence $f_n(x)$ constructed converges to $f(x)$ everywhere with the possible exception of discontinuities of the function $f(x)$. The set of these discontinuities is at most countable. Hence, again applying the diagonal process, we can extract from the sequence $f_n(x)$ a subsequence which converges even at the discontinuities of $f(x)$. We have thus extracted from the given family of non-decreasing functions a sequence which converges at every point of $[a, b]$ as required.

In the general case when the functions $f(x) \in K$ are not necessarily non-decreasing we express each of them in the form

$$f(x) = V(x) - G(x),$$

where $V(x)$ is the total variation of $f(x)$ over the closed interval $[a, x]$. The functions $V(x)$ are non-decreasing and, by condition (2), bounded; the functions $G(x)$ have the same properties. By what we have proved there exists a sequence $f_n(x) \in K$ for which the functions $V_n(x)$ converge at every point of the interval $[a, b]$. From this sequence we can extract a subsequence $f_n\ (x)$ for which the functions $G_{nk}(x)$ converge at every point of $[a, b]$; but then the $f_{nk}(x)$ also converge at every point of $[a, b]$. The proof of theorem 2 is therefore complete.

5. *The Stieltjes Integral for Several Variables*

For definiteness we shall discuss the case of two variables x, y, varying over the square $\Delta_0 = \{a \leqq x \leqq b, a \leqq y \leqq b\}$. We shall call an "interval" in the plane any set of points (x, y) where each of the coordinates runs through some interval (any of the five types given in art. 1) on its own axis. Let us suppose that for each interval Δ in the square Δ_0 a function $\varrho\,\Delta \geqq 0$ is given, satisfying the condition of total additivity: if $\Delta_1, \Delta_2, \ldots, \Delta_n, \ldots$ are mutually disjoint, and their union Δ is also an interval, then

$$\varrho\,\Delta = \varrho\,\Delta_1 + \cdots + \varrho\,\Delta_n + \cdots$$

As above, we shall the function $\varrho\,\Delta$ *the Stieltjes measure*. We partition the square Δ_0 into a finite number of disjoint intervals $\Delta_1, \ldots, \Delta_n$; a function $h(x, y)$ equal to a constant h_j on the set Δ_j is said to be a step function and its integral is defined by the formula

$$I\,\varrho\,h = \sum_j h_j\,\varrho\,\Delta_j.$$

Applying the process already described many times above, we can extend the integral $I\,\varrho$ over a wide class L_ϱ of functions which are then said to be *summable in the Lebesgue–Stieltjes sense with respect to the measure* ϱ. The simplest way of ascertaining the compass of the class and validating the construction is to apply a method, similar to that used in Section 4 for Lebesgue measure, of setting up a correspondence with linear measure. The method is as follows. From the measure $\varrho\,\Delta$ we construct two functions, each of one variable:

$$F_1(x) = \varrho\,\Delta_x, \quad F_2(y) = \varrho\,\Delta_y.$$

Here Δ_x, Δ_y denote the regions of the plane the points (ξ, η) of which satisfy the inequalities $\xi \leqq x$, $\eta \leqq y$ respectively. The functions $F_1(x)$, $F_2(y)$ are non-decreasing and therefore possess at most countable sets of discontinuities. Let $x_1, x_2, \ldots, x_n, \ldots$ be all the discontinuities of $F_1(x)$. We can construct on the x-axis a set of numbers of the form $x_0 + \frac{p}{2^q}$ (x_0 fixed; p, q integers) not containing any of the points $x_1, \ldots, x_n, \ldots$; this follows from the fact that the set $\left\{x_n + \frac{p}{2^q}\right\}$ for all $n, p, q = 1, 2, \ldots$ is countable and consequently does not contain all the points of the axis; as x_0 we take any point not contained in it. Similarly we construct on the y-axis a set of numbers $y_0 + \frac{p}{2^q}$ not containing any of the discontinuities of $F_2(y)$. The straight lines $x = x_0 + \frac{p}{2^q}$, $y = y_0 + \frac{p}{2^q}$ ($p = 0, \pm 1, \pm 2, \ldots$; q fixed) form a net—a partition of the plane into squares of side $\frac{1}{2^q}$, the boundaries of which do not have positive measure. We can now set up a correspondence between the squares of the plane net and the intervals on the axis, with one modification just as we did in Section 4: there we mapped a square onto an interval of length equal to the area of the square, but now we shall map it onto an interval of length equal to the ϱ-measure of the square. Of course we have to take care that inclusion relations are preserved.

Just as in Section 4 we satisfy ourselves that the map constructed is one–one up to a set of ϱ-measure zero in the plane and a set of Lebesgue measure zero on the line. To every ϱ-measurable set in the plane will correspond a Lebesgue-measurable set on the line with the same Lebesgue measure. We observe certain peculiarities of this correspondence. A point (x_0, y_0) in the plane, of positive ϱ-measure, goes over into an entire interval on the line of corresponding length. The aggregate of such intervals will be at most countable and we denote them by $\delta_1, \ldots, \delta_n, \ldots$ On the other hand a square of the plane net of measure zero maps onto a solitary point of the x-axis.

To a step function on the plane which takes constant values on the squares of the net, corresponds a step function on the line, constant on each of the intervals $\delta_1, \delta_2, \ldots$ The integrals of corresponding step functions—one with respect to plane ϱ-measure, the other with respect to linear Lebesgue measure—are equal. By means of a limiting passage through the specified linear step functions we can obtain all the summable functions that are constant on the intervals $\delta_1, \delta_2, \ldots, \delta_n, \ldots$; the corresponding limiting passage in the plane will yield the totality of functions summable with respect to the measure ϱ. To each property of the Lebesgue integral on the line which relates to functions constant on the intervals $\{\delta_n\}$ corresponds some property of the Lebesgue–Stieltjes integral with measure ϱ. Thus the given correspondence absolves us from the need to prove specially for the Lebesgue-Stieltjes integral all the theorems that we proved (Chapter IV and VI) for the Lebesgue integral; all these theorems carry over automatically to the Lebesgue–Stieltjes integral.

6. *The Generating Function in the Case of Several Variables*

Let (ξ, η) be an arbitrary point of the square Δ_0 and let $\Delta_{\xi\eta}$ be the interval determined by the inequalities $n \leq \xi$, $y \leq \eta$. We put

$$F(\xi, \eta) = \varrho\, \Delta_{\xi\eta}.$$

In accordance with the formulae analogous to (2)–(6) of Section 5, art. 2, the function $F(\xi, \eta)$ allows the recovery of the measure $\varrho \Delta$ for every interval Δ and is therefore said be the generating function of the measure $\varrho \Delta$. It is "non-decreasing" in the sense that for $\xi \leqq \xi', \eta \leqq \eta'$ we have

$$F(\xi, \eta) \leqq F(\xi', \eta'),$$

and is also "continuous above", i.e.

$$F(\xi + 0, \eta + 0) = \lim_{\substack{x \to \xi + 0 \\ y \to \eta + 0}} F(x, y) = F(\xi, \eta).$$

Conversely, every function that is non-decreasing and continuous above in the sense indicated can serve as the generating function for some totally additive measure.

The Stieltjes integral of a function $\varphi(x, y)$, constructed from the generating function $F(\xi, \eta)$ is written in the form

$$\int_a^b \int_a^b \varphi(x, y) \, \mathrm{d} F(x, y).$$

In place of a non-negative measure $\varrho \Delta$ we can consider a measure $\varrho \Delta$ *of bounded variation* and of arbitrary sign; this means that for any partition of the base square Δ_0 into intervals $\Delta_1, \ldots, \Delta_m$ we have

$$|\varrho \Delta_1| + \cdots + |\varrho \Delta_m| < C \quad (1)$$

where C is some fixed constant. Correspondingly we can consider a generating function $F(x, y)$ of *bounded variation* in place of a non-decreasing generating function; this implies that the measure $\varrho \Delta$ constructed from $F(x, y)$ in accordance with the general rules must satisfy inequality (1). Of course the assumption of continuity above, which ensures the total additivity of the measure $\varrho \Delta$, must be retained in the more general case.

7. Applications of the Stieltjes Integral in Analysis

The Stieltjes integral has numerous applications. In this paragraph we shall give the derivation of three formulae from three different branches of mathematical analysis. One further formula—the representation of a positive-definite function—will be given in Chapter VII, Section 7.

1. *Linear Functionals on the Space $C(a, b)$*

The simplest continuous linear functional on the space $C(a, b)$ of all continuous functions $\varphi(x)$ on the closed interval $[a, b]$ is the value of the function $\varphi(x)$ at a fixed point $x = \xi$. It turns out that the general form of a linear functional on this space is obtained as a "Stieltjes combination" of such simple functionals; we have the following theorem:

THEOREM 1. (F. Riesz). *Every continuous linear functional $f[\varphi]$ on the space $C(a, b)$ can be written in the form*

$$f(\varphi) = \int_a^b \varphi(\xi)\, \mathrm{d}F(\xi), \tag{2}$$

where $F(\xi)$ is some function of bounded variation, continuous to the right.

The proof will be carried out in several stages.

I. Let us consider a linear space K of bounded functions $\varphi(x)$ defined on some set X. We suppose that a linear functional $f[\varphi]$ on K is given, satisfying the condition

$$|f[\varphi]| \leqq C \sup |\varphi(x)| \tag{2}$$

where C is a fixed constant.

We claim that *for every bounded increasing sequence $\varphi_n(x) \in K$ the corresponding sequence $f[\varphi_n]$ converges.* For, given $\varphi \in K$, we can write

$$|f[\varphi]| = \pm f[\varphi] = f[\pm\varphi]$$

with the appropriate choice of sign. For a given bounded (by M, say) increasing sequence $\varphi_n(x) \in K$ we can form the series

$$|f[\varphi_2 - \varphi_1]| + |f[\varphi_3 - \varphi_2]| + \cdots + |f[\varphi_{n+1} - \varphi_n]| + \cdots.$$

With the reservation mentioned the nth partial sum of this series can be written in the form

$$f[\pm(\varphi_2 - \varphi_1)] + f[\pm(\varphi_3 - \varphi_2)] + \cdots + f[\pm(\varphi_{n+1} - \varphi_n)]$$
$$= f[\pm(\varphi_2 - \varphi_1) \pm (\varphi_3 - \varphi_2) \pm \cdots \pm (\varphi_{n+1} - \varphi_n)].$$

But

$$|\pm(\varphi_2 - \varphi_1) \pm \cdots \pm (\varphi_{n+1} - \varphi_n)|$$
$$\leqq (\varphi_2 - \varphi_1) + \cdots + (\varphi_{n+1} - \varphi_n) = \varphi_{n+1} - \varphi_1 \leqq 2M,$$

so that in virtue of (2)

$$|f[\varphi_2 - \varphi_1]| + \cdots + |f[\varphi_{n+1} - \varphi_n]| \leqq 2MC$$

and consequently the series

$$f[\varphi_2 - \varphi_1] + f[\varphi_3 - \varphi_2] + \cdots + f[\varphi_{n+1} - \varphi_n] + \cdots$$

converges; but this implies the convergence of the sequence $f[\varphi_n]$.

The following query naturally arises: is the limiting value of $f[\varphi_n]$ for two distinct increasing sequences φ_n that converge to the same function φ the same?

We confine ourselves here to the following simple assertion: *starting from given sequences* $\varphi_n \nearrow \varphi, \psi_n \nearrow \psi$ *we construct the strictly increasing sequences* $\bar{\varphi}_n = \varphi_n - 1/n$, $\bar{\psi}_n = \psi_n - 1/n$; *if for any n there exist k, m such that* $\bar{\psi}_m > \bar{\varphi}_n$, $\bar{\varphi}_k > \bar{\psi}_n$, *then* $\lim f[\varphi_n] = \lim f[\psi_n]$. For in this case we can construct the new sequence

$$\varphi_1 < \bar{\psi}_{n1} < \bar{\varphi}_{n2} < \bar{\psi}_{n3} < \cdots,$$

which also converges to φ; by what we have proved the values of the functional f on this sequence have a limit; but this means that the numbers $f[\varphi_n]$, $f[\psi_n]$ tend to one and the same limit.

The same thing applies of course to decreasing sequences.

Such a situation always occurs, for example, in the case where K is the space C of all continuous functions on a closed interval or on a closed bounded set in p-dimensional space. In fact let us fix n and let m increase without bound. We shall suppose that for each m the set E_m on which $\psi_m - 1/m \leqq \varphi_n - 1/n$ is non-empty. The closed sets E_m form a decreasing sequence ($E_1 \supset E_2 \supset \cdots$) and if they are all non-empty, they will contain a common point x_0. Proceeding to the limit as $m \to \infty$ in the inequality

$$\psi_m(x_0) - \frac{1}{m} \leqq \varphi_n(x_0) - \frac{1}{n}$$

we get

$$\varphi(x_0) \leqq \varphi_n(x_0) - \frac{1}{n}$$

or

$$\varphi_n(x_0) \geqq \varphi(x_0) + \frac{1}{n},$$

which is impossible.

Thus in the case of the space C the functional f can be uniquely defined on the class C^+ of bounded functions which are limits of increasing sequences of continuous functions. The equation $f[\varphi + \psi] = f[\varphi] + f[\psi]$ obviously remains valid on the class C^+. Let us form the class R of differences $g = \varphi - \psi$, $\varphi \in C^+$, $\psi \in C^+$, and put $f[g] = f[\varphi] - f[\psi]$. This definition leads to a unique result (cf. Chapter IV, Section 2). The functional f remains additive and homogeneous over the whole class R. If the class K contains $\max\{\varphi, \psi\}$, $\min\{\varphi, \psi\}$ together with the functions φ, φ, inequality (2) continues to hold on the class R; more precisely, $g \in R$ implies

$$|f[g]| \leqq C \sup |g(x)|. \tag{3}$$

For let

$$g = \varphi - \psi, \quad \varphi_n \nearrow \varphi, \quad \psi_n \nearrow \psi \quad (\varphi_n, \psi_n \in K).$$

We put $\sup |\varphi(x) - \psi(x)| = \mu$. If $\sup |\varphi_n(x) - \psi_n(x)| \leqq \mu$ for all n, the required inequality (3) is obtained by proceeding to the limit in the inequality

$$|f[\varphi_n - \psi_n]| \leqq C \sup |\varphi_n(x) - \psi_n(x)|,$$

which holds since $\varphi_n - \psi_n \in K$. In the general case we replace the function φ_n by φ_n by means of the formula

$$\bar{\varphi}_n = \max[\psi_n - \mu, \min(\varphi_n, \psi_n + \mu)],$$

which "truncates" φ_n at the limits $\psi_n \pm \mu$. Together with φ_n, ψ_n, the functions $\bar{\varphi}_n$ are also increasing and tend to φ. But now $|\bar{\varphi}_n - \psi_n| \leqq \mu$ and our assertion is therefore correct.

II. As the space K let us take the space $C(a, b)$ of all continuous functions $\varphi(x)$ on the closed interval $[a, b]$. According to I the given continuous linear functional $f[\varphi]$ has a unique extension to some class of discontinuous functions; without describing this class completely, we observe only that it contains the characteristic functions of all intervals on the line. We define the function

$$F(\xi) = f[\chi_{[a,\xi]}(x)],$$

where $\chi_{[a,\xi]}(x)$ is equal to 1 for $a \leqq x \leqq \xi$ and 0 for $\xi < x \leqq b$. We shall now show that $F(\xi)$ is of bounded variation. Let $a = \xi_0 < \xi_1 < \cdots < \xi_n = b$ be some partition of $[a, b]$; we obtain a bound for the sum

$$\sum_{j=0}^{n-1} |F(\xi_{j+1}) - F(\xi_j)|.$$

Evidently

$$\begin{aligned} |F(\xi_{j+1}) - F(\xi_j)| = \pm[F(\xi_{j+1}) - F(\xi_j)] &= \pm f[\chi_{[\xi_j, \xi_{j+1}]}(x)] \\ &= f[\pm\chi_{(\xi_j, \xi_{j+1})}(x)] \end{aligned}$$

and

$$\sum_{j=0}^{n-1} |F(\xi_{j+1}) - F(\xi_j)|| = f\left[\sum_{j=0}^{n-1} \pm \chi_{(\xi_j,\, \xi_{j+1})}(x)\right].$$

The absolute value of the function in square brackets is at most 1; hence in virtue of inequality (3)

$$\sum_{j=0}^{n-1} |F(\xi_{j+1}) - F(\xi_j)| \leqq C,$$

and it follows that $F(x)$ is of bounded variation.

We shall show further that $F(x)$ is continuous to the right, i.e. for any $\xi < b$ and any sequence $\xi_n \searrow \xi$

$$F(\xi) = \lim F(\xi_k). \tag{4}$$

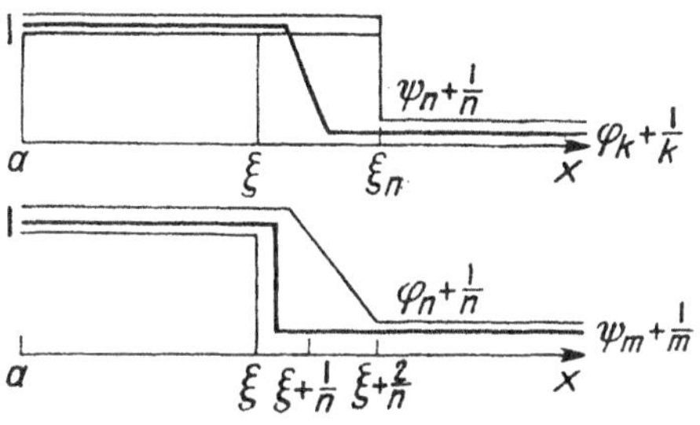

FIG. 13

The sequence $\psi_n(x) = \chi_{[a,\,\xi_n]}(x)$ is decreasing and tends to the function $\psi(x) = \chi_{[a,\xi]}(x)$. The numbers $f[\psi_n]$, $f[\psi]$ are determined uniquely; in particular, $f[\psi]$ is the limit of the numbers $f[\varphi_n]$, where $\varphi_n(x)$ is a continuous function, equal, say, to 1 for $n \leqq \xi + 1/n$ and to 0 for $x \geqq \xi - 2/n$ and linear over the interval $[\xi + 1/n, \xi + 2/n]$.

To establish the required relation $f[\psi_n] \to f[\psi]$, it is sufficient to show that for any n we can find k, m such that

$$\psi_m + \frac{1}{m} < \varphi_n + \frac{1}{n}, \quad \varphi_k + \frac{1}{k} < \psi_n + \frac{1}{n}.$$

But by an elementary geometrical construction (Fig. 13) it is easily seen that for a given n the required k, m always exists. The relation (4) therefore holds.

We observe in particular that if the functional $f[\varphi]$ is non-negative, i.e. takes values $\geqq 0$ for functions $\varphi(x) \geqq 0$, then this property is preserved under the extension of the functional to the specified class of discontinuous functions. Since we have $\chi_{[a,\xi]}(x) \leqq \chi_{[a,\eta]}(x)$ for $\xi < y$, it follows that in this case $F(\xi) \leqq F(y)$, i.e. the function $F(\xi)$ is non-decreasing.

The function $F(x)$, as a function of bounded variation, can act as integrating function for the Stieltjes integral. We then have

$$f[\chi_{[a,\xi]}(x)] = F(\xi) = \int_a^\xi \mathrm{d}F(x) = \int_a^b \chi_{[a,\xi]}(x)\,\mathrm{d}F(x), \qquad (5)$$

which holds not only for the characteristic functions of intervals, but for all step functions; and since every continuous function is the limit of a uniformly convergent sequence of step functions, we get on proceeding to the limit

$$f[\varphi] = \int_a^b \varphi(x)\,\mathrm{d}F(x),$$

valid now for any continuous function $\varphi(x)$. The theorem is proved.

III. Let us see how we can generalise this theorem to the case of functions of several independent variables.

In accordance with art. 1 a functional $f[\varphi]$ on the space C of continuous functions (for simplicity, of two variables x, y) on the square $a \leqq x \leqq b$, $a \leqq y \leqq b$† can be continued over a class of discontinuous functions containing the characteristic functions of all rectangles. We define the function of two variables

$$F(\xi, \eta) = f[\chi_{[a,a,\xi,\eta]}(x, y)],$$

where $\chi_{[a,a,\xi,\eta]}(x, y)$ is the characteristic function of the rectangle $[a \leqq x \leqq \xi, a \leqq y \leqq \eta] = D_{\xi,\eta}$. By the same procedure as in II it can be shown that the function $F(\xi, \eta)$ is of bounded variation and is continuous above (cf. the end of Section 6) and that consequently it is an integrating function for some Stieltjes integral.

† The general case where the functions $\varphi(x, y)$ vary over an arbitrary bounded closed set F also reduces to this case. The set F can be enclosed in a square Q of the given form and every continuous function can be continued, without increasing the upper bound of its modulus, into a continuous function on Q. Conversely, since every continuous function on Q is also continuous on F, we can regard the functional F as defined on the whole set $C(Q)$.

Further, since

$$f[\chi_{[a,a,\xi,\eta]}(x, y)] = F(\xi, \eta) = \iint\limits_{D\xi\eta} \mathrm{d}F(x, y)$$

$$= \int\limits_a^b \int\limits_a^b \chi_{[a,a,\xi,\eta]}(x, y)\,\mathrm{d}F(x, y),$$

we get by taking linear combinations and by completion with respect to the metric of C that, for any continuous function

$$f[\varphi] = \int\limits_a^b \int\limits_a^b \varphi(x, y)\,\mathrm{d}F(x, y).$$

Similar reasoning can be applied to the case of any number of variables.

2. *Absolutely Monotonic Functions.*

An infinitely differentiable function $f(x)$, defined on the closed interval $a \leqq x \leqq b$ $(-\infty \leqq a, b \leqq \infty)$, is said to be *absolutely monotonic* if it is non-negative together with all its derivatives

$$f^{(n)}(x) \geqq 0 \quad (n = 0, 1, 2, \ldots).$$

Examples of absolutely monotonic functions are positive constants and functions of the form $e^{\alpha x}$ $(\alpha > 0)$. It is found that if the interval $[a, b]$ is infinite, every absolutely monotonic function is a "Stieltjes combination" of the simple absolutely monotonic functions $e^{\alpha x}$. For definiteness we confine ourselves to the half-line $-\infty < x < 0$.

THEOREM 2. (S. N. Bernstein). *Every function $f(x)$ that is absolutely monotonic for $x < 0$ can be expressed in the form*

$$f(x) = C + \int\limits_0^\infty e^{\alpha x}\,\mathrm{d}F(\alpha), \tag{1}$$

where $F(x)$ is some bounded non-decreasing function.

Alternatively the constant C can be included in the integral if the function $F(\alpha)$ is given an additional saltus at $\alpha = 0$.

Before we prove the theorem†, we give an account of some of the properties of absolutely monotonic functions. Since $f^n(x) \geqq 0$,

† After B. I. Korenblum, *Advances in Mathematical Sciences*, 1951, Vol. 6, No. 4.

and is non-decreasing, the limit $\theta_n = \lim_{x \to -\infty} f^n(x)$ exists; evidently $\theta_0 \geqq 0$, $\theta_1 = \theta_2 = \cdots = 0$. We claim further that $x \to -\infty$ the functions $f^n(x)$ tend to zero so rapidly that all the integrals

$$I_n = \int_{-\infty}^{0} (-x)^n f^{(n+1)}(x)\,\mathrm{d}x,$$

converge, and that the value of the integral I_n is

$$I_n = M\,n!,$$

where $M = f(0) - f(-\infty)$. For, for any $n \geqq 1$ and $x < 0$, we have

$$f^{(n)}(x) \leqq \frac{2}{|x|}\int_{x}^{x/2} f^{(n)}(\xi)\,\mathrm{d}\xi \leqq \frac{2}{|x|}\left[f^{(n-1)}\left(\frac{x}{2}\right) - f^{(n-1)}(x)\right],$$

so that as $x \to -\infty$ each derivative decreases more rapidly than its predecessor, the difference being at least of the first order (in powers of x); since $f(x/2) - f(x) \to 0$, we have $f^n(x)\,x^n \to 0$ for any η. Hence, integrating by parts,

$$\int_{-\infty}^{0} \frac{(-x)^n}{n!} f^{(n+1)}(x)\,\mathrm{d}x = \frac{(-x)^n}{n!} f^{(n)}(x)\Big|_{-\infty}^{0} + \int_{-\infty}^{0} \frac{(-x)^{n-1}}{(n-1)!} f^{(n)}(x)\,\mathrm{d}x$$

and the extra-integral term vanishes. Successive integrations give us finally

$$\int_{-\infty}^{0} \frac{(-x)^n}{n!} f^{(n+1}(x)\,\mathrm{d}x = \int_{-\infty}^{0} f'(x)\,\mathrm{d}x = f(0) - f(-\infty) = M,$$

as required.

We now turn to the proof of theorem 2. By Dirichlet's formula

$$f(x) - f(-\infty) = \int_{-\infty}^{x} f'(\xi)\,\mathrm{d}\xi = \frac{1}{n!}\int_{-\infty}^{x} (x-\xi)^n f^{(n+1)}(x)\,(\xi)\,\mathrm{d}\xi$$

$$= \frac{1}{n!}\int_{-\infty}^{x} \xi^n \left(1 - \frac{x}{\xi}\right)^n f^{(n+1)}(\xi)\,\mathrm{d}\xi.$$

The substitution $\xi = -\eta\, t$ gives

$$f(x) - f(-\infty) = \frac{(-1)^{n+1}}{n!} \int_{-x/n}^{\infty} n^n t^n \left(1 + \frac{x}{n\,t}\right)^n n^{n+1)} f^{(n+1)}(-n\,t)\, n\, \mathrm{d}t$$

$$= \int_{-x/n}^{\infty} \left(1 + \frac{x}{n\,t}\right)^n \mathrm{d}F_n(t),$$

where

$$F_n(t) = \frac{(-1)^{n+1} n^{2n+2}}{n!} \int_0^t \tau^n f^{(n+1)}(-n\,\tau)\, \mathrm{d}\tau. \tag{2}$$

We claim that the functions $F_n(t)$ are uniformly bounded for all $t \geqq 0$. For substituting n for $-n\,\tau$ in (2) we get

$$0 \leqq F_n(t) = \frac{n^{2n+2}}{n!} \int_{-nt}^{0} \frac{\eta^n}{n^n} f^{(n+1)}(\eta) \frac{1}{n^{n+1}} \frac{\mathrm{d}\eta}{n}$$

$$\leqq \frac{1}{n!} \int_{-\infty}^{0} f^{(n+1)}(\eta)\, |\eta|^n\, \mathrm{d}\eta = M.$$

We define the functions

$$\varphi_n(t, x) = \begin{cases} \left(1 + \dfrac{x}{n\,t}\right)^n, & t \geqq -\dfrac{x}{n} > 0, \\[2ex] 0, & 0 \leqq t \leqq -\dfrac{x}{n}. \end{cases}$$

As $n \to \infty$ these functions tend uniformly in t ($t \geqq 0$) to the limit

$$\varphi(t, x) = e^{x/t}.$$

We observe that as $t \to \infty$ the exponential increases to the value 1 in virtue of the hypothesis $x < 0$.

By Helly's second theorem (Section 6) we can extract from the sequence of non-decreasing functions $F_n(t)$ a subsequence which converges everywhere; by the first theorem, on the convergence of Stieltjes integrals (cf. notes 1 and 2 following it), we have

$$\int_0^{\infty} \varphi_n(t, x)\, \mathrm{d}F_n(t) \to \int_0^{\infty} \varphi(t, x)\, \mathrm{d}F(t).$$

Hence for $x < 0$

$$f(x) = f(-\infty) + \int_0^\infty e^{x/t}\, \mathrm{d}F(t);$$

substituting α for $1/t$, we get the required formula (1).

3. *The Map of the Unit Disc into the Right Half-plane*

We shall find here the general form of a function $w = f(z)$, analytic on the disc $|z| < 1$ and having a non-negative real part, i.e. a function which maps the disc $|z| < 1$ into the right half-plane. Examples of such functions are the constant $\alpha + i\beta$, $\alpha \geqq 0$, and the function

$$f_t(z) = \frac{e^{it} + z}{e^{it} - z}$$

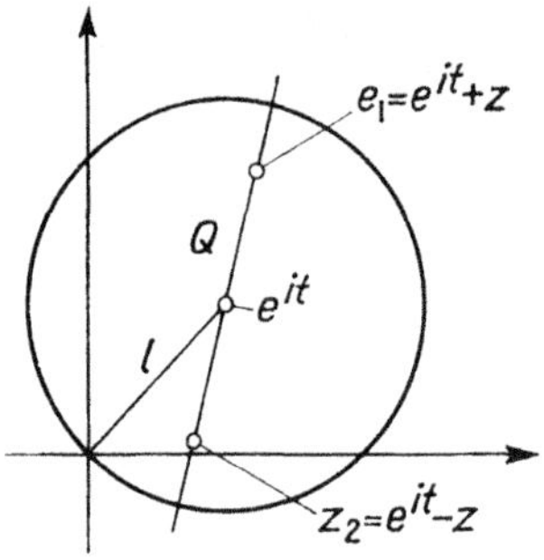

FIG. 14

with arbitrary real t. For, for a given t and $|z| \leqq 1$, the points $z_1 = e^{it} + z$, $z_2 = e^{it} - z$ lie in the closed disc Q of unit radius, centre e^{it}, on a diameter (Fig. 14). The whole diameter subtends an angle $\pi/2$ at the origin (which lies on the boundary of Q); the segment contained between z_1 and z_2 therefore subtends an angle $\alpha \leqq \pi/2$. Hence $|\arg f_t(z)| = |\arg z_1 - \arg z_2| \leqq \pi/2$ and it follows that $Re\, f(z) \geqq 0$.

It is found that every function $w = f(z)$ that is analytic on the disc $|z| < 1$ and maps it into the right half-plane is a "Stieltjes combination" of the given simple functions; we have the following theorem:

THEOREM 3. (G. Herglotz). *Every analytic function on the disc* $|z| < 1$ *with a non-negative real part can be expressed in the form*

$$f(z) = i\beta + \int_0^{2\pi} \frac{e^{it} + z}{e^{it} - z} \, dF(t), \tag{1}$$

where β *is a real number and* $F(t)$ *a non-decreasing function.*

Proof.† The analytic function $f(z)$ is well known to be expressible in the disc $|z| \leqq r < 1$ in terms of the boundary values of its real part $u(z)$ by Schwarz' formula

$$f(z) = \frac{1}{2\pi} \int_0^{2\pi} \frac{r\,e^{it} + z}{r\,e^{it} - z} \, u(r\,e^{it}) \, dt + i\beta .$$

This integral can be written in the form

$$f(z) = \int_0^{2\pi} \frac{r\,e^{it} + z}{r\,e^{it} - z} \, dF_r(t) + i\beta ,$$

where

$$F_r(t) = \frac{1}{2\pi} \int_0^{t} u(r\,e^{i\tau}) \, d\tau$$

is a non-decreasing function of t. By the mean-value theorem for harmonic functions we have further

$$F_r(t) \leqq F_r(2\pi) = \frac{1}{2\pi} \int_0^{2\pi} u(r\,e^{i\tau}) \, d\tau = \frac{1}{2\pi} u(0),$$

so that the family of functions $F_r(t)$ is uniformly bounded for all $r < 1$. As $r \to 1$ the functions $(r\,e^{it} + z)/(r\,e^{it} - z)$ ($|z| < 1$ fixed) converge uniformly in t to the function $(e^{it} + z)/(e^{it} - z)$. By theorem 2 of Section 6 we can extract from the sequence $F_r(t)$ $(r \to 1)$ a subsequence which converges everywhere to some non-

† After N. I. Akhiezer and I. M. Glazman (*Theory of Linear operators in Hilbert Space,* Ungor, New York, 1961). For Schwarz' formula, see e.g. A. I. Markushevich, *The theory of analytic functions,* Hindustan Pub., Dehli, 1963, ch. 6.

decreasing function $F(t)$; applying theorem 1 of Section 6, we get

$$f(z) = \int_0^{2\pi} \frac{r\,e^{it} - z}{r\,e^{it} + z} \mathrm{d}F_r(t) + i\beta \to \int_0^{2\pi} \frac{e^{it} - z}{e^{it} + z} \mathrm{d}F(t) + i\beta,$$

as asserted.

Note. A constant $\alpha > 0$ can also be represented by formula (1); for this it is sufficient to put $F(t) = \alpha t$.

8. Differentiation of Functions of Sets

1. The most general formulation of theorems on the differentiation of set functions is achieved by abstracting from the properties of the primary set with respect to which differentiation is carried out.

Let there be given some abstract set X and some family L of real functions $f(x)$ defined on X. The family L is assumed to be a linear space under the usual operations of addition and scalar multiplication, containing every constant and together with each function $f(x)$ its modulus $|f(x)|$. It follows that $f \in L$ implies $f^+ \in L, f^- \in L$ and $f \in L, g \in L$ implies $\max(f, g) \in L, \min(f, g) \in L$.

Further let an "integral" I be given on L, in other words, a linear functional with the properties (a)–(g) listed below:

(a) *$I\varphi \geqq 0$ whenever $\varphi(x) \geqq 0$.*

This implies that $I\varphi \leqq I\psi$ for $\varphi \leqq \psi$ and that $|I\varphi| \leqq I(|\varphi|)$.

(b) *If the monotone increasing sequence φ_n converges to the function φ and the $I\varphi_n$ are bounded, then $\varphi \in L$ and $I\varphi = \lim I\varphi_n$.*

A function $\varphi(x)$ for which $I(|\varphi|) = 0$ is said to be I-equivalent to zero and a set on which it is non-zero is said to be a set of I-measure zero.

(c) *Any function $\varphi(x)$ that is non-zero only on a set of I-measure zero belongs to the space L, and $I\varphi = 0$.*

(d) *The space L_1 of classes of I-equivalent functions is a complete normed space with the norm*

$$\|\varphi\| = I(|\varphi|).$$

From (d) it can be inferred that the set of $\varphi \in L$ for which $|\varphi|^2 \in L$ constitutes a complete Hilbert space with the scalar product

$$(\varphi, \psi) = I(\varphi, \psi).$$

(e) *There exists a set L_0 of bounded functions which is dense in the space L_I.*

The set of all functions $\varphi \in L_I$ that are limits of increasing sequences of functions $\varphi_n \in L_0$ is denoted by L_I^+.

(f) *Every function $\varphi \in L_I$ is the difference of two functions belonging to the class L_I^+.*

Since we have sets of measure zero, it is natural to define the concept "almost everywhere". For instance, a sequence of functions converges almost everywhere if it converges at all points of X except perhaps on a set of measure zero. We shall call the limit of a sequence of functions $\varphi_n(x) \in L_I$ which converges almost everywhere an I-measurable function.

(g) *The product of two I-measurable functions is an I-measurable function: the quotient $1/\varphi$ is an I-measurable function if the denominator φ is I-measurable and vanishes at most on a set of I-measure zero.*

Just as in Chapter IV it can be shown that an I-measurable function bounded in absolute value by an I-summable function (i.e. by a function $\varphi \in L_I$) is itself I-summable. In particular every bounded I-measurable function is I-summable.

Examples of the class L_I are afforded by the spaces of functions which are integrable with respect to the measures of Lebesgue or Stieltjes over a closed interval on the line or a region in n-dimensional space.

2. Our problem is to compare two integrals I, J which satisfy the conditions postulated. It is assumed that the aggregate L_I of functions φ integrable in the sense of the integral I (briefly I-integrable) and the aggregate L_J of functions ψ integrable in the sense of the functional J (J-integrable) are defined on one and the same set X and have an intersection L_0 (functions which are both I- and J-integrable) dense in L_I with respect to the metric of L_I and in L_J with respect to the metric of L_J.

We shall say that the integral I is absolutely continuous relative to the integral J if for any $\psi \in L_0$ we have $I\psi = 0$ whenever $\psi \geqq 0$, $J\psi = 0$.

For example, let I be the Stieltjes integral over the closed interval $[a, b]$ with an absolutely continuous non-negative integrating function $F(x)$, and let J be the Lebesgue integral over the same interval. As the set L_0 we take the set of all bounded measurable functions. If $\psi \in L_0$, $\psi \geqq 0$, $J\psi = 0$, then, as we know, the function ψ

vanishes almost everywhere; but then also

$$I\,\psi = \int_a^b \psi\,\mathrm{d}F = \int_a^b \psi\,F'(x)\,\mathrm{d}x = 0,$$

i.e. the integral I is absolutely continuous relative to J in the sense just explained. In this case, as we saw in Section 5, the I-integrable functions φ are characterised by the property that the product of a function φ by some fixed J-integrable function g $(= F'(x))$ is again a J-integrable function.

An analogous result holds in the general case; we shall prove the following fundamental theorem:

THEOREM (Radon–Nikodym). *A necessary and sufficient condition for the integral I to be absolutely continuous relative to the integral J is that there exist a J-integrable function ψ_0 such that its product with any I-integrable function φ is again a J-integrable function and*

$$I\,\varphi = J(\varphi\,\psi_0).$$

Proof†. The sufficiency of the condition is obvious: if $\psi \in L_0$, $\psi \geqq 0$, $J\,\psi = 0$, then the set on which $\psi > 0$ is of J-measure zero; by axiom (c) the function $\psi\psi_0 \in L_J$, $J(\psi\,\psi_0) = 0$, and consequently $I\,\psi = J(\psi\,\psi_0) = 0$. We must show that the condition is necessary for the absolute continuity of I relative to J.

Let us consider first the case when $I \leqq J$, i.e. $I\,\varphi \leqq J\,\varphi$ for any $\varphi(x) \geqq 0$, $\varphi \in L_0$. In this case every set of J-measure zero will be of I-measure zero. Every J-measurable function, as the limit of a J-almost everywhere convergent sequence $\varphi_n \in L_0$, will also be I-measurable.

We claim further that in this case the integral I can be defined over *all* functions $\varphi \in L_J$.

For we can form a sequence $\varphi_n \in L_0$ which converges to a given function $\varphi \in L_J$ in the metric of L_J such that

$$J(|\varphi_n - \varphi|) \to 0.$$

But then

$$|I\,\varphi_n - I\,\varphi_m| \leqq I(|\varphi_n - \varphi_m|) \leqq J(|\varphi_n - \varphi_m|) \to 0,$$

so that the sequence φ_n is fundamental with respect to the norm of L_I. We put $I\,\varphi = \lim I\,\varphi_n$. It is easily seen that this definition is unique and that for $\varphi \geqq 0$ it is always the case that $I\,\varphi \leqq J\,\varphi$.

† After F. Riesz.

We claim moreover that the functional I, extended as indicated over the whole space L_J, is a bounded functional on any space $L_J^p (p \geqq 1)$ with respect to the norm of that space. For we have for any $\varphi \in L_J^p$

$$I(|\varphi|^p) \leqq J(|\varphi|^p),$$

so that the functional I is bounded on the unit sphere of the space L_J^p. We shall consider only the value $p = 2$. In accordance with the theorem on the general form of a bounded linear functional in Hilbert space (Chapter V, Section 2) there exists a function $g \in L_J^2$ such that for any $\varphi \in L_J^2$

$$I\varphi = J(g\varphi). \tag{1}$$

We claim that the function g is included between the bounds $0 \leqq g(x) \leqq 1$ almost everywhere in the sense of J-measure. For, putting $\varphi = g^-$ in (1), we get

$$I g^- = J(g g^-) = J(-(g^-)^2) = -J((g^-)^2),$$

and since $I g^- \geqq 0$, $J((g^-)^2) \geqq 0$, we have

$$I g^- = J((g^-)^2) = 0,$$

so that the set on which $g^-(x) \geqq 0$ is of J-measure zero. Thus we have $g(x) \geqq 0$ almost everywhere in the sense of J-measure. The second inequality $g(x) \leqq 1$ is obtained by substituting J–I for the functional I in the foregoing argument.

We have established equation (1) for all $\varphi \in L_J^2$. We shall now carry it over to all functions $\varphi \in L_I$. It is sufficient to consider functions of the class L_I^+. By hypothesis every function $\varphi \in L_I^+$ can be expressed as the limit of an increasing sequence of functions $\varphi_n \in L_0$. For the functions φ_n equation (1) holds:

$$I\varphi_n = J(g\varphi_n).$$

The relation $\varphi_n \nearrow \varphi$ implies that $g\varphi_n \nearrow g\varphi$ and in virtue of property (b) the function $g\varphi$ belongs to L_J, so that

$$J(g\varphi) = \lim J(g\varphi_n) = \lim I\varphi_n = I\varphi.$$

Thus for any $\varphi \in L_I$ we have $g\varphi \in L_J$ and $J(g\varphi) = I\varphi$. The converse result also holds in the following form: if φ is I-measurable and $g\varphi \in L_J$, then $\varphi \in L_I$ and $I\varphi = J(g\varphi)$†. For the proof we

† If we include axioms connecting measurable functions with measurable sets, as in the Lebesgue case, the requirement for φ to be I-measurable becomes superfluous (cf. Section 5, art. 3).

put $\varphi_N = \min\{|\varphi|, N\}$. The function φ_N is bounded and I-measurable and therefore I-summable, and by what we have proved $I\varphi_N = J(g\varphi_N)$. Since $J(g\varphi)_N \leqq J(g|\varphi|)$, the numbers $I\varphi_N$ are bounded; it follows that $|\varphi| = \lim \varphi_N$ belongs to L_I, as required.

We have proved the Radon–Nikodym theorem in the case $I \leqq J$. We proceed now to the general case.

Let I, J be arbitrary non-negative functionals satisfying the conditions of art. 1 with I absolutely continuous relative to J. We define the functional $K = I + J$. Since $I \leqq K$, $J \leqq K$, there exist functions k, l belonging to the space L_K^2 and included between the bounds 0, 1, such that for any $\varphi \in L_I$

$$I\varphi = K(k\varphi) \tag{2}$$

and for any $\psi \in L_J$

$$J\psi = K(l\psi). \tag{3}$$

For $\varphi = \psi \in L_0$ we get

$$I\varphi + J\varphi = K\varphi = K[(k + l)\varphi],$$

and it follows, since L_0 is dense in L_K, that for any $\varphi \in L_K$

$$K\varphi = K[(k + l)\varphi].$$

We claim that almost everywhere in the sense of K-measure

$$k + l = 1. \tag{4}$$

To see this we take as φ the characteristic function e_0 of the set $E_0 = \{k + l < 1\}$. We get

$$K e_0 = K[(k + l)e_0]$$

or

$$K[(1 - (k + l)e_0] = 0. \tag{5}$$

But the function $[1 - (k + l)]e_0$ is non-negative; in virtue of (5) it vanishes almost everywhere in the sense of K-measure. Hence the inequality $k + l < 1$ can hold at most on a set of K-measure zero, and similarly for the inequality $k + l > 1$. Thus (4) holds almost everywhere in the sense of K-measure, as asserted.

It follows that for any J-summable function ψ the product $(1 - k)\psi$ is K-summable and we have

$$J\psi = K[(1 - k)\psi]. \tag{6}$$

Equation (6) holds whenever ψ is J-measurable and $(1 - k)\psi$ is K-summable; as we saw above, it follows in this case that ψ is

J-summable. Taking as φ in (2) and as ψ in (3) the characteristic function e of the set Z on which $k(x) = 1$, we get

$$I\,e = K(k\,e) = K(e),$$
$$J\,e = K((1-k)\,e) = K(0) = 0.$$

We see that the set Z has J-measure zero. Since I is absolutely continuous relative to J, we have $I\,e = 0$ and $K\,e = 0$; thus the set Z has zero I-measure and K-measure. Now let φ be any function in the space L_I. For the function ψ given by the condition

$$k\,\varphi = (1-k)\,\psi$$

we have

$$\psi = \frac{k}{1-k}\,\varphi = \frac{1}{1-k}\,k\,\varphi.$$

The function $k\,\varphi \in L_K$ here is K-measurable. The coefficient $1/(1-k)$ is also a K-measurable function since the domain on which the denominator vanishes has K-measure zero. Hence the function ψ is I-measurable and therefore J-measurable. It follows since $(1-k\,\psi = k\,\varphi \in L_K$, that ψ is J-summable and

$$I\,\varphi = K(k\,\varphi) = K((1-k)\,\psi) = J\,\psi = J\left(\frac{k}{1-k}\,\varphi\right). \qquad (7)$$

Putting $\varphi = 1$ in (7), we find that the function $\psi_0 = k/(1-k)$ is J-integrable, which completes the proof of the Radon–Nikodym theorem.

Concluding Remark

The connection between differentiation and integration described by us in Sections 1–3 was first discovered by Lebesgue (1902) and constitutes one of the most important achievements of the Lebesgue theory of integration. Somewhat earlier (1894) T. Stieltjes (Dutch mathematician, 1856–1894), while engaged in a study of the theory of continued fractions, arrived at a new concept of the integral, now known as the Riemann–Stieltjes integral. With the work of F. Riesz, who in 1909 obtained with the aid of the Stieltjes integral the general representation of a linear functional on the space of continuous functions, the Stieltjes integral began to infiltrate widely into very diverse branches of analysis: the general theory of measure developed hand-in-hand, progressing gradually from the line to many-dimensional space and then to abstract sets. In just such general form measure theory and integral theory proved to be useful tools in problems of higher analysis such as the harmonic analysis of groups, the theory of random processes, dynamic systems, and others. Recommended literature: P. R. Halmos, *Measure Theory,* Van Nostrand, N. Y. (1950).

CHAPTER VII

THE FOURIER TRANSFORM

1. On the Convergence of Fourier Series

1. The development of functions in Fourier series is useful in many analytical problems. In the simplest case, that of the closed interval $-\pi \leqq x \leqq \pi$, this development is given in complex terms by an expression of the form

$$\varphi(x) = \sum_{m=-\infty}^{\infty} a_m \, \mathrm{e}^{i m x}. \tag{1}$$

The development in Fourier series appears more frequently than other possible developments for the following reasons. First, the functions e^{imx} are orthogonal for distinct m [in the metric of the complex Hilbert space $L_2(-\pi, \pi)$], so that (1) is a development over an orthogonal basis. Secondly, the functions $\mathrm{e}^{imx} = u_m(x)$ remain well-behaved analytically (entire analytic functions) when continued with period 2π over the whole axis, and satisfy simple functional equations such as

$$u_m(x + \xi) = u_m(x) \, u_m(\xi)$$

or

$$u'_m(x) = i \, m \, u_m(x).$$

Thirdly, the coefficients of the development (1) can be calculated from the simple formula

$$a_m = \frac{1}{2\pi} \int_{-\pi}^{\pi} \varphi(\xi) \, \mathrm{e}^{-i m \xi} \, \mathrm{d}\xi. \tag{2}$$

Various formulations are possible for the question as to whether the Fourier series (1) converges. We can enquire first of all whether it converges at a given point x_0. Then we can consider its convergence in different norms. We shall narrow down the second question in art. 2; here we concern ourselves with the first question, viz the convergence in the usual numerical sense of the Fourier series at a particular point.

We prove the following theorem, which gives a sufficient condition for the convergence of the Fourier series (1) to the value $\varphi(x)$ at a given point x_0.

THEOREM 1. *If $\varphi(x)$ is a summable function and the integral*

$$\int_{-\varepsilon}^{\varepsilon} \frac{|\varphi(x_0+h)-\varphi(x_0)|}{|h|}\,\mathrm{d}h,$$

converges, then the partial sums of the Fourier series of $\varphi(x)$ converge at $x = x_0$ to the value $\varphi(x_0)$.

The condition that the ratio $[\varphi(x+t)-\varphi(x)]/t$ be summable for $|t| < \delta$ is known as *Dini's condition*. It is satisfied, for example, if $\varphi(x)$ satisfied the *Lipschitz condition of order* α:

$$|\varphi(x+t)-\varphi(x)| \leqq C|t|^{\alpha} \quad (0 < \alpha \leqq 1).$$

In particular if φ has a finite derivative at the point x (or even just finite derived numbers, Chapter VI, Section 1), the Lipschitz condition of order 1 is satisfied and hence the numbers $s_n(x)$ converge to $\varphi(x)$.

Proceeding to the proof of the theorem, we begin by transforming the expression for the partial sum $s_n(x)$ of the series (1). We have:

$$s_n(x) = \sum_{-n}^{n} a_k\,\mathrm{e}^{ikx} = \frac{1}{2\pi}\sum_{-n}^{n}\int_{-\pi}^{\pi}\varphi(\xi)\,\mathrm{e}^{ik(x-\xi)}\,\mathrm{d}\xi.$$

We make the substitution $x - \xi = -t$. We shall suppose that the function $\varphi(\xi)$ is continued from the interval $[-\pi, \pi]$ to the whole line as a periodic function of period 2π; then we can replace the limits $-\pi, \pi$ by the new limits of integration $-\pi - x, \pi - x$, giving

$$\sum_{-n}^{n} a_k\,\mathrm{e}^{ikx} = \frac{1}{2\pi}\int_{-\pi}^{\pi}\varphi(x+t)\sum_{-n}^{n}\mathrm{e}^{ikt}\,\mathrm{d}t.$$

Summing the geometric progression, we get

$$\sum_{-n}^{n} e^{ikt} = \frac{e^{int} - e^{-i(n+1)t}}{1 - e^{-it}} = \frac{e^{i(n+\frac{1}{2})t} - e^{-i(n+\frac{1}{2})t}}{e^{i\frac{t}{2}} - e^{-i\frac{t}{2}}}$$

$$= \frac{\sin(n+\frac{1}{2})t}{\sin t/2}$$

and thus

$$s_n(x) = \frac{1}{2\pi}\int_{-\pi}^{\pi} \varphi(x+t)\frac{\sin(n+\frac{1}{2}t)}{\sin\frac{t}{2}}\,dt. \quad (3)$$

The function

$$D_n(t) = \frac{1}{2\pi}\frac{\sin(n+\frac{1}{2})t}{\sin\frac{t}{2}}$$

is called *Dirichlet's kernel.* If we put $\varphi(x) \equiv 1$, then evidently $s_n(x) \equiv 1$ for any n; in this case formulae (3) gives

$$\frac{1}{2\pi}\int_{-\pi}^{\pi} \frac{\sin(n+\frac{1}{2})t}{\sin\frac{t}{2}}\,dt = 1.$$

The difference $s_n(x) - \varphi(x)$ can now be written in the form

$$s_n(x) - \varphi(x) = \frac{1}{2\pi}\int_{-\pi}^{\pi} [\varphi(x+t) - \varphi(x)]\frac{\sin(n+\frac{1}{2})t}{\sin\frac{t}{2}}\,dt. \quad (4)$$

We want to ascertain under what conditions $s_n(x)$ tends to $\varphi(x)$ or, what is the same thing, under what conditions the integral (4) tends to zero. For this we prove a lemma:

LEMMA 1. *If* $\varphi(x)$ *is a summable function on the closed interval* $[a, b]$ *then the integrals*

$$\int_a^b \varphi(x)\sin\lambda x\,dx, \int_a^b \varphi(x)\cos\lambda x\,dx$$

tend to zero as $\lambda \to \infty$.

Proof. First let $\varphi(x)$ be the characteristic function of an interval $(c, d) \subset [a, b]$. Then

$$\int_a^b \varphi(x) \sin \lambda x \, \mathrm{d}x = \int_c^d \sin \lambda x \, \mathrm{d}x = \frac{\cos \lambda c - \cos \lambda d}{\lambda} \to 0.$$

Any step function $h(x)$ is a linear combination of characteristic functions of intervals, hence the assertion of the lemma also holds for step functions. If now $\varphi(x)$ is an arbitrary summable function, then for a given $\varepsilon > 0$ we find a step function $h(x)$ such that

$$\int_a^b |\varphi(x) - h(x)| \, \mathrm{d}x < \frac{\varepsilon}{2},$$

and $\lambda_0 > 0$ such that for $|\lambda| > \lambda_0$

$$\left| \int_a^b h(x) \sin \lambda x \, \mathrm{d}x \right| < \frac{\varepsilon}{2}.$$

Then for these values of λ

$$\left| \int_a^b \varphi(x) \sin \lambda x \, \mathrm{d}x \right| \leqq \int_a^b |\varphi(x) - h(x)| \, \mathrm{d}x + \left| \int_a^b h(x) \sin \lambda x \, \mathrm{d}x \right| < \varepsilon,$$

which gives the required result. The proof with $\cos \lambda x$ as multiplier is similar.

In particular we get the result:

The Fourier coefficient a_n of any integrable function $\varphi(x)$ tends to zero as $n \to \infty$.

We turn now to the integral (4)

$$\frac{1}{2\pi} \int_{-\pi}^{\pi} [\varphi(x + t) - \varphi(x)] \frac{\sin (n + \frac{1}{2}) t}{\sin \frac{t}{2}} \, \mathrm{d}t = I_n.$$

Let us suppose that for a given value of x the summable function $\varphi(x)$ is defined and finite and the ratio

$$\frac{\varphi(x + t) - \varphi(x)}{t}$$

is integrable with respect to t over $|t| \leqq \delta$ and consequently over the whole interval $-\pi \leqq t \leqq \pi$. Then the function

$$\frac{\varphi(x+t)-\varphi(x)}{\sin\frac{t}{2}} = \frac{\varphi(x+t)-\varphi(x)}{t} \frac{t}{\sin\frac{t}{2}}$$

is integrable between the limits $-\pi < t < \pi$ and lemma 1 can be applied to the integral I_n; we thus get the required result that I_n tends to zero as $n \to \infty$, and theorem 1 is proved.

Note. In a number of cases Dini's condition can be weakened, but it cannot be totally rejected if convergence of the Fourier series is to be preserved. There exist even continuous functions for which the Fourier series diverges at isolated points (cf. art. 3). A. N. Kolmogoroff has constructed a summable function the Fourier development of which diverges at every point.† There has so far been no solution of the problem posed in 1915 by N. N. Luzin: is the Fourier development of a function $f \in L_2$ convergent almost everywhere?

A condition for the uniform convergence of a Fourier series can be formulated in the same terms.

THEOREM 1′. *If the function $\varphi(x)$ is bounded and summable on some set $E \subset [-\pi, \pi]$ and Dini's condition is satisfied uniformly, i.e. corresponding to any $\varepsilon > 0$ there exists $\delta > 0$ such that for all $x \in E$*

$$\int_{-\delta}^{\delta} \frac{|\varphi(x+t)-\varphi(x)|}{|t|} \, dt < \varepsilon$$

then the Fourier series of $\varphi(x)$ converges to $\varphi(x)$ uniformly on the set E.

For the proof we use a strengthened version of lemma 1.

LEMMA 1. *The relation*

$$\lim_{\lambda \to \infty} \int_a^b f(t) \sin \lambda t \, dt = 0$$

is realised uniformly on any set B of functions $f(t)$ summable on $[a, b]$ that is compact with respect to the metric of $L_1(a, b)$.

† Cf. A. Zygmund, *Trigonometric Series*, Stechert, N.Y. (1935), Chapter 8.

For, given $\varepsilon > 0$, we can construct in $L_1(a, b)$ a finite $\varepsilon/2$-net for B; let this consist of the functions $f_1(t), \ldots, f_m(t)$. By lemma 1 we can find λ_0 such that for $\lambda > \lambda_0$

$$\left| \int_a^b f_j(t) \sin \lambda t \, dt \right| < \frac{\varepsilon}{2} \quad (j = 1, 2, \ldots, m).$$

If now $f(t) \in B$ is any function and for some j we have $\| f(t) - f_j(t) \| < \varepsilon/2$, then for $\lambda > \lambda_0$

$$\left| \int_a^b f(t) \sin \lambda t \, dt \right| \leq \int_a^b |f(t) - f_j(t)| \, dt + \left| \int_a^b f_j(t) \sin \lambda t \, dt \right|$$

$$< \frac{\varepsilon}{2} + \frac{\varepsilon}{2} = \varepsilon,$$

as required.

We proceed to the proof of theorem 1′. Given $\varepsilon > 0$ we find $\delta > 0$ such that for all $x \in E$

$$\int_{-\delta}^{\delta} \frac{|\varphi(x + t) - \varphi(x)|}{|t|} \, dt < \frac{\varepsilon}{3}. \tag{5}$$

Then

$$s_n(x) - \varphi(x) = \frac{1}{2\pi} \int_{-\pi}^{\pi} \frac{\varphi(x + t) - \varphi(x)}{t} \, \frac{t}{\sin \dfrac{t}{2}} \sin n\, (+ \tfrac{1}{2} t) \, dt$$

$$= \frac{1}{2\pi} \int_{-\delta}^{\delta} + \frac{1}{2\pi} \int_{\delta}^{\pi} \frac{1}{2\pi} \int_{-\pi}^{-\delta}. \tag{6}$$

Since $\dfrac{t}{2\pi \sin t/2} < 1$ and in virtue of (5), the first term does not exceed $\varepsilon/3$ for any $x \in E$. To obtain bounds for the remaining terms we show that the functions

$$f_x(t) = \frac{1}{2\pi} \frac{\varphi(x + t) - \varphi(x)}{\sin \dfrac{t}{2}},$$

as functions of $t \in [\delta, \pi]$ with parameter $x \in E$, form a compact set B in the space $L_1(\delta, \pi)$. Let $x_n \in E$ be any sequence of points; we can suppose that the x_n tend to some point x_0 and that the values $\varphi(x_n)$ tend to some number c_0. Then in the metric of $L_1(\delta, \pi)$ we get $\varphi(x_n + t) \to \varphi(x_0 + t)$, which means that

$$\|f_{x_n}(t) - f_{x_0}(t)\| \leq \frac{1}{2\pi}\left\|\frac{\varphi(x_n + t) - \varphi(x_0 + t)}{\left|\sin\frac{t}{2}\right|}\right\| + \frac{1}{2\pi}\left\|\frac{\varphi(x_n) - c_0}{\left|\sin\frac{t}{2}\right|}\right\|$$

$$\leq \frac{1}{2\pi}\frac{1}{\left|\sin\frac{\delta}{2}\right|}[\|\varphi(x_n + t) - \varphi(x_0 + t)\| + \pi|\varphi(x_n) - c_0|] \to 0,$$

i.e. the sequence $f\,x_n(t)$ is fundamental in $L_1(\delta, \pi)$.

Thus the set B is compact. By lemma 1′ we can find λ_0 such that for $n > \lambda_0$

$$\left|\frac{1}{2\pi}\int_\delta^\pi \frac{\varphi(x + t) - \varphi(x)}{\sin\frac{t}{2}}\sin\left(n + \tfrac{1}{2}\right)t\,\mathrm{d}t\right| < \frac{\varepsilon}{3}$$

for all $x \in E$. The last term in (6) can be treated similarly. We see as a result that for sufficiently large n the quantity $|s_n(x) - \varphi(x)|$ becomes $< \varepsilon$ for all $x \in E$, which proves theorem 1′.

COROLLARY. *If all the derived numbers of a summable function $\varphi(x)$ are bounded by some constant K on an interval $[\alpha, \beta] \subset [-\pi, \pi]$, then the Fourier series of $\varphi(x)$ converges uniformly on any closed interval $[\alpha', \beta']$ such that $\alpha < \alpha' < \beta' < \beta$.*

For we have for $x \in [\alpha', \beta']$ and all $|t| \leq \min(\beta - \beta', \alpha' - \alpha)$

$$|\varphi(x + t) - \varphi(x)| \leq K|t|$$

so that $\varphi(x)$ is bounded on $[\alpha', \beta']$ and Dini's condition is satisfied uniformly.

For instance, if a summable function $\varphi(x)$ vanishes on the closed interval $[\alpha, \beta]$, its Fourier series converges to zero uniformly on any closed interval $[\alpha', \beta']$ interior to $[\alpha, \beta]$.

Problems. 1. If a function $\varphi(x)$ is continuous at $x = x_0$ and is of bounded variation in the neighbourhood of x_0, its Fourier development converges to $\varphi(x_0)$ for $x = x_0$.

Hint. It is sufficient to consider a non-decreasing $\varphi(x)$. Use the second mean value theorem:

$$\int_0^h \varphi(t)\, g(t)\, dt = \varphi(h) \int_\xi^h g(t)\, dt \quad 0 < \xi < h,$$

and the fact that $\int\limits_{n\xi}^{nh} \frac{\sin t}{t}\, dt$ is uniformly bounded.

2. Without the assumption that $\varphi(x)$ is continuous in problem 1, the Fourier development converges to $1/2\,[\varphi(x_0 + 0) + \varphi(x_0 - 0)]$.

Hint. Since the Dirichlet kernel is even,

$$\int_{-\pi}^{\pi} \varphi(x+t)\, D_n(t)\, dt = \int_0^{\pi} [\varphi(x+t) + \varphi(x-t)]\, D_n(t)\, dt.$$

2. We now turn our attention to questions concerning the convergence of the Fourier series

$$\sum_{m=-\infty}^{\infty} a_m\, e^{imx} \tag{1}$$

in the norms of different functional spaces. We begin by recalling some well-known facts in connection with the convergence of Fourier series. It is proved in elementary courses of analysis that every function $\varphi(x)$ that is continuous on $[-\pi, \pi]$ is piecewise-smooth, and satisfies the condition $\varphi(\pi) = \varphi(-\pi)$ (thus ensuring the continuity of the 2π-periodic continuation of $\varphi(x)$ over the whole axis) has a Fourier series development which converges absolutely and uniformly, in particular, that is, in the norm of the space $C(-\pi, \pi)$. On the other hand, we saw in Chapter V that every function $\varphi(x)$ square-summable on $[-\pi, \pi]$ is the sum of a Fourier series (1) which converges to $\varphi(x)$ in the quadratic mean [i.e. in the metric of the space $L_2(-\pi, \pi)$].

Series in the functions e^{imx} can also be constructed and studied in other normed spaces of functions on $[-\pi, \pi]$. There are too many such spaces for us to consider them all; we shall confine ourselves to an important class of spaces, which contains the majority of those used in analytic applications.

Definition. A normed space R of functions defined on the closed interval $[-\pi, \pi]$ is said to be a *homogeneous functional space* if the following conditions are satisfied.

(1) All functions $\varphi(x) \in R$ are summable and the convergence $\varphi_n(x) \to \varphi(x)$ in the norm of R implies the convergence

$\varphi_n(x) \to \varphi(x)$ in the norm of $L_1(-\pi, \pi)$, i.e. the relation

$$\int_{-\pi}^{\pi} |\varphi_n(x) - \varphi(x)| \, dx \to 0.$$

(2) If a function $\varphi(x)$ is continued over the whole axis of x as a periodic function of period 2π, then for any real h the functions $\varphi(x+h)$—the displacements of $\varphi(x)$—will be well-defined. All such displacements are required to belong to the space R together with the functions $\varphi(x)$ and the norm is to be invariant under translation:

$$\|\varphi(x+h)\| = \|\varphi(x)\| \quad \text{for any } h. \tag{2}$$

(3) The space R contains all trigonometric polynomials—linear combinations of the functions e^{imx}, and their aggregate forms a set everywhere dense in R.

Conditions (1)–(3) are satisfied by many of the functional spaces with which we are familiar. They hold for example in the spaces $L_p(-\pi, \pi)$ for all $p \geqq 1$. In the space $C(-\pi, \pi)$ of all continuous functions on $[-\pi, \pi]$ condition (2) is not satisfied since a continuous function $\varphi(x)$ with $\varphi(\pi) = \varphi(-\pi)$ ceases to be continuous as a periodic continuation over the whole axis and $\varphi(x+\pi)$, for example, is not contained in $C(-\pi, \pi)$. But if in place of the whole space $C(-\pi, \pi)$ we consider only the subspace $\hat{C}(-\pi, \pi)$ determined by the condition $\varphi(-\pi) = \varphi(\pi)$, conditions (1)–(3) will all be satisfied. The analogous subspace $\hat{D}_n(-\pi, \pi)$ of the space $D_n(-\pi, \pi)$ determined by the conditions $\varphi(-\pi) = \varphi(\pi)$, $\varphi'(-\pi) = \varphi'(\pi), \ldots, \varphi^{(n)}(-\pi) = \varphi^{(n)}(\pi)$ also satisfies conditions (1)–(3).

We shall establish first of all the form of the Fourier series in a homogeneous space R. Strictly speaking we require only condition (1) here.

LEMMA 2. *If for some function $\varphi(x)$ which belongs to a homogeneous space R the development*

$$\varphi(x) = \sum_{-\infty}^{\infty} a_m \, e^{imx} \tag{3}$$

converges in the norm of R, then the usual Fourier coefficients of $\varphi(x)$ are given by

$$a_m = \frac{1}{2\pi} \int_{-\pi}^{\pi} \varphi(\xi) \, e^{-im\xi} \, d\xi \tag{4}$$

Proof. By condition (1) we have

$$\int_{-\pi}^{\pi} \left| \sum_{-n}^{n} a_m e^{imx} - \varphi(x) \right| dx \to 0.$$

But then for every fixed k

$$\int_{-\pi}^{\pi} \left\{ \left| e^{-ikx} \sum_{-n}^{n} a_m e^{imx} - e^{-ikx} \varphi(x) \right\} dx \to 0,$$

and hence in virtue of the orthogonality of the functions e_{imx} over $[-\pi, \pi]$

$$\int_{-\pi}^{\pi} \varphi(x) e^{-ikx} dx = \lim_{n\to\infty} \int_{-\pi}^{\pi} e^{-ikx} \sum_{-n}^{n} a_m e^{imx} dx = a_k \cdot 2\pi,$$

which yields formula (4).

The following property of functions which belong to a homogeneous space plays an important part in the subsequent discussion.

Lemma 3. *Every function $\varphi(x)$ that belongs to a homogeneous space R is continuous in the norm under displacement: to any $\varepsilon > 0$ corresponds $\delta > 0$ such that for $|h| < \delta$ we have $\|\varphi(x+h) - \varphi(x)\| < \varepsilon$.*

Proof. We denote by Q the totality of functions $\varphi(x) \in R$ which are continuous in the norm under displacement. Evidently the set Q is a subspace of R: it contains together with functions $\varphi(x)$, $\psi(x)$ every linear combination of them. We shall show that Q is closed in the norm. Let $\varphi_n \to \varphi$, where $\varphi_n \in Q$. For a given $\varepsilon > 0$ we find n such that $\|\varphi - \varphi_n\| < \varepsilon/3$. Then we choose δ in the condition for continuity of $\varphi_n(x)$ under displacement so that $\|\varphi_n(x+h) - \varphi_n(x)\| < \varepsilon/3$ for $|h| < \delta$. By condition (2) we shall have at the same time $\|\varphi(x+h) - \varphi_n(x+h)\| < \varepsilon/3$. Hence

$$\|\varphi(x+h) - \varphi(x)\| \leq \|\varphi(x+h) - \varphi_n(x+h)\| + \|\varphi_n(x+h) - \varphi_n(x)\| + \|\varphi_n(x) - \varphi(x)\| < \varepsilon,$$

so that the function $\varphi(x)$ is also continuous under displacement. Finally we observe that each of the functions e^{imx} is continuous under displacement since

$$\|e^{im(x+h)} - e^{imx}\| = |e^{imh} - 1| \, \|e^{imx}\| - 0 \text{ for } |h| \to 0.$$

Thus the set Q contains all trigonometric polynomials and is closed; it follows from condition (3) that $Q = R$, and the lemma is proved.

We now formulate our basic problems.

A. *We are given a function $\varphi(x)$ contained in a homogeneous space R and its Fourier coefficients are calculated in accordance with formula* (4). *Does the Fourier series $\sum_{-\infty}^{\infty} a_m e^{imx}$ converge to $\varphi(x)$ in the norm of the space R?*

The answer to this question is in general negative: there exist homogeneous spaces, moreover ones with which we are very familiar such as $\hat{C}(-\pi, \pi)$ and $L_1(-\pi, \pi)$, in which the Fourier series is found to be divergent for some functions in every case.

Such an answer evokes a natural desire:

B. *To find a procedure, as simple and uniform as possible, which will allow a function $\varphi(x)$ to be deduced from its Fourier series despite the possible divergence of the latter.*

In attempting to fulfil this desire we obtain the following result:

THEOREM 2. *The arithmetic means*

$$\sigma_p(x) = \frac{s_0(x) + s_1(x) + \cdots + s_{p-1}(x)}{p}$$

of the partial sums $s_n(x) = \sum_{-n}^{n} a_m e^{imx}$ of the Fourier series of any function $\varphi(x) \in R$ converge in the norm to $\varphi(x)$ as $p \to \infty$.

3. In this paragraph we shall satisfy ourselves that the spaces $\hat{C}(-\pi, \pi)$ $L_1(-\pi, \pi)$ really do contain functions whose Fourier series do not converge in the norm of the space. We begin by showing that the integrals

$$D_n = \int_{-\pi}^{\pi} |D_n(t)|\, dt,$$

and

$$D_n(t) = \frac{1}{2\pi} \frac{\sin\left(n + \frac{1}{2}\right) t}{\sin \frac{t}{2}},$$

increase without bound as $n \to \infty$. The graph of the function

$D_n(t)$ is given in Fig. 15. At points t for which $(n + 1/2)\,t = \pi/2$, $3\pi/2$, ...,$(k + 1/2)\,\pi$, ..., the quantity $|\sin\,(n + n\;1/2)\,t|$ is equal to unity, while in the intervals

$$|(n + \tfrac{1}{2})\,t - (k + \tfrac{1}{2})\,\pi| < \frac{\pi}{3} \quad (k = 0, 1, 2, \ldots)$$

it exceeds 1/2. In the same intervals $\sin t/2$ does not exceed

$$\sin \frac{(k + \tfrac{1}{2})\,\pi + \dfrac{1}{3}\pi}{2\,(n + \tfrac{1}{2})} < \frac{(k + 1)\,\pi}{2\pi}.$$

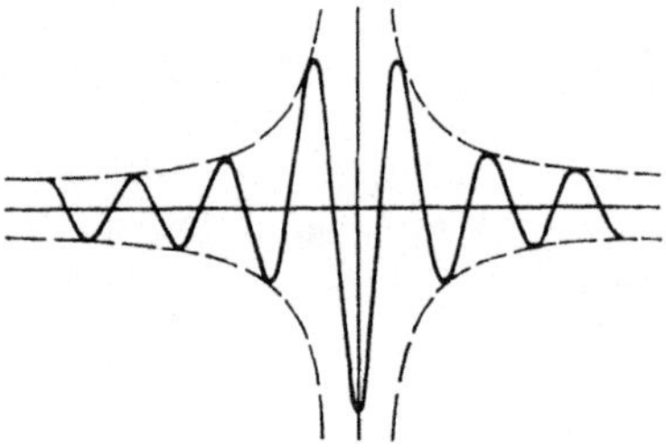

FIG. 15

Hence the integral of $|D_n(t)|$, taken only over the intervals specified, exceeds

$$\frac{2\pi}{3\,(n + \tfrac{1}{2})} \sum_{k=0}^{n} \frac{\tfrac{1}{2}}{\dfrac{(k + 1)\,\pi}{2n}} = \frac{2n}{3\,(n + \tfrac{1}{2})} \sum_{k=0}^{n} \frac{1}{k + 1} \to \infty.$$

In this fact lies the primary reason why the Fourier series does not, in general, converge in the spaces $\hat{C}(-\pi, \pi)$, $L_1(-\pi, \pi)$. Let us consider on the space $\hat{C}(-\pi, \pi)$ the functionals

$$\Phi_n[\varphi] = \frac{1}{2\pi} \int_{-\pi}^{\pi} \varphi(t) \frac{\sin\,(n + \tfrac{1}{2})\,t}{\sin \dfrac{t}{2}}\,dt = \int_{-\pi}^{\pi} \varphi(t)\,D_n(t)\,dt,$$

which give the value of the partial sum $s_n(x)$ of the Fourier series for $\varphi(x)$ at $x = 0$. Each of the functionals $\Phi_n(\varphi)$ is bounded on the unit sphere of the space $\hat{C}(-\pi, \pi)$, but they do not have a *common* bound on this sphere. For if we take as $\varphi(t)$ continuous functions which approximate to $+1$ on intervals where $D_n(t)$ is positive and to -1 on intervals where $D_n(t)$ is negative [such functions do not exceed 1 in absolute value and therefore belong to the unit sphere of $C(-\pi, \pi)$], we shall obtain arbitrarily large numerical values of $\Phi(\varphi)$ (by taking n sufficiently large). We claim that there exists a function $\varphi_0(x)$ for which the values of the functionals $\Phi_n(\varphi_0)$ are unbounded (so that the corresponding Fourier series diverges for $x = 0$). This is a consequence of the following general lemma of functional analysis:

LEMMA 4. *If a sequence of linear functionals Φ_n on a complete normed space R is unbounded on the unit sphere $|\varphi| \leqq 1$, there exists an element φ_0 for which the values $\Phi_n(\varphi_0)$ are unbounded.*

The proof of this lemma will be found in the Supplement, Section 2 (p. 465).

Hence, *there exists a continuous function whose Fourier development is divergent at $x = 0$.*

With minor technical modifications, the same argument can be carried through for space $L_1(-\pi, \pi)$. We assume that the partial sums $S_n \varphi$ of the Fourier development of any $\varphi \in L_1$ converge in the norm to the element φ. An even weaker assumption can be made, namely that the numbers $\|S_n \varphi\|$ are bounded for any $\varphi \in L_1$. We can now assert that *the numbers $\|S_n \varphi\|$ are bounded by the same constant K for all $\|\varphi\| \leqq 1$.* If this were not the case, we could apply the same lemma of functional analysis and find an element φ_0 such that the numbers $\|S_n \varphi_0\|$ are unbounded. Now let $\mu(x)$ be any bounded measurable function, say not exceeding 1 in absolute value. We have the inequality

$$\left|\int_{-\pi}^{\pi} S_n \varphi(x)\,\mu(x)\,\mathrm{d}x\right| \leqq \int_{-\pi}^{\pi} |S_n \varphi(x)|\,\mathrm{d}x = \|S_n \varphi\| \leqq K.$$

Introducing the explicit expression for operator S_n, we get

$$\left|\int_{-\pi}^{\pi}\int_{-\pi}^{\pi} \varphi(x+t)\,D_n(t)\,\mu(x)\,\mathrm{d}t\,\mathrm{d}x\right|$$

$$= \left|\int_{-\pi}^{\pi}\int_{-\pi}^{\pi} \varphi(t)\,D_n(x-t)\,\mu(x)\,\mathrm{d}x\,\mathrm{d}t\right| \leqq K. \quad (1)$$

But it is easily shown that this inequality cannot be fulfilled for all n, any $\varphi \in L_1$ with $\|\varphi\| \leqq 1$ and any measurable $\mu(x)$ with

$|\mu(x)| \leqq 1$. For we can write (1) in the form

$$\left| \int_{-\pi}^{\pi} \varphi(t) \, M_n(t) \, \mathrm{d}t \right| \leqq K, \tag{2}$$

where $M_n(t) = \int_{-\pi}^{\pi} D_n(x-t) \, \mu(x) \, \mathrm{d}x$ is obviously a continuous function of t. We put $\varphi(t)$ equal to $1/2\varepsilon$ for $|t| \leqq \varepsilon$ and 0 outside this interval; we get $\|\varphi\| = 1$ and the integral in (2) becomes the mean of the function $M_n(t)$ over the interval $|t| \leqq \varepsilon$. Proceeding to the limit as $\varepsilon \to 0$, we get

$$|M_n(0)| = \left| \int_{-\pi}^{\pi} D_n(x) \, \mu(x) \, \mathrm{d}x \right| \leqq K.$$

We have not yet defined the function $\mu(x)$. Let us put it equal to $+1$ for $D_n(x) > 0$ and -1 for $D_n(x) < 0$; then we get

$$\int_{-\pi}^{\pi} |D_n(x)| \, \mathrm{d}x \leqq K.$$

But we have already shown that such an inequality cannot hold for all n. Hence *there exists an element $\varphi_0 \in L_1$ such that the partial sums $S_n \varphi_0$ do not converge to φ_0 in the norm of $L_1(-\pi, \pi)$.*

Of course our results do not preclude the convergence $S_n \varphi \to \varphi$ from holding for every element φ in particular homogeneous spaces. But in this event the property is peculiar to the space under consideration. We know, for example, that the space $L_2(-\pi, \pi)$ has this property. There is also a theorem due to M. Riesz which establishes an analogous property for any space $L_p(-\pi, \pi)$ with $p > 1$†.

Note. Theorem 2 ceases to hold if we do not assume that space R is homogeneous; it can happen, moreover, that no linear combinations of partial sums of the Fourier development of $\varphi(x)$ converge to $\varphi(x)$ in the norm of R.

We take as an example the space R of functions $\varphi(x)$, continuous for $|x| \leqq \frac{\pi}{2}$, belonging to L_2 for $|x| \leqq \pi$ and with the norm

$$\|\varphi\| = \max_{|x| \leqq \pi/2} |\varphi(x)| + \int_{-\pi}^{\pi} |\varphi(x)| \, \mathrm{d}x.$$

† Cf. A. Zygmund, *Trigonometric Series*, Stechert, N. Y. (1935), Chapter 7.

This space obviously satisfies conditions (1) and (3) for a homogeneous space, but does not satisfy condition (2). The function $\varphi_0(x)$, equal to $+\pi/4$ for $|x| \leqq \frac{\pi}{2}$, and $-\pi/4$ at the remaining points of the interval $[-\pi, \pi]$, belongs to R and has the formal Fourier development

$$\cos x - \frac{1}{3}\cos 3x + \frac{1}{5}\cos 5x - \cdots$$

All the terms vanish at $x = \pm\frac{\pi}{2}$. Hence any linear combination of partial sums also vanishes at $x = \pm\frac{\pi}{2}$. Since $\varphi_0(x)$ itself is equal to $-1/2$ at $x = \pm\frac{\pi}{2}$, whilst convergence in the norm of R requires in particular uniform convergence in the closed interval $\left[-\frac{\pi}{2}, \frac{\pi}{2}\right]$, it is clear that there can be no linear combinations of partial sums of the Fourier development that converge to $\varphi_0(x)$ in the norm of space R.

4. Our primary device in proving theorem 2 will be the *integration of continuous abstract functions with values in a normed space R.* In this paragraph we shall give the relevant definitions and essential elements of the theory.

Let $f(t)$ denote an element of a complete normed space R, which depends on a real parameter t, or what is the same thing, a function of the parameter t with values in the space R. Such functions are said to be *abstract*. We shall say that $f(t)$ is continuously dependent on the parameter t at $t = \tau$ if whenever $t \to \tau$

$$\|f(t) - f(\tau)\| \to 0.$$

An abstract function $f(t)$ which is continuously dependent on t at any $t = \tau$ in the closed interval $a \leqq t \leqq b$ is said to be a *continuous abstract function of t on* $[a, b]$.

The following propositions, which are the natural generalisations of well-known elementary theorems of analysis, are easily proved by means of the usual arguments using the compactness of the closed interval:

(a) An abstract function $f(t)$, continuous on the interval $[a, b]$, is bounded in the norm, so that $\|f(t)\| < M$ for all t.

(b) An abstract function $f(t)$, continuous on the interval $[a, b]$, is uniformly continuous on it: for any $\varepsilon > 0$ there exists $\delta > 0$ such that $\|f(t') - f(t'')\| < \varepsilon$ whenever $|t' - t''| < \delta$.

(c) A sequence of abstract functions $f_n(t)$ is said to converge to an abstract function $f(t)$ *uniformly on* $[a, b]$ if for any $\varepsilon > 0$ there exists $N = N(\varepsilon)$ such that for $n > N$

$$\max_t \|f_n(t) - f(t)\| < \varepsilon.$$

The limit $f(t)$ of a uniformly convergent sequence of continuous functions $f_n(t)$ is also a continuous function.

We define further the Riemann integral of a function $f(t)$. Let π denote a partition $a = t_0 < t_1 < \cdots < t_n = b$ of the interval $[a, b]$ and let $\Delta t_j = t_{j+1} - t_j$. We shall call the quantity $\mathrm{d}(\pi) = \max \Delta t_j$ the parameter of the partition π. We form the integral sum

$$S_\pi = \sum_{j=0}^{n-1} f(t_j) \Delta t_j. \tag{1}$$

The quantity S_π is evidently an element of the same space R. We claim that when the partition π is indefinitely refined, i.e. as $\mathrm{d}(\pi) \to 0$, the integral sum S_π tends to a uniquely determined element If of R, said to be *the Riemann integral of the abstract function* $f(t)$.

We prove the existence of the integral.

LEMMA 5. *Let* $\varepsilon > 0$ *be given and let* $\delta > 0$ *be chosen so that* $|f(t') - f(t'')| < \varepsilon/2$ *whenever* $|t' - t''| < \delta$; *then for* $\mathrm{d}(\pi) < \delta$ *the sums* (1) *differ by at most* $\varepsilon(b - a)$.

We consider first the integral sums

$$s = \sum_{j=0}^{n-1} f(t_j) \Delta t_j, \quad s' = \sum_{k=0}^{m-1} f(t_k) \Delta t_k,$$

where the partition of the primary interval corresponding to the second sum is a refinement of the partition corresponding to the first sum. In this case each term $f(t_j) \Delta t_j$ of the first sum is replaced in the refinement by a quantity of the form

$$f(t_{j1}) \Delta t_{j1} + \cdots + f(t_{jr}) \Delta t_{jr}.$$

By hypothesis each of the quantities $f(t_j), \ldots, f(t_j)$ can be replaced here by $f(t_j)$ with an error in the norm of less than $\varepsilon/2$:

$$f(t_{j1}) = f(t_j) + h_1, \ldots, f(t_{jr}) = f(t_j) + h_r, \ \|h_i\| < \frac{\varepsilon}{2}.$$

Hence

$$\|f(t_{j1}) \Delta t_{j1} + \cdots + f(t_{jr}) \Delta t_{jr} - f(t_j) \Delta t_j\| \leq \sum_{i=1}^{r} \|h_i\| \Delta t_{ji} < \frac{\varepsilon}{2} \Delta t_j,$$

and therefore

$$\|s - s'\| < \sum \frac{\varepsilon}{2} \Delta t_j = \frac{\varepsilon}{2}(b - a).$$

Now let s_1, s_2 be any two integral sums, with the sole restriction that the elements of the corresponding partitions do not exceed the given δ. We form the integral sums corresponding to the partition obtained by superposing the former partitions. Then by what we have proved

$$\|s - s_1\| < \frac{\varepsilon}{2}(b - a), \quad \|s - s_2\| < \frac{\varepsilon}{2}(b - a),$$

from which we get the required result

$$\|s_1 - s_2\| < \varepsilon(b - a).$$

It is easily deduced from lemma 5 that as the interval $[a, b]$ undergoes indefinitely refined partitions the sums (1) tend to a limit. For let π_n be an arbitrary sequence of partitions with $\mathrm{d}(\pi_n) \to 0$. By what we have proved the corresponding integral sums s_n form a fundamental sequence; we denote its limit in R by If. Any other sequence of integral sums s_n' with $\mathrm{d}(\pi_n) \to 0$ has the same limit, since, as we have shown, $\|s_n - s_n'\| \to 0$. We shall call the element If *the integral of the function* $f(t)$ *over the interval* $[a, b]$.

The integral of an abstract function possesses the usual properties of an integral

$$I(f + g) = If + Ig; \tag{2}$$

$$I(\alpha f) = \alpha I f; \tag{3}$$

$$\text{if} \quad \|f\| \leq M, \quad \text{then} \quad \|If\| \leq M(b - a). \tag{4}$$

The product of a continuous abstract function $f(t)$ by a real continuous function $\beta(t)$ is again a continuous abstract function. If in addition $\beta(t) \geq 0$, $|f| \leq M$, then

$$\|I(\beta f)\| \leq M \int_a^b \beta(t)\,\mathrm{d}t. \tag{5}$$

All these properties are proved by taking limits in the relevant integral sums.

An important example is furnished by abstract functions of a parameter t which take values in a normed space R of ordinary functions of an argument x, so that $f(t) = \varphi(x, t)$.

It is evident from the definition of the integral of an abstract function that in this case the operation If and the ordinary integration of the function $\varphi(x, t)$ with respect to the variable t lead to one and the same result.

5. In this article we shall prove theorem 2: *the arithmetic means of the partial sums of the Fourier series of any function $\varphi(x)$ that belongs to a homogeneous space R converge to $\varphi(x)$ in the norm of R.*

Before carrying out the proof, we shall obtain an expression for the arithmetic means of the partial sums of the Fourier series.

We put

$$\sigma_n(x) = \frac{s_0(x) + s_1(x) + \cdots + s_{n-1}(x)}{n},$$

where $s_m(x) = \sum_{-m}^{m} a_k e^{ikx}$ is a partial sum of the Fourier series of the function $\varphi(x)$. As we saw above

$$s_m(x) = \frac{1}{2\pi} \int_{-\pi}^{\pi} \varphi(x + t) \frac{\sin(m + \frac{1}{2})\, t}{\sin \frac{t}{2}} \mathrm{d}t,$$

hence

$$\sigma_n(x) = \frac{1}{2\pi\, n} \int_{-\pi}^{\pi} \varphi(x - t) \sum_{m=0}^{n-1} \frac{\sin(m + \frac{1}{2})\, t}{\sin \frac{t}{2}} \mathrm{d}t.$$

The sum under the integral sign is easily calculated if we multiply the numerator and denominator of each term by $\sin t/2$:

$$\sum_{m=0}^{n-1} \frac{\sin(m + \frac{1}{2})\, t \sin \frac{t}{2}}{\sin^2 \frac{t}{2}} = \sum_{m=0}^{n-1} \frac{\cos m\, t - \cos\,(m + 1)\, t}{2 \sin^2 \frac{t}{2}}$$

$$= \frac{1 - \cos n\, t}{2 \sin^2 \frac{t}{2}} = \frac{\sin^2 \frac{n}{2} t}{\sin^2 \frac{t}{2}}.$$

Thus we get:

$$\sigma_n(x) = \frac{1}{2\pi n} \int\limits_{-\pi}^{\pi} \varphi(x+t) \frac{\sin^2 \frac{n}{2} t}{\sin^2 \frac{t}{2}} \mathrm{d}t. \tag{1}$$

The function

$$F_n(t) = \frac{1}{2\pi n} \frac{\sin^2 \frac{n}{2} t}{\sin^2 \frac{t}{2}}$$

is known as Fejer's kernel. In contrast with Dirichlet's kernel Fejer's kernel is non-negative. Further, if $\varphi(x) \equiv 1$, then $s_n(x) = 1$, $\sigma_n(x) \equiv 1$, and (1) implies that

$$\int\limits_{-\pi}^{\pi} F_n(t)\,\mathrm{d}t = 1.$$

We turn to the proof of theorem 2. We show first that it holds for any trigonometric polynomial

$$\varphi(x) = \sum_{-n}^{n} a_m \, e^{imx}.$$

We express the arithmetic mean of the first p partial sums of the Fourier series of $\varphi(x)$ in the form

$$\begin{aligned} \sigma_p(x) &= \frac{s_0(x) + \cdots + s_n(x)}{p} + \frac{s_{n+1}(x) + \cdots + s_p(x)}{p} \\ &= \frac{s_0(x) + \cdots + s_n(x)}{p} + \frac{p-n}{p} \varphi(x), \end{aligned}$$

since for $q > n$ the partial sums $s_q(x)$ coincide with the function $\varphi(x)$ itself. As $p \to \infty$ the first term tends to zero and the second to $\varphi(x)$; hence as $p \to \infty$, we have $\sigma_p(x) \to \varphi(x)$, as required.

We now consider the general case. Equation (1) gives $\sigma_n(x)$ as the result of operating on an element $\varphi \in R$ with a linear integral operator, say A_n, with Fejer's kernel. We apply inequality (5) of art. 4 to the integral on the right-hand side, regarded as the integral of a continuous abstract function of t with values in the space R; we get the bound

$$\|\sigma_n(x)\| = \|A_n\varphi\| \leq \|\varphi\| \int_{-\pi}^{\pi} F_n(t)\,dt = \|\varphi\|.$$

This means that the norm of the operator A_n does not exceed 1 for any n. We now employ the following simple lemma:

LEMMA 6. *Let a sequence of linear operators A_n bounded in the norm by a fixed constant K be given on a normed space R. If the relation $A_n\varphi \to \varphi$ holds for the elements φ that belong to some everywhere dense set $Q \subset R$, then it holds for all elements $\varphi \in R$.*

With the aid of this lemma the proof of theorem 2 is quickly completed, for we have to establish the relation $A_n\varphi \to \varphi$, where the A_n are operators with Fejer's kernel. But we have seen that the norms of these operators are bounded by the number 1 and that the relation $A_n\varphi \to \varphi$ holds for trigonometric polynomials, which are everywhere dense in the homogeneous space R by hypothesis. By lemma 6, $A_n\varphi \to \varphi$ for all $\varphi \in R$, and theorem 2 is proved.

It remains for us to prove lemma 6. Let $\varphi \in R$ be any element and let $\varepsilon > 0$ be given. We find an element $\varphi_\varepsilon \in Q$ such that $\|\varphi - \varphi_\varepsilon\| < \varepsilon$ and a number N such that $\|A_n\varphi_\varepsilon - \varphi\| < \varepsilon$ for all $n > N$. Then for $n > N$ we have

$$\begin{aligned}\|A_n\varphi - \varphi\| &\leq \|A_n\varphi - A_n\varphi_\varepsilon\| + \|A_n\varphi_\varepsilon - \varphi_\varepsilon\| + \|\varphi_\varepsilon - \varphi\| \\ &\leq K\varepsilon + \varepsilon + \varepsilon,\end{aligned}$$

and it follows that $A_n\varphi \to \varphi$ as $n \to \infty$.

In applying theorem 2 to the space $\hat{C}(-\pi, \pi)$ we obtain the following result: *every function $\varphi(x)$ that is continuous on the interval $[-\pi, \pi]$ and satisfies the condition $\varphi(-\pi) = \varphi(\pi)$ is the limit of the uniformly convergent sequence of arithmetic means of partial sums of its Fourier series.* Theorem 2 was first proved in this special form by L. Fejer in 1905.

Applied to the space $L_1(-\pi, \pi)$ theorem 2 leads to an important uniqueness property:

If all the Fourier coefficients of a summable function $\varphi(x)$ vanish, the function $\varphi(x)$ itself vanishes (almost everywhere). For it follows from the condition of the theorem that all the terms in the Fourier series of $\varphi(x)$ vanish; but then all the $s_n(x)$ vanish, likewise all the $\sigma_n(x)$, and consequently in the norm of L_1

$$\varphi(x) = \lim_{n\to\infty} \sigma_n(x) = 0.$$

The same property is alternatively expressed as follows:

If all the coefficients of two integrable functions $\varphi(x)$, $\psi(x)$ coincide in pairs, the functions themselves coincide almost everywhere.

For the proof it is sufficient to form the difference $f(x) = \varphi(x) - \psi(x)$; all its Fourier coefficients vanish by hypothesis and therefore $f(x)$ vanishes almost everywhere.

Problems. 1. A point x_0 is said to be a generalised Dini point for a summable function $\varphi(x)$ if the integral

$$\int_{-\varepsilon}^{\varepsilon} \frac{|\varphi(x_0+t)-c|}{|t|}\,dt$$

converges for some c.

Show that the Fourier series of the function φ converges to the value c at a generalised Dini point.

2. A point x_0 is said to be a (regular) Dini point for a summable function $\varphi(x)$ if the integral

$$\int_{-\varepsilon}^{\varepsilon} \frac{|\varphi(x_0+t)-\varphi(x_0)|}{|t|}\,dt$$

converges.

Show that almost all generalised Dini points are regular Dini points.

Hint. Show that every generalised Dini point that is also a Lebesgue point for the function $\varphi(x)$ is a regular Dini point.

3. If $f(t)$ is a continuous abstract function ($a \leq t \leq b$), then

$$\lim_{\lambda\to\infty} \left\| \int f(t)\sin \lambda t\,dt \right\| = 0.$$

4. Show that the terms of the Fourier series of a function $\varphi(x)$ which belongs to a homogeneous space R tend to zero in the norm of R.

Hint. Interpret the Fourier series term

$$a_n e^{inx} = \frac{1}{2\pi}\int_{-\pi}^{\pi} \varphi(t)\,e^{in(x-t)}\,dt = \frac{1}{2\pi}\int_{-\pi}^{\pi} \varphi(x+t)\,e^{int}\,dt$$

as the integral of a continuous abstract function. Apply the result of problem 3.

5. If a sequence $\{y_n\}$ of elements of a normed space R converges in the norm to an element y, the sequence of arithmetic means $s_n = \dfrac{y_1+\cdots+y_n}{n}$ also converges in the norm to y.

6. Prove that at every Lebesgue point of a summable function the arithmetic means of its Fourier series converge to the value of the function. Deduce the uniqueness theorem (cf. above).

Hint. Put $u(t) = |\varphi(x+t) - \varphi(x)|$, $U'(t) = u(t)$; then

$$\int_0^\delta u(t)\frac{\sin^2\frac{n}{2}t}{t^2}\,dt = \int_0^{\frac{1}{n}} + \int_{\frac{1}{n}}^\delta = I_1 + I_2$$

$$I_1 \leqq \left(\frac{n}{2}\right)^2 \int_0^{\frac{1}{n}} u(t)\,dt < \frac{\varepsilon n}{4} \quad \text{(Lebesgue point)},$$

$$I_2 \leqq \int_{\frac{1}{n}}^\delta \frac{u(t)}{t^2}\,dt = \frac{1}{t^2}U(t)\Big|_{\frac{1}{n}}^\delta + 2\int_{\frac{1}{n}}^\delta \frac{1}{t^3}U(t)\,dt < \frac{U(\delta)}{\delta^2}$$

$$+ 2\varepsilon \int_{\frac{1}{n}}^\delta \frac{dt}{t^2} < \frac{\varepsilon}{\delta} + 2\varepsilon n_\varepsilon$$

$$\frac{1}{n}(I_1 + I_2) < \frac{\varepsilon}{4} + \frac{\varepsilon}{n\delta} + 2\varepsilon < 4\varepsilon \text{ for } n > \frac{1}{\delta}.$$

7. Show that condition (3) (p. 367) can be replaced by the result of lemma 3 (p. 368) in the definition of homogeneous function space.

8. Show that every function of bounded variation, continuous relative to displacement in a norm equal to the total variation (see problem 6, p. 298), is absolutely continuous.

9. Prove the theorem: if a subset A of a homogeneous space R of functions $\varphi(x)$, $-\pi \leqq x \leqq \pi$, is uniformly bounded, i.e. $\|\varphi\| \leqq C_1$, and equicontinuous, i.e. given any $\varepsilon < 0$, we can find $\delta < 0$ so that

$$\|\varphi(x+h) - \varphi(x)\| < \varepsilon \quad \text{for} \quad |h| < \delta,$$

then A is compact in R (cf. Arzela's theorem, Chapter II, Section 7, problem 5) (S. B. Stechkin).

Hint. For sufficiently large n the $\sigma_n(x)$ form a compact ε-net in A relative to R (Chapter II, Section 7).

2. The Fourier Transform

1. When we wish to exhibit a periodic function $\varphi(x)$ of period 2π in the form of a superposition of pure harmonic waves, we have recourse to the Fourier series

$$\varphi(x) = \sum_{-\infty}^{\infty} a_n e^{inx}. \tag{1}$$

If we are dealing with a function of period $2\pi l$, the corresponding Fourier series acquires the form

$$\varphi(x) = \sum_{-\infty}^{\infty} a_n e^{in\frac{x}{l}}, \tag{2}$$

where the coefficients a_n are defined by the formula

$$a_n = \frac{1}{2\pi l} \int_{-\pi l}^{\pi l} \varphi(\xi) e^{-in\frac{\xi}{l}} d\xi. \tag{3}$$

Formula (3) is obtained by multiplying (2) by $e^{-in\frac{x}{l}}$ and integrating with respect to x between the limits $-\pi l$ and πl.

It follows from (2) and (3) that

$$\varphi(x) = \sum_{-\infty}^{\infty} \frac{1}{2\pi l} \int_{-\pi l}^{\pi l} \varphi(\xi) e^{\frac{in}{l}(x-\xi)} d\xi. \tag{4}$$

It is natural to try to effect the limiting passage $l \to \infty$ in (4) with the object of representing a quite arbitrary function $\varphi(x)$ defined on the whole axis $-\infty < x < \infty$ as a superposition of harmonic waves. The formal passage to the limit $l \to \infty$ leads to the formula

$$\varphi(x) = \int_{-\infty}^{\infty} d\sigma \left\{ \frac{1}{2\pi} \int_{-\infty}^{\infty} \varphi(\xi) e^{i\sigma(x-\xi)} d\xi \right\}, \tag{5}$$

where the symbol σ denotes the continuous argument derived from the discrete argument $\sigma_n = n/l$. Thus the required formula for the development of $\varphi(x)$ in harmonic waves must be of the form

$$\varphi(x) = \int_{-\infty}^{\infty} \psi(\sigma) e^{i\sigma x} d\sigma, \tag{6}$$

where

$$\psi(\sigma) = \frac{1}{2\pi} \int_{-\infty}^{\infty} \varphi(\xi) e^{-i\sigma\xi} d\xi. \tag{7}$$

The function $\psi(\sigma)$ defined by the formula (7) is said to be the *Fourier transform* (or *Fourier integral*) of the function $\varphi(x)$; formula (6) is said to be the *inversion* formula of the Fourier transform or the *inverse Fourier transform*. The inverse Fourier transform (6) differs from the direct transform (7) only in the sign of the exponent and the coefficient $1/2\pi$. Sometimes the Fourier transform is written in the form

$$\psi(\sigma) = \int_{-\infty}^{\infty} \varphi(\xi)\, e^{i\sigma\xi}\, d\xi; \tag{8}$$

when the inversion formula assumes the form

$$\varphi(x) = \frac{1}{2\pi} \int_{-\infty}^{\infty} \psi(\sigma)\, e^{-i\sigma\xi}\, d\sigma. \tag{9}$$

To confer the maximum symmetry on the direct and inverse transforms, it is a common practice to define the direct transform by the formula

$$\psi(\sigma) = \frac{1}{\sqrt{2\pi}} \int_{-\infty}^{\infty} \varphi(\xi)\, e^{i\sigma\xi}\, d\xi; \tag{10}$$

when the inversion formula assumes the form

$$\varphi(x) = \frac{1}{\sqrt{2\pi}} \int_{-\infty}^{\infty} \psi(\sigma)\, e^{-i\sigma\xi}\, d\sigma. \tag{11}$$

Whatever the notation employed, it is evident that in every case the Fourier transform is a linear transformation: it carries the sum of the functions $\varphi_1(x)$, $\varphi_2(x)$ into the sum of $\psi_1(\sigma)$, $\psi_2(\sigma)$ and the product of the function $\varphi(x)$ by a number λ into the product of $\psi(\sigma)$ by the same number λ.

We shall adhere to the definition (7) and the inversion formula (6).

2. Instead of proving the validity of the limiting passage to formula (5), we shall show immediately that (6) follows from (7) under certain hypotheses concerning the function $\varphi(x)$.

The first hypothesis is, naturally, that *$\varphi(x)$ is integrable over the whole axis* $-\infty < x < \infty$. This ensures the existence of the integral (7) for any value of σ, $-\infty < \sigma < \infty$.

The first consequence of this hypothesis is that *the function $\psi(\sigma)$ is bounded, is continuous for all σ, and tends to* 0 as $|\sigma| \to \infty$. The first assertion derives from the inequality

$$|\psi(\sigma)| \leq \int_{-\infty}^{\infty} |\varphi(\xi)|\, d\xi.$$

It also follows from this inequality that a sequence of functions $\varphi_n(x)$ which converges in the metric of the space $L_1(-\infty, \infty)$ carries over under the Fourier transform to a sequence of functions $\psi_n(\sigma)$ which converges uniformly over the axis $-\infty < \sigma < \infty$.

We verify the second and third assertions first of all for the characteristic function of an interval (c, d). In this case

$$\psi(\sigma) = \int_c^d e^{-i\sigma x} dx = \frac{e^{-i\sigma c} - e^{-i\sigma d}}{i\sigma}$$

and the expression obtained shows that $\psi(\sigma)$ is continuous and tends to zero as $|\sigma| \to \infty$. Since any step function $h(x)$ is a linear combination of characteristic functions of intervals, the second and third assertions hold for all step functions. Finally, any summable function $\varphi(x)$ is a limit (with respect to the metric of $L_1(-\infty, \infty)$) of step functions. By what has been proved, its Fourier transform $\psi(\sigma)$ is a limit (in the sense of uniform convergence over the σ-axis) of continuous functions which vanish at infinity. But then $\psi(\sigma)$ itself is a continuous function and vanishes at infinity, as required.

We now revert to the proof of formula (6). We begin by considering the finite integral

$$\varphi_N(x) = \frac{1}{2\pi} \int_{-N}^{N} \psi(\sigma)\, e^{i\sigma x} d\sigma = \frac{1}{2\pi} \int_{-N}^{N} \left\{ \int_{-\infty}^{\infty} \varphi(\xi)\, e^{i\sigma(x-\xi)}\, d\xi \right\} d\sigma .$$

The interior integral converges uniformly in the parameter σ, hence we can invert the order of integration:

$$\varphi_N(x) = \frac{1}{2\pi} \int_{-\infty}^{\infty} \varphi(\xi) \left\{ \int_{-N}^{N} e^{i\sigma(x-\xi)}\, d\sigma \right\} d\xi$$

$$= \frac{1}{2\pi} \int_{-\infty}^{\infty} \varphi(\xi) \frac{e^{iN(x-\xi)} - {}^{iN(x-\xi)}}{i(x-\xi)} d\xi$$

$$= \frac{1}{\pi} \int_{-\infty}^{\infty} \varphi(\xi) \frac{\sin N(x-\xi)}{x-\xi} d\xi = \frac{1}{\pi} \int_{-\infty}^{\infty} \varphi(x+t) \frac{\sin Nt}{t} dt. \qquad (1)$$

The last transformation is effected by means of the substitution $x - \xi = -t$. We shall show that if $\varphi(x)$ *satisfies Dini's condition*

$$\int_{-\delta}^{\delta} \frac{|\varphi(x+t) - \varphi(x)|}{|t|} \, dt < \infty \text{ for some } \delta > 0.$$

then as $N \to \infty$ *the function* $\varphi_N(x)$ *tends to* $\varphi(x)$. For the proof we recall that†

$$\frac{1}{\pi} \int_{-\infty}^{\infty} \frac{\sin N t}{t} \, dt = 1.$$

Hence the difference $\varphi_N(x) - \varphi(x)$ can be expressed in the form

$$\varphi_N(x) - \varphi(x) = \frac{1}{\pi} \int_{-\infty}^{\infty} [\varphi(x+t) - \varphi(x)] \frac{\sin N t}{t} \, dt.$$

We divide the integral into two parts:

$$\int_{-\infty}^{\infty} = \int_{|t| \leqq T} + \int_{|t| \geqq T} .$$

The second term can be written in the form

$$\int_{|t| \geqq T} \varphi(x+t) \frac{\sin N t}{t} \, dt - \varphi(x) \int_{|t| \geqq T} \frac{\sin N t}{t} \, dt,$$

and it is clear that for a given x and sufficiently large T it becomes arbitrarily small independently of the value of N, $N > 1$ say.

For the first term we have

$$\int_{-T}^{T} \frac{\varphi(x+t) - \varphi(x)}{t} \sin N t \, dt,$$

and since the function $\varphi(x+t) - \varphi(x)/|t|$ is summable over the interval specified (Dini's condition!), it tends to zero as N increases by lemma 2 of Section 1. Hence

$$\lim \varphi_N(x) = \varphi(x),$$

as required.

† Cf. for example A. Ya. Khinchin, *A Course in Mathematical Analysis*, Chapter 26, Section III, Gordon and Breach, New York, 1961.

Thus if the function $\varphi(x)$ is summable and satisfies Dini's condition, it is given in terms of its Fourier transform $\psi(\sigma)$ by formula (6).

We emphasise that in general the integral (6) is not absolutely convergent and cannot be defined by the formula

$$\lim_{N_1, N_2 \to \infty} \int_{N_1}^{N_2},$$

where N_1, N_2 tend independently to infinity.

3. We give a few examples on the calculation of the Fourier transform.

(1) To find the Fourier transform of the function

$$\varphi(x) = \frac{1}{(x-\lambda)^m}, \tag{1}$$

where m is a natural number and λ a non-real constant; let, say, $Im\,\lambda > 0$.

The integral

$$\psi(\sigma) = \int_{-\infty}^{\infty} \frac{e^{-i\sigma x}}{(x-\lambda)^m} dx \tag{2}$$

is absolutely convergent for $m > 1$, but for $m = 1$ converges conditionally in the sense

$$\lim_{N\to\infty} \int_{-N}^{N}.$$

For any $m \geqq 1$ it is conveniently calculated by the method of contour integration. For $\sigma > 0$ we consider a contour in the plane $z = x + i\,y$ formed by the segment $-N \leqq x \leqq N$ of the x-axis and the semi-circle in the lower half-plane on this segment as diameter (Fig. 16). The function $e^{-i\sigma z} = e^{-i\sigma x} e^{\sigma y}$ is bounded in the

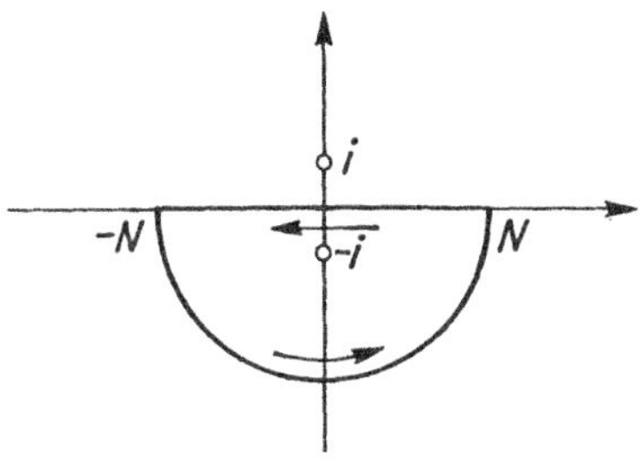

FIG. 16

lower half-plane for $\sigma > 0$, and the integral round the semi-circle tends to zero in virtue of the well-known Jordan's lemma†. Since the singularity of the integrand is located in the upper half-plane for $\sigma > 0$, we get $\psi(\sigma) = 0$.

To use Jordan's lemma when $\sigma < 0$ we have to take the semi-circle in the upper half-plane, and by the residue theorem we get

$$\psi(\sigma) = 2\pi i \operatorname{Res} \left.\frac{e^{-i\sigma z}}{(z-\lambda)^m}\right|_{z=\lambda}.$$

The given residue is easily evaluated if we develop the function $e^{-i\sigma z}$ in a Taylor series in powers of $z - \lambda$:

$$e^{-i\sigma z} = e^{-i\sigma(z-\lambda)}\, e^{-i\sigma\lambda} = e^{-i\sigma\lambda} \sum_{n=0}^{\infty} \frac{[-i\sigma(z-\lambda)]^n}{n!}.$$

The residue is given by the coefficient of $(z-\lambda)^{-1}$; hence

$$\psi(\sigma) = 2\pi i\, e^{-i\sigma\lambda} \frac{(-i\sigma)^{m-1}}{(m-1)!}.$$

Thus for $Im\,\lambda > 0$

$$\psi(\sigma) = \begin{cases} 0 & \text{for} \quad \sigma > 0, \\ 2\pi i\, e^{-i\sigma\lambda} \dfrac{(-i\sigma)^{m-1}}{(m-1)!} & \text{for} \quad \sigma < 0. \end{cases} \tag{3}$$

Similarly for $Im\,\lambda < 0$ we find

$$\psi(\sigma) = \begin{cases} -2\pi i\, e^{-i\sigma\lambda} \dfrac{(-i\sigma)^{m-1}}{(m-1)!} & \text{for} \quad \sigma > 0, \\ 0 & \text{for} \quad \sigma < 0. \end{cases} \tag{4}$$

Any rational function that has no singularities on the real axis and vanishes at infinity can be developed in simple fractions of the form $A/(x-\lambda)^m$, where $Im\,\lambda \neq 0$. Hence the formulae obtained enable us to write down the Fourier transforms of any such rational function. It is easily seen that the functions $\psi(\sigma)$ given by (3) and (4) decrease exponentially as $|\sigma| \to \infty$; hence the Fourier transform of any such rational function decreases exponentially as $|\sigma| \to \infty$.

† V. I. Smirnov, *Course of Higher Mathematics*, 1951, Vol. III, part 2, p. 232 (Pergamon, London, 1964).

(2) To find the Fourier transform of the function

$$\varphi(x) = e^{-ax^2} \quad (a > 0).$$

The expression

$$\varphi(\sigma) = \int_{-\infty}^{\infty} e^{-ax^2} e^{-i\sigma x}\, dx$$

is the integral of the analytic function $e^{-az^2 - i\sigma z}$, $z = x + iy$ over the real axis. Since

$$\left| e^{-a(x+iy)^2 - i\sigma(x+iy)} \right| = e^{-ax^2 + ay^2 + \sigma y},$$

the integrand tends to zero uniformly in y as $x \to \pm\infty$ in any horizontal strip $|y| \leq y_0$. Hence by Cauchy's theorem we can integrate over any parallel line in the z-plane without altering the result:

$$\psi(\sigma) = \int_{-\infty}^{\infty} e^{-a(x+iy)^2} e^{-i\sigma(x+iy)}\, dx$$

$$= \int_{-\infty}^{\infty} e^{-ax^2 + ay^2 + \sigma y - 2aixy - i\sigma x}\, dx = e^{ay^2 + \sigma y} \int_{-\infty}^{\infty} e^{-ax^2 - ix(2ay + \sigma)}\, dx.$$

We put $y = -\sigma/2a$; then we have $ay^2 + \sigma y = -\sigma^2/4a$ and by a well-known formula†

$$\psi(\sigma) = e^{-\frac{\sigma^2}{4a}} \int_{-\infty}^{\infty} e^{-ax^2}\, dx = -e^{-\frac{\sigma^2}{4a}} \sqrt{\frac{\pi}{a}}.$$

In particular, for $\varphi(x) = e^{-x^2/2} (a = 1/2)$ we get $\psi(\sigma) = \sqrt{2\pi}\, e^{-\sigma^2/2}$, a function of the same form, differing from the original function only by the factor $\sqrt{2\pi}$.

Problem. Complete the empty places in the table

№	$\varphi(x)$	$\psi(\sigma)$
1	$\dfrac{1}{x^2 + a^2}$	
2		$\dfrac{1}{\sigma + \sigma_0 + i\tau_0}$
3		$\dfrac{\sin a\sigma}{\sigma}$

№	$\varphi(x)$	$\psi(\sigma)$
4		$\dfrac{\sin^2 a\sigma}{\sigma}$
5	$\dfrac{\sin^2 ax}{x^2}$	

† Cf., for example, V. I. Smirnov, *Course of Higher Mathematics*, Vol. II, Chapter III, Section 8, art. 78.

Answer:

$$\psi_1(\sigma) = \frac{\pi}{\sigma} e^{-a|\sigma|}, \ \varphi_2(x) = e^{x(\tau_0 - i\sigma_0)} \text{ for } x < 0 \text{ and } 0 \text{ for } x > 0;$$

$\varphi_3(x) = \frac{1}{2}$ for $|x| < a$ and 0 for $|x| > a$; $\varphi_4(x) = \frac{i}{4}$ for $0 < x < 2a$, $-\frac{i}{4}$ for $-2a < x < 0$, 0 for $|x| > 2a$; $\psi_5(\sigma) = \pi\left(a - \frac{|\sigma|}{2}\right)$ for $|\sigma| < 2a$ and 0 for $|\sigma| < 2a$.

4. We now consider the question of the arithmetic means of Fourier integrals in the same way as we considered them in Section 1 in relation to Fourier series. In place of the arithmetic mean of n partial sums of the Fourier series, we naturally consider the integral mean

$$\sigma_N(x) = \frac{1}{N} \int_0^N \varphi_\nu(x)\, d\nu. \tag{1}$$

Substituting the value of $\varphi_\nu(x)$ given by formula (1) of art. 2, we find:

$$\begin{aligned}
\sigma_N(x) &= \frac{1}{\pi N} \int_0^N \left\{ \int_{-\infty}^{\infty} \varphi(x+t) \frac{\sin \nu t}{t} dt \right\} d\nu \\
&= \frac{1}{\pi N} \int_{-\infty}^{\infty} \frac{\varphi(x+t)}{t} \left\{ \int_0^N \sin \nu t \, d\nu \right\} dt \\
&= \frac{1}{N\pi} \int_{-\infty}^{\infty} \frac{\varphi(x+t)}{t} \, \frac{1 - \cos Nt}{t} dt \\
&= \frac{2}{\pi N} \int_{-\infty}^{\infty} \varphi(x+t) \frac{\sin^2 \frac{N}{2} t}{t^2} dt.
\end{aligned}$$

The expression

$$F_N(t) = \frac{2}{\pi N} \frac{\sin^2 \frac{N}{2} t}{t^2}$$

is called *Fejer's kernel for the Fourier integral*. It possesses the following properties:

(a) $F_N(t) \geqq 0$;

(b) $\int_{-\infty}^{\infty} F_N(t)\,\mathrm{d}t = 1;$

(c) $\int_{|t| \geqq \delta} F_N(t)\,\mathrm{d}t \to 0 \quad \text{as} \quad N \to \infty \quad \text{for any fixed} \quad \delta > 0.$

Inequality (a) is obvious; equation (b) is deduced from the equation

$$\frac{1}{\pi} \int_{-\infty}^{\infty} \frac{\sin \nu t}{t}\,\mathrm{d}t = 1 \tag{2}$$

by integrating with respect to the parameter ν from ε to N and then letting $\varepsilon \to 0$†.

Relation (c) follows from the inequality

$$\int_{|t| \geqq \delta} F_N(t)\,\mathrm{d}t \leqq \frac{2}{\pi N} \int_{|t| \geqq \delta} \frac{\mathrm{d}t}{t^2} = \frac{4}{\pi N \delta}.$$

Equation (b) implies the relation

$$\sigma_N(x) - \varphi(x) = \int_{-\infty}^{\infty} [\varphi(x+t) - \varphi(x)]\, F_N(t)\,\mathrm{d}t. \tag{3}$$

We shall further consider the convergence of the arithmetic means of the Fourier integral in different normed spaces. Modifying somewhat the definitions of art. 1, which relate to functions on the interval $[-\pi, \pi]$ we shall call a normed space R of functions $\varphi(x)$, $-\infty < x < \infty$ *homogeneous* if it satisfies the following conditions:

(1) all functions $\varphi(x) \in R$ are summable over $(-\infty, \infty)$ and the convergence $\varphi_n \to \varphi$ in R implies the convergence $\varphi_n \to \varphi$ in the norm of $L_1(-\infty, \infty)$;

(2) all translations $\varphi(x+h)$ are contained in R together with $\varphi(x)$ and

$$\|\varphi(x+h)\| = \|\varphi(x)\| \quad \text{for any real } h;$$

(3) the norm in R is continuous under translation, i.e.

$$\lim_{h \to 0} \|\varphi(x+h) - \varphi(x)\| = 0.$$

† The validity of integrating the integrand with respect to the parameter, over the interval $\varepsilon \leqq \nu \leqq N$ is ensured by the uniform convergence in ν of the integral (2) over the region. The convergence is not uniform for $0 < \nu \leqq N$.

We prove the following theorem:

THEOREM: *If a function $\varphi(x)$ belongs to a homogeneous space R, then the arithmetic means $\sigma_N(x)$ of its Fourier integral also belong to R and* $\lim_{N\to\infty} \sigma_N(x) = \varphi(x)$ *in the norm of R.*

The proof of this theorem, just like that of the analogous theorem in Section 1, will be based on the integration of abstract functions with values in the space R. We have also to consider improper integrals of abstract functions; we give the relevant definitions.

Improper integrals of abstract functions. Let us suppose that an abstract function $f(t)$ with values in the normed space R is defined and continuous on the half-line (a, ∞). We define the improper integral

$$\int_a^\infty f(t)\,\mathrm{d}t \tag{4}$$

as the limit of the proper integral

$$\int_a^b f(t)\,\mathrm{d}t$$

as $b \to \infty$, provided the limit exists.

In particular, if the ordinary improper integral

$$\int_a^\infty \|f(t)\|\,\mathrm{d}t, \tag{5}$$

is finite, the improper integral (4) exists. For in this case, for any b', b''

$$\left\| \int_{b'}^{b''} f(t)\,\mathrm{d}t \right\| \leq \int_{b'}^{b''} \|f(t)\|\,\mathrm{d}t \to 0 \quad \text{as} \quad b' \to \infty,\ b'' \to \infty,$$

so that any sequence of proper integrals

$$\int_a^{b_n} f(t)\,\mathrm{d}t \quad (b_n \to \infty)$$

satisfies Cauchy's criterion; the limit (4) exists since R is complete. If the integral (5) exists, the integral (4) is said to be absolutely convergent.

The improper integrals $\int\limits_{-\infty}^{a}$, $\int\limits_{-\infty}^{\infty}$ are similarly defined.

Letting $b \to \infty$ in the inequality

$$\left\| \int\limits_a^b f(t)\,\mathrm{d}t \right\| \leq \int\limits_a^b \|f(t)\|\,\mathrm{d}t$$

we get

$$\left\| \int\limits_a^\infty f(t)\,\mathrm{d}t \right\| \leq \int\limits_a^\infty \|f(t)\|\,\mathrm{d}t. \tag{6}$$

The analogous results hold for absolutely convergent integrals of the other two types.

We now proceed to the proof of the theorem.

The integral (3) can be regarded as an improper integral of the abstract function $f(t) = [\varphi(x+t) - \varphi(x)]\,F_N(t)$, which has a value in R for each value of t. By hypothesis the function $f(t)$ is continuous. The integral (3) converges absolutely in virtue of the inequality

$$\|[\varphi(x+t) - \varphi(x)]\,F_N(t)\| \leqq 2\|\varphi(x)\|\,F_N(t) \in L_1(-\infty, \infty).$$

In particular, we get $\sigma_N(x) - \varphi(x) \in R$; consequently $\sigma_N(x) \in R$.

Further, for a given $\varepsilon > 0$ there exists $\delta > 0$ such that $\|\varphi(x+t) - \varphi(x)\| < \varepsilon/2$ whenever $|t| < \delta$. Using properties (b)–(e) of Fejer's kernel, we find

$$\begin{aligned}\|\sigma_N(x) - \varphi(x)\| \leqq & \int\limits_{|t|\leqq\delta} \|\varphi(x+t) - \varphi(x)\|\,F_N(t)\,\mathrm{d}t + \\ & + \int\limits_{|t|\geqq\delta} \|\varphi(x+t) - \varphi(x)\|\,F_N(t)\,\mathrm{d}t \leqq \\ \leqq \max_{|t|\leqq\delta} \|\varphi(x+t) - \varphi(x)\| & \int\limits_{-\infty}^{\infty} F_N(t)\,\mathrm{d}t + 2\|\varphi(x)\| \int\limits_{|t|\geqq\delta} F_N(t)\,\mathrm{d}t.\end{aligned}$$

The first term does not exceed $\varepsilon/2$ for any N, while the second becomes $< \varepsilon/2$ for sufficiently large $N > N_0$. Hence for $N > N_0$ we get

$$\|\sigma_N(x) - \varphi(x)\| < \varepsilon,$$

which completes the proof.

We shall apply the theorem to the space $R = L_1(-\infty, \infty)$. Of course we have to verify that $L_1(-\infty, \infty)$ is a homogeneous space. Conditions 1 and 2 are satisfied here in an obvious way. To verify condition 3, we remark the following:

(a) every characteristic function of an interval is continuous under displacement in the metric of L_1 (this is easily verified directly);

(b) the set $Q \subset L_1$ of all functions continuous under displacement is a closed subspace in L_1 (the proof is similar to that of lemma 3, Section 1).

If we combine (a) and (b) and recall that linear combinations of characteristic functions (i.e. step functions) are everywhere dense in space L_1, we find that $Q = L_1$. In other words, every function $\varphi(x) \in L_1$ is continuous under displacement, as required.

All in all, we have obtained the theorem:

The arithmetic means of the Fourier integral of any summable function $\varphi(x)$ converge to $\varphi(x)$ on the line $-\infty < x < \infty$ in the metric of $L_1(-\infty, \infty)$.

As a corollary we get the following uniqueness theorem for the Fourier transform.

If the Fourier transform $\psi(\sigma)$ of a summable function $\varphi(x)$ vanishes for all σ, then $\varphi(x) \equiv 0$ (almost everywhere).

For then $\psi(\sigma) \equiv 0$, $\varphi_\nu(x) \equiv 0$, $\sigma_N(x) \equiv 0$, and therefore $\varphi(x) \equiv \lim \sigma_N(x) = 0$.

As a further example we consider the space $CL_1(-\infty, \infty)$ of all uniformly continuous functions $\varphi(x)$ summable over $(-\infty, \infty)$ with the norm

$$\|\varphi(x)\| = \max_{-\infty < x < \infty} |\varphi(x)| + \int_{-\infty}^{\infty} |\varphi(x)|\, dx.$$

We leave it to the reader to verify that conditions 1–3 for a homogeneous space are satisfied. Applying theorem 2, *the arithmetic means of the Fourier integral of any uniformly continuous summable function $\varphi(x)$, $-\infty < x < \infty$, converge to $\varphi(x)$ uniformly over the whole axis in the metric of $CL_1(-\infty, \infty)$.* This is the generalisation of Fejer's theorem to the case of the Fourier integral.

3. The Fourier Transform (continued)

In this paragraph and subsequently we shall denote the Fourier operator by the symbol F:

$$F[\varphi(x)] = \int_{-\infty}^{\infty} \varphi(x)\, e^{-i\sigma x}\, dx.$$

The operator F is, as we know, a linear operator with the inverse

$$F^{-1}[\psi(\sigma)] = \frac{1}{2\pi} \int_{-\infty}^{\infty} \psi(\sigma)\, e^{i\sigma x}\, d\sigma.$$

1. *The Fourier Transform and the Operation of Differentiation*

Let us suppose that an absolutely integrable function $\varphi(x)$ is absolutely continuous in the neighbourhood of any point and that its derivative is also integrable over the line $-\infty < x < \infty$. We shall explain how the Fourier transforms of $\varphi(x)$ and its derivative are related. We observe that in virtue of the hypothesis that $\varphi'(x)$ is integrable the function

$$\varphi(x) = \varphi(0) + \int_0^x \varphi'(\xi)\, d\xi$$

has a limit as $x \to \infty$; this limit can only be zero since otherwise $\varphi(x)$ would not be integrable. The same thing applies to the case $x \to -\infty$. Integrating by parts, we get:

$$F[\varphi'] = \int_{-\infty}^{\infty} \varphi'(x)\, e^{-ix\sigma}\, dx = \varphi(x)\, e^{-ix\sigma} \Big|_{-\infty}^{\infty} + i\sigma \int_{-\infty}^{\infty} \varphi(x)\, e^{-ix\sigma}\, dx.$$

Our observation shows that the first term vanishes and we have

$$F[\varphi'] = i\sigma F[\varphi].$$

In other words differentiation of the function $\varphi(x)$ corresponds to multiplication of the function $\psi(\sigma) = F[\varphi]$ by $i\sigma$. If the derivatives of $\varphi(x)$ up to order m are integrable, repetition of the process yields the results

$$F[\varphi^{(k)}(x)] = (i\sigma)^k F[\varphi] \quad (k = 0, 1, 2, \ldots, m).$$

Since $F[\varphi^k(x)]$ is the Fourier transform of an integrable function, and hence is bounded as a function of τ (and even tends to zero as $|\sigma| \to \infty$), we have the bound for $F[\varphi(x)]$:

$$F[\varphi] = \frac{|F[\varphi^{(k)}(x)]|}{|\sigma|^k} \leqq \frac{C}{|\sigma|^k}.$$

Thus, *the more integrable derivatives $\varphi(x)$ has, the faster its Fourier transform tends to zero at infinity.*

In particular, if $\varphi(x)$ is reasonably smooth, its Fourier transform $\psi(\sigma)$ will also be an absolutely integrable function. It is evident from inequality (1) that the existence of φ, φ' and φ'' (in L_1) is sufficient for this. We can confine ourselves to requiring the existence of φ, φ', with the additional condition that they belong to L_2, and not merely to L_1. In fact, as we shall see in Section 6, $\varphi \in L_2$, $\varphi' \in L_2$ implies $\sigma\,\psi(\sigma) \in L_2$, so that

$$|\psi(\sigma)| = \left|\sigma\,\psi(\sigma)\cdot\frac{1}{\sigma}\right| \leqq \frac{1}{2}\left\{|\sigma\,\psi(\sigma)|^2 + \frac{1}{\sigma^2}\right\}$$

is an integrable function, as required.

For any differential operator $P(\mathrm{d}/\mathrm{d}x)$ of order $\leqq m$

$$F\left[P\left(\frac{\mathrm{d}}{\mathrm{d}x}\right)\varphi\right] = P(i\,\sigma)\,F[\varphi].$$

A linear differential equation in $\varphi(x)$ over the x-axis carries into an algebraic equation in $\psi(\sigma)$ over the σ-axis. This suggests new possibilities for solving differential equations. But since the application of this method requires that the equations be linear and have constant coefficients, it is inadequate in general for the solution of ordinary differential equations (especially since we are restricted to the class of functions integrable over the whole line). For partial differential equations, however, it is found to be useful. We shall give an illustration in art. 3 in the example on thermal conductivity.

2. *The Fourier Transform and the Resultant*

Let $\varphi_1(\sigma)$, $\psi(\sigma)$ be the Fourier transforms of absolutely integrable functions $\varphi_1(x)$, $\varphi_2(x)$; we should like to know what function has the product $\psi_1(\sigma)$, $\psi_2(\sigma)$ as its Fourier transform. We have

$$\psi_1(\sigma)\,\psi_2(\sigma) = \int_{-\infty}^{\infty} \varphi_1(\xi)\,\mathrm{e}^{-i\sigma\xi}\,\mathrm{d}\xi \int_{-\infty}^{\infty} \varphi_2(\eta)\,\mathrm{e}^{-i\sigma\eta}\,\mathrm{d}\eta$$

$$= \int_{-\infty}^{\infty}\int_{-\infty}^{\infty} \varphi_1(\xi)\,\varphi_2(\eta)\,\mathrm{e}^{-i\sigma(\xi+\eta)}\,\mathrm{d}\xi\,\mathrm{d}\eta,$$

with the double integral absolutely convergent (cf. the note to Fubini's theorem, Chapter IV, Section 5, art. 2). To obtain a single exponent we make the substitution $\eta = x - \xi$; then we get

$$\psi_1(\sigma)\,\psi_2(\sigma) = \int_{-\infty}^{\infty} \varphi_1(\xi) \left\{ \int_{-\infty}^{\infty} \varphi_2(x-\xi)\, e^{-i\sigma x}\, dx \right\} d\xi$$

$$= \int_{-\infty}^{\infty} e^{-i\sigma x} \left\{ \int_{-\infty}^{\infty} \varphi_1(\xi)\,\varphi_2(x-\xi)\, d\xi \right\} dx, \qquad (1)$$

the inversion of the order of integration being valid in virtue of Fubini's theorem. The integral

$$\varphi(x) = \int_{-\infty}^{\infty} \varphi_1(\xi)\,\varphi_2(x-\xi)\, d\xi,$$

which exists (again by one of the assertions of Fubini's theorem) for almost all x and is absolutely integrable with respect to x (by another of its assertions), is said to be the *convolute* or the *resultant* of the functions $\varphi_1(x)$, $\varphi_2(x)$. Formula (1) shows that *the product of the functions $\psi_1(\sigma)$, $\psi_2(\sigma)$ is the Fourier transform of the resultant of $\varphi_1(x)$, $\varphi_2(x)$.*

The convolute of $\varphi_1(x)$, $\varphi_2(x)$ is denoted by $\varphi_1 * \varphi_2$. This is a commutative and associative operation since it is carried under Fourier transformation into the commutative and associative operation $\psi_1\psi_2$.

Problems. 1. Let $e_a(x)$ be the characteristic function of the interval $0 < x < a$. Find the resultant

$$e_b(x) * \frac{e_{a+h}(x) - e_a(x)}{h}.$$

Answer. Cf. Fig. 17

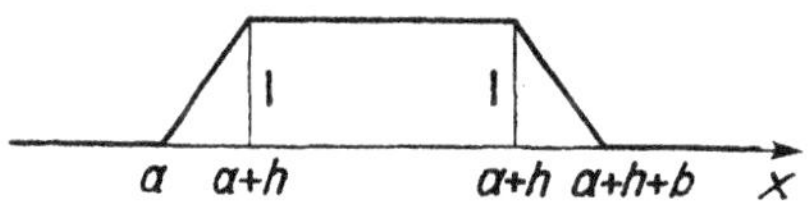

FIG. 17

2. Prove that for any $\varphi(x) \in L_1(-\infty, \infty)$

$$\lim_{h\to 0} \varphi(x) * \frac{e_{a+h}(x) - e_a(x)}{h} = \varphi(x-a)$$

(the limit being in the sense of the metric of $L_1(-\infty, \infty)$).

Hint. Prove (2) for the functions $\varphi(x) = e_b(x)$ and their linear combinations. In the passage to the limit use the boundedness of the norm of the second factor.

3. If A is a closed subspace in $L_1(-\infty, \infty)$ which contains together with any function $\varphi(x)$ all its displacements $\varphi(x-h)$, then for any function $\psi(x) \in L_1(-\infty, \infty)$ it will contain the resultant of $\psi(x)$ with $\varphi(x)$.

Hint. The resultant is a limit of linear combinations of displacements.

3. *Application of the Fourier Transform to the Solution of the Thermal Conductivity Equation*

We shall find the solution of the thermal conductivity equation

$$\frac{\partial u(x,t)}{\partial t} = \frac{\partial^2 u(x,t)}{\partial x^2}, \tag{1}$$

$(-\infty < x < \infty, t \geqq 0)$ which coincides with a given function $u_0(x)$ at $t = 0$. The physical significance of the problem specified consists in determining the temperature of a one-dimensional homogeneous continuum (an infinite rod) at any instant $t > 0$ from its known temperature at time $t = 0$. We shall make the following assumptions:

(a) the functions $u(x, t)$, $u_x(x, t)$, $u_{xx}(x, t)$ are integrable with respect to x for $-\infty < x < \infty$ and any fixed $t \geqq 0$;

(b) in every interval $0 \leqq t \leqq T$ the function $u_t = (x, t)$ has an integrable majorant:

$$|u_t(x,t)| \leqq \Phi(x), \quad \int_{-\infty}^{\infty} \Phi(x)\,\mathrm{d}x < \infty.$$

We shall pass from equation (1) into Fourier transforms by multiplying by $\mathrm{e}^{-i\sigma x}$ and integrating with respect to x from 0 to ∞. Condition (b) allows us to write

$$\int_{-\infty}^{\infty} u_t(x,t)\,\mathrm{e}^{-i\sigma x}\,\mathrm{d}x = \frac{\partial}{\partial t}\int_{-\infty}^{\infty} u(x,t)\,\mathrm{e}^{-i\sigma x}\,\mathrm{d}x = v_t(\sigma, t),$$

where

$$v(\sigma, t) = \int_{-\infty}^{\infty} u(x,t)\,\mathrm{e}^{-i\sigma x}\,\mathrm{d}x$$

is the Fourier transform of the required solution $u(x, t)$.

By condition (a) and the results of art. 1

$$F[u_{xx}(x,t)] = -\sigma^2 F[u] = -\sigma^2 v(\sigma, t).$$

Consequently we obtain the ordinary differential equation

$$v_t(\sigma, t) = -\sigma^2 v(\sigma, t).$$

We have to find a solution of this equation which reduces at $t = 0$ to

$$v_0(\sigma) = F[u_0(x)] = \int_{-\infty}^{\infty} u_0(x)\, e^{-i\sigma x}\, dx.$$

The required solution is evidently of the form

$$v(\sigma, t) = e^{-\sigma^2 t}\, v_0(\sigma).$$

We know (cf. Section 2, ex. 2 with $a = 1/4t$) that

$$e^{-\sigma^2 t} = F\left[\frac{1}{2\sqrt{\pi t}}\, e^{-\frac{x^2}{4t}}\right].$$

By the resultant formula (art. 2)

$$v(\sigma, t) = F\left[\frac{1}{2\sqrt{\pi t}}\, e^{-\frac{x^2}{4t}}\right] F[u_0] = F\left[\frac{1}{2\sqrt{\pi t}}\, e^{-\frac{x^2}{4t}} * u_0(x)\right],$$

and since $v(\sigma, t) = F[u(x, t)]$, we have finally

$$u(x, t) = \frac{1}{2\sqrt{\pi t}}\, e^{-\frac{x^2}{4t}} * u_0(x) = \frac{1}{2\sqrt{\pi t}} \int_{-\infty}^{\infty} e^{-\frac{\xi^2}{4t}}\, u_0(x - \xi)\, d\xi.$$

This form of the solution is known as *Poisson's integral.*

4. *The Relation Between the Decrease of a Function $\varphi(x)$ as $|x| \to \infty$ and the Smoothness of its Fourier Transform*

We know that the Fourier transform $\psi(\sigma)$ of an absolutely integrable function $\varphi(x)$ is a bounded continuous function of σ, $-\infty < \sigma < \infty$, which tends to zero as $|\sigma| \to \infty$. Let us suppose now that not only $\varphi(x)$ but also $x\,\varphi(x)$ is an integrable function over the line $-\infty < x < \infty$. Then we can assert that the function $\psi(\sigma)$ is differentiable. For formal differentiation of the Fourier integral

$$\int_{-\infty}^{\infty} \varphi(x)\, e^{-i\sigma x}\, dx = \psi(\sigma)$$

with respect to the parameter σ yields the integral

$$-i \int_{-\infty}^{\infty} x\,\varphi(x)\, e^{-i\sigma x}\, dx,$$

which converges absolutely and uniformly in σ. By a well-known theorem on the differentiation of a uniformly convergent integral the function $\psi(\sigma)$ is differentiable and

$$\psi'(\sigma) = -i\int_{-\infty}^{\infty} x\,\varphi(x)\,e^{-i\sigma x}\,dx.$$

We obtain the significant formula

$$i\,F'[\varphi] = F[x\,\varphi], \tag{1}$$

which shows that the operation of multiplication by x carries under Fourier transformation into the operation $i\,d/d\sigma$. As the Fourier transform of an integrable function, the function $\psi'(\sigma)$ is again continuous and bounded and tends to zero as $|\sigma| \to \infty$. If the functions $x\,\varphi(x)$, $x^2\,\varphi(x)$, ..., $x^m\,\varphi(x)$ are all integrable together with $\varphi(x)$ over the x-axis, the process of differentiation can be continued; we get that the function $\psi(\sigma) = F[\varphi]$ has derivatives up to order m which are continuous and bounded and tend to zero as $|\sigma| \to \infty$, and

$$i^k\,F^{(k)}[\varphi] = F[x^k\,\varphi] \quad (k = 0, 1, \ldots, m). \tag{2}$$

For an arbitrary polynomial $P(x)$ of degree $\leq m$ we obtain the formula

$$P\left(i\,\frac{d}{d\sigma}\right)F[\varphi] = F[P(x)\,\varphi]. \tag{3}$$

If all the products $x^m\,\varphi(x)$ $(m = 0, 1, 2, \ldots)$ are integrable, the function $F[\varphi] = \psi(\sigma)$ has derivatives with respect to σ of *all* orders, each one being continuous and tending to zero as $|\sigma| \to \infty$. We see that the stronger the conditions we impose on the decrease of $\varphi(x)$, the smoother is the function $\psi(\sigma)$ we obtain.

In this context we can specify an important class of functions which maps into itself under Fourier transformation, of course with the argument x replaced by σ. Let us consider the set S_x of infinitely differentiable functions $\varphi(x)$ which for all k, $q = 0$, $1, 2, \ldots$ satisfy an inequality

$$|x^k\,\varphi^{(q)}(x)| \leq C_{kq}, \tag{4}$$

where C_{kq} is a constant depending on the choice of the function φ. We denote by S_σ the class of the same functions $\psi(\sigma)$ of the argument σ.

We observe first that each of the functions $x^k \varphi^{(q)}(x)$ is not only bounded but also integrable over the line, since in addition to (4) we have the inequality

$$|x^{k+2} \varphi^{(q)}(x)| \leqq C_{k+2,q}, \quad |x^k \varphi^{(q)}(x)| \leqq \frac{C_{k+2,q}}{x^2}.$$

Let $\psi(\sigma) = F[\varphi]$ be the Fourier transform of a function $\varphi(x) \in S_x$. By what we have proved the function $\psi(\sigma)$ is infinitely differentiable and

$$i^q \psi^{(q)}(\sigma) = F[x^q \varphi(x)].$$

Further, the function $x^q \varphi(x)$ is infinitely differentiable together with $\varphi(x)$ and all its successive derivatives are integrable, since by Leibnitz' formula they can be expressed linearly in terms of the integrable functions $x^j \varphi^{(q-j)}(x)$. Hence the functions

$$(i\sigma)^k \psi^{(q)}(\sigma) = (-i)^q F[(x^q \varphi(x))^{(k)}],$$

being the Fourier transforms of integrable functions, are bounded for all k, q. Thus *if* $\varphi(x) \in S_x$, *then* $\psi(\sigma) \in S_\sigma$. Conversely let a function $\psi(\sigma) \in S_\tau$ be given. We construct the function

$$\varphi(x) = \frac{1}{2\pi} \int_{-\infty}^{\infty} \psi(\sigma)\, e^{i\sigma x}\, d\sigma.$$

The function $2\pi\varphi(-x)$ is evidently the Fourier transform of $\psi(\sigma)$ and is therefore contained in S_x. But then obviously $\varphi(x) \in S_x$. By the inversion formula

$$\psi(\sigma) = \frac{1}{2\pi} \int_{-\infty}^{\infty} 2\pi\varphi(-x)\, e^{i\sigma x}\, dx = \int_{-\infty}^{\infty} \varphi(x)\, e^{-i\sigma x}\, dx,$$

so that $\psi(\sigma)$ is the Fourier transform of $\varphi(x)$. Thus *each function* $\psi \in S_\sigma$ *is the Fourier transform of a function* $\varphi \in S_x$. *The class* S_x *maps under Fourier transformation onto the whole class* S_σ. We can express this fact symbolically by the equation

$$F[S_x] = S_\sigma.$$

Let us see how the smoothness properties of the functions $\psi(\sigma)$ can be improved by imposing further restrictions on the behaviour of the functions $\varphi(x)$ at infinity.

Let the product $\varphi(x)\, e^{b|x|}$ be integrable, where $b > 0$ is a fixed constant. We can assert in this case that the Fourier transform $\psi(\sigma)$ of $\varphi(x)$ is not only infinitely differentiable but analytic. For the Fourier integral

$$\psi(\sigma) = \int_{-\infty}^{\infty} \varphi(x)\, e^{-i\sigma x}\, dx$$

is now defined not only for real σ, but also for certain complex σ; if we put $s = \sigma + i\tau$ (σ, τ real), we get

$$\psi(\sigma + i\tau) = \int_{-\infty}^{\infty} \varphi(x)\, e^{-i\sigma x}\, e^{\tau x}\, dx = \int_{-\infty}^{\infty} \varphi(x)\, e^{-isx}\, dx,$$

and the integral obtained converges for $|\tau| \leqq b$, i.e. over a complete horizontal strip of the s-plane. We have obtained a function of the complex variable s which is analytic at every interior point of this strip; for, differentiating formally with respect to s, we get

$$\int_{-\infty}^{\infty} \varphi(x)\, e^{-isx}(-ix)\, dx.$$

The integral obtained converges uniformly in some neighbourhood of the point s (provided it lies within the specified strip) and is therefore the derivative of the function $\psi(s)$. The function $\psi(s)$ is bounded over the whole strip, since

$$|\psi(s)| \leqq \int_{-\infty}^{\infty} |\varphi(x)|\, e^{|\tau||x|} dx \leqq \int_{-\infty}^{\infty} |\varphi(x)|\, e^{b|x|} dx.$$

It follows, in particular, that to a sequence of functions $\varphi_n(x)$ which converges in the norm $\|\varphi\| = \int_{-\infty}^{\infty} |\varphi(x)|\, e^{b|x|}\, dx$ corresponds a sequence $\psi_n(s)$ which converges uniformly over the whole strip $|\tau| \leqq b$.

We can assert further that as $\sigma \to \pm\infty$ the function $\psi(s) = \psi(\sigma + i\tau)$ tends to zero uniformly in τ, $|\tau| \leqq b$. For this is the case for the Fourier transform of the characteristic function of an interval (α, β):

$$\psi(s) = \int_{\alpha}^{\beta} e^{-isx}\, dx = \frac{e^{-is\alpha} - e^{-is\beta}}{is},$$

since the numerator of the ratio obtained is bounded for $|\tau| \leqq b$. The transition to the general case is effected by means of the usual limiting process with step functions.

We observe that in virtue of the last property the integration in the inversion formula can be taken, not only over the real axis, but over any parallel line that lies in the prescribed strip of the s-plane, so that

$$\varphi(x) = \frac{1}{2\pi} \int_{-\infty}^{\infty} \psi(\sigma)\, e^{i\sigma x}\, d\sigma = \frac{1}{2\pi} \int_{-\infty}^{\infty} \psi(\sigma + i\tau)\, e^{i(\sigma + i\tau)x}\, d\sigma .$$

Let us make the further assumption that the product of $\varphi(x)$ with $e^{b|x|}$ is integrable for *any* b. Then the function $\psi(s)$ is defined and analytic on any strip $|\tau| \leqq b$, i.e. it is an entire analytic function; by what we have proved it is bounded over any strip $|\tau| \leqq b$ (the bound depending on b) and tends uniformly to zero as $0 \to \pm\infty$. In the inversion formula we can integrate over any line parallel to the axis of abscissae.

We can impose a still more severe restriction on the behaviour of $\varphi(x)$ at infinity by requiring that its product with the function $e^{b|x|^p}$, $p > 1$, be integrable. It can be shown (though we shall not dwell here on the proof) that in this case the entire function $\psi(s)$ will satisfy a bound of the form

$$|\psi(\sigma + i\tau)| \leqq C\, e^{a|\tau|^{p'}}, \quad \text{where} \quad \frac{1}{p} + \frac{1}{p'} = 1 .$$

The function $\psi(s)$ is then said to have an *exponential order of growth* $\leqq p$ in the s-plane.

The numbers p, p' both exceed 1; but they vary in opposite directions and when p increases without bound, p' approaches 1.

Let us suppose finally that the product of $\varphi(x)$ with any increasing function of $|x|$ is integrable. This property is possessed by finite functions (which vanish almost everywhere outside some interval $|x| \leqq a$) and, as is easily seen, only by such functions. So let us put $\varphi(x)$ equal to zero for $|x| \geqq a$. Then the Fourier transform

$$\psi(s) = \int_{-a}^{a} \varphi(x)\, e^{-isx}\, dx$$

is an entire analytic function of s; it admits the following bound in the s-plane:

$$|\psi(\sigma + i\tau)| \leqq \int_{-a}^{a} |\varphi(x)| \, e^{|\tau||x|} dx \leqq C \, e^{a|\tau|},$$

where $C = \int_{-a}^{a} |\varphi(x)| \, dx$; the function $\psi(x)$ is said to be an entire function of at most *first order growth of type* $\leqq a$. Thus the more rapidly the function $\varphi(x)$ decreases at infinity, the "smoother" its Fourier transform $\psi(\sigma)$ becomes. Beginning with continuous functions $\psi(\sigma)$, we have progressed through finitely differentiable, infinitely differentiable, analytic-in-a-strip, and analytic-in-the-plane function before arriving at first-order analytic functions of finite type. This is the limit of smoothness for functions tending to zero on both sides of the real axis (we know that the Fourier transforms of integrable functions always have this latter property); it is known that there do not exist entire analytic functions distinct from zero which are bounded over the real line and have a slower rate of growth in the plane than $e^{a|\tau|}$ for all $a > 0$†.

5. We give one of the simplest applications of the theorems proved. Let $\varphi_0(x) \in L_2(a, b)$ $(-\infty \leqq a, b \leqq \infty)$ be a function which is non-zero almost everywhere and satisfies the inequality

$$|\varphi_0(x)| \leqq C \, e^{-a|x|}$$

for some positive a. We shall show that *the system of functions* $\varphi_n(x) = x^n \varphi_0(x)$ $(n = 0, 1, 2, \ldots)$ *is complete in the space* L_2 (a, b) *in the sense that the linear combinations of these functions form a set which is everywhere dense in that space.*

For if this were not the case, there would exist, by the orthogonal complement theorem (Chapter V, Section 2, art. 8), a function $f(x) \in L_2(a, b)$, not identically zero, orthogonal, to all the functions $\varphi_n(x)$, so that for any $n = 0, 1, 2, \ldots$

$$\int_{a}^{b} f(x) \, x^n \varphi_0(x) \, dx = 0. \tag{5}$$

The product of the functions $f(x)\,\varphi_0(x)$ with any function $e^{\beta|x|}$, $\beta < \alpha$ is integrable; hence continuing $f(x)\,\varphi_0(x)$ if necessary as

† Cf. for example, A. I. Markushevich, *Theory of Analytic functions*, State Tech. Pub. Dept. 1950, Chapter 6, Section 3.

identically zero over the remaining part of the line $(-\infty, \infty)$, we get that its Fourier transform $g(s)$ is analytic in the strip $|\tau| < \alpha$.

Since in accordance with formula (2)

$$i^n g^{(n)}(\sigma) = \int_a^b f(x)\, x^n \varphi_0(x)\, e^{-i\sigma x}\, dx,$$

it follows from (5) that $g^{(n)}(0) = 0$ for all $n = 0, 1, 2, \ldots$ Hence $g(\sigma) \equiv 0$ everywhere in virtue of the analyticity of $g(s)$. By the uniqueness theorem proved at the end of Section 2, we have also $f(x)\varphi_0(x) \equiv 0$ (almost everywhere) and consequently $f(x) \equiv 0$ (almost everywhere), in contradiction to the construction. Thus the system $x^n \varphi_0(x)$ is complete in $L_2(-a, b)$ as required.

For example, taking $a = -\infty$, $b = \infty$, $\varphi_0(x) = e^{-x^2}$, we verify the completeness of the system of Hermite functions $x^n e^{-x^2}$ (Chapter V, Section 2), and with $a = 0$, $b = \infty$, $\varphi_0(x) = e^{-x}$, the completeness of the Laguerre functions (Chapter V, Section 2).

4. The Laplace Transform

1. Let there be given a function $\varphi(x)$ (possibly non-integrable itself) whose product with $e^{-\gamma x}$ is integrable for some real γ. Then the Fourier transform of $\varphi(x)$, which may not exist in our original sense, is found to exist for certain complex s:

$$\psi(s) = \int_{-\infty}^{\infty} \varphi(x)\, e^{-isx}\, dx = \int_{-\infty}^{\infty} \varphi(x)\, e^{-ix\sigma} e^{x\tau}\, dx,$$

in particular, on the line $\tau = \gamma$. We see that on this line $\psi(s)$ is the Fourier transform of the integrable function $\varphi(x)\, e^{x\tau}$.

The most important instance of this kind occurs under the conditions

$$\left.\begin{aligned} |\varphi(x)| &< C\, e^{ax} \quad \text{for} \quad x > 0, \\ \varphi(x) &= 0 \qquad\;\; \text{for} \quad x < 0. \end{aligned}\right\} \qquad (1)$$

Hence the Fourier transform

$$\psi(s) = \int_0^{\infty} \varphi(x)\, e^{x\tau} e^{-ix\sigma}\, dx = \int_0^{\infty} \varphi(x)\, e^{-ixs}\, dx \qquad (2)$$

exists for all $\tau < -\alpha$, i. e. in the halfplane of the s-plane bounded above by the line $\tau = -\alpha$. As we know already, we can carry out the integration in the inversion formula, over any horizontal line $\tau = -\gamma$ that lies below the line $\tau = -\alpha$:

$$\varphi(x) = \frac{1}{2\pi} \int\limits_{-i\gamma-\infty}^{-i\gamma+\infty} \psi(s)\, e^{isx}\, ds. \tag{3}$$

We introduce a change of variable $is = p$ in formulae (2) and (3). When s varies over the half-plane $Im\, s < -\alpha$, p varies over the half-plane $\text{Re}\, p > \alpha$. The function

$$\Phi(p) \equiv \psi(s) = \int\limits_0^\infty \varphi(x)\, e^{-px}\, dx$$

is defined and analytic in the half-plane $\text{Re}\, p > \alpha$; on each vertical line of this half-plane it tends to zero as $\text{Im}\, p \to \pm\infty$, uniformly over any finite interval of variation of $\text{Re}\, p$. The inversion formula (3) acquires the form

$$\varphi(x) = \frac{1}{2\pi} \int\limits_{\gamma-i\infty}^{\gamma+i\infty} \Phi(p)\, e^{px} \frac{dp}{i} = \frac{1}{2\pi i} \int\limits_{\gamma-i\infty}^{\gamma+i\infty} \Phi(p)\, e^{px}\, dp (\gamma > \alpha).$$

The function $\Phi(p)$ is said to be the Laplace transform of the function $\varphi(x)$ [which is subject to the condition (1)]. As we have seen, the Laplace transform is distinguished from the Fourier transform (considered in the complex plane) only by a rotation of 90° in the domain of the complex argument.

The following simple theorem gives general sufficient (but by no means necessary) conditions for a given function $\Phi(p)$ to be the Laplace transform of some function $\varphi(x)$ which satisfies conditions (1).

THEOREM 1. *If a function $\Phi(p)$ satisfies the conditions:*

(a) *$\Phi(p)$ is analytic in the half-plane* $\text{Re}\, p > \gamma_0$

(b) *$\Phi(p)$ admits the bound* $(p = p_1 + i\, p_2)$

$$|\Phi(p_1 + i\, p_2)| \le B(p_2), \quad \text{where} \quad \int\limits_{-\infty}^{\infty} B(p_2)\, dp_2 = B < \infty,$$

then it is the Laplace transform of a function $\varphi(x)$ which vanishes for $x < 0$ and satisfies an inequality of the form

$$|\varphi(x)| < C\, e^{\gamma_0 x}.$$

for $x > 0$.

Proof. We define a function $\varphi(x)$ by means of the formula

$$\varphi(x) = \frac{1}{2\pi i} \int_{\gamma - i\infty}^{\gamma + i\infty} \Phi(p)\, e^{pt}\, dp \quad (\gamma > \gamma_0). \tag{4}$$

Arguing as usual with the aid of Cauchy's formula and using properties (a), (b) it is easily verified that the integral (4) is independent of γ. At the same time we have the bound

$$|\varphi(x)| \leqq \frac{1}{2\pi} \int_{-\infty}^{\infty} |\Phi(\gamma + i\, p_2)|\, e^{\gamma x}\, dp_2 \leqq \frac{B}{2\pi} e^{\gamma x}.$$

For $x > 0$, letting γ tend to γ_0, we obtain the bound

$$|\varphi(x)| \leqq C\, e^{\gamma_0 x};$$

for $x < 0$, letting γ tend to $+\infty$, we get $\varphi(x) \equiv 0$.

If we express formula (4) in the form

$$\varphi(x) = \frac{1}{2\pi i} \int_{-\infty}^{\infty} \Phi(p_1 + i\, p_2)\, e^{(p_1 + i p_2)x}\, i\, dp_2$$

$$= \frac{e^{p_1 x}}{2\pi} \int_{-\infty}^{\infty} \Phi(p_1 + i\, p_2)\, e^{i p_2 x}\, dp_2,$$

we see that $2\pi\, \varphi(-x)\, e^{p_1 (x)}$ is the Fourier transform with respect to the variable p_2 of the absolutely integrable function $\Phi(p_1 + i\, p_2)$ (p_1 fixed). By the inversion formula

$$\Phi(p_1 + i\, p_2) = \frac{1}{2\pi} \int_{-\infty}^{\infty} 2\pi\, \varphi(-x)\, e^{(p_1 + i p_2)x}\, dx = \int_{-\infty}^{\infty} \varphi(x)\, e^{-px}\, dx,$$

so that $\Phi(p_1 + i\, p_2)$ is in fact the Laplace transform of the function $\varphi(x)$.

2. The Laplace transform is often used to solve both ordinary and partial differential equations which describe regulating processes; in such problems the unknown function $f(t)$ vanishes for $t < 0$, and for $t > 0$ must be a solution of some equation which satisfies certain initial conditions at $t = 0$.

We begin by considering the ordinary differential equation

$$a_0\, y^{(n)}(t) + a_1\, y^{(n-1)}(t) + \cdots + a_n\, y(t) = b(t) \tag{1}$$

with prescribed values

$$y(0) = y_0,$$
$$y'(0) = y_1,$$
$$\dots\dots\dots\dots$$
$$y^{(n-1)}(0) = y_{n-1}.$$

We multiply equation (1) by e^{-pt} and integrate with respect to t from 0 to ∞. Denoting the Laplace transform of $y(t)$ by

$$Y(p) = \int_0^\infty y(t)\, e^{-pt}\, dt$$

and integrating by parts, we get

$$\left.\begin{aligned}
\int_0^\infty y'(t)\, e^{-pt}\, dt &= y(t)\, e^{-pt}\Big|_0^\infty + p\int_0^\infty y(t)\, e^{-pt}\, dt \\
&= -y_0 + p\, Y(p), \\
\int_0^\infty y''(t)\, e^{-pt}\, dt &= y'(t)\, e^{-pt}\Big|_0^\infty + p\int_0^\infty y'(t)\, e^{-pt}\, dt \\
&= -y_1 + p(-y_0 + p\, Y(p)) = -y_1 - p\, y_0 + p^2\, Y(p), \\
&\dots\dots\dots\dots\dots\dots\dots\dots\dots\dots \\
\int_0^\infty y^{(n)}(t)\, e^{-pt}\, dt &= y^{(n-1)}(t)\, e^{-pt}\Big|_0^\infty + p\int_0^\infty y^{(n-1)}(t)\, e^{-pt}\, dt \\
&= -y_{n-1} + p(-y_{n-2} - p\, y_{n-3} - \cdots + p^{n-1}\, Y(p)) \\
&= -y_{n-1} - p\, y_{n-2} - \cdots + p^n\, Y(p).
\end{aligned}\right\} \quad (2)$$

Multiplying each of the equations (2) by the corresponding coefficient a_k and adding, we obtain an equation of the form

$$R_0(p) + R(p)\, Y(p) = B(p),$$

where $R_0(p)$ is a polynomial of not higher than the $(n-1)$th degree in p, $R(p)$ is a polynomial of the nth degree in p, and $B(p)$ is the Laplace transform of the function $b(t)$. We thus obtain a purely algebraic equation for the unknown function $Y(p)$. Solving it, we find

$$Y(p) = \frac{B(p) - R_0(p)}{R(p)},$$

and the required solution is given by the inversion formula

$$y(t) = \frac{1}{2\pi i} \int_{\gamma - i\infty}^{\gamma + i\infty} \frac{B(p) - R_0(p)}{R(p)} \mathrm{e}^{pt} \,\mathrm{d}p. \qquad (3)$$

To evaluate the integral (3), we resort as a rule to contour integration and the theory of residues, as we did when evaluating the Fourier integral of rational functions. We observe that the function e^{pt} is bounded in the left half-plane ($\operatorname{Re} p < \gamma$) for $t > 0$ but not in the right; hence any semicircles which form part of the contour must be taken to the left of the line $\operatorname{Re} p = \gamma$ and not to the right. We can take as γ any number that satisfies the requirement that all the singularities of the function $R(p)$ lie to the left of the line $\operatorname{Re} p = \gamma$.

Example. Let us consider the second-order equation

$$a_0 y'' + a_1 y' + a_2 y = b \sin k t, \quad y_0 = 0, \quad y_1 = 0,$$

with complex conjugate (non-real) characteristic roots $\lambda = \alpha + i\beta$, $\bar{\lambda} = \alpha - i\beta$, where $\alpha < 0$.

In electrical engineering an equation of this form describes forced oscillations in a resistance inductance capacity circuit under the action of a constraining e.m.f of frequency k. Under Laplace transformation it gives rise to the equation

$$(a_0 p^2 + a_1 p + a_2)\, Y(p) = \int_0^\infty b \sin k t\, \mathrm{e}^{-pt} \,\mathrm{d}t = \frac{b k}{k^2 + p^2}.$$

Solving this equation, we find

$$Y(p) = \frac{b k}{(a_0 p^2 + a_1 p + a_2)(k^2 + p^2)}.$$

By the inversion formula

$$y(t) = \frac{b k}{2\pi i} \int_{\gamma - i\infty}^{\gamma + i\infty} \frac{\mathrm{e}^{pt} \,\mathrm{d}p}{(a_0 p^2 + a_1 p + a_2)(k^2 + p^2)}.$$

We put

$$f(p) = \frac{\mathrm{e}^{pt}}{(a_0 p^2 + a_1 p + a_2)(k^2 + p^2)}.$$

The function $f(p)$ has four simple poles in the unextended plane at the points $\pm i\,k, \alpha + i\,\beta$. As γ we can take any positive number. To evaluate the integral we append to the line Re $p = \gamma$ an infinite semicircle in the left half-plane; then by the residue theorem

$$y(t) = b\,k\{\operatorname{Res} f(p)|_{p=ik} + \operatorname{Res} f(p)|_{p=-ik} + \operatorname{Res} f(p)|_{p=\alpha+i\beta} + \\ + \operatorname{Res} f(p)|_{p=\alpha-i\beta}$$

The residue at each point is calculated in accordance with the general formula for simple poles

$$\operatorname{Res} \frac{A(p)}{B(p)}\bigg|_{p=p_0} = \frac{A(p_0)}{B'(p_0)}.$$

Hence we get

$$y(t) = b\,k\left[\frac{e^{(\alpha+i\beta)t}}{(\lambda^2 + k^2)\,2i\,\beta\,a_0} - \frac{e^{(\alpha-i\beta)t}}{(\lambda^2 + k^2)\,2i\,\beta\,a_0}\right.$$

$$\left. + \frac{e^{ikt}}{(-a_0\,k^2 + a_1\,i\,k + a_2)\,2i\,k} - \frac{e^{-ikt}}{(-a_0\,k^2 - a_1\,i\,k + a_2)\,2i\,k}\right].$$

The resulting process is the superposition of a periodic oscillation with the frequency of the external force and a damping oscillation with the natural frequency of the system; the rate of damping is defined by α, i.e. the abscissae of the characteristic roots.

For $\alpha = 0$, $\beta = k$ resonance occurs. In this case the original equation is of the form

$$y'' + k^2\,y = b \sin k\,t$$

and the solution becomes

$$y(t) = \frac{b\,k}{2\pi\,i}\int_{\gamma-i\infty}^{\gamma+i\infty} \frac{e^{pt}\,dp}{(p^2 + k^2)^2}.$$

The points $p = \pm i\,k$ are second-order poles of the integrand. Evaluating the residues in accordance with the general rules (for multiple poles), we find

$$y(t) = b\,k\left[e^{ikt}\left(-\frac{t}{4k^2} + \frac{1}{4i\,k^3}\right) + e^{-ikt}\left(-\frac{t}{4k^2} - \frac{1}{4i\,k^3}\right)\right]$$

$$= -\frac{b\,t}{2k}\cos k\,t - \frac{b}{2k^2}\sin k\,t.$$

The amplitude of the resulting oscillation increases without bound.

3. The same methods are applicable to partial differential equations.

If the Laplace transformation with respect to t carried an ordinary equation into an algebraic equation in the unknown function, then in the transform of an equation which contains derivatives not only with respect to t but also with respect to variables x, y, ..., derivatives with respect to t will have vanished while derivatives with respect to x, y, ... remain. For a large number of independent variables, the simplification achieved is of course slight, but in the case of two independent variables t, x, the Laplace transform method can be applied to great effect.

As an example, let us consider the thermal conductivity equation $\partial u/\partial t = \partial^2 u/\partial x^2$ over a finite interval $0 \leqq x \leqq l$ with initial and boundary conditions $u_x(0, t) = 0$, $u(l, t) = u_1$, $(x, 0) = u_0$. In physical terms, these conditions mean that no heat is lost through the end $x = 0$, while a constant temperature u_1 is maintained at $x = l$; at the initial instant the temperature is constant and equal to u_0.

For the solution of the problem we apply the Laplace transform with respect to t, i.e. we pass from the function $u(x, t)$ to the function

$$v(x, p) = \int_0^\infty e^{-pt} u(x, t) \, \mathrm{d}t .$$

For the function $v(x, p)$ we obtain the equation

$$\frac{\mathrm{d}^2 v(x, p)}{\mathrm{d}x^2} - p\, v(x, p) = -u_0$$

with the conditions

$$v_x(0, p) = 0, \quad v(l, p) = \frac{u_1}{p} .$$

This second-order equation has the solution

$$v(x, p) = \frac{u_0}{p} + \frac{u_1 - u_0}{p} \, \frac{\cosh x \sqrt{p}}{\cosh l \sqrt{p}}$$

and hence

$$u(x, t) = u_0 + \frac{u_1 - u_0}{2\pi i} \int_{\gamma - i\infty}^{\gamma + i\infty} e^{pt} \frac{\cosh x \sqrt{p}}{\cosh l \sqrt{p}} \, \frac{\mathrm{d}p}{p} .$$

The integrand is a well-defined function of p with poles at $p_0 = 0$ and $p_n = \pi^2/l^2\,(n - 1/2)^2$ $(n = 1, 2, \ldots)$. We shall show that its integral is equal to the sum of its residues at all these poles. For this we consider the semicircle T_n in the left half-plane with centre the origin and radius $n^2 = \pi^2/l^2$; it passes between two adjacent poles and we shall show that the ratio $\dfrac{\cosh x\sqrt{p}}{\cosh l\sqrt{p}}$ is bounded over its whole length; then by Jordan's lemma, as $n \to \infty$ the integral along T_n tends to zero and the whole integral (1) reduces, as usual, to the sum of the residues.

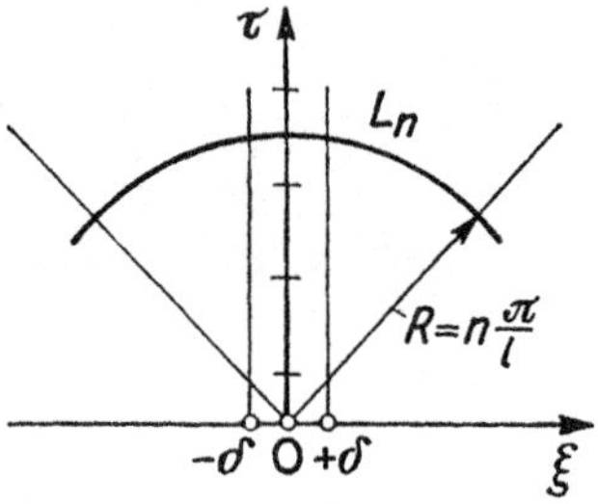

FIG. 18

Instead of considering the ratio $\dfrac{\cosh x\sqrt{p}}{\cosh l\sqrt{p}}$ on the semicircle T_n, where $|p| = n^2\,(\pi^2/l^2)$ we can replace $\sqrt{p}$ by ζ, and consider the ratio $\dfrac{\cosh x\zeta}{\cosh l\zeta}$ on the quarter-circumference of the circle L_n of radius $n\,(\pi/l)$ where the argument varies from $\pi/4$ to $3\pi/4$ (Fig. 18). Putting $\zeta = \xi + i\tau$, we have $\tau > 0$, $|\xi| \leqq \tau$, and

$$\left|\frac{\cosh x\zeta}{\cosh l\zeta}\right|^2 = \left|\frac{\cosh x(\xi + i\tau)}{\cosh l(\xi + i\tau)}\right|^2$$

$$= \left|\frac{\cosh x\xi \cos x\tau + i \sinh x\xi \sin x\tau}{\cosh l\xi \cos l\tau + i \sinh l\xi \sin l\tau}\right|^2$$

$$= \frac{\cosh^2 x\xi \cos^2 x\tau + \sinh^2 x\xi \sin^2 x\tau}{\cosh^2 l\xi \cos^2 l\tau + \sinh^2 l\xi \sin^2 l\tau}$$

$$\leqq \frac{\cosh^2 l\xi}{\cosh^2 l\xi \cos^2 l\tau + \sinh^2 l\xi \sin^2 l\tau}. \tag{1}$$

If $|\xi| \geqq \delta$, then on the circle L_n we have $|\tau - n(\pi/l)| < \varepsilon$ for sufficiently large n, and consequently $\cos^2 l\tau > 1 - \eta$, where ε, η are arbitrarily small; hence

$$\left|\frac{\cosh x\zeta}{\cosh l\zeta}\right|^2 \leq \frac{\cosh^2 l\xi}{(1-\eta)\cosh^2 l\xi} = \frac{1}{1-\eta}. \tag{2}$$

If $|\xi| \geqq \delta$, we substitute $\sinh^2 l\xi$ for $\cosh^2 l\xi$ in the denominator of (1); then we get

$$\left|\frac{\cosh x\zeta}{\cosh l\zeta}\right|^2 \leq \frac{\cosh^2 l\xi}{\sinh^2 l\xi} = \coth^2 l\xi \leqslant \coth^2 l\delta. \tag{3}$$

Inequalities (2) and (3) show that the ratio $\dfrac{\cosh x\sqrt{p}}{\cosh l\sqrt{p}}$ is bounded by a fixed constant on the given circles. Hence the integral reduces, as we said, to the sum of the residues. For the pole $p = 0$ the residue is 1, and at the pole $p_n = -\pi^2/l^2\,(n - 1/2)^2$ the residue is easily computed to be

$$\frac{(-1)^n 4}{\pi(2n-1)} e^{-\frac{\pi^2}{l^2}\left(n-\frac{1}{2}\right)^2 t} \cos\left(n - \frac{1}{2}\right)\frac{\pi x}{l}.$$

Finally we obtain the solution in the form of a series

$$u(x,t) = u_0 +$$

$$+ \frac{4}{\pi}(u_1 - u_0)\sum_{n=1}^{\infty} \frac{(-1)^n}{2n-1} e^{-\frac{\pi^2}{l^2}\left(n-\frac{1}{2}\right)^2 t} \cos\left(n - \frac{1}{2}\right)\frac{\pi x}{l}.$$

Problems. 1. Complete the empty places in the table

№	$y(t)$	$Y(p)$
1	t^n	
2	$\cos at$	
3	$\sinh at$	
4		$\dfrac{p}{p^2 - a^2}$
5		$\dfrac{a}{p^2 + a^2}$

Answer. $Y_1(p) = \dfrac{n!}{p^{n+1}}$, $Y_2(p) = \dfrac{p}{a^2 + p^2}$, $Y_3(p) = \dfrac{a}{p^2 - a^2}$, $y_4(t) = \cosh at$, $y_5(t) = \sin at$.

2. Solve the equation $g^{(IV)} + 4y''' + 4y'' = 0$ under the conditions $y_0 = 0$, $y_1 = 1$, $y_2 = 2$, $y_3 = 3$.

Answer. $4y(t) = -9 + 15t + 9e^{-2t} + 7t\,e^{-2t}$.

3. Solve the system of equations

$$y'' - 3y' + y + z' - z = 0$$
$$-y' + y + z'' - 5z' + 4z = 0$$

subject to the conditions $y_0 = y_1 = z_1 = 0$, $z_0 = 1$.

Answer. $4y = e^t - e^{3t} + 2t\,e^{3t}$, $4z = 5e^t - e^{3t} - 2t\,e^{3t}$.

4. Solve the thermal conductivity equation

$$\frac{\partial u}{\partial t} = \frac{\partial^2 u}{\partial x^2} \quad (x \geqq 0,\ t \geqq 0)$$

subject to the conditions $u(x, 0) = 0$, $u(0, t) = a \cos \omega t$.

Answer:

$$u(x, t) = a e^{-x\sqrt{\frac{\omega}{2}}} \cos\left(\omega t - x\sqrt{\frac{\omega}{2}}\right) - \frac{a}{\pi}\int_0^\infty e^{-\xi t} \sin x\sqrt{\xi}\,\frac{\xi\, d\xi}{\xi^2 + \omega^2}$$

Note. A set of examples on the Laplace transform and its application in problems of mathematical physics can be found in: H. S. Carslaw and J. C. Jaeger: *Operational Methods in Applied Mathematics*, Oxford (1941); I. N. Sneddon, *Fourier Transforms*, McGraw-Hill (1951).

5. Quasi-analytic Classes of Functions

1. The Laplace transform method can also be successfully applied to the solution of theoretical problems. As one such application we give an account here of the fundamental theorem of the theory of quasi-analytic classes†.

It is known that a function $f(x)$ of a real variable x, though infinitely differentiable in a neighbourhood of a point x_0, need nevertheless not be analytic, i.e. capable of development in a Taylor series in a neighbourhood of the point. But if the successive derived functions of $f(x)$ do not increase too rapidly, viz. if

$$\max_{|x - x_0| < \delta} |f^{(n)}(x)| \leqq C M^n n!, \tag{1}$$

then the analyticity of the function $f(x)$ in a neighbourhood of the point x_0 is guaranteed. For the remainder in Taylor's formula

$$R_n(x) = f(x) - f(x_0) - (x - x_0) f'(x_0) - \cdots$$
$$- \frac{(x - x_0)^{n-1}}{(n-1)!} f^{(n-1)}(x_0)$$
$$= \frac{(x - x_0)^n}{n!} f^{(n)}(x_1) \quad (x_0 - \delta < x_1 < x_0 + \delta)$$

† From S. Mandelbrojt: *Series de Fourier et Classes quasi-analytiques de fonctions*, Gauthier-Villars (1935).

in this case admits the bound $|R_n(x)| \leqq cM^n|x - x_0|^n$ and tends to zero for $|x - x_0| < 1/M$; hence in the interval $|x - x_0| < 1/M$ the function $f(x)$ is the sum of its Taylor series. Applying Cauchy's formula for the derivatives of an analytic function, it is easily verified that, conversely, the analyticity of $f(x)$ in a neighbourhood of the point x_0 entails the fulfilment of condition (1). Let $m_0, m_1, \ldots, m_n, \ldots$ be an arbitrary sequence of positive numbers. We form the class $C_{\langle m_n \rangle}$ of functions $f(x)$ which are defined on the line $-\infty < x < \infty$ and satisfy inequalities

$$|f^{(n)}(x)| \leqq C\, M^n\, m_n \quad (n = 0, 1, 2 \ldots),$$

where C, M are constants which may depend on the choice of the function f. If the numbers m_n increase more rapidly than $n!$, the class $C_{\langle m_n \rangle}$ can include non-analytic functions. But, as was shown by A. Denjoy in 1921, if $\sum\limits_{n=1}^{\infty} \left(\frac{1}{\sqrt[n]{m_n}}\right) = \infty$, the class $C_{\langle m_n \rangle}$ possesses the following remarkable property: *if two functions $f(x)$, $g(x)$ belonging to the class $C_{\langle m_n \rangle}$ coincide at some point x_0 together with all their derivatives, they coincide identically for all values of x.* For analytic functions this property has been well-known since Cauchy's time.

The classes $C_{\langle m_n \rangle}$ in which the coincidence of two functions together with all their derivatives implies their coincidence everywhere have been termed *quasi-analytic classes*. In 1926 T. Carleman gave a complete description of quasi-analytic classes; a somewhat simpler formulation was proposed by A. Ostrovsky in 1930. In Ostrovsky's formulation Carleman's theorem reads as follows:

THEOREM 1. *Let us put*

$$T(r) = \sup_{n \geqq 0} \frac{r^n}{m_n}. \tag{2}$$

Then a necessary and sufficient condition for the class $C_{\langle m_n \rangle}$ to be quasi-analytic is that

$$\int_1^{\infty} \frac{\log T(r)}{r^2}\, dr = \infty. \tag{3}$$

For example, let $m_n = n^{n\alpha}$, where α is fixed. Then it is easily calculated that

$$T(r) \sim r^{1-\alpha};$$

the integral (3) converges for $\alpha > 1$ and diverges for $\alpha \leqq 1$. In accordance with Carleman's theorem the class $C_{\langle n^{n\alpha} \rangle}$ is quasi-analytic for $\alpha \leqq 1$ (as we saw above, it may even be composed of analytic functions) and non-quasi-analytic for $\alpha > 1$.

There exist quasi-analytic classes which are not composed solely of analytic functions. It can be shown that the function $f(x) = \sum T^{-1}(n) \cos n x$ is contained in the class $C_{\langle m_n \rangle}$ and is not analytic if $\frac{\sqrt[n]{m_n}}{n} \to \infty$; hence for $m_n = n!\,(\log n)$, say, there exist non-analytic functions in the quasi-analytic class $C_{\langle m_n \rangle}$.

2. In this article we shall use the Laplace transform to reduce the problem of quasi-analytic classes to another problem, which relates to analytic functions in a half-plane.

Let us suppose that the class $C_{\langle m_n \rangle}$ is non-quasi-analytic. This means that it contains functions $f(x)$, $g(x)$ which coincide at $x = x_0$ together with all their derivatives but do not coincide everywhere. Without loss of generality we can suppose that $x_0 = 0$ and $f(x) \not\equiv g(x)$ for $x > 0$; we can always comply with these conditions by means of translation and the substitution of $-x$ for x, i.e. through operations which can be carried out within the class $C_{\langle m_n \rangle}$. We consider further the function $\varphi(x)$ equal to $f(x) - g(x)$ for $x \geqq 0$ and equal to 0 for $x < 0$; evidently it also belongs to the class $C_{\langle m_n \rangle}$. Since it vanishes for $x < 0$ and is bounded for $x > 0$, it possesses a Laplace transform

$$\Phi(p) = \int_0^\infty \varphi(x)\, e^{-px}\, dx, \tag{1}$$

which is analytic in the half-plane $\operatorname{Re} p > 0$.

Let us see what properties the function $\Phi(p)$ possesses. Integrating (1) n times by parts, we get

$$p^n \Phi(p) = \int_0^\infty \varphi^{(n)}(x)\, e^{-px}\, dx,$$

which yields the bound

$$|p^n \Phi(p)| \leqq C M^n m_n \int_0^\infty e^{-px}\, dx = C M^n m_n \frac{1}{|p|} \leqq C_1 M^n m_n$$

for $|p| > \gamma > 0$. Conversely, let there exist in the half-plane $\operatorname{Re} p > \gamma$ an analytic function $\Phi(p) \not\equiv 0$ which satisfies inequalities of the form

$$|p^n \Phi(p)| \leqq CM^n m_n \quad (n = 0, 1, 2, \ldots).$$

It is clear that $\frac{(\Phi(p))}{p} e^{-\gamma x}$ satisfies the conditions of theorem 1 of Section 4; as the integrable majorant required by condition (b) we can take, say, $Cm_0 \frac{1}{|p|^2}$. In virtue of this theorem the function $\varphi(x)$ defined by the equation

$$\varphi(x) = \frac{1}{2\pi i} \int_{\gamma - i\infty}^{\gamma + i\infty} \frac{\Phi(p)}{p^2} e^{(p-\gamma)x} \, dp, \tag{2}$$

vanishes for $x < 0$. Since $\Phi(p) \not\equiv 0$, we have $\varphi(x) \not\equiv 0$ for $x > 0$. Moreover $\varphi(x)$ has derivatives of all orders and

$$|\varphi^{(n)}(x)| = \frac{1}{2\pi} \left| \int_{\gamma - i\infty}^{\gamma + i\infty} \frac{\Phi(p)}{p^2} (p - \gamma)^n e^{(p-\gamma)x} \, dp \right|$$

$$\leqq \frac{CM^n m_n}{2\pi} \int_{\gamma - i\infty}^{\gamma + i\infty} \left| \frac{p - \gamma}{p} \right|^n \frac{|dp|}{|p^2|}$$

$$\leqq \frac{C}{2\pi} M^n m_n \int_{\gamma - i\infty}^{\gamma + i\infty} \frac{|dp|}{|p^2|} = C' M^n m_n.$$

We see that $\varphi(x)$ belongs to the class $C_{\langle m_n \rangle}$. Since $\varphi(x) = 0$ for $x < 0$ and $\varphi(x) \not\equiv 0$, the class $C_{\langle m_n \rangle}$ is not quasi-analytic. Thus *the problem of quasi-analytic classes is equivalent to the problem* ("Watson's problem") *of the existence of a function* $\Phi(p) \not\equiv 0$ *which is analytic in the right half-plane and satisfies inequalities of the form*

$$|p^n \Phi(p)| \leqq CM^n m_n \quad (n = 0, 1, 2, \ldots).$$

3. The inverse transformation $p = 2\gamma/s$ takes the half-plane $\operatorname{Re} p > \gamma$ into the disc $|s - 1| < 1$, and Watson's problem reduces to the following: under what conditions, imposed on the sequence m_n, does there exist in the disc $|s - 1| < 1$ an analytic function $F(s) \not\equiv 0$ which satisfies inequalities of the form

$$|F(s)| \leqq CM^n m_n |s|^n ? \tag{3}$$

Let us suppose that such a function $F(s)$ exists. We can find ϱ such that $F(\varrho) \neq 0$, $|F(\varrho + \varrho\, e^{i\theta})| < 1$ for all real θ, and $s = 0$ is the only zero of $F(s)$ on the circle $s = \varrho + \varrho\, e^{i\theta}$. All the subsequent constructions will extend over the disc $|s - \varrho| \leqq \varrho$. In virtue of the inequalities (3)

$$|F(\varrho + \varrho\, e^{i\theta})| \leqq CM^n\, m_n\, \varrho^n |\, 1 + e^{i\theta}|^n = CM^n\, m_n \left| 2\varrho \cos\frac{\theta}{2} \right|^n .$$

Taking the minimum over n on the right-hand side, we get

$$F|(\varrho + \varrho\, e^{i\theta})| \leqq \frac{C}{\dfrac{1}{\max\limits_n M^n\, m_n \left| 2\varrho \cos\dfrac{\theta}{2} \right|^n}}$$

and by the definition of the function $T(r)$

$$|F(\varrho + \varrho\, e^{i\theta})| \leqq \frac{C}{T\left(\dfrac{1}{2M\,\varrho \left|\cos\dfrac{\theta}{2}\right|}\right)},$$

so that

$$\log|F(\varrho + \varrho e^{i\theta})| \leqq \log C - \log T\left(\frac{1}{2M\,\varrho \left|\cos\dfrac{\theta}{2}\right|}\right).$$

The following theorem is well known in the theory of analytic functions (we shall give a proof of it in art. 5): if the function $f(z)$ is analytic in the disc $|z - z_0| < h$, non-zero at $z = z_0$, at most 1 in absolute value, continuous over the closed disc $|z - z_0| \leqq h$ and has a single zero on the circle $|z - z_0| = h$, then the integral

$$-\int_0^{2\pi} \log|f(z_0 + h\, e^{i\theta})|\, d\theta$$

has a finite value.

Applying this theorem to the case we are considering, we find that the function

$$\log T\left(\frac{1}{2M\,\varrho \left|\cos\dfrac{\theta}{2}\right|}\right) \leqq \log C - \log|F(\varrho + \varrho\, e^{i\theta})|$$

has a finite integral with respect to θ between the limits 0 and 2π. If we make the substitution

$$2M\varrho\cos\frac{\theta}{2}=\frac{1}{r},$$

we see that the integral

$$\int_a^\infty \frac{\log T(r)}{r^2}\frac{1}{\sqrt{M^2\varrho^2-\frac{1}{4r^2}}}\,\mathrm{d}r,$$

converges and with it the integral

$$\int_a^\infty \frac{\log T(r)}{r^2}\,\mathrm{d}r. \tag{4}$$

Thus if the class $C_{\langle m_n\rangle}$ is not quasi-analytic, the integral (4) converges. We have established the sufficiency of Carleman's condition in theorem 1

4. Proceeding to the proof of the necessity of Carleman's condition, let us assume that the integral (4) converges. Then the integral

$$\int_0^{2\pi}\log T\left(\frac{1}{2M\varrho\left|\cos\frac{\theta}{2}\right|}\right)\mathrm{d}\theta$$

also converges and we can therefore construct Poisson's integral

$$G(r\,\mathrm{e}^{i\varphi})=\frac{1}{2\pi}\int_0^{2\pi}\log T\left(\frac{1}{2\left|\cos\frac{\theta}{2}\right|}\right)\frac{1-r^2}{1-2r\cos(\theta-\varphi)+r^2}\,\mathrm{d}\theta,$$

which represents a function harmonic on the disc $r<1$. We put $G(s-1)=P(s)$ and denote by $Q(s)$ the conjugate harmonic function on the disc $|s-1|<1$. Further, let

$$F(s)=\mathrm{e}^{-\{P(s)+i\,Q(s)\}}.$$

We claim that the function $F(s)$ satisfies the inequalities

$$|F(s)|\leqq m_n|s|^n\quad(n=0,1,2,\ldots). \tag{5}$$

For the inequalities (5) are equivalent to the inequalities

$$e^{-P(s)} \leq m_n |s|^n \quad (n = 0, 1, 2, \ldots)$$

or

$$-G(s) = -P(s+1) \leq \log m_n + n \log |s+1|. \tag{6}$$

Both terms on the right can be expressed in the form of Poisson integrals:

$$\log m_n = \frac{1}{2\pi} \int_0^{2\pi} \frac{\log m_n (1 - r^2)}{1 - 2r \cos s(\theta - \varphi) + r^2} d\theta,$$

$$n \log |s+1| = \frac{1}{2\pi} \int_0^{2\pi} \frac{n \log |e^{i\theta} + 1| (1 - r^2)}{1 - 2r \cos(\theta - \varphi) + r^2} d\theta,$$

and hence inequality (6), which is subject to proof, can be expressed in the form

$$\int_0^{2\pi} \log \left\{ T\left(\frac{1}{2\left|\cos \frac{\theta}{2}\right|} \right) m_n |1 + e^{i\theta}|^n \right\} \frac{1 - r^2}{1 - 2r \cos(\theta - \varphi) + r^2} d\theta \geq 0. \tag{7}$$

But $|1 + e^{i\theta}| = 2 |\cos \theta/2|$; since

$$T(r) = \sup_{n \geq 0} \frac{r^n}{m_n},$$

we have for every n

$$T(r) \geq \frac{r^n}{m_n}, \quad T(r) m_n r^n \geq 1,$$

and hence the integrand in (7) is non-negative. It follows that (7) is true, and therefore inequality (5) is satisfied; in accordance with art. 1, the class $C_{\langle m_n \rangle}$ is not quasi-analytic. This completes the proof of Carleman's theorem.

5. In this article we shall prove the theorem from the theory of functions of a complex variable that we used in art. 3.

THEOREM. *If the function $f(z)$ is analytic in the disc $|z - z_0| < h$, non-zero at $z = z_0$ at most 1 in absolute value, continuous over the closed disc $|z - z_0| \leq h$ and has a single zero on the circle $|z - z_0| = h$*

at the point z^*, then the integral

$$-\int_0^{2\pi} \log |f(z_0 + h\, e^{i\theta})|\, d\theta$$

is finite.

Proof. Without loss of generality we can put $z_0 = 0$, $h = 1$, $z^* = 1$. In the disc $|z| \leq r < 1$ the function $f(z)$ is analytic and can have only a finite number of zeros $z_1, \ldots, z_m$; we assume that the boundary $|z| = r$ is free from zeros.

Consider the closed contour C shown in Fig. 19, composed of arcs of the circle $|z| = r$ taken in the positive direction, circles C_k $(k = 1, 2, \ldots, m)$ of very small radius ε taken in the negative

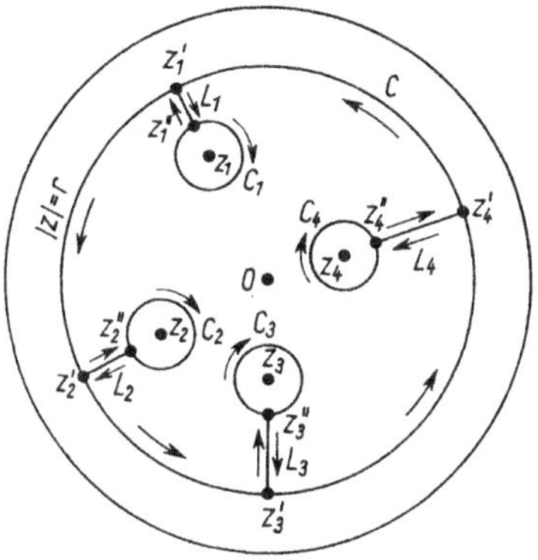

FIG. 19

direction, and lines $L_k = [z_k'', z'_k]$ joining the curves specified and taken twice in opposite directions. The function $\log f(z)$ is analytic inside the contour C and the value of $\log f(0)$ can be expressed by Cauchy's formula

$$\log f(0) = \frac{1}{2\pi i} \int_C \log f(z) \frac{dz}{z}. \tag{1}$$

Let us consider the part of the contour formed by the circle C_j of radius ε with centre z_j taken in the negative direction. The part of the integral (1) that is taken round C_j has the form

$$\frac{1}{2\pi i} \int_{C_j} \log f(z) \frac{i\, \varepsilon\, e^{i\theta}\, d\theta}{z_j + \varepsilon\, e^{i\theta}}. \tag{2}$$

If k_j denotes the multiplicity of the zero z_j, then $f(z) = (z - z_0)\, k_j f_j(z)$ where $f_j(z_j) \neq 0$, and

$$|\log f(z)| = |\log (z - z_j) k_j\, f_j(z)| = |k_j \log (z - z_j) + \log f_j(z)|$$
$$\leq k_j |\log |z - z_j| + 2\pi k_j| + |\log f_j(z)| \leq k_j |\log \varepsilon| + c_j.$$

This bound shows that the integrand in (2) becomes arbitrarily small as $\varepsilon \to 0$ and therefore all the integrals round the circles C_j tend to zero as $\varepsilon \to 0$.

In a negative circuit of the point z_j the function $\log f(z) = \log |f(z)| + i \arg f(z)$ acquires the increment $-2\pi k_j i$, hence the integral over the part of the contour formed by the segment L_j taken twice in opposite directions is equal to

$$k_j \int_{L_j} \frac{dz}{z} = k_j [\log z'_j - \log z''_j].$$

On each successive arc of the circle $|z| = r$ the function $\log f(z)$ acquires an increment $-2\pi k_j i$ relative to the preceding arc which gives a corresponding increment in the whole integral (1) of the form

$$k_j \int_{z_j''}^{z''_{j+1}} i\, d\theta,$$

which is evidently purely imaginary. With all this in view we can detach the real part in equation (1) as $\varepsilon \to 0$; we get

$$\log |f(0)| = \sum_{j=1}^{m} k_j \log |z_j'| + \frac{1}{2\pi} \int_0^{2\pi} \log |f(r\, e^{i\theta})|\, d\theta;$$

and since $|z_j'| < 1$, $\log |z_j'| < 0$,

$$\frac{1}{2\pi} \int_0^{2\pi} \log |f(r\, e^{i\theta})|\, d\theta \geq \log |f(0)|,$$

or, what is the same thing,

$$-\frac{1}{2\pi} \int_0^{2\pi} \log |f(r\, e^{i\theta})|\, d\theta \leq -\log |f(0)|.$$

By hypothesis there is a single zero on the circle $|z| = 1$ at the point $z^* = 1$.

Choosing an arbitrary $\delta > 0$, we evidently have

$$-\frac{1}{2\pi}\int_{\delta}^{2\pi-\delta}\log|f(r\,e^{i\theta})|\,d\theta \leqq -\frac{1}{2\pi}\int_{0}^{2\pi}\log|f(r\,e^{i\theta})|\,d\theta$$

$$\leqq -\log|f(0)|.$$

Keeping δ fixed, we allow r to approach 1; we get

$$-\frac{1}{2\pi}\int_{\delta}^{2\pi-\delta}\log|f(e^{i\theta})|\,d\theta \leqq -\log|f(0)|.$$

This inequality holds for any $\delta > 0$. In the limit as $\delta \to 0$ we get the integral

$$-\frac{1}{2\pi}\int_{0}^{2\pi}\log|f(e^{i\theta})|\,d\theta$$

exists. The proof is complete.

6. The Fourier Transform in the Class $L_2(-\infty, \infty)$

1. A function $\varphi(x)$ whose square is integrable over the whole x-axis is not, in general, so integrable itself (e.g. $1/\sqrt{x^2+1}$) and so, in the ordinary sense, it has no Fourier transform. We shall show, nevertheless, that the following proposition holds (replacing the theorem on the Fourier transform in the class L_1):

Theorem. (Plancherel, 1910). *For any function $\varphi(x) \in L_2(-\infty, \infty)$ the integral*

$$\psi_N(\sigma) = \int_{-N}^{N}\varphi(x)\,e^{-ix\sigma}\,dx \tag{1}$$

represents a function belonging to the space $L_2(-\infty, \infty)$ (over σ). As $N \to \infty$ the function $\psi_N(\sigma)$ has a certain limit $\psi(\sigma)$ in the metric of $L_2(-\infty, \infty)$, with

$$\int_{-\infty}^{\infty}|\psi(\sigma)|^2\,d\sigma = 2\pi\int_{-\infty}^{\infty}|\varphi(x)|^2\,dx. \tag{2}$$

If in addition $\varphi(x)$ belongs to $L_1(-\infty, \infty)$ then $\psi(\sigma)$ is its ordinary Fourier transform. Hence, even in the general case (when

$\varphi(x) \notin L_1(-\infty, \infty)$) $\psi(\sigma)$ is said to be the Fourier transform of $\varphi(x)$.

Proof. Consider functions $\varphi_1(x)$, $\varphi_2(x)$ in the class S_x (Section 3, art. 4) and let $\psi_1(\sigma)$, $\psi_2(\sigma)$ be their Fourier transforms, which belong to the class S_σ. Then

$$\int_{-\infty}^{\infty} \varphi_1(x)\overline{\varphi_2(x)}\,\mathrm{d}x = \int_{-\infty}^{\infty}\left\{\frac{1}{2\pi}\int_{-\infty}^{\infty}\psi_1(\sigma)\mathrm{e}^{i\sigma x}\,\mathrm{d}\sigma\right\}\overline{\varphi_2(x)}\,\mathrm{d}x$$

$$= \frac{1}{2\pi}\int_{-\infty}^{\infty}\psi_1(\sigma)\left\{\int_{-\infty}^{\infty}\mathrm{e}^{i\sigma x}\overline{\varphi_2(x)}\,\mathrm{d}x\right\}\mathrm{d}\sigma$$

$$= \frac{1}{2\pi}\int_{-\infty}^{\infty}\psi_1(\sigma)\overline{\left\{\int_{-\infty}^{\infty}\varphi_2(x)\,\mathrm{e}^{-i\sigma x}\,\mathrm{d}x\right\}}\mathrm{d}\sigma = \frac{1}{2\pi}\int_{-\infty}^{\infty}\psi_1(\sigma)\overline{\psi_2(\sigma)}\,\mathrm{d}\sigma.$$

The inversion of the order of integration at the third stage is valid in virtue of the absolute convergence of the double integral

$$\int_{-\infty}^{\infty}\int_{-\infty}^{\infty}|\psi_1(\sigma)|\,|\varphi_2(x)|\,\mathrm{d}x\,\mathrm{d}\sigma.$$

In particular, for $\varphi_1(x) = \varphi_2(x) = \varphi(x), \psi_1(\sigma) = \psi_2(\sigma) = \psi(\sigma)$, we get:

$$\int_{-\infty}^{\infty}|\psi(\sigma)|^2\,\mathrm{d}\sigma = 2\pi\int_{-\infty}^{\infty}|\varphi(x)|^2\,\mathrm{d}x. \tag{3}$$

Further, let $\varphi(x)$ be a square-summable function which vanishes for $|x| \geqq a$. We can form a sequence of functions $\varphi_n(x) \in S$ which vanish for $|x| \geqq a$ and converge to $\varphi(x)$ in the metric of $L_2(-a, a)$. Since by Cauchy–Bunyakowsky inequality

$$\int_{-a}^{a}|f(x)|\,\mathrm{d}x \leqq \sqrt{\int_{-a}^{a}1\,\mathrm{d}x}\sqrt{\int_{-a}^{a}|f(x)|^2\,\mathrm{d}x}$$

$$= \sqrt{2a}\sqrt{\int_{-a}^{a}|f(x)|^2\,\mathrm{d}x},$$

for any function $f \in L_2(-a, a)$, we have

$$\int_{-a}^{a}|\varphi(x) - \varphi_n(x)|\,\mathrm{d}x \leqq \sqrt{2a}\sqrt{\int_{-a}^{a}|\varphi(x) - \varphi_n(x)|^2\,\mathrm{d}x} \to 0,$$

so that $\varphi_n(x)$ also converges to $\varphi(x)$ in the metric of $L_1(-a, a)$; and since both $\varphi_n(x)$ and $\varphi(x)$ vanish for $|x| \geqq a$, this convergence holds in the metric of $L_1(-\infty, \infty)$. But then the Fourier transforms $\psi_n(\sigma)$ of the functions $\varphi_n(x)$ converge uniformly for $-\infty < \sigma < \infty$ to the Fourier transform $\psi(\sigma)$ of the function $\varphi(x)$. In addition the functions $\psi_n(\sigma)$ form a sequence which is fundamental in the metric of $L_2(-\infty, \infty)$, since by what we have proved

$$\int_{-\infty}^{\infty} |\psi_n(\sigma) - \psi_m(\sigma)|^2 \,\mathrm{d}\sigma = 2\pi \int_{-\infty}^{\infty} |\varphi_n(x) - \varphi_m(x)|^2 \,\mathrm{d}x.$$

It follows that $\psi(\sigma) = \lim \psi_n(\sigma)$ belongs to $L_2(-\infty, \infty)$ and let

$$\int_{-\infty}^{\infty} |\psi(\sigma)|^2 \,\mathrm{d}\sigma = \lim_{n\to\infty} \int_{-\infty}^{\infty} |\psi_n(\sigma)|^2 \,\mathrm{d}\sigma =$$

$$= 2\pi \lim_{n\to\infty} \int_{-\infty}^{\infty} |\varphi_n(x)|^2 \,\mathrm{d}x = 2\pi \int_{-\infty}^{\infty} |\varphi(x)|^2 \,\mathrm{d}x.$$

In effect the limits of the last two integrals $-a, a$.

Finally let $\psi(x)$ be any function in $L_2(-\infty, \infty)$ and let $\varphi_N(x)$ be equal to $\varphi(x)$ for $|x| \leqq N$ and 0 for $|x| \geqq N$. By what we have proved the Fourier transform $\psi_N(\sigma)$ of the function $\varphi_N(x)$ belongs to $L_2(-\infty, \infty)$, and

$$\int_{-\infty}^{\infty} |\psi_N(\sigma)|^2 \,\mathrm{d}\sigma = 2\pi \int_{-\infty}^{\infty} |\varphi_N(x)|^2 \,\mathrm{d}x, \tag{4}$$

while also

$$\int_{-\infty}^{\infty} |\psi_N(\sigma) - \psi_M(\sigma)|^2 \,\mathrm{d}\sigma = 2\pi \int_{-\infty}^{\infty} |\varphi_N(x) - \varphi_M(x)|^2 \,\mathrm{d}x. \tag{5}$$

But the sequence $\varphi_N(x)$ converges in the metric of $L_2(-\infty, \infty)$ to $\varphi(x)$ and is therefore fundamental; it follows from equation (5) that the sequence $\psi_N(\sigma)$ is also fundamental. Putting $\psi(\sigma) = \lim \psi_N(\sigma)$, it follows from (4) that

$$\int_{-\infty}^{\infty} |\psi(\sigma)|^2 \,\mathrm{d}\sigma = 2\pi \lim_{N\to\infty} \int_{-\infty}^{\infty} |\psi_N(\sigma)|^2 \,\mathrm{d}\sigma$$

$$= 2\pi \lim_{N\to\infty} \int_{-\infty}^{\infty} |\varphi_N(\sigma)|^2 \,\mathrm{d}x = 2\pi \int_{-\infty}^{\infty} |\varphi(x)|^2 \,\mathrm{d}x. \tag{6}$$

Finally if $\varphi(x)$ belongs also to $L_1(-\infty, \infty)$, we shall have for the function $\varphi_N(x)$:

$$\int_{-\infty}^{\infty} |\varphi_N(x) - \varphi(x)|\, dx \to 0,$$

whence it follows that $\psi_N(\sigma)$ converges uniformly to the (ordinary) Fourier transform of $\varphi(x)$. But since the $\psi_N(\sigma)$ converge in the mean to $\psi(\sigma)$, this transform is none other than $\psi(\sigma)$ itself.

This completes the proof of Plancherel's theorem.

A somewhat more general relation than (6) is easily verified, namely, if $\varphi_1(x)$, $\varphi_2(x)$ are functions in $L_2(-\infty, \infty)$ and $\psi_1(\sigma)$, $\psi_2(\sigma)$ are their Fourier transforms, then

$$\int_{-\infty}^{\infty} \psi_1(\sigma)\overline{\psi_2(\sigma)}\, d\sigma = 2\pi \int_{-\infty}^{\infty} \varphi_1(x)\overline{\varphi_2(x)}\, dx.$$

For the proof it is sufficient to apply Plancherel's theorem to the function $\varphi_1(x) + \varphi_2(x)$ and compare the results on the right—and left hand sides.

2. The relationships between smoothness and rate of decrease of the Fourier transform of a function, obtained in Section 3 for integrable functions, still hold for square summable functions.

Suppose first that the square summable function $\varphi(x)$ is locally absolutely continuous and its derivative $\varphi'(x)$ is also square summable. Then *the Fourier transform of $\varphi'(x)$ is the derivative of the Fourier transform $\psi(\sigma)$ of $\varphi(x)$ on $i\sigma$.*

For $\varphi(x)$ now has a limit as $x \to \infty$, since

$$|\varphi(x)|^2 = \varphi(x)\overline{\varphi(x)} = \varphi(0)\overline{\varphi}(0) + \int_0^x \varphi(\xi)\overline{\varphi'(\xi)}\, d\xi + \\ + \int_0^x \varphi'(\xi)\overline{\varphi(\xi)}\, d\xi,$$

and $\varphi\bar{\varphi}'$, $\bar{\varphi}\varphi'$ are integrable over an infinite interval; and obviously, the limit can only be zero.

We use this fact to construct a sequence of finite absolutely continuous functions $\varphi_\nu(x)$, so that

$$\varphi_\nu(x) \to \varphi(x),$$
$$\varphi_\nu'(x) \to \varphi'(x).$$

in space $L_2(-\infty, \infty)$. We actually take as $\varphi_\nu(x)$ a continuous function equal to $\varphi(x)$ on the interval $(-\nu, \nu)$, 0 outside the interval $U_\nu = (-\nu - |\varphi(-\nu)|, \nu + |\varphi(\nu)|)$ and linear in the two remaining intervals; since $\varphi(\nu) \to 0$ as $\nu \to \pm\infty$, $\varphi_\nu(x)$ tends to $\varphi(x)$ in the metric of $L_2(-\infty, \infty)$. Further, $\varphi'_\nu(x)$ coincides with $\varphi'(x)$ in the interval $(-\nu, \nu)$, is equal to 0 outside U_ν and is equal to ± 1 in the two remaining intervals; obviously, $\varphi'_\nu(x)$ tends to $\varphi'(x)$ in the metric of $L_2(-\infty, \infty)$.

By Plancherel's theorem,

$$F[\varphi_\nu(x)] = \psi_\nu(\sigma) \to F[\varphi(x)] = \psi(\sigma);$$
$$F[\varphi'_\nu(x)] = i\sigma\psi_\nu(\sigma) \to F(\varphi'(x)] = \psi_1(\sigma),$$

in space $L_2(-\infty, \infty)$, and since the sequence $i\,\sigma\,\psi_\nu(\sigma)$ evidently tends to the function $i\,\sigma\,\psi(\sigma)$, we get

$$\psi_1(\sigma) = F[\varphi'(x)] = i\,\sigma\,\psi(\sigma),$$

as required.

Suppose conversely that $x\,\varphi(x)$ is square summable along with $\varphi(x)$. We show that *the Fourier transform* $\psi(\sigma)$ *of* $\varphi(x)$ *is locally absolutely continuous and* $F[x\,\varphi(x)] = i\,\psi'(\sigma)$. We put $\varphi_\nu(x) = \varphi(x)$ for $|x| \leq \nu$ and $\varphi_\nu(x) = 0$ for $|x| > \nu$; then $\varphi_\nu(x) \to \varphi(x)$ in the metric of $L_2(-\infty, \infty)$. Obviously, $x\,\varphi_\nu(x) \to x\,\varphi(x)$ also in the metric of $L_2(-\infty, \infty)$. Let $F[x\,\varphi(x)] = i\,g(\sigma)$; we get

$$\psi_\nu(\sigma) = F[\varphi_\nu(x)] \to F[\varphi(x)] = \psi(\sigma),$$
$$i\,\psi'_\nu(\sigma) = F[x\,\varphi_\nu(x)] \to F[x\,\varphi(x)] = i\,g(\sigma).$$

By the lemma of art. 6, Section 3, Chapter IV a subsequence $\psi_{\nu_k}(\sigma)$, convergent almost everywhere to $\psi(\sigma)$, can be extracted from the sequence $\psi_\nu(\sigma)$. Let e.g. $\psi(\sigma_0) = \lim \psi_{\nu_k}(\sigma_0)$. We have further

$$\int_{\sigma_0}^{\sigma} [\psi'_{\nu_k}(\xi) - \psi'_{\nu_m}(\xi)]\,d\xi = [\psi_{\nu_k}(\sigma) - \psi_{\nu_m}(\sigma)] - [\psi_{\nu_k}(\sigma_0) - \psi_{\nu_m}(\sigma_0)];$$

on the other hand,

$$\int_{\sigma_0}^{\sigma} |\psi'_{\nu_k}(\xi) - \psi'_{\nu_m}(\xi)|\,d\xi \leq \sqrt{|\sigma - \sigma_0|}\,\sqrt{\int_{\sigma_0}^{\sigma} |\psi'_{\nu_k}(\xi) - \psi'_{\nu_m}(\xi)|^2\,d\xi} \to 0,$$

and it follows that the sequence $\psi_{\nu_k}(\sigma)$ is uniformly convergent in any finite interval. But now $\psi(\sigma) = \lim \psi_{\nu_k}(\sigma)$ is a continuous function. Also,

$$\psi(\sigma) - \psi(\sigma_0) = \lim \int_{\sigma_0}^{\sigma} [\psi_{\nu_k}(\xi)]\, d\xi = \int_{\sigma_0}^{\sigma} g(\xi)\, d\xi,$$

and it follows that $\psi(\sigma)$ is locally absolutely continuous and $\psi'(\sigma) = g(\sigma)$, as we asserted.

3. *A Theorem of Wiener and Paley*

If a square-summable function $\varphi(x)$ vanishes outside the interval $[-b, b]$, its Fourier transform $\psi(\sigma)$, in addition to being also square-summable, can be continued analytically in the plane $s = \sigma + i\tau$. For the expression

$$\psi(s) = \int_{-b}^{b} \varphi(x)\, e^{-isx}\, dx$$

is defined for all complex $s = \sigma + i\tau$. It represents an analytic function of s and satisfies the bound

$$|\psi(s)| \leqq \int_{-b}^{b} |\varphi(x)|\, e^{|\tau x|} dx \leqq C\, e^{b|\tau|} \leqq C\, e^{b|s|}.$$

An entire analytic function $\psi(s)$ which satisfies an inequality of the form

$$|\psi(s)| \leqq C\, e^{b|s|},$$

is said to be *a function of exponential type* $\leqq b$. We see that the Fourier transform of a square-summable function which vanishes for $|x| > b$ is an entire function of exponential type $\leqq b$. We shall devote this article to establishing the converse.

THEOREM (Wiener–Paley). *If an entire function $\psi(s)$ of exponential type $\leqq b$ is square-summable over the real axis, it is the Fourier transform of a square-summable function $\varphi(x)$ which vanishes outside the interval $[-b, b]$.*

Before proving the theorem we shall obtain bounds for the coefficients of the Taylor development of entire functions of exponential type.

It is well known that the coefficients of the Taylor development of an analytic function

$$\psi(s) = \sum_0^\infty a_n s^n,$$

are given by Cauchy's formula:

$$a_n = \frac{1}{2\pi i} \int_{|s|=r} \frac{\psi(s)}{s^{n+1}} \,\mathrm{d}s \qquad (n = 0, 1, 2, \ldots).$$

If $\psi(s)$ is an entire function of exponential type $\leqq b$, we obtain bounds for the numbers a_n in the form

$$|a_n| \leqq C \frac{e^{br}}{r^n} \qquad (n = 0, 1, 2, \ldots).$$

Taking the minimum over r, we get the inequalities

$$|a_n| \leqq C \left(\frac{eb}{n}\right)^n \qquad (n = 0, 1, 2, \ldots). \tag{*}$$

We show by starting from the inequality (*) that the partial sums of the Taylor series of $\psi(\sigma)$ have a common majorant of the form $C_\varepsilon e^{(b+\varepsilon)r}$ for any $\varepsilon > 0$. For

$$\frac{eb|s|}{n} \leqq \frac{1}{2}$$

as from a certain N, which can be taken between the numbers $2e\,b|s|$ and $2e\,b|s| + 1$, and consequently

$$\sum_N^\infty |a_n s^n| \leqq C \sum_N^\infty \left(\frac{eb|s|}{n}\right)^n < C \sum_N^\infty \frac{1}{2^n} \leqq C.$$

On the other hand, $(e\,b|s|/n)^n$, regarded as a function of n, attains at $n = b|s|$ the maximum value $\mathrm{e}^{b|s|}$, as is easily verified by differentiation. Hence

$$\sum_0^\infty |a_n s^n| = \sum_0^{N-1} |a_n s^n| + \sum_N^\infty |a_n s^n| \leqq C(2eb|s| + 1)\, e^{b|s|} + C \leqq C_\varepsilon e^{(b+\varepsilon)|s|}$$

for any $\varepsilon > 0$, as required.

We now turn to the proof of the theorem itself. Let $\psi(s) = \sum_0^\infty a_n s^n$ be an entire function of exponential type $\leqq b$, square-summable over the real axis, and let $\varphi(x)$ be its (inverse) Fourier transform; by Plancherel's theorem, we have for any function $u(x) \in L_2(-\infty, \infty)$ and its Fourier transform $v(\sigma) \in L_2(-\infty, \infty)$

$$\int_{-\infty}^{\infty} \psi(\sigma)\, v(\sigma)\, d\sigma = 2\pi \int_{-\infty}^{\infty} \varphi(x)\, u(x)\, dx.$$

Let us suppose that $v(\sigma)$ vanishes outside the interval $[-c, c]$ with the result that $u(x)$ is an entire analytic function of the argument $z = x + iy$. Then the series

$$\psi(\sigma)\, v(\sigma) = \sum_{n=0}^{\infty} a_n \sigma^n v(\sigma) \tag{1}$$

converges in the metric of the space $L_2(-\infty, \infty)$. Hence we obtain the integral

$$\int_{-\infty}^{\infty} \psi(\sigma)\, v(\sigma)\, d\sigma$$

by term-by-term integration of (1); we get

$$\int_{-\infty}^{\infty} \psi(\sigma) v(\sigma) d\sigma = \sum_{n=0}^{\infty} a_n \int_{-\infty}^{\infty} \sigma^n v(\sigma) d\sigma = \sum_{n=0}^{\infty} a_n i^n u^{(n)}(0). \tag{2}$$

Equation (2) also holds under more general hypotheses, when $v(\sigma)$ is not necessarily a finite function but merely decreases with sufficient rapidity for $\sum_{n=0}^{\infty} a_n \sigma^n v(\sigma)$ to remain convergent in the metric of the space $L_2(-\infty, \infty)$. Since, by what has been proved, the partial sums of the Taylor series of the function $\psi(\sigma)$ share a common majorant of the form $C_\varepsilon e^{(b+\varepsilon)|\sigma|}$ for any $\varepsilon > 0$, it is enough for the function $v(\sigma)$ to decrease as $|\sigma| \to \infty$ more rapidly than $e^{-c|\sigma|}$, $c > b$.

We denote by E_c the class of functions $v(\sigma)$ which satisfy the inequality

$$|v(\sigma)| \leqq C e^{-c|\sigma|} \qquad (c > 0).$$

We shall show that if the function $v(\sigma)$ belongs to E_c, then each of the equations

$$w'(\sigma) \pm c_1 w(\sigma) = v(\sigma) \quad (0 < c_1 < c)$$

has a solution in the class E_{c_1}. For, taking the " + " sign, we can take as solution

$$w(\sigma) = e^{-c_1\sigma} \int_{-\infty}^{\sigma} e^{c_1\lambda} v(\lambda)\, d\lambda .$$

It is clear that the integral on the right-hand side exists for all σ. Further, as $\sigma \to +\infty$ we have

$$|w(\sigma)| \leqq e^{-c_1\sigma} \int_{-\infty}^{\infty} e^{c_1\lambda} |v(\lambda)|\, d\lambda \leqq C_1 e^{-c_1\sigma},$$

and as $\sigma \to -\infty$

$$|w(\sigma)| \leqq e^{-c_1\sigma} C \int_{-\infty}^{\sigma} e^{c_1\lambda} e^{c\lambda}\, d\lambda = Ce^{-c_1\sigma} \frac{e^{(c_1+c)\sigma}}{c_1 + c} \leqq C_2 e^{-c|\sigma|} \leqq C_2 e^{-c_1|\sigma|}$$

so that $w(\sigma) \in E_{c_1}$. Similarly for the " – " sign, we take as solution

$$w(\sigma) = e^{c_1\sigma} \int_{\sigma}^{\infty} e^{-c_1\lambda} v(\lambda)\, d\lambda .$$

Now let us consider, independently of the foregoing constructions, the series

$$\sum_{n=0}^{\infty} a_n\, i^n\, u^{(n)}(0), \tag{3}$$

where the numbers a_n satisfy the inequalities $|a_n| < C\,(e\,b\,n^{-1})^n$ and $u(x + i\,y)$ is an arbitrary function, analytic in the disc $|z| \leqq b$. We show that the series (3) is convergent. Every function $u(z)$ that is analytic for $|z| \leqq b$ is defined and analytic on some disc $|z| \leqq b + \varepsilon$, and by Cauchy's formula

$$u^{(n)}(0) = \frac{n!}{2\pi i} \int_{|z|=b+\varepsilon} \frac{u(z)\, dz}{z^{n+1}},$$

hence

$$|u^{(n)}(0)| \leqq \frac{Cn!}{(b + \varepsilon)^n}. \tag{4}$$

Substituting $C_1 n^{n+1/2}\ e^{-n}$ for $n!$; in accordance with Stirling's formula we get

$$|u^{(n)}(0)| \leqq C_2 \left[\frac{n}{e(b + \varepsilon)}\right]^n \tag{5}$$

and therefore

$$\sum_{n=0}^{\infty} |a_n|\, |u^{(n)}(0)| \leqq \sum_{n=0}^{\infty} C_3 \left(\frac{e\,b}{n}\right)^n \left(\frac{n}{e(b+\varepsilon)}\right)^n$$
$$= C_4 \sum_{n=0}^{\infty} \left(\frac{b}{b+\varepsilon}\right)^n < \infty, \tag{6}$$

as required.

Consider now the functional

$$\Phi(u) = 2\pi \int_{-\infty}^{\infty} \varphi(x)\, u(x)\, \mathrm{d}x = \int_{-\infty}^{\infty} \psi(\sigma)\, v(\sigma)\, \mathrm{d}\sigma,$$

which is defined over functions $u(x) \in L_2(-\infty, \infty)$; its value over functions $u(x)$ which belong to the class E_c, $c > b$, is given by formula (2):

$$\Phi(u) = \sum_{n=0}^{\infty} a_n\, i^n\, u^{(n)}(0).$$

As we have shown, this formula can be taken to define it over the totality of functions that are analytic in the disc $|z| \leqq b$. In addition, the functional $\Phi(u)$ is continuous in the following sense: if the functions $u_m(z)$ are defined and analytic on a disc $|z| \leqq b + \varepsilon$ and converge uniformly to zero as $m \to \infty$ over the whole disc, then $\Phi(u_m) \to 0$. This follows from equations (4)–(6), where we can evidently substitute the quantity $\max\limits_{|z| \leqq b+\varepsilon} |u(z)|$ for the constant C.

We must show that the function $\varphi(x)$—the Fourier transform of the given function $\psi(\sigma)$—vanishes for $|x| \geqq a$. Let $\varepsilon/2 < \varepsilon_1 < < \varepsilon_2 < \cdots < \varepsilon_m < \cdots \to \varepsilon$ and put

$$u_m(z) = \frac{\prod\limits_{k=1}^{m} \left(\dfrac{z}{b+\varepsilon_k}\right)^2}{\prod\limits_{k=1}^{m} \left[1 + \left(\dfrac{z}{b+\varepsilon_k}\right)^2\right]}\, u_0(z),$$

where $u_0(z) \in L_2(-\infty, \infty)$ is some entire function whose Fourier transform $v_0(\sigma)$ belongs to the class $E_{b+\varepsilon}$.

Resolving the coefficient of $u_0(z)$ into partial fractions, we obtain a representation of $u_m(z)$ in the form

$$u_m(z) = A_0\, u_0(z) + \sum_{k=1}^{m} \frac{A_k}{1 + \dfrac{i\,z}{b+\varepsilon_k}}\, u_0(z) + \sum_{k=1}^{m} \frac{B_k}{1 - \dfrac{i\,z}{b+\varepsilon_k}}\, u_0(z).$$

Let

$$u_k^+(z) = \left(1 + \frac{i\,z}{b + \varepsilon_k}\right)^{-1} u_0(z),$$

$$u_k^-(z) = \left(1 - \frac{i\,z}{b + \varepsilon_k}\right)^{-1} u_0(z),$$

and let $v_k^+(\sigma)$, $v_k^-(\sigma)$ be the corresponding Fourier transforms; by formula (3) of Section 3, art. 4, we have

$$v_0(\sigma) = v_k^+(\sigma) + \frac{d}{d\sigma}\,\frac{1}{b + \varepsilon_k}\,v_k^+(\sigma),$$

$$v_0(\sigma) = v_k^-(\sigma) - \frac{d}{d\sigma}\,\frac{1}{b + \varepsilon_k}\,v_k^-(\sigma).$$

Since the function $v_0(\sigma)$ belongs, by hypothesis, to the class $E_{b+\varepsilon}$ and the numbers $b + \varepsilon_j$ are less than $b + \varepsilon$, by what we have proved v_k^+, v_k^- belong to the class $E_{b+\varepsilon_k}$; it follows that the function $v_m(\sigma) = F[u_m(x)]$ belongs to the class $E_{b+1/2\varepsilon}$. Hence at the function $v_m(\sigma)$ the functional (2) assumes the value

$$\Phi(u_m) = \int_{-\infty}^{\infty} \psi(\sigma)\,v_m(\sigma)\,d\sigma = \sum_{n=0}^{\infty} a_n\,i^n\,u_m^{(n)}(0).$$

As $m \to \infty$ the sequence of functions $u_m(z)$.

(a) converges to 0 uniformly over the disc $|z| \leq b + \varepsilon/2$,

(b) converges in the metric of $L_2(-\infty, \infty)$ to a function equal to $u_0(x)$ for $|x| > b + \varepsilon$ and equal to 0 for $|x| < b + \varepsilon$.

In virtue of what we have proved $\Phi(u_m) \to 0$. At the same time

$$\Phi(u_m) = \int_{-\infty}^{\infty} \varphi(x)\,u_m(x)\,dx \to \int_{|x| \geqq b+\varepsilon} \varphi(x)\,u_0(x)\,dx.$$

It follows that

$$\int_{|x| \geqq b+\varepsilon} \varphi(x)\,u_0(x)\,dx = 0.$$

We can write this equation in the form

$$\int_{-\infty}^{\infty} \varphi_0(x)\,u_0(x)\,dx = 0,$$

where $\varphi_0(x)$ is equal to $\varphi(x)$ for $|x| > b + \varepsilon$ and to 0 for $|x| < b + \varepsilon$. The function $u_0(x)$ here is any entire function whose Fourier transform belongs to the class $E_{b+\varepsilon}$. Since the aggregate of such functions is complete in the space $L_2(-\infty, \infty)$†, we have $\varphi_0(x) \equiv 0$ (almost everywhere), and hence $\varphi(x) \equiv 0$ almost everywhere that $|x| > b + \varepsilon$; since $\varepsilon > 0$ is arbitrary, $\varphi(x)$ vanishes for $|x| > b$ almost everywhere, as required.

Problems. 1. Let E denote the totality of functions $\psi(\sigma)$ which are entire functions of the argument $s = \sigma + i\tau$ of exponential type $\leq b$ and such that

$$\int_{-\infty}^{\infty} |\psi(\sigma)|^2 \, d\sigma = 1 . \tag{*}$$

Let G be a bounded measurable set on the σ axis. Show that

$$\theta(G) \equiv \sup_{\psi \in E} \int_G |\psi(\sigma)|^2 \, d\sigma < 1 .$$

Hint. Use the Wiener-Paley theorem to show that the functions $\psi(\sigma + i\tau \in E$ are uniformly bounded in any circle. Show by Cauchy's formula that the first derivatives of the functions $\psi(s)$ are bounded in any circle. By Arzela's theorem (Chapter II, Section 7, problem 5), the set E is compact in any circle Q. If there exists a sequence $\psi_\nu \in E$ for which $\int_G |\psi_\nu(\sigma)|^2 \, d\sigma \to 1$, on choosing a subsequence from it, uniformly convergent in $Q \supset 2G$, we obtain the equations for the limit function $\psi\sigma)$:

$$\int_G |\varphi(\sigma)|^2 \, d\sigma = 1 , \qquad \int_{Q-G} |\varphi(\sigma)|^2 \, d\sigma = 0 , \tag{*}$$

whence $\psi(\sigma) \equiv 0$ in $Q - G$, so that $\psi(\sigma) \equiv 0$, which contradicts (*).

2. (continued). Show that

$$\theta(G) \leq 2b \, \mu \, G$$

or any set G of finite measure.

Hint. Apply the Cauchy–Bunyakovsky inequality to the expression for $\psi(\sigma)$ in terms of $\Phi(x)$.

3. (continued). Show that the result of problem 1 still holds for any set G of finite measure (B. P. Paneyakh).

Hint. $G = G_1 + G_2$, where G_1 is bounded, $2b \, \mu \, G_2 < 1$. If $\theta(G) = 1$, we can find as in problem 1 a sequence $\psi_\nu(\sigma)$ convergent to zero uniformly outside G, and hence on G_1, so that $\int_{G_1} |\psi_\nu(\sigma)|^2 \, d\sigma \to 0$; but now $\theta(G_2) = 1$, which contradicts the result of problem 2.

† For example, we can take as the function $u_0(x)$ the Hermite function $x^n e^{-x^2}$ (cf. Section 3, art. 5).

4. Show that the Fourier transform $\psi(s)$ of a function $\Phi(x) \in L_2(-\infty, \infty)$, equal to zero for $x < 0$, is characterised by the condition: $\psi(\sigma + i\tau)$ is analytic for $\tau < 0$ and $\int_{-\infty}^{\infty} |\psi(\sigma + i\tau)|^2 \, d\sigma \leqq C$ for all $\tau \leqq 0$.

Hint. Use Plancherel's theorem for $\tau < 0$.

4. *The Uncertainty Principle in Quantum Mechanics*

In investigating the motion of a material particle M in quantum mechanics the quantities sought are not the coordinates of the particle and its velocity, as in classical mechanics, but the probability distributions of these quantities. For simplicity we shall consider the one-dimensional case. Then the functions sought are the following:

(1) *The position function* $\varphi(x)$. This function is defined on the whole line $-\infty < x < \infty$ and satisfies the condition

$$\int_{-\infty}^{\infty} |\varphi(x)|^2 \, dx = 1; \tag{1}$$

it determines the probability that (at a given instant) the material particle M will be situated in the interval (α, β) in accordance with the formula

$$\text{Prob}\,\{x \in (\alpha, \beta)\} = \int_{\alpha}^{\beta} |\varphi(x)|^2 \, dx.$$

(2) *The momentum function* $\psi(p)$. This function is defined on the line $-\infty < p < \infty$ and satisfies the condition

$$\int_{-\infty}^{\infty} |\psi(p)|^2 \, dp = 2\pi. \tag{2}$$

It determines the probability that the magnitude of the momentum of the particle (the product of its mass and its velocity) will lie in the interval (γ, δ) in accordance with the formula

$$\text{Prob}\,\{p \in (\gamma, \delta)\} = \frac{1}{2\pi} \int_{\gamma}^{\delta} |\psi(p)|^2 \, dp.$$

One of the fundamental axioms of quantum mechanics (we shall not go into its physical significance) consists in the hypothesis that the momentum function is the Fourier transform of the position function:

$$\psi(p) = \int_{-\infty}^{\infty} \varphi(x)\, e^{-ipx}\, dx. \tag{3}$$

If we know the position function, we can write down the "most probable" value (the mathematical expectation) of the position of the particle:

$$\xi = \int_{-\infty}^{\infty} x\,|\varphi(x)|^2\, dx.$$

It can be assumed that $\xi = 0$, since we can always effect a translation along the x-axis. We observe that the quantity $|\psi(p)|$ is invariant under such translation since for any $\varphi(x)$

$$\int_{-\infty}^{\infty} \varphi(x-h)\, e^{-ipx}\, dx = e^{-iph} \int_{-\infty}^{\infty} \varphi(x)\, e^{-ipx}\, dx. \tag{4}$$

Similarly we can also assume that the mathematical expectation of the momentum

$$\eta = \frac{1}{2\pi} \int_{-\infty}^{\infty} p\,|\psi(p)|^2\, dp \tag{5}$$

is equal to zero. An estimate of the scatter of the quantity x is given by the mean square deviation (the dispersion)

$$\delta_x^2 = \int_{-\infty}^{\infty} x^2\,|\varphi(x)|^2\, dx. \tag{6}$$

The smaller δ_x^2, the greater the likelihood that the point M is actually situated close to the origin. Similarly the scatter estimate δ_p^2 is given by the mean square deviation

$$\delta_p^2 = \frac{1}{2\pi} \int_{-\infty}^{\infty} p^2\,|\psi(p)|^2\, dp. \tag{7}$$

It is naturally assumed that the functions $\varphi(x)$, $\psi(p)$ are such that the integrals (6), (7) exist. Hence, by what was said in art. 2, the functions $\varphi'(x)$, $p'(\sigma)$ exist and are square summable. The function $x\,\varphi(x)$, $\bar{\varphi}'(x)$ is integrable in the first degree, whilst the integrable function $x\,\varphi\,\bar{\varphi}$, which has the integrable derivative

$\varphi\bar{\varphi} + x\varphi\bar{\varphi} + x\varphi\bar{\varphi}'$, vanishes at infinity. We shall establish the inequality

$$\delta_x^2 \delta_p^2 \geqq \tfrac{1}{4}, \tag{8}$$

which is said to be *the uncertainty relation*. It shows that the more accurately we know the position of the particle (i.e. the smaller δ_x) the less accurately we know the magnitude of its momentum (i.e. the greater δ_p), and conversely, so that the simultaneous existence of functions $\varphi(x)$, $\psi(p)$ determining with great precision both the position of the particle and the magnitude of its momentum is impossible.

For the proof of the relation (8) we consider the integral

$$I(\alpha) = \int_{-\infty}^{\infty} |\alpha x \varphi(x) + \varphi'(x)|^2 \,\mathrm{d}x$$

where α is a real parameter. Using the result $|z|^2 = z\bar{z}$, we find

$$I(\alpha) = \int_{-\infty}^{\infty} (\alpha x \varphi + \varphi')(\alpha x \bar{\varphi} + \bar{\varphi}')\,\mathrm{d}x$$

$$= \alpha^2 \int_{-\infty}^{\infty} x^2 |\varphi|^2 \,\mathrm{d}x + \alpha \int_{-\infty}^{\infty} x(\varphi\bar{\varphi}' + \varphi'\bar{\varphi})\,\mathrm{d}x + \int_{-\infty}^{\infty} |\varphi'|^2 \,\mathrm{d}x.$$

In addition

$$\int_{-\infty}^{\infty} x^2 |\varphi|^2 \,\mathrm{d}x = \delta_x^2,$$

$$\int_{-\infty}^{\infty} x(\varphi\bar{\varphi}' + \varphi'\bar{\varphi})\,\mathrm{d}x = \int_{-\infty}^{\infty} x(\varphi\bar{\varphi})'\,\mathrm{d}x$$

$$= x\varphi\bar{\varphi}\Big|_{-\infty}^{\infty} - \int_{-\infty}^{\infty} |\varphi|^2 \,\mathrm{d}x = -1,$$

and by Plancherel's theorem, since $F[\varphi'] = i\,p\,\psi(p)$

$$\int_{-\infty}^{\infty} |\varphi'(x)|^2 \,\mathrm{d}x = \frac{1}{2\pi} \int_{-\infty}^{\infty} p^2 |\psi(p)|^2 \,\mathrm{d}p = \delta_p^2.$$

Hence

$$I(\alpha) = \alpha^2 \delta_x^2 - \alpha + \delta_p^2. \tag{9}$$

Since by construction $I(\alpha) \geqq 0$, the quadratic trinomial (9) does not have real zeros, and consequently

$$1 - 4\delta_x^2 \delta_p^2 \leqq 0,$$

which is equivalent to the required inequality

$$\delta_x^2 \delta_p^2 \geqq \tfrac{1}{4},$$

7. The Fourier–Stieltjes Transform

The formula for the Fourier transform of an absolutely integrable function,

$$\psi(\sigma) = \int_{-\infty}^{\infty} e^{-ix\sigma} \varphi(x)\, dx \tag{1}$$

can be expressed in the form of a Stieltjes integral

$$\psi(\sigma) = \int_{-\infty}^{\infty} e^{-ix\sigma}\, d\Phi(x),$$

where

$$\Phi(x) = \int_{-\infty}^{x} \varphi(\xi)\, d\xi$$

is an absolutely continuous function of bounded variation on the line $-\infty < x < \infty$.

It is possible to consider general integrals of the form

$$\psi(\sigma) = \int_{-\infty}^{\infty} e^{-ix\sigma}\, d\Phi(x), \tag{2}$$

where $\Phi(x)$ is now an arbitrary function of bounded variation on the line $-\infty < x < \infty$. An integral of the form (2) is said to be a *Fourier–Stieltjes integral.*

The function $\psi(\sigma)$ defined by the integral (2) is bounded:

$$|\psi(\sigma)| \leqq \left|\int_{-\infty}^{\infty} e^{-ix\sigma} d\Phi(x)\right| \leqq \int_{-\infty}^{\infty} |d\Phi(x)| = \mathrm{V}_{-\infty}^{\infty}[\Phi].$$

It is also continuous; for

$$|\psi(\sigma') - \psi(\sigma'')| = \int_{-A}^{A} [e^{-ix\sigma'} - e^{-ix_{\sigma}''}]\, d\Phi(x) +$$
$$+ \int_{|x| \geqq A} [e^{-ix\sigma'} - e^{-ix\sigma''}]\, d\Phi(x);$$

the second integral becomes arbitrarily small for sufficiently large A independently of σ' and σ'', while the first, for any chosen A, becomes arbitrarily small for a sufficiently small difference $|\sigma' - \sigma''|$.

But in contrast with the Fourier–Lebesgue integral (1), the Fourier–Stieltjes integral (2) does not, generally speaking, tend to zero as $|\sigma| \to \infty$. For example, if $\Phi(x)$ corresponds to a unit mass concentrated at the point x_0, then

$$\psi(\sigma) = \int_{-\infty}^{\infty} e^{-ix\sigma}\, \mathrm{d}\Phi(x) = e^{-ix_0\sigma}$$

is a periodic function of σ.

Every periodic function $\psi(\sigma)$ which has a Fourier series development

$$\psi(\sigma) = \sum_{-\infty}^{\infty} a_n\, e^{in\sigma} \tag{3}$$

for which the series of coefficients is absolutely convergent,

$$\sum_{-\infty}^{\infty} |a_n| < \infty,$$

can be expressed in the form of a Fourier–Stieltjes integral: for this a generating function must be chosen which is piecewise-constant and has a saltus a_n at each point $x = n$. The same is true of a more general class of functions, obtained from (3) by replacing the exponent in σ by $i\,\lambda_n\,\sigma$, where λ_n is an arbitrary sequence of real numbers; these functions belong to the class of so-called *almost periodic functions*.

We shall demonstrate how the Fourier–Stieltjes integral can be applied in proving a theorem which has applications in probability theory.

We shall call a measurable function $\psi(\sigma)$ $(-\infty < \sigma < \infty)$ positive–definitive if for any continuous function $u(x)$ and any a, b

$$\int_a^b \int_a^b \psi(\sigma - \eta)\, u(\sigma)\, \overline{u(\eta)}\, \mathrm{d}\sigma\, \mathrm{d}\eta \geqq 0. \tag{4}$$

An example of a positive–definite function is the function $e^{i\sigma x}$ with x fixed; for

$$\int_a^b \int_a^b e^{-i(\sigma-\eta)x}\, u(\sigma)\, \overline{u(\eta)}\, \mathrm{d}\sigma\, \mathrm{d}\eta = \int_a^b e^{-i\sigma x}\, u(\sigma)\, \mathrm{d}\sigma \int_a^b e^{i\eta x}\, \overline{u(\eta)}\, \mathrm{d}\eta$$

$$= \left|\int_a^b e^{-i\sigma x}\, u(\sigma)\, \mathrm{d}\sigma\, \right|^2 \geqq 0.$$

It is found that any continuous positive–definite function $\psi(\sigma)$ can be expressed as a "Stieltjes combination" of the functions $e^{-ix\sigma}$; we have the following theorem:

THEOREM (S. Bochner, A. Ya. Khinchin, 1932). *Every continuous positive–definite function $\psi(\sigma)$ can be expressed in the form*

$$\psi(\sigma) = \int_{-\infty}^{\infty} e^{-ix\sigma}\, \mathrm{d}\Phi(x),$$

where $\Phi(x)$ is a bounded non-decreasing function.

Proof. We consider first the case when the positive–definite function $\psi(\sigma)$ has, in addition, a continuous derivative. Putting $u(\sigma) = e^{i\xi\sigma}$, $a = 0$, $b = n$ in (4), we get

$$\int_0^n \int_0^n \psi(\sigma - \eta)\, e^{i\xi\sigma}\, e^{-i\xi\eta}\, \mathrm{d}\sigma\, \mathrm{d}\eta \geqq 0$$

or, replacing $\sigma - \eta$ by λ,

$$\int_{-n}^{n} \psi(\lambda)\, e^{i\lambda\xi}\left(1 - \frac{|\lambda|}{n}\right) \mathrm{d}\lambda = f_n(\xi) \geqq 0. \tag{5}$$

We can interpret (5) as the Fourier transform of the function

$$\Psi_n(\lambda) = \begin{cases} \psi(\lambda)\left(1 - \dfrac{|\lambda|}{n}\right) & (|\lambda| \leqq n), \\ 0 & (|\lambda| > n). \end{cases}$$

Since, by hypothesis, the function $\psi(\sigma)$ is differentiable, the function $\Psi_n(\lambda)$ is piecewise-smooth and by a theorem of Section 2 (cf. p. 382) the inversion formula

$$\Psi_n(\lambda) = \frac{1}{2\pi} \int_{-\infty}^{\infty} e^{-i\lambda\xi} f_n(\xi)\, \mathrm{d}\xi. \tag{6}$$

holds for all λ.

In addition

$$\frac{1}{2\pi} \int_{-\infty}^{\infty} f_n(\xi)\, \mathrm{d}\xi = \Psi_n(0) = \psi(0).$$

Introducing the monotone function

$$\Phi_n(x) = \frac{1}{2\pi} \int_{-\infty}^{x} f_n(\xi)\, \mathrm{d}\xi,$$

We can write the integral (6) in the form of a Stieltjes integral

$$\Psi_n(\lambda) = \int_{-\infty}^{\infty} e^{-i\lambda x} \, d\Phi_n(x). \tag{7}$$

As $n \to \infty$ the left-hand side obviously tends to the limit $\psi(\lambda)$. By Helly's theorem (Chapter VI, Section 6, art. 4) the functions $\Phi_n(x)$ form a sequence which contains an everywhere convergent subsequence; with a suitable renumbering, we can assume that the sequence $\Phi_n(x)$ itself converges everywhere to some function $\Phi(x)$, which is also non-decreasing and varies within the same limits O and $\psi(O)$. If we were now to apply the theorem of Helly (Chapter VI, Section 6, art. 3) which ensures the legality of a passage to the limit under the Stieltjes integral sign, we should get that

$$\psi(\lambda) = \lim_{n \leftarrow \infty} \int_{-\infty}^{\infty} e^{-i\lambda x} \, d\Phi_n(x) = \int_{-\infty}^{\infty} e^{-i\lambda x} \, d\Phi(x),$$

and the proof would be complete.

When using Helly's theorem, we have to note the fact the interval of integration is infinite, and the function $e^{-i\lambda x}$ is not continuous at infinity. In accordance with note 3 after Helly's theorem, we must therefore verify that condition (*) is fulfilled: given any $\varepsilon > 0$, an $N = N(\varepsilon)$ can be found such that

$$\operatorname*{Var}_{|x| \geqq N} \Phi_n(x) \leqq \varepsilon.$$

for all n.

For this, we apply the following lemma:

LEMMA. *Given a family of functions $\Psi_\alpha(\lambda)$ of the form*

$$\Psi_\alpha(\xi) = \int_{-\infty}^{\infty} e^{-i\lambda x} \, d\Phi_\alpha(x),$$

where $\Phi_\alpha(x)$ are non-decreasing functions of bounded variation, if the family is equicontinuous for $\lambda = 0$, i.e. if, given any $\varepsilon > 0$, a $\delta > 0$ can be found such that

$$|\Psi_\alpha(\lambda) - \Psi_\alpha(0)| < \varepsilon,$$

for all α and $|\lambda| < \delta$, the functions $\Phi_\alpha(x)$ satisfy condition (*) (with n replaced by α).

Proof of the lemma. It follows in particular from the equicontinuity condition that, for $|h| < \delta$,

$$\left|\Psi_\alpha(0) - \frac{1}{2h}\int_{-h}^{h} \Psi_\alpha(\lambda)\,\mathrm{d}\lambda\right| = \frac{1}{2h}\left|\int_{-h}^{h} [\Psi_\alpha(0) - \Psi_\alpha(\lambda)]\,\mathrm{d}\lambda\right|$$

$$\leqq \frac{1}{2h}\int_{-h}^{h} |\Psi_\alpha(0) - \Psi_\alpha(\lambda)|\,\mathrm{d}\lambda < \varepsilon.$$

Further, having found h, we can find A such that $\left|\frac{\sin hx}{hx}\right| \leqq 1/2$ for $|x| > A$ (it is sufficient to put $A = 1/h$); we then get

$$\varepsilon > \Psi_\alpha(0) - \frac{1}{2h}\int_{-h}^{h}\Psi_\alpha(\lambda)\,\mathrm{d}\lambda = \int_{-\infty}^{\infty}\mathrm{d}\Phi_\alpha(x) - \int_{-\infty}^{\infty}\left\{\frac{1}{2h}\int_{-h}^{h}\mathrm{e}^{i\lambda x}\,\mathrm{d}\lambda\right\}\mathrm{d}\Phi_\alpha(x)$$

$$= \int_{-\infty}^{\infty}\left[1 - \frac{\sin hx}{hx}\right]\mathrm{d}\Phi_\alpha(x) \geqq \int_{|x|\geqq A}\left[1 - \frac{\sin hx}{hx}\right]\mathrm{d}\Phi_\alpha(x) \geqq \frac{1}{2}\int_{|x|\geqq A}\mathrm{d}\Phi_\alpha(x),$$

and condition (*) is fulfilled, as required.

We return to the proof of the theorem. We see that it remains to verify the equicontinuity at $\lambda = 0$ of the set of functions $\Psi_n(\lambda)$ figuring in equation (7). But each of these functions is obtained from a fixed continuous function $\psi(\lambda)$ by multiplying by $1 - (|\lambda|/n)$, which tends uniformly to 1 as $n \to \infty$. The equicontinuity condition is clearly satisfied in this case. We have thus proved the Bochner–Khinchin theorem, at any rate for the case when $\psi(\sigma)$ is differentiable.

Now let the positive definite function $\psi(\sigma)$ be merely continuous. We form the *symmetric mean* of $\psi(\sigma)$:

$$\psi^h(\sigma) \equiv S_h\,\psi(\sigma) \equiv \frac{1}{2h}\int_{-h}^{h}\psi(\sigma + \xi)\,\mathrm{d}\xi = \frac{1}{2h}\int_{-h}^{h}\psi(\sigma - \xi)\,\mathrm{d}\xi$$

$$= \frac{1}{2h}\int_{\sigma+h}^{\sigma+h}\psi(\eta)\,\mathrm{d}\eta.$$

The function $\psi^h(\sigma)$ evidently has a continuous derivative. We show that the double symmetric mean of the function $\psi(\sigma)$

$$S_h S_h \psi(\sigma) = \frac{1}{4h^2}\int_{-h}^{h}\int_{-h}^{h} \psi(\sigma + \xi + \eta)\,d\xi\,d\eta \tag{8}$$

is again a positive-definite function. For

$$\iint S_h S_h \psi(\sigma - \tau)\, u(\sigma)\,\overline{u(\tau)}\,d\sigma\,d\tau$$

$$= \frac{1}{4h^2}\iiiint \psi(\sigma - \tau + \xi + \eta)\, u(\sigma)\,\overline{u(\tau)}\,d\sigma\,d\tau\,d\xi\,d\eta$$

$$= \frac{1}{4h^2}\iiiint \psi(\sigma' - \tau')\, u(\sigma' - \xi)\,\overline{u(\tau' + \eta)}\,d\sigma'\,d\tau'\,d\xi\,d\eta$$

$$= \iint \psi(\sigma' - \tau')\left\{\frac{1}{2h}\int_{-h}^{h} u(\sigma' - \xi)\,d\xi\right\}\left\{\frac{1}{2h}\int_{-h}^{h} \overline{u(\tau' + \eta)}\,d\eta\right\} d\sigma'\,d\tau'$$

$$= \iint \psi(\sigma' - \tau')\, v(\sigma')\,\overline{v(\tau')}\,d\sigma'\,d\tau' \geqq 0,$$

where

$$v(\sigma) = \frac{1}{2h}\int_{-h}^{h} u(\sigma - \xi)\,d\xi = \frac{1}{2h}\int_{-h}^{h} u(\sigma + \eta)\,d\eta.$$

Applying the theorem in the form already proved to the double symmetric mean $\psi^{hh}(\sigma) = S_h S_h \psi(\sigma)$ of the function $\psi(\sigma)$, we get

$$\psi^{hh}(\sigma) = \int_{-\infty}^{\infty} e^{-ix\sigma}\,d\Phi_n(x), \tag{9}$$

where $\Phi_h(\sigma)$ is a non-decreasing function which varies between 0 and $\psi^{hh}(\sigma)$. The family of functions $\psi^{hh}(\sigma)$ is uniformly continuous at $\sigma = 0$, for we have

$$|\psi^{hh}(\sigma) - \psi^{hh}(0)|$$

$$\leqq \frac{1}{4h^2}\int_{-h}^{h}\int_{-h}^{h} |\psi(\sigma + \xi + \eta) - \psi(\xi + \eta)|\,d\xi\,d\eta$$

$$\leqq \max_{\substack{|\xi| \leq h \\ |h| \leq \eta}} |\,\psi(\sigma + \xi + \eta) - \psi(\xi + \eta)|,$$

and the question reduces to the uniform continuity of the function $\psi(\sigma)$ itself in a neighbourhood of the point zero.

We now let h tend to 0. The functions $\psi^{hh}(\sigma)$ tend to $\psi(\sigma)$. The sequence $\Phi_h(x)$ contains a subsequence, which we can take to be the sequence $\Phi_h(x)$ itself, which converges everywhere to a non-decreasing function $\Phi(x)$ whose values range between 0 and $\Psi(\sigma)$. Using again the lemma proved above, we can once more apply Helly's theorem to the interval $-\infty < x < \infty$ and thus obtain from (9) the equation

$$\psi(\sigma) = \int_{-\infty}^{\infty} e^{-ix\sigma} \mathrm{d}\,\Phi(\sigma). \tag{10}$$

This completes the proof of the Bochner–Khinchin theorem.

Note. A more searching analysis reveals that the requirement that the positive-definite function of the Bochner–Khinchin theorem be continuous is superfluous. The theorem holds under the supposition that the positive-definite function $\psi(\sigma)$ is *measurable*; the representation (10) is then found to hold almost everywhere. It follows as a corollary that every measurable positive-definite function can be made continuous by altering its values on a set of measure zero.

8. The Fourier Transform in the Case of Several Independent Variables

In problems of mathematical physics we have to deal with the Fourier transform of functions of several variables. In this paragraph we consider the simplest properties of this transform.

Let $\varphi(x) = \varphi(x_1, \ldots, x_n)$ be an integrable function of n variables $x_1, \ldots, x_n$ defined on the whole of the n-dimensional space R_n. *The Fourier transform* of the function $\varphi(x)$ is defined as the function

$$\begin{aligned}\psi(\sigma) &= \psi(\sigma_1, \ldots, \sigma_n)\\ &= \underbrace{\int \cdots \int}_{n} e^{-i(x_1\sigma_1 + \cdots + x_n\sigma_n)} \varphi(x_1, \ldots, x_n)\, \mathrm{d}x_1, \ldots, \mathrm{d}x_n \end{aligned} \tag{1}$$

or in symbolic notation

$$\psi(\sigma) = \int_{R_n} e^{-i(x\sigma)} \varphi(x)\, \mathrm{d}x.$$

If $\varphi(x)$ is the product of functions $\varphi_1(x), \ldots, \varphi_n(x_n)$, each of which is integrable with respect to its own variable, then the

n-ple integral (1) reduces to the product of the n simple integrals:

$$\psi(\sigma) = \int_{-\infty}^{\infty} \varphi_1(x_1) e^{-i x_1 \sigma_1} \mathrm{d}x_1 \ldots \int_{-\infty}^{\infty} \varphi_n(x_n) e^{-i x_n \sigma_n} \mathrm{d}x_n = \psi_1(\sigma_1) \ldots \psi_n(\sigma_n),$$

where $\psi_k(\sigma_k)$ is the usual Fourier transform of the function $\varphi_k(x_k)$.

In the general case the multiple integral (1) can be expressed, by Fubini's theorem, in the form of a repeated integral

$$\psi(\sigma) = \int_{-\infty}^{\infty} \Big\{ \cdots \Big\{ \int_{-\infty}^{\infty} \Big\{ \int_{-\infty}^{\infty} \psi(x_1, \ldots, x_n) e^{-i x_1 \sigma_1} \mathrm{d}x_1 \Big\} e^{-i x_2 \sigma_2} \mathrm{d}x_2 \Big\} \cdots$$

$$\cdots \Big\} e^{-i x_n \sigma_n} \mathrm{d}x_n .$$

Each of the curly brackets determines the Fourier transform with respect to one coordinate with the others held fixed. Inverting each of these operations in succession, we get formally

$$\varphi(x) = \frac{1}{2\pi} \int_{-\infty}^{\infty} \Big\{ \cdots \frac{1}{2\pi} \int_{-\infty}^{\infty} \frac{1}{2\pi} \int_{-\infty}^{\infty} \psi(\sigma_1, \ldots, \sigma_n) e^{i x_n \sigma_n} \mathrm{d}\sigma_n \Big\} \times$$

$$\times e^{i x_{n-1} \sigma_{n-1}} \mathrm{d}\sigma_{n-1} \cdots \Big\} e^{i x_1 \sigma_1} \mathrm{d}\sigma_1$$

Or in the form of an n-tuple integral

$$\varphi(x) = \varphi(x_1, \ldots, x_n) = \frac{1}{(2\pi)^n} \int \cdots \int \psi(\sigma_1, \ldots, \sigma_n) e^{i(x,\sigma)} \mathrm{d}\sigma_1 \ldots \mathrm{d}\sigma_n . \quad (2)$$

Since in general the function $\psi(\sigma)$ is not absolutely integrable over R_n, formula (2) can only be meaningful if we specify a procedure for evaluating the integral on the right-hand side. We shall give several possible interpretations of the integral (2) below.

THEOREM. *Let us suppose that the function* $\varphi(x) = \varphi(x_1, \ldots, x_n)$ *satisfies the conditions*

$$|\varphi(x_1 + t_1, x_2, \ldots, x_n) - \varphi(x_1, x_2, \ldots, x_n)| \leqq C |t_1|^{\alpha}, \quad (3_1)$$

$$|\varphi(x_1, x_2 + t_2, x_3, \ldots, x_n) - \varphi(x_1, x_2, \ldots, x_n)| \leqq C(x_1) |t_2|^{\alpha}, \quad (3_2)$$

. . . .

$$|\varphi(x_1, x_2, \ldots, x_n + t_n) - \varphi(x_1, x_2, \ldots, x_n)|$$
$$\leqq C(x_1, \ldots, x_{n-1}) |t_n|^{\alpha}, \quad (3_n)$$

$$\int_{-\infty}^{\infty} C(x_1) \mathrm{d}x_1 < \infty, \cdots, \int_{-\infty}^{\infty} \cdots \int_{-\infty}^{\infty} C(x_1, \ldots, x_{n-1}) \mathrm{d}x_1 \ldots \mathrm{d}x_{n-1} < \infty$$

$$(0 < \alpha \leqq 1).$$

Then formula (2) is valid if it is understood as the result of successive passages to the limit as $N_1 \to \infty, \ldots, N_n \to \infty$:

$$\varphi(x) = \frac{1}{(2\pi)^n} \lim_{N_1\to\infty} \int_{-N_1}^{N_1} \Big\{ \cdots \lim_{N_{n-1}\to\infty} \int_{-N_{n-1}}^{N_{n-1}} \Big\{ \lim_{N_n\to\infty} \int_{-N_n}^{N_n} \psi(\sigma_1 \ldots \sigma_n) \times e^{i x_n\sigma_n} \mathrm{d}\sigma_n \Big\} e^{i x_{n-1}\sigma_{n-1}} \mathrm{d}\sigma_{n-1} \ldots \Big\} e^{i x_1\sigma_1} \mathrm{d}\sigma_1. \tag{4}$$

Proof. We put

$$\varphi_1(\sigma_1, x_2, \ldots, x_n) = \int_{-\infty}^{\infty} \varphi(x_1, x_2, \ldots, x_n)\, e^{-i x_1\sigma_1} \mathrm{d}x_1.$$

By Fubini's theorem the function $\varphi(x_1, \ldots, x_n)$ is summable with respect to x_1 for almost all $x_2, \ldots, x_n$.

It follows from condition (3) and the theorem of Section 2, art. 2 that the inversion formula

$$\varphi(x_1, \ldots, x_n) = \lim_{N_1\to\infty} \frac{1}{2\pi} \int_{-N_1}^{N_1} \varphi_1(\sigma_1, x_2, \ldots, x_n)\, e^{i x_1\sigma_1} \mathrm{d}\sigma_1,$$

holds. The function $\varphi_1(\sigma_1, x_2, \ldots, x_n)$ is summable with respect to x_2 for almost all $x_3, \ldots, x_n$ and satisfies the condition

$$|\varphi_1(\sigma_1, x_2 + t_2, \ldots, x_n) - \varphi_1(\sigma_1, x_2, \ldots, x_n)|$$
$$\leqq \int_{-\infty}^{\infty} |\varphi(x_1, x_2 + t_2, \ldots, x_n) - \varphi(x_1, x_2, \ldots, x_n)|\, \mathrm{d}x_1$$
$$\leqq |t_2|^\alpha \int_{-\infty}^{\infty} C(x_1)\, \mathrm{d}x_1 \leqq C_2 |t_2|^\alpha,$$

in virtue of (3_2). Hence for the function

$$\varphi_2(\sigma_1, \sigma_2, x_3, \ldots, x_n) = \int_{-\infty}^{\infty} \varphi_1(\sigma_1, x_2, \ldots, x_n)\, e^{-i x_2\sigma_2} \mathrm{d}x_2$$

the inversion formula

$$\varphi_1(\sigma_1, x_2, \ldots, x_n) = \lim_{N_2\to\infty} \frac{1}{2\pi} \int_{-N_2}^{N_2} \varphi_2(\sigma_1, \sigma_2, x_2, \ldots, x_n)\, e^{i x_2\sigma_2} \mathrm{d}\sigma_2,$$

holds and therefore

$$\varphi(x_1, \ldots, x_n) = \lim_{N_1 \to \infty} \frac{1}{2\pi} \int_{-N_1}^{N_1} \left\{ \lim_{N_2 \to \infty} \frac{1}{2\pi} \int_{-N_2}^{N_2} \varphi_2(\sigma_1, \sigma_2, x_3, \ldots, x_n) e^{i x_2 \sigma_2} d\sigma_2 \right\} e^{i x_1 \sigma_1} d\sigma_1.$$

Continuing the process, we arrive ultimately at the required formula (4).

The conditions of the theorem are satisfied, for example, if the integrable function $\varphi(x_1, \ldots, x_n)$ has partial derivatives $\partial\varphi/\partial x_1$, $\partial\varphi/\partial x_2, \ldots, \partial\varphi/\partial x_n$, of which the first is bounded by a constant, the second is an integrable function of x_1, the third is an integrable function of x_1 and x_2, etc.

There exist alternative procedures for reducing the n-tuple integral to a repeated one. Let us consider the expression of the Fourier transform in spherical coordinates. In spherical coordinates the integration is taken first for a fixed $r > 0$ over the sphere of radius r, centre the origin, then with respect to r from 0 to ∞. We denote by $d\omega$ an element of the unit sphere Ω_1; then an element of the sphere Ω_r of radius r will be represented by $r^{n-1}\, d\omega$. The expression (1) is transformed into

$$\psi(\sigma) = \int_0^\infty \left\{ \int_{\Omega_r} e^{-i\varrho r \cos\theta} \varphi(r; \omega)\, d\omega \right\} r^{n-1}\, dr.$$

Here ω denotes direction from the origin of coordinates, or, if we like, a point of the unit sphere; $\varphi(r, \omega)$ is the value of the function $\varphi(x)$ at the intersection of the directed line ω and the sphere Ω_r; ϱ is $|\sigma|$, and θ is the angle between the vectors x and σ.

This formula acquires a particularly simple form when the function φ depends only on r, i.e. is spherically symmetric. In this case $\varphi(r, \omega) = \varphi(r)$ and we have

$$\psi(\sigma) = \int_0^\infty \left\{ \int_{\Omega_r} e^{-i\varrho r \cos\theta} d\omega \right\} \varphi(r)\, r^{n-1}\, dr. \tag{5}$$

The enclosed integral can be completely determined. We consider first the case $n = 3$; then, taking the direction of the vector σ as polar axis, we have:

$$d\omega = \sin\theta\, d\theta\, d\alpha,$$

where α is the polar angle in the plane orthogonal to the vector σ; as a result we get

$$\int_{\Omega_r} e^{-i\varrho r\cos\theta}\sin\theta\, d\theta\, d\alpha = \int_0^{2\pi}\left\{\int_0^{\pi} e^{-i\varrho r\cos\theta}\sin\theta\, d\theta\right\} d\alpha = \frac{2\pi}{i\varrho r} e^{-i\varrho r\cos\theta}\Big|_0^{\pi}$$
$$= 4\pi\frac{\sin\varrho r}{\varrho r}$$

and hence

$$\psi(\sigma) = \frac{4\pi}{\varrho}\int_0^{\infty}\varphi(r)\, r\sin\varrho r\, dr.$$

In the general case, with x arbitrary, the enclosed integral in (5) can be expressed in terms of Bessel functions† in the form

$$\int_{\Omega_r} e^{-i\varrho r\cos\theta} d\omega = (2\pi)^{\frac{n}{2}-1}(\varrho r)^{1-\frac{n}{2}} J_{\frac{n}{2}-1}(\varrho r) \tag{6}$$

and therefore a complete representation of the Fourier transform of a spherically symmetric function is given by

$$\psi(\sigma) = \left(\frac{2\pi}{\varrho}\right)^{\frac{n}{2}-1}\int_{-\infty}^{\infty}\varphi(r) J_{\frac{n}{2}-1}(\varrho r)\, r^{\frac{n}{2}}\, dr. \tag{7}$$

From formula (7) we can draw a somewhat startling conclusion: for $n \geqq 3$ the Fourier transform of a spherically symmetric function in n-dimensional space is a differentiable function (for $\varrho \neq 0$), and furthermore the order of its differentiability increases together with n. For a function $\varphi(r)$ is integrable over as n-dimensional space if the integral

$$\int_{-\infty}^{\infty} |\varphi(r)|\, r^{n-1}\, dr$$

exists. In the integral (7) each formal differentiation with respect to ϱ increases the exponent of r in the integrand by one; since, for large r, $|J_{n/2-1}(\varrho r)| \leqq C r^{-1/2}$, multiplication of the integrand by r is permissible so long as the total exponent of r does not exceed $n - 1/2$, i.e. at least $[n - 1/2 - n/2] = [(n-1)/2]$ times.

† Cf. for example, R. Courant and D. Hilbert, *Methods of Mathematical Physics*, Interscience, N.Y. (1953), Vol. 1, Chapter VII.

Consequently the function $\psi(\sigma)$ has derivatives (with respect to ϱ) at least up to order $[(n-1)/2]$. For $n = 3$ the function $\psi(\sigma)$ is differentiable at least once.

Let us consider the Fourier integral inversion formula

$$\varphi(x) = \frac{1}{(2\pi)^n} \int \cdots \int_{R_n} \psi(\sigma)\, e^{i(\sigma, x)}\, \mathrm{d}\sigma$$

and try to assign a meaning to it by first integrating over a sphere $|\sigma| \leqq R$ and then letting R tend to infinity. We put

$$\varphi_R(x) = \frac{1}{(2\pi)^n} \int \cdots \int_{|\sigma| \leqq R} \psi(\sigma)\, e^{i(\sigma, x)}\, \mathrm{d}\sigma .$$

Substituting for $\psi(\sigma)$ from (1) and reversing the order of integration with respect to x and σ (which we can do in virtue of the absolute and uniform convergence with respect to σ of the n-tuple integral (1)), we get

$$\varphi_R(x) = \frac{1}{(2\pi)^n} \int_{R_n} \varphi(\xi) \left\{ \int_{|\sigma| \leqq R} e^{i(\xi - x, \sigma)}\, \mathrm{d}\sigma \right\} \mathrm{d}\xi = \frac{1}{(2\pi)^n} \int_{R_n} \varphi(\xi)\, H_R(\xi)\, \mathrm{d}\xi .$$

We transform the enclosed integral to spherical coordinates in accordance with formula (6):

$$\begin{aligned} H_R(\xi) &= \int_{|\sigma| \leqq R} e^{i(\xi - x, \sigma)}\, \mathrm{d}\sigma = \int_0^R \varrho^{n-1} \left\{ \int_{\Omega_\varrho} e^{i|\xi - x|\varrho \cos\theta} \mathrm{d}\omega \right\} \mathrm{d}\varrho \\ &= \int_0^R \varrho^{n-1} (2\pi)^{\frac{n}{2}-1} (\varrho\, r)^{1-\frac{n}{2}} J_{\frac{n}{2}-1}(\varrho\, r)\, \mathrm{d}\varrho \\ &= (2\pi)^{\frac{n}{2}-1}\, r^{1-\frac{n}{2}} \int_0^R \varrho^{\frac{n}{2}} J_{\frac{n}{2}-1}(\varrho\, r)\, \mathrm{d}\varrho \quad (r = |x - \xi|) \end{aligned}$$

Thus substituting τ for $\varrho\, r$, we get

$$H_R(\xi) = (2\pi)^{\frac{n}{2}-1}\, r^{-n} \int_0^{Rr} \tau^{\frac{n}{2}} J_{\frac{n}{2}-1}(\tau)\, \mathrm{d}\tau .$$

One of the results of the theory of Bessel functions is

$$\tau^{\frac{n}{2}} J_{\frac{n}{2}-1}(\tau) = \frac{\mathrm{d}}{\mathrm{d}\tau} \left(\tau^{\frac{n}{2}} J_{\frac{n}{2}}(\tau)\right),$$

hence

$$H_R(\xi) = (2\pi)^{\frac{n}{2}-1} r^{-n} (R r)^{\frac{n}{2}} J_{\frac{n}{2}}(R r) = (2\pi)^{\frac{n}{2}-1} |\xi - x|^{-\frac{n}{2}} R^{\frac{n}{2}} J_{\frac{n}{2}}(R r).$$

For $\varphi_R(x)$ we get the expression

$$\varphi_R(x) = (2\pi)^{-\frac{n}{2}-1} R^{\frac{n}{2}} \int \varphi(\xi) |\xi - x|^{-\frac{n}{2}} J_{\frac{n}{2}}(R|\xi - x|)\, \mathrm{d}\xi$$

$$= (2\pi)^{-\frac{n}{2}-1} R^{\frac{n}{2}} \int \varphi(\xi + x) |\xi|^{-\frac{n}{2}} J_{\frac{n}{2}}(R|\xi|)\, \mathrm{d}\xi. \tag{8}$$

In the last integral we first integrate with respect to ξ over the sphere $|\xi| = r$. We put

$$\Phi(r, x) = \frac{1}{\omega_n} \int_{\Omega} \varphi(x + \omega r)\, \mathrm{d}\omega. \tag{9}$$

This quantity is the mean value of the function φ on a sphere of radius r, centre x. The integral (8) now reduces to the form

$$\varphi_R(x) = C R^{\frac{n}{2}} \int_0^\infty \Phi(r, x)\, r^{\frac{n}{2}-1} J_{\frac{n}{2}}(R r)\, \mathrm{d}r$$

$$= C \int_0^\infty \Phi\left(\frac{\tau}{R}, x\right) J_{\frac{n}{2}}(\tau)\, \tau^{\frac{n}{2}-1}\, \mathrm{d}\tau.$$

Let us put

$$f(\tau) = \begin{cases} \Phi(x, \tau)\, \tau^{\frac{n}{2}-1} & \text{for } n \text{ even.} \\ \Phi(x, \tau)\, \tau^{\frac{n}{2}-\frac{1}{2}} & \text{for } n \text{ odd;} \end{cases}$$

then, denoting the integer $n/2 - 1$ or $n/2 - 1/2$ by m, we get

$$\varphi_R(x) = C R^m \int_0^\infty f\left(\frac{\tau}{R}\right) J_m^*(\tau)\, \mathrm{d}\tau, \tag{10}$$

where

$$J_m^*(\tau) = \begin{cases} J_{\frac{n}{2}}(\tau) & \text{for } n \text{ even,} \\ \dfrac{J_{\frac{n}{2}}(\tau)}{\sqrt{\tau}} & \text{for } n \text{ odd.} \end{cases}$$

It is known that the function $J_{n/2}(\tau)$ is bounded as $\tau \to 0$, and at infinity has the form

$$J_{\frac{n}{2}}(\tau) = a_0 \frac{e^{i\frac{n}{2}\tau}}{\sqrt{\tau}} + H(\tau),$$

where $H(\tau)$ is absolutely integrable. It follows that the integral of $J_{n/2}(\tau)$ is defined and finite on $(0, \infty)$ (and converges conditionally as $\tau \to \infty$). We put

$$J^1_{\frac{n}{2}}(x) = \int_x^\infty J_{\frac{n}{2}}(\tau)\, d\tau.$$

The asymptotic behaviour of the function $J^1_{n/2}(x)$ as $x \to \infty$ is identical with that of $J_{\frac{n}{2}}(x)$. For integrating by parts, we find that for any $r > 0$

$$\int_x^\infty \frac{e^{ia\tau}}{\tau^r} d\tau = \frac{e^{iar}}{ia\tau^r}\Bigg|_x^\infty - \int_x^\infty \frac{re^{ia\tau}\, d\tau}{ia\tau^{r+1}} = -\frac{e^{iax}}{iax^r} + \theta \frac{C}{x^{r+1}} \, (|\theta| \leqq 1),$$

which ensures that the second term is absolutely integrable as $x \to \infty$. As $x \to 0$, $J^1_{n/2}(x)$ is bounded in virtue of the convergence of the integral $\int\limits_0^\infty J_{n/2}(x)\, dx$. We can construct functions

$$J^2_{\frac{n}{2}}(x) = \int_x^\infty J^1_{\frac{n}{2}}(\tau)\, d\tau, \quad J^3_{\frac{n}{2}}(x) = \int_x^\infty J^2_{\frac{n}{2}}(\tau)\, d\tau, \ \ldots;$$

they all possess the same asymptotic properties as $J_{n/2}(x)$. The same applies to the function $J^*_m(x)$.

After these preparatory remarks we formulate a theorem.

THEOREM (S. Bochner, 1932). *If the function* $f(\tau) = \tau^m \Phi(\tau, x)$ (cf. (9)) *is bounded, continuous, and absolutely integrable (from* 0 *to* ∞) *and together with its derivatives up to order m is of bounded variation, then for all x*

$$\varphi(x) = \lim_{R\to\infty} \varphi_R(x) = \frac{1}{(2\pi)^n} \lim_{R\to\infty} \int\limits_{|\sigma| \leqq R} \psi(\sigma)\, e^{i(\sigma, x)}\, d\sigma. \quad (11)$$

Proof. Integrating (10) by parts, we get

$$\int_0^\infty f\left(\frac{\tau}{R}\right) J^*_m(\tau)\, d\tau = f\left(\frac{\tau}{R}\right) J^{*1}_m(\tau)\Bigg|_0^\infty - \frac{1}{R}\int_0^\infty f'\left(\frac{\tau}{R}\right) J^{*1}_m(\tau)\, d\tau.$$

Since the function $f(\tau)$ has a zero of order not less than m at $\tau = 0$ and is bounded as $x \to \infty$, the first term on the right vanishes. Repeated integration by parts gives

$$(-1)^m R^m \int_0^\infty f\left(\frac{\tau}{R}\right) J_m^*(\tau)\, d\tau = \int_0^\infty f^{*(m)}\left(\frac{\tau}{R}\right) J_m^{*m}(\tau)\, d\tau .$$

We split the resulting integral into two parts:

$$\int_0^\infty = \int_0^N + \int_N^\infty .$$

In virtue of general theorems on the limiting passage under the integral sign, we have as $R \to \infty$

$$\int_0^N f^{(m)}\left(\frac{\tau}{R}\right) J_m^{*m}(\tau)\, d\tau \to \int_0^N f^{(m)}(0)\, J_m^{*m}(\tau)\, d\tau = C_m f^{(m)}(0).$$

The second integral admits the bound

$$\left|\int_N^\infty f^{(m)}\left(\frac{\tau}{R}\right) J_m^{*m}(\tau)\, d\tau\right| \le \left| C \int_N^\infty f^{(m)}\left(\frac{\tau}{R}\right) \frac{e^{ia\tau}}{\tau^p}\, d\tau\right| + \left|\int_N^\infty f^{(m)}\left(\frac{\tau}{R}\right) H(\tau)\, d\tau\right| ,$$

where the function $H(\tau)$ is absolutely integrable, and $p = 1/2$ or 1. Since $f^{(m)}(x)$ is of bounded variation, we can write

$$f^{(m)}(x) = A(x) - B(x),$$

where $A(x)$, $B(x)$ are non-negative bounded non-decreasing functions. Then

$$\left|\int_N^\infty f^{(m)}\left(\frac{\tau}{R}\right) \frac{e^{ia\tau}}{\tau^p}\, d\tau\right| \le \left|\int_N^\infty A\left(\frac{\tau}{R}\right) \frac{e^{ia\tau}}{\tau^p}\, d\tau\right| + \left|\int_N^\infty B\left(\frac{\tau}{R}\right) \frac{e^{ia\tau}}{\tau^p}\, d\tau\right|$$

$$\le A\left(\frac{N}{R}\right) \frac{1}{N^p\, 2\pi a} + B\left(\frac{N}{R}\right) \frac{1}{N^p\, 2\pi a} \to 0$$

as $N \to \infty$ independently of the value of R.

It follows that

$$R^m \int_0^\infty f\left(\frac{\tau}{R}\right) j_m^*(\tau)\, \mathrm{d}\tau \to f^{(m)}(0)\, C_m.$$

Thus

$$\lim \varphi_R(x) = f^{(m)}(0)\, C_m' = \varphi(x)\, C_m'' \tag{12}$$

for some constant C_m''. But since the relation (12) holds, for example, for the function e^{-x^2} with $C_m'' = 1$, the constant C_m'' is equal to unity for all $\varphi(x)$, and the theorem is proved.

Note. The conditions of the theorem are satisfied, for example, if we suppose that the function $\varphi(x)$ itself has derivatives up to order $m = [n/2 - 1]$, which are absolutely integrable over the whole space. This follows immediately from formula (9) if we consider that the integrals

$$\int_0^\infty \frac{1}{\omega_n} \left\{ \int_\Omega \frac{\partial^k \varphi(x + \omega r)}{\partial x_1^{k_1} \ldots \partial x_n^{k_n}} \mathrm{d}\omega \right\} r^{n-1}\, \mathrm{d}r$$

exist.

Let us consider now the question of the summability of the n-tuple Fourier integral by the method of arithmetic means. We proceed from the expression

$$\varphi_\varrho(x_1, \ldots, x_n) = \frac{1}{(2\pi)^n} \int_{-\varrho}^{\varrho} \cdots \int_{-\varrho}^{\varrho} \psi(\sigma_1, \ldots, \sigma_n)\, e^{i(x,\sigma)}\, \mathrm{d}\sigma$$

$$= \frac{1}{(2\pi)^n} \int \cdots \int \varphi(\xi_1, \ldots, \xi_n) \left\{ \int \cdots \int e^{i(x-\xi,\sigma)}\, \mathrm{d}\sigma \right\} \mathrm{d}\xi$$

$$= \frac{1}{\pi^n} \int \cdots \int \varphi(\xi_1, \ldots, \xi_n) \prod \frac{\sin(x_j - \xi_j)}{x_j - \xi_j}\, \mathrm{d}\xi$$

to the mean over the interval $0 \leqq \varrho \leqq R$:

$$\sigma_R(x_1, \ldots, x_n) = \frac{1}{R^n} \int_0^R \varphi_\varrho(x_1, \ldots, x_n)\, \mathrm{d}\varrho$$

$$= \frac{1}{\pi^n} \frac{1}{R^n} \int \varphi(x + t) \prod_{j=1}^{n} \frac{\sin^2 R \dfrac{t_j}{2}}{t_j^2}\, \mathrm{d}t_1, \ldots, \mathrm{d}t_n.$$

The Fejer kernel

$$\Phi(R, t) = \prod_{j=1}^{n} \frac{\sin \frac{R}{2} t_j}{t_j^2}$$

is the product of n one-dimensional Fejer kernels of the form $(\sin^2 Rt/2)t^2$. It possesses the following properties:

(a) $\Phi(R, t) \geq 0;$

(b) $\int \cdots \int \Phi(R, t)\, \mathrm{d}t = 1;$

(c) $\int \cdots \int_{|t| \geq \delta} \Phi(R, t)\, \mathrm{d}t \to 0 \quad \text{as} \quad R \to \infty.$

The last is equivalent to the limiting relation

$$\int_{\substack{|t_j| \leq \delta_j \\ j=1,2,\ldots,n}} \Phi(R, t)\, \mathrm{d}t = \prod_{j=1}^{n} \int_{|t_j| \leq \delta_j} \frac{\sin^2 \frac{R t_j}{2}}{t_j^2} \mathrm{d}t_j \to 1,$$

which holds since each of the one-dimensional kernels has the given property.

Just as in the single variable case, it follows from properties (a)–(c) that

$$\varphi_R(x) \to \varphi(x), \tag{13}$$

if the function $\varphi(x)$ is continuous.

More generally: if $\varphi(x) = \varphi(x_1, \ldots, x_n)$ belongs to some normed space $E \subset L_1(R)$ and is continuous relative to displacement in that space, i.e.

$$\lim_{t \to 0} \|\varphi(x_1 + t_1, \ldots, x_n + t_n) - \varphi(x_1, \ldots, x_n)\| = 0,$$

then relation (13) holds in the norm of the space:

$$\|\varphi(x) - \varphi_R(x)\| \to 0.$$

As in the case of one variable, this last theorem reduces to a uniqueness theorem for the Fourier transform when applied to the space of all integrable functions: *if the Fourier transforms $\psi_1(\sigma)$, $\psi_2(\sigma)$ of two integrable functions $\varphi_1(x)$, $\varphi_2(x)$ coincide identically, then $\varphi_1(x)$, $\varphi_2(x)$ coincide almost everywhere.*

A class S (Section 3) can also be defined in the case of n independent variables; it is composed of the infinitely differentiable functions $\varphi(x_1, \ldots, x_n)$ for which, for any $k_1, \ldots, k_n, q_1, \ldots, q_n$, the quantities

$$\left| x_1^{k_1} \cdots x_n^{k_n} \frac{\partial^{q_1+q_1+\cdots+q_n}\partial}{\partial x_1^{q_1} \cdots \varphi\, x_n^{q_n}} \right|$$

are bounded over the whole space. Again as in the case of one variable this class carries into itself under Fourier transformation.

Finally, the whole L_2-theorem of (Section 6) extends to the case of n-variables: if the function $\varphi(x)$ is square-summable over R_n, then its Fourier transform, defined by the formula

$$\psi(\sigma) = \lim_{N\to\infty} \int_{-N}^{N} \cdots \int_{-N}^{N} \varphi(x)\, e^{-i(x,\sigma)}\, dx \tag{17}$$

(the limit being taken in the metric of the space of square-summable functions of σ), exists and belongs to L_2, and

$$\int \cdots \int |\psi(\sigma)|^2\, d\sigma = (2\pi)^n \int \cdots \int |\varphi(x)|^2\, dx. \tag{18}$$

There is also an analogue of the Wiener–Paley theorem, and of the Bochner–Khinchin theorem on the representation of positive-definite functions (Section 7).

Problems. 1. Prove that the Fourier transform of the function

$$e^{-\sum_{j=1}^{n}\sum_{k=1}^{n} a_{jk}x_j x_k} \tag{19}$$

where the exponent is the negation of a positive-definite quadratic form, is equal to

$$\frac{\pi^{\frac{n}{2}}}{\sqrt{D}}\, e^{\frac{D(\sigma)}{4D}}, \tag{20}$$

where

$$D = \det \|a_{jk}\|, \quad D(\sigma) = \begin{vmatrix} 0 & \sigma_1 & \cdots & \sigma_n \\ \sigma_1 & a_{11} & \cdots & a_{1n} \\ \cdots & \cdots & \cdots & \cdots \\ \sigma_n & a_{n1} & \cdots & a_{nn} \end{vmatrix} \tag{21}$$

Hint. Reduce the quadratic form to canonical form by means of an orthogonal transformation.

2. Find the Fourier transform of the function $e^{-ar}\left(r = \sqrt{\sum_{j=1}^{n} x_j^2}\right)$.

Answer:

$$F(e^{-ar}) = \frac{2^n \pi^{\frac{n+1}{2}} \Gamma\left(\frac{n+1}{2}\right)}{\left(\sqrt{a^2+\varrho^2}\right)^{\frac{n+1}{2}}} \qquad \left(\varrho = \sqrt{\sum_{j=1}^{n} \sigma_j^2}\right)$$

Concluding Remark

Fourier series and integrals have their origin as a means of solving problems in mathematical physics in the works of J. B. Fourier (French mathematician 1768–1830) and were systematised in his book *The Analytic Theory of Heat* (1822). The Laplace transform was applied by Euler (1737) to the solution of differential equations; P. S. Laplace (French mathematician, astronomer, and physicist 1749–1827) developed it and made extensive use of it in his book *The Analytic Theory of Probability* (1812). The Fourier and Laplace transforms have become one of the primary tools of mathematical physics of the nineteenth and twentieth centuries. The problem of representing arbitrary functions in Fourier series, which captured the attention of the foremost mathematicians of these centuries, did much to promote the emergence of a theory of functions of a real variable. The Lebesgue integral and the related equation $F[L_2(-\infty, \infty)] = L_2(-\infty, \infty)$ rendered the Fourier transform indispensable in constructing the basic concepts of theoretical physics (quantum mechanics). The thirties saw important new advances in the theory of the Fourier integral—in particular, the proofs of the Wiener–Paley (Section 6, art. 3) and Bochner–Khinchin (Section 7) theorems. In contemporary investigations the technique of Fourier transformation is of increasing importance, and in particular, has enabled an approach to be made to the solution of fundamental problems in the general theory of linear partial differential equations with constant coefficients. Recommended literature: A. Zygmund, *Trigonometric Series*, Stechert, N.Y. (1935); E. C. Titchmarsh, *Introduction to the Theory of Fourier Integrals*, Oxford (1937); B. van der Pol and H. Bremmer, *Operational Calculus*, Cambridge (1950); D. I. Blochintsev, *Foundations of Quantum Mechanics*, State Tech. Pub.

Dept., (1949); I. M. Gelfand and G. E. Shilov, *Generalised Functions*, Nos. 1–3, Fizmatgiz (1958) (2nd ed., No. 1, 1960), I. M. Gelfand and N. Ya. Vilenkin, *Generalised Functions*, No. 4, Fizmatgiz (1961) (being translated, Academic Press, New York).

SUPPLEMENT

1. Further Remarks on Sets

The ideas and propositions that have made up Chapter 1 of our book are sometimes referred to as "Naive set theory". The fact is that even the early stages of the development of set theory revealed paradoxes and contradictions indicating that no advance is possible without clear axioms and rules of proof. In particular, many doubts have been expressed about the so-called "axiom of choice", or its equivalent, Zorn's lemma (see below). This subject became pressing, inasmuch as the axiom of choice proved necessary for the development of the more advanced branches of analysis itself and of other branches of mathematics. As from the twenties, several systems of set theory have been proposed, with axioms and rules of proof that avoid the classical paradoxes; unfortunately, self-consistency has not been proved for any of them. K. Gödel (1938) showed that association of the axiom of choice with one of them, the "Neumann–Bernays system", cannot lead to inconsistencies provided the system itself (without the axiom of choice) is consistent. On the other hand, Spekker (1933) showed that the axiom of choice does not hold in another system, the "Quine system", which admits of greater freedom in the treatment of sets. Our future constructions will relate to sets in the Neumann–Bernays sense; we shall not adduce here the relevant axioms and the reader must take our word that the constructions are carried out within the limits of the system.

Definition. A set A is said to be *partially ordered* if a comparison relation $\leqq$ (less than or equal to) is defined for certain pairs of its elements (these pairs being described as *comparable*) with fulfilment of the following conditions:

(1) $x \leqq x$ for any $x \in A$;

(2) if $x \leqq y$, $y \leqq x$, then $x = y$;

(3) if $x \leqq y$, $y \leqq z$, then $x \leqq z$.

We can regard any set A as partially ordered if the comparison relation $x \leqq y$ is assumed to be defined only for $y = x$. This case is trivial, and it seems natural to describe such a set as non-ordered. The opposite will be the case if any two elements x, y of A are comparable, so that the relations $x \leqq y$ or $y \leqq x$ always hold. In this case A is simply said to be *ordered*, or *linearly ordered*.

Examples. 1. The set of integers is linearly ordered with the usual definition of the sign $\leqq$. The same is true for the set of real numbers.

2. The set of points (x, y) of the plane is partially ordered if we assume that $(x_1, y_1) \leqq (x_2, y_2)$ provided $y_1 = y_2$, $x_1 \leqq x_2$.

3. The system $A, B, \ldots$ of subsets of a given set E is partially ordered if $A \leqq B$ implies the inclusion $A \subset B$.

Every subset of a partially ordered set is also partially ordered, with the same comparison relation. Generally speaking, linearly ordered subsets can be distinguished in a partially ordered set: for instance, the subset of points (x, y) on a horizontal line in example 2. A linearly ordered subset B of a partially ordered subset A is said to be *maximal* if no wider linearly ordered subset $B_1 \supset B$ exists. For example, in the partially ordered set of all points of a plane (example 2) the set of all points of any horizontal line is a maximal linearly ordered subset.

The following axiom plays an essential part in what follows.

Axiom (Zorn, 1935). *Every partially ordered set contains a maximal linearly ordered subset.*

Let us introduce two important new concepts.

Let B be a subset of a partially ordered set A. The set B is said to be *bounded* in A if an element $b_0 \in A$ exists such that $b \leqq b_0$ for all $b \in B$; every element $b_0 \in A$ satisfying this condition is described as an *upper bound* of the set B. An element x of a partially ordered set A is said to be *maximal* if there exists no element $y \neq x$ greater than x.

In the trivial partially ordered set mentioned above, every element x is maximal. At the same time, a partially ordered set may have no maximal element at all (like the set of integers).

The following theorem gives a sufficient condition for the existence of maximal elements in a partially ordered set.

Theorem 1 (Zorn). *If every linearly ordered subset B of a partially ordered set A is bounded, A contains a maximal element.*

Proof. Let B_0 be a maximal linearly ordered subset of A; by Zorn's axiom, such a subset exists. Let b_0 be an upper bound of the set B_0. We claim that b_0 is a maximal element. For, if there were to exist an element $a_0 \in A$ greater than b_0, the set $B_0 + a_0$ would again be linearly ordered and larger than B_0, which contradicts the fact that B_0 is maximal.

The proof of a theorem of great importance in set theory, called the theorem of choice, is based on theorem 1. Briefly, this theorem asserts that, given a system of sets $\{A_\alpha\}$, a set Z exists which contains precisely one point of each of the sets A_α.

THEOREM 2 (theorem of choice). *Let there be a set A_α for every index α of some set Λ. Then there exists a set of elements a_α such that, whatever the fixed α of Λ, the element a_α belongs to the set A_α.*

Proof. We consider the set $\mathfrak{U}$ of all functions x_α that are defined on subsets of the set Λ and take values in the set A_α. Every fixed element y_α of A_α determines one such function, which is defined only for one value of α. Thus $\mathfrak{U}$ is non-empty. A comparison relation can be set up between the functions x_α by taking $x_\alpha \leqq y_\alpha$ if y_α is defined in every case for the same values of α as x_α (and possibly for other values of α), whilst $y_\alpha = x_\alpha$ on the domain of definition of the function x_α. We take any linearly ordered subset $\mathfrak{U}_0 \supset \mathfrak{U}$. Let Λ_0 be the union of all subsets of the set Λ on which the functions $x_\alpha \in \mathfrak{U}_0$ are given. Certain of the functions $x_\alpha \in \mathfrak{U}_0$ are defined at every point $\alpha_0 \in \Lambda_0$, whilst the fact that they can be compared implies that they take the same value $x_{\alpha 0}$ for $\alpha = \alpha_0$. This uniquely defined value defines a single-valued function y_α on Λ_0. This latter can clearly be compared with all the functions $x_\alpha \in \mathfrak{U}_0$, and in fact $x_\alpha \leqq y_\alpha$. We see that the linearly ordered subset $\mathfrak{U}_0$ is bounded in $\mathfrak{U}$. By theorem 1, $\mathfrak{U}$ contains a maximal element a_α. We claim that the function a_α is defined on the whole set Λ. For, if an element $\alpha_0 \in \Lambda$ were not contained in its domain of definition, we could use any element $y_0 \in A_{\alpha_0}$ to widen its domain of definition by adding the value y_0 at the point α_0, and a_α would not be a maximal element of $\mathfrak{U}$. Hence a_α is the required function.

Many constructions of analysis are based, explicitly or implicitly, on the theorem of choice. For example, the proof of the existence of a non-measurable set (p. 167) in fact requires the employment of this theorem. Applications of it can be traced in more elementary discussions. Take, for instance, the proof of the equivalence of two familiar definitions of the continuity of a function $f(x)$ at a point x_0 of a metric space (the "ε–δ-definition" and the "lim-definition"). The second definition is usually proved equivalent to the first by *reductio ad absurdum*: it is assumed that the first definition is not satisfied; given $\varepsilon > 0$, we assign a sequence of values $\delta_n \to 0$ to δ and choose points x_n in the corresponding δ_n-neighbourhoods of the point x_0 such that $|f(x_0) - f(x_n)| > \varepsilon$, and thus obtain a contradiction with the second definition. It will be seen use has been made of the possibility of an arbitrary choice during the argument.

Long before Zorn, the theorem of choice was stated as an axiom by Zermelo (1904). Zermelo deduced from the "axiom of choice" a theorem—at first sight somewhat paradoxical—to the effect that every set can be "well ordered". To see the meaning of this result we compare the order properties of three linearly ordered sets: the set A_1 of all real numbers, the set A_2 of all real non-negative numbers and the set A_3 of all natural numbers. The set A_1 does not possess a first (least) element. The set A_2 has a first element 0 but no immediately consecutive element. The set A_3 has a first element 1 and an immediately consecutive element 2, etc.

The term "immediately consecutive" can be replaced by the more precise "least consecutive". We say that a linearly ordered set A is *well ordered* if every non-empty subset $C \subset A$ has a least element. We denote the least element of a well ordered set A by 1, the immediately consecutive element by 2, the next by 3, 4 etc. If the set A is not finite this process leads to the construction of a countable subset $B = (1, 2, ..., n, ...)$. B may not coincide with the whole of A, in which case there is an element immediately following B, which we denote by ω. We denote the element immediately following $B + \omega$ by $\omega + 1$; the next elements are naturally denoted by $\omega + 2, \omega + 3, ...$ The elements $\omega + \omega = 2\omega$, $3\omega, ..., \omega^2 = \omega \cdot \omega$ then make their appearance and so on. A canonical system of numbering the elements of a well ordered set is thus defined. But this numbering limits us to obtaining each time no more than a finite or countable subset, and it is difficult to imagine a non-countable well ordered set. We can now proceed to an effective statement of Zermelo's theorem:

A comparison relation $\leqq$ can be introduced into any set A in such a way that it is well ordered.

Zermelo's theorem may be deduced from Zorn's as follows.

Given any set A, let $\mathfrak{U}$ be the system of subsets of A that can be well ordered. If a subset $B \subset A$ can be well ordered in several, say p, different ways, it appears p times in the system $\mathfrak{U}$. This system is not empty, since it contains, for example, all single-element subsets of A. We introduce partial orderedness into the system $\mathfrak{U}$ by assuming that $B_1 \leqq B_2$ if $B_1 \subset B_2$, the order relation is the same on B_1 as on B_2, and all the elements of $B_2 - B_1$ are greater than any element of B_1 (so that B_1 is so to speak the origin of B_2). We show that the condition of Zorn's theorem is satisfied. Let $\mathfrak{U}_0 \supset \mathfrak{U}$ be a linearly ordered system. By hypothesis, there is a unique well orderedness for all elements of a set of the system $\mathfrak{U}_0$. This will also hold on the union B of all subsets of $\mathfrak{U}_0$. The union B, together with the comparison relation holding on it, belongs to U_0, so that U_0 is bounded. By Zorn's theorem, U contains a maximal element A_0. The element A_0 is a subset of A admitting of well orderedness. We claim that $A_0 = A$; for, if there were an element $z \in A$ not belonging to A_0, we could consider the subset $A_0 + z$ and regard z as following all the elements of A_0; but in this case A_0 would not be a maximal well ordered subset. Hence $A_0 = A$, and the theorem is proved.

2. Theorems on Linear Functionals

The following theorem enables homeomorphic linear functionals to be constructed in a linear space. Its "complex" statement originates from Hahn (1927) (although Hahn himself only considered real spaces), and the "real" statement (which imposes less conditions on the functional $p(x)$) from Banach (1929).

Theorem (Hahn–Banach).

(a) *let $p(x)$ be a functional given in a real linear space E and satisfying the conditions*

$$p(x + y) \leqq p(x) + p(y), \quad p(\lambda x) = \lambda p(x) \quad \text{for} \quad \lambda \geqq 0. \quad (1)$$

Further, let $f(x)$ be a given linear functional in a subspace $E_0 \subset E$, satisfying the inequality

$$f(x) \leqq p(x). \quad (2)$$

We claim that there exists a linear functional $f^*(x)$ in space E that coincides with $f(x)$ on E_0 and satisfies everywhere on E the ine-

quality

$$f^*(x) \leqq p(x). \tag{3}$$

In other words, we can continue $f(x)$ from E_0 on to all of E whilst retaining inequality (2).

(b) *let $p(x)$ be a functional given in a complex linear space E and satisfying the conditions*

$$p(x + y) \leqq p(x) + p(y), \quad p(\lambda x) = |\lambda|\, p(x) \tag{1'}$$

for any complex λ.

Further, let $f(x)$ be a given linear functional on the subspace $E_0 \subset E$, satisfying the inequality

$$|f(x)| \leqq p(x). \tag{2'}$$

We claim that there exists a linear functional $f^*(x)$ in E, coinciding with $f(x)$ on E_0 and satisfying everywhere on E the inequality

$$|f^*(x)| \leqq p(x). \tag{3'}$$

In other words, the functional $f(x)$ can be continued from E_0 to all of E whilst retaining inequality (2').

Proof. We denote elements of the subspace E_0 by y. We show first that the functional f can be extended from E_0 on to a subspace E_1 with one more dimension.

More precisely, what we do is this: we take any vector x_0 not belonging to E_0 and show that it is possible to construct a linear functional $f_1(x)$, defined on the subspace E_1 of all linear combinations of vectors $y \in E$ and the vector x_0, in such a way that $f_1(x)$ coincides with $f(x)$ on E_0 and is bounded on E_1 by the same functional $p(x)$ (in absolute value in the complex case).

We first perform the construction for the real case. We define f_1 in accordance with the natural formula

$$f_1(x) = f_1(y + \lambda x_0) = f(y) + \lambda f_1(x_0), \tag{4}$$

where the number $f_1(x_0)$ is defined in such a way that condition (2) is satisfied. Condition (2) can now be written as

$$f_1(y) = f(y) + \lambda f_1(x_0) \leqq p(y + \lambda x_0). \tag{5}$$

When $\lambda > 0$ we can write this as

$$f(y/\lambda) + f_1(x_0) \leqq p(y/\lambda + x_0)$$

or

$$f(y_1) + f_1(x_0) \leqq p(y_1 + x_0), \; y_1 = y/\lambda. \tag{6}$$

When $\lambda > 0$ we put $\mu = -\lambda$ and obtain

$$f(y) - \mu f_1(x_0) \leqq p(y - \mu x_0),$$

or, dividing by μ and denoting $y_2 = y/\mu$,

$$f(y_2) - f_1(x_0) \leqq p(y_2 - x_0). \tag{7}$$

Conversely, if the number $f_1(x_0)$ satisfies inequalities (6) and (7) for any $y_1 \in E_0$, $y_2 \in E_0$, then (5) is also satisfied for any $y \in E_0$ and any λ. Inequalities (6), (7) can be written as

$$f(y_2) - p(y_2 - x_0) \leqq f_1(x_0) \leqq p(y_1 + x_0) - f(y_1).$$

We see that the solution of our problem depends on the relationship between the numbers

$$\alpha = \sup_{y_2 \in E_0} \{f(y_2) - p(y_2 - x_0)\}$$

and

$$\beta = \inf_{y_1 \in E_0} \{p(y_1 + x_0) - f(y_1)\}.$$

If $\alpha \leqq \beta$, the problem is soluble; if $\alpha > \beta$, it could prove impossible to find a number $f_1(x_0)$ satisfying condition (2). Let us show that always, in fact, $\alpha \leqq \beta$. We need to show that

$$f(y_2) - p(y_2 - x_0) \leqq p(y_1 + x_0) - f(y_1),$$

for any y_1, y_2 of E_0, or what is the same thing,

$$f(y_1) + f(y_2) \leqq p(y_1 + x_0) + p(y_2 - x_0). \tag{8}$$

We show that (8) in fact holds. We have

$$\begin{aligned} f(y_1) + f(y_2) &= f(y_1 + y_2) \leqq p(y_1 + y_2) \\ &= p(y_1 + x_0 + y_2 - x_0) \leqq p(y_1 + x_0) + p(y_2 - x_0), \end{aligned}$$

as required. Thus the required number $f_1(x_0)$ exists, so that the required continuation of the functional f is possible.

In the complex case we resolve the functional $f(y)$ into real and imaginary components:

$$f(y) = g(y) + i\,h(y).$$

The functionals $g(y)$ and $h(y)$ are real and linear in E_0, regarded as a real space, and are bounded along with the functional f by the same functional $p(x)$. We also have

$$f(i\,y) = i\,f(y) = g(i\,y) + i\,h(i\,y) = i\,g(y) - h(y),$$

and $h(y) = -g(i\,y)$, whence

$$f(y) = g(y) - i\,g(i\,y).$$

By what we have proved, the functional $g(y)$ can be continued whilst retaining inequality (2′) on to the real subspace $E_{1/2}$ of real linear combinations of vectors $y \in E_0$ and the vector x_0, and then, whilst still retaining the inequality, on to the subspace $E_1 \supset E_{1/2}$ of real linear combinations of vectors $y \in E_0$, x_0 and $i\,x_0$. We put

$$f(x) = g(x) - i\,g(i\,x)$$

on E_1. We get an extension of the functional f from E_0 onto E_1. Let us verify that it remains a linear functional in the complex space E_1. It is sufficient to verify that $f(i\,x) = i\,f(x)$. For

$$\begin{aligned} f(i\,x) &= g(i\,x) - i\,g(-x) \\ &= i[-i\,g(i\,x) - g(-x)] = i[g(x) - i\,g(i\,x)] = i\,f(x) \end{aligned}$$

It remains to show that inequality (3′) is satisfied for the functional f in E_1. Given x_0, we choose a real number θ such that $e^{i\theta} f(x)$ is a real non-negative number. Then $e^{i\theta} f(x) = f(e^{i\theta} x) = g(e^{i\theta} x)$ and consequently,

$$|f(x)| = |e^{i\theta} f(x)| = g(e^{i\theta} x) \leqq p(e^{i\theta} x) = p(x),$$

as required.

Thus a given functional can always be extended—in the real or complex case—from a given subspace $E_0 \neq E$ to a larger space E_1 whilst retaining inequality (3) or (3′). We show that an extension of the functional f exists on to the whole space E whilst retaining the relevant inequality. We apply Zorn's theorem of Section 1, and consider the system $\mathfrak{U}$ of the subspaces of E on which the required extension is possible. It will be assumed that a given subspace E_α appears as many times in $\mathfrak{U}$ as there are possible methods of extending the functional f onto E_α. The whole family $\mathfrak{U}$ can be partially ordered by taking $E_\alpha \leqq E_\beta$ if $E_\alpha \subset E_\beta$ and the values of the functional f, extended on to E_α and E_β, coincide on E_α. Let $\mathfrak{U}_0$ be a linearly ordered subsystem of $\mathfrak{U}$. The functional f is uniquely defined and satisfies inequality (3) or (3′) on the union E_ω of all subspaces E_α belonging to $\mathfrak{U}$. Consequently E_ω itself belongs to $\mathfrak{U}$, and obviously, is an upper bound for all the $E_\alpha \in \mathfrak{U}_0$. Hence $\mathfrak{U}_0$ is bounded. By Zorn's theorem, the system contains a maximal element E^*. The linear functional f can be extended on to the subspace E^* whilst retaining inequality (3) or (3'). Suppose $E^* \neq E$; then f could be extended on to a larger subspace, which contradicts the fact that E^* is maximal. Hence $E^* = E$, and the theorem is proved.

Some corollaries of the Hahn–Banach theorem may be mentioned.

1. The norm of the element x, or a multiple of it, can be taken as the functional $p(x)$ (in both the real and the complex case). We now find, in particular, that *a functional, satisfying the inequality*

$$|f(x)| \leqq C\|x\|$$

on the subspace $E_0 \subset E$*—and hence having a norm not exceeding* C *on* E_0*—can be extended to all of* E *whilst preserving this norm.* This is the most commonly encountered corollary of the Hahn–Banach theorem.

2. *There exists for every element* $x_0 \neq 0$ *of a normed space* E *a linear functional* f, *defined on all* E, *of norm* 1, *and such that* $f(x_0) \neq 0$. For, putting $f(\lambda x_0) = \lambda\|x_0\|$, we get a linear functional with norm 1, defined on the one-dimensional subspace generated by the element x_0. We continue it onto all E whilst preserving the norm and obtain the required functional.

3. There exists for any closed subspace $E_0 \neq E$ and element $x_0 \in E_0$ a linear functional f with norm 1, equal to 0 on E_0 and such that $f(x_0) \neq 0$.

For we can define the functional f on the subspace E_1 composed of all vectors of the form $x = y + \lambda x_0$ ($y \in E_0$, λ is any number) by means of the formula

$$f(x) = c\lambda,$$

where c is a positive constant. We have on E_1:

$$\|f\| = \sup_{y \in E_0} \frac{c|\lambda|}{\|y + \lambda x_0\|} = \sup_{y \in E_0} \frac{c}{\left\|\dfrac{y}{\lambda} + x_0\right\|}$$

$$= \frac{c}{\inf\limits_{y \in E_0} \left\|\dfrac{y}{\lambda} + x_0\right\|} = \frac{c}{d},$$

where $d = \inf\limits_{y \in E_0} \|y/\lambda + x_0\|$ is the distance from the element x_0 to the subspace E_0; d is positive, since E_0 is closed. It may be seen that, if we put $c = d$, a functional with norm 1 is obtained on E_1. On continuing it further onto all E whilst preserving the norm, the required functional is obtained.

Note 1. An alternative statement of corollary 3 is: *there exists for any subspace $E_0 \not\subset E$ and element x_0, not belonging to the closure of E_0, a linear functional with norm 1, equal to 0 on E_0 and such that $f(x_0) \neq 0$.*

Note 2. The functional $p(x)$, figuring in the complex statement of the Hahn–Banach theorem, is "almost" a norm; it differs from the norm in that it can vanish, not only for $x \neq 0$, but also on an entire manifold. We shall call such a functional a *semi-norm*. In the real case $p(x)$ differs still further from the norm in its properties; it can even take negative values.

Note 3. The Hahn-Banach theorem can be applied with great advantage in certain constructions of analysis. For example, when deducing the general form of a continuous linear functional in space $C(a, b)$ (Chapter VI, Section 7, art. 1), we could have avoided the argument about continuing the functional onto a larger space (stage I) and applied the theorem directly instead.

It is easily verified in this case, as in stage II, that the resulting function $F(\xi) = f[\chi_{a,\xi}(x)]$ is of bounded variation. But it will not in general be true that this function is continuous from the right, as is required for total additivity of the measure. Though this fact may not be important for a single variable, since the integrating function can be "improved" at discontinuities by making it continuous from the right (art. 2 Section 6), the possibility of such improvement is a much more complicated matter in the case of several variables.

Let us mention some facts in connection with sequences of linear functionals.

Theorem (Banach and Steinhaus, 1927). *If the values of linear functionals $f_1(x), \ldots, f_n(x), \ldots$ are bounded on every fixed element $x \in E$, they are uniformly bounded on the unit sphere of space E, in other words, the norms of functionals $f_n(x)$ have a common bound.*

Proof. If the sequence of linear functionals $f_n(x)$ is not bounded in the unit sphere $\|x\| \leqq 1$, it is evidently not bounded in any sphere $\|x\| \leqq r$; and further, it is not bounded in any sphere $U(x_0, r) = \{x: \|x - x_0\| \leqq 1\}$, since if the numbers $f_n(x)$, $f_n(x_0)$ were bounded for $x \in U(x_0, r)$, the numbers $f_n(x - x_0) = f_n(x) - f_n(x_0)$ would also be bounded, which is impossible, since $x - x_0$ runs over the sphere of radius r, centre the origin. We note this and choose an element x_1 in the unit sphere, $\|x_1\| < 1$, on which one of the functionals f_n, say f_1, exceeds 1 in absolute value:

$$|f_1(x_1)| > 1.$$

Since f_1 is a continuous functional, there exists a sphere $\|x - x_1\| \leqq r_1$, lying wholly inside the initial sphere $\|x\| \leqq 1$, in which the inequality

$$|f_1(x)| > 1$$

is satisfied. We find an element x_2 inside this sphere, and a functional f_2, such that

$$|f_2(x_2)| > 2,$$

then choose a new sphere $\|x - x_2\| \leqq r_2$, lying in the previous one, at every point of which

$$|f_2(x)| > 2.$$

On proceeding thus, we get a sequence of nested spheres of radii $r_1, r_2, \ldots, r_n, \ldots$, tending to zero. The inequalities

$$|f_1(x_0)| > 1, |f_2(x_0)| > 2, \ldots, |f_n(x_0)| > n, \ldots$$

hold at the common point of all these spheres (which exists by virtue of the completeness of space R and the lemma of Chapter II, Section 5), i.e. the numbers $f_n(x_0)$ are unbounded, which contradicts the hypothesis.

COROLLARY. *If a sequence of continuous linear functionals $f_1(x), \ldots, f_n(x), \ldots$ is convergent at every $x \in E$, the limiting functional $f(x) = \lim f_n(x)$ is also linear and continuous.*

For, the linear properties of $f(x)$ are got by a passage to the limit in the equation $f_n(\alpha x + \beta y) = \alpha f_n(x) + \beta f_n(y)$, whilst it follows from Banach's theorem that the values of this functional are bounded in the unit sphere.

Problems. 1. A linear functional f is given on a subspace E_0 of a linear space E (without a norm). Show that it can be continued linearly onto all of E.

Hint. Use the scheme of proof of the Hahn–Banach theorem (without bothering about the value of $f(x_0)$).

2. Show that there exists a linear functional, defined everywhere in an infinite-dimensional normed space, for which

$$\sup_{|x| \leqq 1} |f(x)| = \infty.$$

Hint. Modify the scheme of proof of the Hahn–Banach theorem in such a way that the norm of the functional is increased by one for every extension "by one dimension".

Start from an arbitrary functional, defined on a one-dimensional space; after a countable number of extensions by one dimension an unbounded functional is obtained. Use problem 1 for the further extension (to the whole of E).

3. Prove that the following spaces of (complex) numerical sequences (with the natural linear operations) are complete:
(a) the space c_0 of sequences tending to zero:

$$x = (\xi_1, \xi_2, \dots, \xi_n, \dots), \quad \lim \xi_n = 0,$$

with the norm

$$\|x\| = \sup_n |\xi_n|;$$

(b) the space l_1 of sequences

$$\alpha = (\alpha_1, \alpha_2, \dots, \alpha_n, \dots), \quad \sum_{n=1}^{\infty} |\alpha_n| < \infty,$$

with the norm

$$\|x\| = \sum_{u=1}^{\infty} |\alpha_n|;$$

(c) the space m of bounded numerical sequences

$$x = (\xi_1, \xi_2, \dots, \xi_n, \dots), \quad \sup_n |\xi_n| < \infty,$$

with the norm

$$\|x\| = \sup_n |\xi_n|.$$

Note. The space c_0 is a closed subset of space m; c_0 is separable, whilst m is not separable (see problem 14 of Chapter II, Section 3), since it possesses a set of elements with the power of a continuum with distance 1 between any two elements (see theorem 3, Chapter I, Section 5).

4. Find the general form (and the norm) of a continuous linear functional in space c_0 and in space l_1 (problem 8 of Section 9).

Hint. Using the notation $e_n = (0, \dots, 0, \underset{n}{\underbrace{1}}, 0, \dots)$, any element of either space can be developed into a series in e_n, convergent in the norm of the space. Hence it is sufficient to find the numbers $f(e_n)$ and discuss their properties.

Answer. A continuous linear functional $f(x)$ is defined in space c_0 by an element $\alpha \in l_1$, so that

$$f(x) = \sum_{n=1}^{\infty} \alpha_n \xi_n, \quad \|f\| = \sum_{n=1}^{\infty} |\alpha_n|$$

for $x = (\xi_1, \dots, \xi_n, \dots) \in c_0$. A continuous linear functional $g(\alpha)$ is given in space l_1 by an element $y = (\eta_1, \dots, \eta_n, \dots) \in m$, so that

$$g(\alpha) = \sum_{n=1}^{\infty} \eta_n \alpha_n, \quad \|g\| = \sup_n |\eta_n|$$

for $\alpha = (\alpha_1, \dots, \alpha_n, \dots) \in l_1$.

Note. We might try to show that any functional in space m can be written in the form

$$f(x) = \sum_{n=1}^{\infty} \beta_n \xi_n, \tag{9}$$

where β_n is a fixed sequence of numbers and $x = (\xi_1, \ldots, \xi_n, \ldots)$. This is not the case, however, We shall see in the next problem that there exists a linear functional $f(x)$ in space m which maps every convergent sequence $x = (\xi_1, \ldots, \xi_n, \ldots)$ onto its limit $\lim \xi_n$. Such a functional cannot have the form (9). For, application of (9) to the element e_n would give $f(e_n) = \beta_n = 0$, so that $f(x) = \sum_{n=1}^{\infty} \beta_n \xi_n = 0$; and this last is false, since e.g. $f(e) = 1$, where $e = (1, 1, \ldots, 1, \ldots)$.

5. Construct on the space m of all bounded complex sequences $x = (\xi_0, \ldots, \xi_n, \ldots)$ with the norm $\|x\| = \sup_n |\xi_n|$ a linear functional with norm 1 that maps every sequence x onto a number ξ, lying in the least convex set that contains all the limit points of the sequence ξ_n (and in particular, maps every convergent sequence onto its limit).

Hint. Let $m_r \subset m$ denote the aggregate of real sequences. Define the seminorm $p(x) = \overline{\lim} |\xi_n|$ on m_r. Take the real linear functional f, equal to 1 on the element $e = (1, 1, \ldots, 1, \ldots)$ and hence having the seminorm 1 on the subspace $\{\lambda e\}$, and continue it on to all m_r whilst preserving the seminorm. Use the relations $|f(x)| \leqq p(x), f(x - \lambda e) = f(x) - \lambda$ to verify that the value of the functional $f(x)$ lies between $\lim \xi_n$ and $\lim \xi_n$. Put $f(x + iy) = f(x) + i f(y)$ for the complex sequence $x + i y$ (prove that the functional obtained is linear). Its value (a complex number) does not lie to the right of the vertical line through the right-hand limit point of the sequence ξ_n. Obtain the required result by replacing x by $e^{i\theta} x$ and applying similar arguments.

6. (continued). Having fixed an element $\alpha = (\alpha_0, \ldots, \alpha_n, \ldots)$ for which $\alpha_j \geqq 0$, $\sum_{j=0}^{\infty} \alpha_j = 1$, we can use it to pass from a bounded sequence $x = (\xi_0, \ldots, \xi_n, \ldots)$ to a new bounded sequence $\alpha^* x = (\eta_0, \ldots, \eta_n, \ldots)$, where $\eta_n = \sum_{j=0}^{\infty} \alpha_j \xi_{j+n} (n = 1, 2, \ldots)$. We shall refer to this operation as an α-operation.

Show that all the limit points of the sequence $\alpha^* x$ lie in the convex envelope of the set of all limit points of the sequence x.

Hint. The number η_n is the "mean" of the numbers $\xi_{0+n}, \xi_{1+n}, \ldots$

7. (continued). Let the condition $\alpha_j \geqq 0$ of problem 6 be replaced by $\sum_{j=0}^{\infty} |\alpha_j| = A < \infty$ (the α_j can now be any complex numbers). Show that the limit points of the sequence $\alpha^* x$ lie in a circle concentric with the circle $Q(x)$ containing all the limit points of the sequence x, and with a radius A times greater than the radius of $Q(x)$.

Hint. It is sufficient to take case when the centre of $Q(x)$ is the coordinate origin.

Note. We shall call the number ξ the *α-limit* of a sequence x if ξ is the ordinary limit of the sequence $\alpha^* x$.

The result of problems 6–7 shows that, if a sequence x has the ordinary limit ξ, it also has an α-limit, equal to ξ for any choice of α; further, if the α-limit exists, it lies in the A-extension of the circle $Q(x)$. In particular, $p(\alpha^* x) \leqq A\, p(x)$.

8. (continued). Improve the structure of the functional f of problem 5 in such a way that every sequence with an α-limit (for a given fixed α) is mapped by the functional f onto the value of this limit.

Hint. The subspace m_α of sequences with zero α-limit does not contain the element e in its closure with respect to the seminorm $p(x)$, since $A\, p(e - x) \geqq p(\alpha^*(e - x)) = p(\alpha^* e - \alpha^* x) = p(e) = 1$. Put $f(x) = 0$ on m_α, $f(e) = 1$ and continue on to all m whilst preserving the seminorm.

9. (continued). Improve the structure of the functional f in such a way that every sequence with an α-limit is mapped, at any rate for one α, by the functional f onto this limit.

Hint. If x has an α-limit, and y a β-limit, $x + y$ has a γ-limit, where

$$\gamma_n = \sum_{j=0}^{n} \alpha_j \beta_{n-j}.$$

The totality m_0 of all elements $x \in m$ with a zero α-limit for some α is therefore *a* subspace.

Note. There exist sequences $x \in m$ that have no α-limit for any α; an example of such a sequence is $(1, 0, \ldots, 0, 1, 0, \ldots, 0, 1, 0, \ldots)$, where the intervals filled continuously by zeros increase indefinitely.

10. (continued). A sequence x is said to be *quasizero* if, given any $\varepsilon > 0$, there exists an element $\alpha = (\alpha_0, \ldots, \alpha_n, \ldots)$, $\alpha_j \geqq 0$, $\sum_{j=0}^{\infty} \alpha_j = 1$, such that $p(\alpha^* x) < \varepsilon$.

Show that the functional f (problem 5) can be chosen so that $f(x) = 0$ for any quasizero sequence,

Hint. The quasizero sequences form a subspace, and $e = (1, \ldots, 1, \ldots)$ does not belong to its closure with respect to the seminorm $p(x)$.

11. (continued). Every sequence of the form $\mu^* x$, where

$$\mu = (\mu_0, \ldots, \mu_n, \ldots), \quad \sum_{n=0}^{\infty} |\mu_n| < \infty, \quad \sum_{n=0}^{\infty} \mu_n = 0,$$

is quasizero.

Hint. Show that the proposition holds for $\mu_0 = (1, -1, 0, 0, \ldots)$ and displacements of it. Express an arbitrary quasizero sequence μ as a linear combination of μ_0, displacements of μ_0, and a sequence z with $p(z) < \varepsilon$.

12. (continued). Show that the functional f formed in problem 10 has the property $f(\alpha^* x) = f(x)$ for any α.

Hint. By the result of problem 11, the sequence $x - \alpha^* x$ is quasizero.

Note. In particular, when $\alpha = (0, 1, \ldots)$, the sequence $\alpha^* x$ is a displacement of the sequence x; hence the value of the functional $f(x)$ remains unchanged on displacement of the sequence x.

13. If the numbers $\alpha_1, \ldots, \alpha_n, \ldots$ are such that the series $\sum_{k=1}^{\infty} \xi_k \alpha_k$ is convergent for any $x = (\xi_1, \ldots, \xi_k, \ldots) \in c_0$, then $\sum_{n=1}^{\infty} |\alpha_n| < \infty$.

Hint. It follows from the hypothesis that the sequence of functionals $f_n(x) = \sum_{k=1}^{n} \xi_k \alpha_k$ on c_0 is convergent for any element $x \in c_0$. Then use the Hahn–Banach theorem and the general form of a linear functional on c_0 (problem 4).

14. The T-limit of a bounded sequence. Let $T = (t_{jk})$, $j = 1, 2, \ldots$; $k = 1, 2, \ldots$, be the infinite matrix satisfying:

(1) $\sum_{k=1}^{\infty} |t_{jk}| \leq C$, C is independent of j;

(2) $\sum_{k=1}^{\infty} t_{jk} = s_j$, $\lim_{j \to \infty} s_j = 1$;

(3) $\lim_{j \to \infty} t_{jk} = 0$ for any $k = 1, 2, \ldots$

(the Toeplitz matrix). Also, let $x = (\xi_1, \ldots, \xi_n, \ldots)$ be a bounded sequence We form the sequence Tx of numbers $t_j(x) = \sum_{k=1}^{\infty} t_{jk} \xi_k$; if it exists, the limit of the sequence Tx is said to be the T-limit of the sequence x. Show that: (a) if the sequence x is convergent (in the usual sense) to the limit ξ, the sequence Tx is also convergent to the limit ξ;

(b) if $T = (t_{jk})$ is any matrix with the property that the numbers $t_j(x) = \sum_{k=1}^{\infty} t_{jk} \xi_k$ exist and are convergent to ξ for any convergent sequence $x = \{\xi_n\}$, $\lim \xi_n = \xi$, T must have properties (1)–(3).

Hint. (a) It is sufficient to take the case $\xi = 0$.
(b) Apply the matrix T to the element $(0, \ldots, 0, 1_k, 0, \ldots)$ and obtain (3). Apply T to the element $(1, \ldots, 1, \ldots)$ and obtain (2). The convergence of every series (1) is deduced from problem 13. The functionals $t_j(x)$ are convergent here for any $x \in c_0$; by the Banach–Steinhaus theorem, their norms have a common bound, which leads to (1).

15. Show that the α-limit (problems 6–12) is a particular case of the T-limit.

Answer. If $\alpha_n = (\alpha_1, \ldots, \alpha_n, \ldots)$, the corresponding T matrix has the form

$$\left\| \begin{matrix} \alpha_1 & \alpha_2 & \alpha_3 & \cdots \\ 0 & \alpha_1 & \alpha_2 & \cdots \\ 0 & 0 & \alpha_1 & \cdots \\ \cdots & \cdots & \cdots & \cdots \\ \cdots & \cdots & \cdots & \cdots \end{matrix} \right\|$$

16. Form the Toeplitz matrix T such that $T\hat{e} = \hat{e} + e$ for

$$\hat{e} = (1, 1, \ldots, 1, \ldots),\ e = (1, -1, 1, -1, \ldots).$$

Answer. For example:

$$T = \begin{Vmatrix} 1+\frac{1}{2} & -\frac{1}{2} & 0 & 0 & 0 & 0 & \dots \\ 0 & 1 & \frac{1}{2} & -\frac{1}{2} & 0 & 0 & \dots \\ 0 & 0 & 1 & 0 & \frac{1}{2} & -\frac{1}{2} & \dots \\ \dots & \dots & \dots & \dots & \dots & \dots & \dots \end{Vmatrix}$$

Note. The existence of such a matrix T shows that there exists no linear functional f on the space m with the properties $f(Tx) = f(x)$, $f(e) = 1$; hence the result of problem 12, relating to α-limits, cannot be carried over to T-limits.

17. Show that the closed subspace of space $L_1(-\infty, \infty)$ that contains some function $\varphi(x) \not\equiv 0$, all its displacements and all its products with exponents $e^{ix\sigma}$, coincides with the whole space $L_1(-\infty, \infty)$.

Hint. Use corollary 3 of the Hahn–Banach theorem, the general form of a continuous linear functional on L_1, and the uniqueness theorem for Fourier transforms.

INDEX

OTHER TITLES IN THE SERIES IN PURE AND APPLIED MATHEMATICS

Vol. 1 WALLACE—*An Introduction to Algebraic Topology*

Vol. 2 PEDOE—*Circles*

Vol. 3 SPAIN—*Analytical Conics*

Vol. 4 MIKHLIN—*Integral Equations*

Vol. 5 EGGLESTON—*Problems in Euclidean Space: Application of Convexity*

Vol. 6 WALLACE—*Homology Theory on Algebraic Varieties*

Vol. 7 NOBLE—*Methods Based on the Wiener—Hopf Technique for the Solution of Partial Differential Equations*

Vol. 8 MIKUSINSKI—*Operational Calculus*

Vol. 9 HEINE—*Group Theory in Quantum Mechanics*

Vol. 10 BLAND—*The Theory of Linear Viscoelasticity*

Vol. 11 KURTH—*Axiomatics of Statistical Mechanics*

Vol. 12 FUCHS—*Abelian Groups*

Vol. 13 KURATOWSKI—*Introduction to Set Theory and Topology*

Vol. 14 SPAIN—*Analytical Quadrics*

Vol. 15 HARTMAN and MIKUSINSKI—*Theory of Lebesgue Measure and Integration*

Vol. 16 KULCZYCKI—*Non-Euclidean Geometry*

Vol. 17 KURATOWSKI—*Introduction to Calculus*

Vol. 18 GERONIMUS—*Polynomials Orthogonal on a Circle and Interval*

Vol. 19 ELSGOLC—*Calculus of Variations*

Vol. 20 ALEXITS—*Convergence Problems of Orthogonal Series*

Vol. 21 FUCHS and LEVIN—*Functions of a Complex Variable, Volume II*

Vol. 22 GOODSTEIN—*Fundamental Concepts of Mathematics*

Vol. 23 KEENE—*Abstract Sets and Finite Ordinals*

Vol. 24 DITKIN and PRUDNIKOV—*Operational Calculus in Two Variables and its Applicationse*

Vol. 25 VEKUA—*Generalized Analytic Functions*

Vol. 26 FASS and AMIR-MOÉZ—*Elements of Linear Spaces*

Vol. 27 GRADSHTEIN—*Direct and Converse Theorems*

Vol. 28 FUCHS—*Partially Ordered Algebraic Systems*

Vol. 29 POSTNIKOV—*Foundations of Galois Theory*

Vol. 30 BERMANT—*A Course of Mathematical Analysis, Part II*

Vol. 31 ŁUKASIEWICZ—*Elements of Mathematical Logic*

Vol. 32 VULIKH—*Introduction to Functional Analysis for Scientists and Technologists*

Vol. 33 PEDOE—*An Introduction to Projective Geometry*

Vol. 34 TIMAN—*Theory of Approximation of Functions of a Real Variable*

Vol. 35 CSÁSZÁR—*Foundations of General Topology*

Vol. 36 BRONSHTEIN and SEMENDYAYEV—*A Guide-Book to Mathematics for Technologists and Engineers*

Vol. 37 MOSTOWSKI and STARK—*Introduction to Higher Algebra*

Vol. 38 GODDARD—*Mathematical Techniques of Operational Research*

Vol. 39 TIKHONOV and SAMARSKII—*Equations of Mathematical Physics*

Vol. 40 McLEOD—*Introduction to Fluid Dynamics*

Vol. 41 MOISIL—*The Algebraic Theory of Switching Circuits*

Vol. 42 OTTO—*Nomography*

Vol. 43 RANKIN—*An Introduction to Mathematical Analysis*

Vol. 44 BERMANT—*A Course of Mathematical Analysis, Part I*

Vol. 45 KRASNOSEL'SKII—*Topological Methods in the Theory of Nonlinear Integral Equations*

Vol. 46 KANTOROVICH and AKILOV—*Functional Analysis in Normed Spaces*

Vol. 47 JONES—*The Theory of Electromagnetism*

Vol. 48 FEJES TÓTH—*Regular Figures*

Vol. 49 YANO—*Differential Geometry on Complex and Almost Complex Spaces*

Vol. 50 MIKHLIN—*Variational Methods in Mathematical Physics*

Vol. 51 FUCHS and SHABAT—*Functions of a Complex Variable and Some of their Applications, Vol. 1*

Vol. 52 BUDAK, SAMARSKII and TIKHONOV—*A Collection of Problems on Mathematical Physics*

Vol. 53 GILES—*Mathematical Foundations of Thermodynamics*

Vol. 54 SAUL'YEV—*Integration of Equations of Parabolic Type by the Method of Nets*

Vol. 55 PONTRYAGIN *et al.*—*The Mathematical Theory of Optimal Processes*

Vol. 56 SOBOLEV—*Partial Differential Equations of Mathematical Physics*

Vol. 57 SMIRNOV—*A Course of Higher Mathematics Vol. I*

Vol. 58 SMIRNOV—*A Course of Higher Mathematics Vol. II*

Vol. 59 SMIRNOV—*A Course of Higher Mathematics Vol. III, Pt.* 1

Vol. 60 SMIRNOV—*A Course of Higher Mathematics Vol. III, Pt.* 2

Vol. 61 SMIRNOV—*A Course of Higher Mathematics Vol. IV*